Foreword

In a 1990 Angewandte article entitled "Organic Synthesis – Where now?" I made the statements [1]: "The primary center of attention for all synthetic methods will continue to shift towards catalytic and enantioselective variants; indeed, it will not be long before such modifications will be available with every standard reaction for converting achiral educts into chiral products" and "The discovery of truly new reactions is likely to be limited to the realm of transition-metal chemistry, which will almost certainly provide us with additional *miracle reagents* in the years to come". It was not very risky to make such predictions, because these developments were already going on in 1990, when the discovery of the Sharpless epoxidation celebrated its 10th anniversary, when the enantioselective hydrogenations with diphosphane–Pd complexes were well established (Kagan, Knowles, Noyori), when highly active and versatile metathesis catalysts for organic synthesis were about to be developed by Schrock and Grubbs, and when chiral Lewis acids, such as those derived from BINOL, TADDOL, diaminocyclohexane or semicorrins were already "on the market".

An almost explosive increase of worldwide activities in the field of stereoselective reactions ("asymmetric synthesis"), especially of catalytic enantioselective transformations, has, however, set in only within the last ten years. A look at any issue of any journal in which organic chemistry results are published is striking evidence for the fact that the center of attention has indeed become enantioselective catalysis by transition-metal complexes (organometallic catalysis), by organic acids or bases/nucleophiles ("organocatalysis"), and by enzymes (biocatalysis). These are exactly the topics of this book with the demanding title "Asymmetric Synthesis – The Essentials", edited by Christmann and Bräse.

The book is unique, mainly for the following two reasons.

1. It is based on a novel concept, containing more than 50 essay-type contributions, of ca. five pages each, covering essentially all current research in the area of stereoselective synthesis. After the title of the essay a general background is given, followed by a description of the type of reaction with illustrative examples, a conclusion with perspectives, a *curriculum vitae* of the main author, and a list of leading references.

2. The editors must be congratulated for having been able to persuade an illustrious group of essentially all the major players in the field of stereoselective

synthesis from around the world to act as authors, and to submit collectively their contributions Surprisingly, there is hardly any overlap between the various essays. It looks to me as if this endeavor must have involved periods of truly Sisyphean efforts by the editors?

The book is a flashlight picture of the state of the art in the field of preparation of enantioenriched compounds. The emphasis is on catalytic processes leading to one of two enantiomers preferentially, directly from achiral precursors. There are some contributions, in which the use of a stoichiometric amount of a chiral additive is described (cf. spartein in Hoppe's work). Diastereoselective transformations for the preparation of enantiopure products (stoichiometrically employed chiral reagents, and the auxiliary approach) are the subject of only a few contributions (cf. Meyers, Evans, Enders, Hoffmann, Kunz, Davis, de Meijere). Chapters by giants in the field of total synthesis of complex natural products (cf. Danishefsky, Nicolaou, Paterson) remind us that, depending upon the degree of linearity, only a few enantioselective steps are involved in such multistep syntheses, while the major task consists of series of functional-group-selective (including protection/deprotection), regioselective and diastereoselective transformations.

There could hardly be more variety of topics, not only in view of the types of reactions dealt with. There are totally practical sections describing just what can be done, and there are more mechanistic essays (cf. non-linear effects by Kagan or autocatalysis by Soai). The techniques covered or mentioned range from immobilization of reagents and catalysts, entire syntheses carried out on solid phase, combinatorial methods, photochemical processes, automated, high-throughput screening of enantiomer ratios, and reactions carried out in water.

With so many authors involved, it is not astonishing that the chemical language, and especially the "language of stereochemistry", in this book is not uniform. The regrettable loss of the term configuration (absolute and relative configuration of stereocenters and E/Z configuration are now commonly called "stereochemistry"), the wrong use of topicity-specifying descriptors (re/si vs. Re/Si) [2], the weird term %de, and 100% yield or ee [3] (there is no analytical method without a detection limit) are only a few examples [4]. This will, however, not really detract from the enormous merits deserved by the editors and authors for having achieved this work, nor will it diminish the great value of the book as an up-to-date overview of the field of stereoselective reactions in Organic Chemistry. Both, experienced researchers and novices will profit by learning from this collection of concise essays about the state of the art and future prospects of the many exciting developments in the center of today's synthetic organic methodology.

Dieter Seebach

Asymmetric Synthesis – The Essentials

Edited by
Mathias Christmann and Stefan Bräse

WILEY-VCH Verlag GmbH & Co. KGaA

The Editors 1004943002

Dr. Mathias Christmann
Institute of Organic Chemistry
RWTH Aachen
Landoltweg 1
52074 Aachen
Germany

Prof. Dr. Stefan Bräse
Institute of Organic Chemistry
Universität Karlsruhe (TH)
Fritz-Haber-Weg 6
76131 Karlsruhe
Germany

■ All books published by Wiley-VCH are carefully produced. Nevertheless, authors, editors, and publisher do not warrant the information contained in these books, including this book, to be free of errors. Readers are advised to keep in mind that statements, data, illustrations, procedural details or other items may inadvertently be inaccurate.

Library of Congress Card No.: Applied for

British Library Cataloguing-in-Publication Data:
A catalogue record for this book is available from the British Library.

Bibliographic information published by the Deutsche Nationalbibliothek
The Deutsche Nationalbibliothek lists this publication in the Deutsche Nationalbibliografie; detailed bibliographic data are available in the Internet at http://dnb.d-nb.de.

© 2007 WILEY-VCH Verlag GmbH & Co. KGaA, Weinheim

All rights reserved (including those of translation into other languages). No part of this book may be reproduced in any form – by photoprinting, microfilm, or any other means – nor transmitted or translated into a machine language without written permission from the publishers. Registered names, trademarks, etc. used in this book, even when not specifically marked as such, are not to be considered unprotected by law.

Printed in the Federal Republic of Germany.
Printed on acid-free paper.

Composition Hagedorn Kommunikation, Viernheim
Printing betz-druck GmbH, Darmstadt
Bookbinding Litges & Dopf Buchbinderei GmbH, Heppenheim

ISBN 978-3-527-31399-0

1. D. Seebach, *Angew. Chem.* **1990**, *102*, 1363; *Angew. Chem. Int. Ed.* **1990**, *29*, 1320.
2. V. Prelog, G. Helmchen, *Angew. Chem.* **1982**, *94*, 614; *Angew. Chem. Int. Ed.* **1982**, *21*, 567; D. Seebach, V. Prelog, *Angew. Chem.* **1982**, *94*, 696; *Angew. Chem. Int. Ed.* **1982**, *21*, 654.
3. I belong to the advocates for the abolishment of the outdated term %*ee*, see: R. E. Gawley, *J. Org. Chem.* **2006**, *71*, 2411.
4. See (a) the glossary in: E. L. Eliel, S. H Wilen, *Stereochemistry of Organic Compounds*, Wiley, New York, **1994**, pp. 1191–1210; (b) G. Helmchen, The Glossary of Problematic Terms, in Houben-Weyl, *Methods of Organic Chemistry, Stereoselective Synthesis*, Workbench Edition, Vol. E 21a, Part A 1., General Aspects, Nomenclature and Vocabulary of Organic Stereochemistry, G.Helmchen, R. W. Hoffmann, J. Mulzer et al. (Eds.), G. Thieme Verlag 1996; (c) G. Quinkert, E. Egert, C. Griesinger, *Aspects of Organic Chemistry*, Ch. 10, Wiley-VCH and Verlag Helv. Chim. Acta, 1996.

Contents

Foreword *V*

Message from the editors *XXV*

List of Abbreviations *XXVII*

Named and Unnamed Reactions *XXXI*

List of Contributors *XXXIII*

Part I

Chiral Auxiliaries in Asymmetric Synthesis *3*
David A. Evans, Günter Helmchen and Magnus Rüping
Background *3*
Results *4*
Summary *8*
CV of David A. Evans *8*
CV of Günter Helmchen *8*
CV of Magnus Rüping *8*
Selected References *9*

Asymmetric Epoxidation of Pentadienols *10*
Reinhard Brückner
Background *10*
Objective: Making Building Blocks for the Synthesis of
  1,3,5,7,...-Polyols from Epoxyalcohols *11*
Results *12*
Asymmetric Epoxidation of a Conjugated Pentadienol *12*
Asymmetric Epoxidation of a Nonconjugated Pentadienol *12*
CV of Reinhard Brückner *14*
References *15*

Asymmetric Synthesis – The Essentials.
Edited by Mathias Christmann and Stefan Bräse
Copyright © 2007 WILEY-VCH Verlag GmbH & Co. KGaA, Weinheim
ISBN: 978-3-527-31399-0

Asymmetric Synthesis using Sulfur–Nitrogen Reagents 16
Franklin A. Davis
Background 16
Results 16
Asymmetric Oxidations using N-Sulfonyloxaziridines 16
Asymmetric Fluorination of Enolates with N-Fluoro-2,10-Camphorsultams 18
Asymmetric Synthesis of Amine Derivatives using Enantiopure
   Sulfinimines (N-Sulfinyl Imines) 18
Outlook 19
CV of Franklin A. Davis 19
Selected Publications 20

Asymmetric Syntheses with SAMP-/RAMP-Hydrazones 21
Dieter Enders and Wolfgang Bettray
Background 21
Strategy 22
Results 23
Application for the Synthesis of Natural Products 24
Other Asymmetric Syntheses from the Enders Group 24
Conclusions and Future Perspectives 24
CV of Dieter Enders 25
Selected Publications 25

Asymmetric Allylation Reactions 27
Reinhard W. Hoffmann
Background 27
Results 27
Type (I) Allylmetallation of Aldehydes 27
Second Generation Allylmetallation Reagents 29
CV of Reinhard W. Hoffmann 30
References 31

Carbohydrates as Chiral Auxiliaries 32
Horst Kunz
Introduction 32
Results 32
Carbohydrate Auxiliaries in Cycloaddition Reactions 32
Conjugate Addition Reactions Directed by Carbohydrate Auxiliaries 33
Glycosylamines in Stereoselective Synthesis 34
Conclusions 35
CV of Horst Kunz 36
Selected Publications 36

The Use of Chiral Oxazolines in early C–C Bond Forming Reactions 37
Albert I. Meyers
Background   37
Results   38
CV of A.I. Meyers   39
Selected Publications   39

Asymmetric Synthesis of Amines and Amino Acids from Amines 40
Shun-Ichi Murahashi and Yasushi Imada
Background   40
Strategy   41
Results   41
Diastereoselective Reactions of Nitrones   41
Enantioselective Reactions of Nitrones   42
Conclusions and Future Perspectives   43
CV of Shun-Ichi Murahashi   43
Selected Publications   44

Part II

Metal-catalyzed Asymmetric Synthesis 47
Masakatsu Shibasaki and Shigeki Matsunaga
Introduction   47
Design of Asymmetric Metal Catalysis   48
Future Prospects for Metal-catalyzed Asymmetric Synthesis   49
CV of Masakatsu Shibasaki   50
CV of Shigeki Matsunaga   50
Selected Publications   50

Catalytic Asymmetric Sulfur Ylide Mediated Epoxidation of Carbonyl Compounds 52
Varinder Aggarwal
Background   52
Results   53
Other Reactions of Sulfur Ylides from the Aggarwal Group   55
Conclusions and Future Perspectives   55
CV of Varinder Aggarwal   56
Selected Publications   56

Asymmetric Baeyer-Villiger Reactions   57
Carsten Bolm
Background   57
Strategy   58
Results   59
Other Research Topics from the Bolm Group   60
CV of Carsten Bolm   60
Selected Publications   61

Planar Chiral Ligands Based on [2.2]Paracyclophanes   62
Stefan Bräse
Background   62
Strategy   63
Results   64
Application for the Synthesis of Natural Products   65
Other Asymmetric Syntheses from the Bräse Group   65
Conclusions and Future Perspectives   65
CV of Stefan Bräse   66
Selected Publications   66

Asymmetric Syntheses of 3-(*trans*-2-Nitrocyclopropyl)alanine and 3-(*trans*-2-Aminocyclopropyl)alanine   67
Armin de Meijere and Oleg V. Larionov
Background   67
Strategy and Results   68
Outlook   70
CV of Armin de Meijere   70
CV of Oleg V. Larionov   70
Selected Publications   71

Copper-Bis(oxazoline) Catalyzed Synthesis of β-Lactams – Enantioselective Reaction of Alkynes with Nitrones   72
David A. Evans, Florian Kleinbeck and Magnus Rüping
Background   72
Results   73
Summary   75
CV of David A. Evans   76
CV of Magnus Rüping   76
CV of Florian Kleinbeck   76
Selected Publications   76

Catalytic Asymmetric Conjugate Addition Reactions of Organometallic Reagents 78
Fernando López and Ben L. Feringa
Background 78
The First Catalytic Enantioselective Approaches 78
Results 79
Cu-catalyzed Enantioselective CA of Organozinc Reagents 79
Highly Enantioselective Cu-catalyzed CA of Grignard Reagents 80
Conclusion and Future Perspectives 81
CV of Fernando López 82
CV of Ben L. Feringa 82
Selected Publications 82

Catalytic Asymmetric Synthesis of Allylic Alcohols via Dynamic Kinetic Resolution 84
Hans-Joachim Gais
Introduction 84
Strategy 84
Results 85
Application to the Synthesis of Natural Products 87
Other Asymmetric Syntheses from the Gais Group 87
Conclusion and Future Perspectives 87
CV of Hans-Joachim Gais 87
Selected Publications 88

Asymmetric Cross-Coupling Reactions 90
Tamio Hayashi
Background 90
Results 90
Asymmetric Cross-coupling of Secondary Alkyl Grignard Reagents 90
Asymmetric Cross-Coupling Forming Axially Chiral Biaryls 92
Conclusions and Future Perspectives 93
CV of Tamio Hayashi 94
Selected Publications 94

Asymmetric Allylic Substitutions 95
Günter Helmchen
Background 95
Results 96
Pd-Catalyzed Allylic Substitutions 96
Ir-Catalyzed Allylic Substitutions 97
Application for the Synthesis of Biologically Active Compounds 97
Other Asymmetric Syntheses from the Helmchen Group 98
Conclusions and Future Perspectives 98
CV of Günter Helmchen 98
Selected Publications 99

Asymmetric Homoaldol Reactions 100
Dieter Hoppe
Background 100
Results 101
Enantioselective Additions via Stereospecific Deprotonation 101
Enantioselective Additions via (–)-Sparteine-mediated α-Deprotonations 102
Enantioselective γ-Deprotonations 103
Synthetic Applications 103
CV of Dieter Hoppe 103
Selected Publications 104

Asymmetric Vinylogous Mukaiyama Aldol Reaction 105
Markus Kalesse and Jorma Hassfeld
Introduction 105
Results 105
Conclusions and Further Perspectives 108
CV of Markus Kalesse 109
CV of Jorma Hassfeld 109
Selected Publications 109

Chiral Lewis Acid Catalysis in Aqueous Media 110
Shu Kobayashi and Chikako Ogawa
Background 110
Results 110
Water-compatible Lewis Acid 110
Asymmetric Aldol Reactions 111
Asymmetric Hydroxymethylation Reactions 112
Asymmetric Mannich-type Reactions 113
CV of Sh Kobayashi 114
CV of Dr. Chikako Ogawa 114
Selected Publications 114

Asymmetric Epoxidation of Non-activated Olefins 116
Kazuhiro Matsumoto and Tsutomu Katsuki
Introduction 116
Results 116
Manganese(salen) Catalyst 116
Ruthenium(salen) Catalyst 117
Titanium(salalen) Catalyst 118
Conclusion 119
CV of Tsutomu Katsuki 120
CV of Kazuhiro Matsumoto 120
Selected Publications 120

Chiral Carbonyl Lewis Acid Complexes in Asymmetric Syntheses *121*
Keiji Maruoka and Takashi Ooi
Background *121*
Results *121*
Conclusions and Future Perspectives *124*
CV of Keiji Maruoka *124*
CV of Takashi Ooi *124*
Selected Publications *125*

ZACA Reaction: Zr-Catalyzed Asymmetric Carboalumination of Alkenes *126*
Ei-ichi Negishi, Bo Liang, Tibor Novak, and Ze Tan
Background and Discovery *126*
Results *127*
ZACA–Pd-Catalyzed Cross-coupling Tandem Processes for the Synthesis of Deoxypolypropionates and Related Compounds *127*
Synthesis of 2-Methyl- or 2-Ethyl-1-alkanols Not Readily Purifiable by Chromatography or Recrystallization *129*
CV of Ei-ichi Negishi *129*
Selected Publications *130*

Bisoxazolines – a Privileged Ligand Class for Asymmetric Catalysis *131*
Andreas Pfaltz
Background *131*
Results *132*
Semicorrins *132*
C_2-Symmetric Bisoxazolines *132*
Boron-bridged Bisoxazolines *134*
Conclusion *134*
CV of Andreas Pfaltz *134*
Selected Publications *135*

Enantioselective Cycloaddition Reactions Catalyzed by Hydrogen Bonding *136*
Viresh H. Rawal and Avinash N. Thadani
Background *136*
Strategy *138*
Results *138*
Application to Other Reactions *139*
Conclusions and Future Perspectives *139*
CV of Viresh Rawal *139*
CV Co-Autor *140*
Selected Publications *140*

Direct Catalytic Asymmetric Aldol-Tishchenko Reaction 141
Masakatsu Shibasaki and Takashi Ohshima
Background 141
Strategy 142
Results 142
Mechanistic Studies 144
Conclusions and Future Perspectives 145
CV of Masakatsu Shibasaki 145
CV of Takashi Ohshima 145
Selected Publications 146

Asymmetric Heck and other Palladium-catalyzed Reactions 147
Lutz F. Tietze and Florian Lotz
Background 147
Enantioselective Heck Reactions with Allylsilanes as the Alkene Component 147
CV of L. F. Tietze 151
CV of F. Lotz 151
Selected Publications 151

Asymmetric Catalysis with Chiral Acid 153
Hisashi Yamamoto
Background 153
Results 154
C_2 Symmetric Chiral Lewis Acid Catalysts 154
Combined Acid Catalyst 154
CV of Hisashi Yamamoto 157
Selected Publications 157

Part III

Biocatalysis and Organocatalysis: Asymmetric Synthesis Inspired by Nature 161
Benjamin List
Introduction 161
The Aldol Reaction as an Example 161
Asymmetric Chemical Aldolizations 162
Conclusions and Outlook 164
CV of Benjamin List 165
References 165

Enantioselective Photochemical Reactions *166*
Thorsten Bach
Background *166*
Concept *167*
Results *167*
Perspectives *169*
CV of Thorsten Bach *170*
Selected Publications *170*

Asymmetric Catalysis via Dynamic Kinetic Resolution *171*
Jan-E. Bäckvall
Background *171*
Combined Enzyme and Metal Catalysis for Efficient DKR *172*
Results *172*
Extension to DKR of Amines *174*
Conclusions and Future Perspectives *174*
CV of Jan-E. Bäckvall *175*
Selected Publications *175*

Catalytic Asymmetric Epoxidation of Enones and Related Compounds *176*
Albrecht Berkessel
General *176*
Electrophilic Catalytic Epoxidation *177*
Nucleophilic Catalytic Epoxidation *177*
Conclusions and Future Perspectives *179*
CV of Albrecht Berkessel *179*
Selected Publications *180*

Kinetic Investigations of the Soai Autocatalytic Reaction *181*
Donna G. Blackmond
Background *181*
Kinetic Studies using Reaction Calorimetry *182*
CV of Donna G. Blackmond *185*
References *185*

Planar-chiral Heterocycles as Enantioselective Organocatalysts *186*
Gregory Fu
Background and Design *186*
Applications *186*
Conclusions and Future Perspectives *189*
CV of Gregory C. Fu *190*
Selected Publications *190*

An Organocatalytic Approach to Optically Active Six-membered Rings *191*
Karl Anker Jørgensen
Background   *191*
Strategy   *192*
Results   *193*
Synthesis of γ- and ε-lactones   *194*
Conclusions and Future Perspectives   *195*
CV of Karl Anker Jørgensen   *195*
Selected Publications   *195*

Non-linear Effects in Asymmetric Catalysis   *196*
Henri B. Kagan
Background   *196*
Enantiomerically Impure Chiral Auxiliaries   *196*
Non-linear Effects   *196*
Results   *197*
Examples of Asymmetric Amplifications   *197*
Kinetic Models   *198*
Applications of Non-linear Effects   *199*
Conclusion and Perspectives   *199*
CV of Henri Kagan   *199*
Selected Publications   *200*

Asymmetric Organocatalysis   *201*
Steven V. Ley
Background   *201*
Results   *202*
Conclusions and Future Perspectives   *204*
CV of Steven V Ley   *205*
Selected Publications   *205*

Directed Evolution of Enzymes for Asymmetric Syntheses   *207*
Manfred T. Reetz
Background   *207*
Strategy and Early Results   *207*
Generalizing the Concept   *209*
New Strategies   *209*
Conclusions and Perspectives   *210*
CV of Manfred T. Reetz   *210*
Selected Publications   *210*

**Asymmetric Autocatalysis and Its Implications
in the Origin of Chiral Homogeneity of Biomolecules** *212*
Kenso Soai and Tsuneomi Kawasaki
Background *212*
Results *213*
Chiral Amplification by Asymmetric Autocatalysis *213*
Asymmetric Autocatalysis as a Link Between the Origin
  of Chirality and Highly Enantioenriched Molecules *214*
Spontaneous Absolute Asymmetric Synthesis *215*
Conclusions and Future Perspectives *215*
CV of Kenso Soai *215*
CV of Tsuneomi Kawasaki *216*
Selected Publications *216*

Asymmetric Synthesis using Deoxyribose-5-phosphate Aldolase *217*
Chi-Huey Wong and William A. Greenberg
Background *217*
Results *218*
Catalytic Mechanism *218*
Synthetic Applications *218*
Application to the Synthesis of Epothilones and Statins *219*
Conclusion *220*
CV of Chi-Huey Wong *221*
CV of William A. Greenberg *221*
Selected Publications *221*

Part IV

Asymmetric Reactions in Total Synthesis *225*
K. C. Nicolaou and Paul G. Bulger
Background *225*
Reactions *225*
Aldol *225*
Alkylation *226*
Diels–Alder *227*
Allylation *230*
Palladium-catalyzed Cross-coupling *231*
Epoxidation *232*
Dihydroxylation/aminohydroxylation *234*
Organocatalytic Reductions *235*
Metal-catalyzed Hydrogenations *236*
Enzymatic *237*
Conclusions *238*
CV of K. C. Nicolaou *238*
CV of Paul G. Bulger *238*
Selected Publications *239*

Ring Rearrangement Metathesis (RRM) in Alkaloid Synthesis 241
Nicole Holub and Siegfried Blechert
Background and Strategy 241
Results 242
Tetraponerines 242
Trans-195A 242
Lasubine II 243
Astrophylline and Dendrochrysine 244
Conclusions and Future Perspectives 245
CV of Siegfried Blechert 245
Selected Publications 245

Asymmetric Synthesis of Biaryls by the 'Lactone Method' 246
Gerhard Bringmann, Tanja Gulder, and Tobias A. M. Gulder
Introduction 246
The Basic Concept of the 'Lactone Method' 246
Application of the 'Lactone Concept' to Natural Product Synthesis 247
Conclusions and Further Perspectives 249
CV of Gerhard Bringmann 249
Selected Publications 250

Asymmetric Synthesis of Merrilactone A 251
Samuel J. Danishefsky
Background 251
Synthetic Strategy 251
Synthesis of Merrilactone A 252
CV of Samuel J. Danishefsky 255
Selected Publications 255

Asymmetric Synthesis of Cyclic Ketal and Spiroaminal-Containing Natural Products 256
Craig J. Forsyth
Background 256
Results 257
Okadaic Acid – Three Spiroketalization Methods 257
Additional Ketals via Double Intramolecular Hetero Michael Additions 258
Synthesis of the Azaspiracid Spiroaminal 258
Conclusion and Future Perspectives 260
CV of Craig J. Forsyth 260
Selected Publications 260

Case Studies at the Metathesis/Asymmetric Synthesis Interface *262*
Alois Fürstner
Background *262*
Results *262*
(−)-Gloeosporone *262*
(−)-Isooncinotine *264*
Conclusions and Future Perspectives *266*
CV of Alois Fürstner *266*
Selected Publications *266*

Asymmetric Synthesis of Amino Acids by Rhodium and Ruthenium Catalysis *267*
Jean Pierre Genet
Introduction *267*
Asymmetric Hydrogenation Reactions *267*
Chiral Diphosphine Ligands and Diversity of Chiral Ru Catalysts *268*
Results *268*
Asymmetric Hydrogenation of Dehydroamino Acids *268*
Asymmetric Syntheses of α-Amino β-Hydroxy Acids (DKR) *269*
Asymmetric Syntheses of Natural Products *269*
Tandem-1,4 addition/Enantioselective Protonation Catalyzed by Rhodium *270*
Conclusions *270*
CV of Jean Pierre Genet *271*
Selected Publications *271*

Asymmetric Syntheses of Pheromones *273*
Kenji Mori
Background *273*
Strategy and Results *274*
Derivation from Optically Active Starting Materials *274*
Enantiomer Separation *274*
Asymmetric Synthesis *275*
Conclusion and Future Perspective *276*
CV of Kenji Mori *276*
Selected Publications *277*

Total Synthesis of Polyketides Using Asymmetric Aldol Reactions *278*
Ian Paterson
Background *278*
Results *278*
Stereocontrol in the Boron Aldol Reactions of Ketones *278*
Application to the Total Synthesis of Altohyrtin A (Spongistatin 1) *279*
Conclusions and Future Perspectives *280*
CV of Ian Paterson *282*
Selected Publications *282*

Asymmetric Synthesis on the Solid Phase 283
Torben Leßmann and Herbert Waldmann
Introduction 283
Results 283
Asymmetric Aldol Reactions 283
Asymmetric Cycloadditions 284
Asymmetric Epoxide Openings 285
Other Reactions 285
Conclusion 286
CV of Herbert Waldmann 286
Selected Publications 286

Part V

Asymmetric Synthesis in Industry 291
Herbert Hugl
Background 291
Industrial Production of Chiral Compounds 291
Outlook 294
CV of Herbert Hugl 294
Selected Publications 294

Industrial Application of Enantioselective Catalysis 296
Hans-Ulrich Blaser
The Potential of Enantioselective Catalysis 296
Hurdles on the Way to an Industrial Process 296
Catalyst Performance 297
Availability and Cost of the Catalyst 297
Development Time 298
Selected Milestones in Industrial Application 298
Conclusions and Future Perspectives 300
CV of Hans-Ulrich Blaser 300
References 300

Crystallization-Induced Diastereoselection for the Synthesis of Aprepitant 301
Karel M. Jos Brands, Philip Pye and Kai Rossen
Background 301
Results 302
Bicyclic Acetal Approach 302
Lactam Approach 303
Conclusion 304
CV of J. Brands 305
CV of P. Pye 305
CV of K. Rossen 305
Selected Publications 305

Combinatorial Methods in Asymmetric Syntheses *306*
Stefan Dahmen
Background *306*
Results *306*
Ligand Screening *306*
Additive Screening *308*
Arylzinc Precursors *309*
CV of Stefan Dahmen *309*
Selected Publications *309*

Biocatalytic Production of Optically Active Amines *311*
Klaus Ditrich
Background *311*
Results *311*
Conclusions and Future Perspectives *315*
CV of Klaus Ditrich *315*
Selected Publications *315*

The Monsanto L-Dopa Process *316*
William S. Knowles
Background *316*
L-DOPA Process *316*
Other Industrial Applications *319*
Conclusions *319*
CV of William Knowles *319*
Selected Publications *320*

Asymmetric Hydrogenation through Metal–Ligand Bifunctional Catalysis *321*
Ryoji Noyori, Takeshi Ohkuma, Christian A. Sandoval, and Kilian Muñiz
Background *321*
Strategy *321*
Results *321*
Catalytic Cycle *323*
CV of Ryoji Noyori *324*
Selected Publications *325*

**Many Ways are Leading to Rome –
Today's Variety of Competing Synthetic Methods in Industry** *326*
Andreas Job and Andreas Stolle
Introduction *326*
Synthetic Concepts *327*
Biochemical Approach *327*
Catalytic Hydrogenation Approach *328*
Chiral Pool Synthesis *329*
Conclusion *329*
CV of Andreas Job *329*
CV of Andreas Stolle *329*
Selected Publications *330*

Index *331*

Author Index *339*

Subject Index *342*

Related Titles *346*

Message from the editors

"[In such cases] the versatility and power of modern synthetic methods, acting on information provided by modern analytical methods and motivated by the economic importance of the target, can be invaluable."[1] This quote by the Nobel laureate Sir John Warcup Cornforth is particularly true for asymmetric synthesis which has been a key target of research in academia and for many industrial applications. While asymmetric synthesis is defined as the intentional construction of enantiomers of a given constitutional isomer by means of chemical reactions, this definition, though simple, does not convey its enormous implications to the life and material sciences. Like not many others, Dieter Enders has shaped the field of asymmetric synthesis from its infancy in the late seventies of the past century up to the present day and one will hardly find a research paper on asymmetric synthesis without his name in the list of references. Most people might connect his name with the SAMP/RAMP hydrazone methodology, but he has also been in the front rank of asymmetric carbene catalysis, asymmetric organometallic chemistry and in particular asymmetric organocatalysis. The editors are grateful to have found an exceptional mentor in Dieter Enders, and on the occasion of his 60th birthday we would like to share his enthusiasm with some friends and colleagues.

In conversations with other colleagues who have accompanied and strongly influenced the development of asymmetric synthesis from the very beginning, we were fascinated to learn about their personal view on this field. Of course, every one of them has her/his own personal perspective and focus.

Our aim in editing this book was to capture the spirit of these informal conversations. After inviting a couple of colleagues first, we were very surprised by the overwhelming positive resonance and only very few colleagues were not able to submit a chapter. Many of the some sixty contributors of this book have been Dieter Enders' personal friends along the way and did not hesitate to give short accounts in order to put their own contributions in the context of asymmetric synthesis. We asked all of them to write very short essays; otherwise we would have ended with a multi-volume set. Additionally, we tried to be as consistent

---

[1] J. W. Cornforth, The Trouble with Synthesis, on the occasion of the Sir Robert Price Lecture 1992. *Aust. J. Chem.* **1993**, *46*, 157–170.

Asymmetric Synthesis – The Essentials.
Edited by Mathias Christmann and Stefan Bräse
Copyright © 2007 WILEY-VCH Verlag GmbH & Co. KGaA, Weinheim
ISBN: 978-3-527-31399-0

throughout the book as possible and did not change the personal style of the authors at all.

We were very proud that one of the pioneers in Asymmetric Synthesis – Professor Dieter Seebach – was able to write a personalized foreword. It was a true pleasure for us to invite him, since he was also a mentor of Dieter Enders.

The book is organized in five chapters reflecting various aspects of asymmetric synthesis. The first chapter "Diastereoselective Asymmetric Synthesis" introduced by David A. Evans, Günther Helmchen and Magnus Rueping deals with the creation of new stereocenters using existing stereochemical information in the molecule or from a chiral auxiliary. Catalysis – the synthesis of chiral molecules with a substoichiometric source of chiral information – is nowadays in most cases the method of choice and the chapter "Metal-Catalyzed Asymmetric Synthesis" is introduced by our Japanese colleagues Masakatsu Shibasaki and Takashi Oshima. Asymmetric catalysis with enzymes and small organic molecules is dealt with in Chapter III ("Enzymatic and Organic Catalysis") and Benjamin List from the Max-Planck Institute at Mülheim wrote the introduction of this chapter. K. C. Nicolaou and his student Paul Bulger explain to us the importance of "Asymmetric Reactions in Total Synthesis" in their detailed introduction to chapter IV. Therein, intriguing applications of the above mentioned methods to the synthesis of complex natural products will be given. Finally, Herbert Hugl, formerly of Lanxess in Germany, has written the last chapter "Asymmetric Synthesis in Industry" outlining the significance of asymmetric synthesis for the production of drugs and fine chemicals.

We would like to thank all distinguished scientists and their coauthors for their rewarding and timely contributions. Grateful acknowledgements are offered to the Wiley-VCH editorial staff, in particular to Dr. Elke Maase, who was a a great help at the very beginning of this project. We also thank Renate Dötzer and Hans-Jochen Schmitt for their professional work during the production process.

Finally, we would like to dedicate this book to Dieter Enders, our mentor and colleague and hope that it serves as a source of inspiration for young chemists.

Happy birthday, Dieter Enders!

Stefan Bräse and Mathias Christmann, May 2006

List of Abbreviations

AA	asymmetric aminohydroxylation
acac	acetylacetonyl
ACE	angiotensin converting enzyme
AD	asymmetric dihydroxylation
API	active pharmaceutical ingredient
BBA	Brønsted acid assisted Brønsted acid
BINAP	2,2'-diphenylphosphino-1,1'-binaphthyl
BINOL	2,2'-dihydroxy-1,1'-binaphthol
BITIANP	2,2'-bis(diphenylphosphino)-3,3'-dibenzo[b]thiophen
BLA	Brønsted acid assisted Lewis acid
BOX	bisoxazoline
BPB	2-[(N-benzylprolyl)amino]benzophenone
Bz	benzoyl
CA	conjugate addition
CALB	*Candida antarctica* lipase B
CAMP	cyclohexyl-*o*-anisylmethylphosphine
CAST	complete active-site saturation test
Cb	N,N-diisopropylcarbamoyl
CBS	Corey-Bakshi-Shibata
Cbz	benzyloxycarbonyl
CD	circular dichroism
CHMO	cyclohexanone monooxygenase
Chx	cyclohexyl
CINV	chemotherapy induced nausea and vomiting
ClMeOBIPHEP	5,5'-dichloro-6,6'-dimethoxy-2,2'-bis(diphenylphosphanyl)-1,1'-biphenyl
CM	cross metathesis
COD	1,5-cyclooctadiene
CPL	circularily polarized light
CTAB	cetyl trimethylammonium bromide
Cy	cyclohexyl
DA	Diels-Alder
DABCO	1,4-diazabicyclo[2.2.2]octane

Asymmetric Synthesis – The Essentials.
Edited by Mathias Christmann and Stefan Bräse
Copyright © 2007 WILEY-VCH Verlag GmbH & Co. KGaA, Weinheim
ISBN: 978-3-527-31399-0

DAIPN	1,1-dianisyl-2-isopropyl-1,2-ethylenediamine
dba	*trans, trans*-dibenzylideneacetone
DBU	1,8-diazabicylo[5.4.0]undec-7-ene
DCM	dichloromethane
De	diastereomeric excess
DERA	deoxyribose-5-phosphate aldolase
DET	diethyl tartrate
DIAD	diisopropyl azodicarboxylate
DIHMA	double intramolecular hetero-Michael addition
DIOP	2,3-*O*-isopropylidene-2,3-dihydroxy-1,4-bis(diphenylphosphino)butane
DiPAMP	1,2-bis(*o*-anisylphenylphosphino)ethane
DIPT	diisopropyl tartrate
DKR	dynamic kinetic resolution
DMAP	4-dimethylaminopyridine
DMDO	2,2-dimethyldioxirane
DME	ethylene glycol dimethyl ether
DMH	dimethylhydrazones
DNA	deoxyribonucleic acid
DOPA	3-(3,4-dihydroxyphenyl)alanine
EDCI	1-(3-dimethylaminopropyl)-3-ethylcarbodiimide hydrochloride
Ee	enantiomeric excess
epPCR	error-prone polymerase chain reaction
EtP_2	$(EtN=P(NMe_2)_2)_2$-$N=P(NMe_2)_3$
EWG	electron withdrawing group
FHPC	5-formyl-4-hydroxy[2.2]paracyclophane
HCd	10,11-dihydrocinchonidine
HDA	hetero-Diels-Alder
HDL	high density lipoprotein
HMG	3-hydroxy-3-hethylglutaryl
HMPA	hexamethylphosphoramide
HMPT	see HMPA
HOBt	1-hydroxybenzotriazole
HOMO	highest occupied molecular orbital
IBAO	isobutylaluminoxan
IPA	isopropyl alcohol
Ipc	isopinocampheyl
KR	kinetic resolution
LAH	lithium aluminium hydride
LBA	Lewis acid assisted Brønsted acid
LDA	lithium diisopropylamide
LDL	low density lipoprotein
LiCHIPA	lithium cyclohexylisopropylamine
LiHMDS	lithium bis(trimethylsilyl)amide
LLA	Lewis acid assisted Lewis acid

LLB	Lewis acid assisted Lewis base
LUMO	lowest unoccupied molecular orbital
MAO	methylaluminoxan
mCPBA	3-chloroperoxybenzoic acid
MEM	2-methoxyethoxymethyl
MMPP	magnesium monoperoxyphthalate
MOIPA	1-methoxyisopropylamine
MRSA	Methicillin-resistant *Staphylococcus aureus*
Ms	methanesulfonyl
MTBE	methyl *tert*-butylether
$NADP^+$	nicotinamide adenine dinucleotide phosphate
NADPH	nicotinamide adenosine dinucleotide phosphate
NCS	*N*-chlorosuccinimide
NLE	nonlinear effect
NMI	1-neomenthylindenyl
NP	natural product
PCC	pyridinium chlorochromate
PDC	pyridinium dichromate
PET	photoinduced electron transfer
PHOX	phosphino oxazoline
Piv	pivaloyl
PMB	*p*-methoxybenzyl
PMP	*p*-methoxyphenyl
PPTS	pyridinium *p*-toluenesulfonate
PTC	phasetranfer catalyst
Py	pyridine
PyBOP	bromotripyrrolidinophosphonium hexafluorophosphate
pyBOX	pyridinebis(oxazoline)
Pyr	pyridine
RAMP	(*R*)-1-amino-2-methoxymethylpyrrolidine
RCM	ring closing metathesis
RE	rare earth
ROM	ring opening metathesis
RRM	ring rearrangement metathesis
SAE	Sharpless asymmetric epoxidation
SAMP	(*S*)-1-amino-2-methoxymethylpyrrolidine
S-C	substrate-catalyst
SES	2-(trimethylsilyl)ethylsulfonyl
SMB	simulated moving bed
TADDOL	2,2-dimethyl-$\alpha,\alpha,\alpha',\alpha'$-tetraphenyl-1,3-dioxolan-4,5-dimethanol
TBAF	tetra-*n*-butylammonium fluoride
TBDPS	*tert*-butyldiphenylsilyl
TBS	*tert*-butyldimethylsilyl
TES	triethylsilyl
Tf	trifluoromethanesulfonyl

TFAA	trifluoroacetic anhydride
THC	Δ^9-tetrahydrocannabinol
TMBTP	4,4'-bis(diphenylphosphino)-2,2',5,5'-tetramethyl-3,3'-bithiophene
TMEDA	N,N,N',N'-tetramethylethylenediamine
TMS	trimethylsilyl
TOF	turnover frequency
TolBINAP	2,2'-bis(di-4-tolylphosphino)-1,1'-binapthyl
TON	turnover number
TPPB	tris(pentafluorophenyl)borane
Ts	p-toluenesulfonyl
UHP	urea/hydrogen peroxide adduct
VANOL	3,3'-diphenyl[2,2'-binaphthalene]-1,1'-diol
VMAR	vinylogous Mukaiyama aldol reaction
WERC	water exchange rate constant
X_c	chiral auxiliary
XylBINAP	2,2'-bis(di-4-xylylphosphino)-1,1'-binapthyl
ZACA	zirconium-catalyzed asymmetric carboalumination

Named and Unnamed Reactions

Aldol reaction   4, 6f, 53, 105, 161f, 164, 192f, 217ff, 225, 278ff
Aldol-Tishchenko reaction   141ff
Aza-Wittig reaction   259
Baeyer-Villiger   52ff, 194, 252f
Brehme reaction   24
Brønsted acid   154ff, 188f
Brønsted base   49
Brown allylation   230
Claisen rearrangement   123
Corey-Bakshi-Shibata reduction (CBS)   235ff, 243, 257
Danishefsky diene   34, 122, 154
Darzens reaction   52
Diels-Alder Reaction   4f, 32, 48f, 73, 85, 98, 133, 136, 139, 154, 156, 168, 194, 227, 229, 252, 285
Enders alkylation   21ff, 226, 228
Ene reaction   48, 155, 197
Erlenmeyer azlacton synthesis   316
Evans Auxiliary   33, 162, 227, 283
Felkin-Anh model   106, 226f, 281
Geissman-Weiss lactone   42
Glorius hydrogenation 264
Grieco oxidation   102
Grignard reagent   16, 35, 78f, 90, 93, 265
Grubbs catalyst   262ff
Hajos-Parrish-Eder-Sauer-Wiechert reaction   21, 163
Heck reaction   147ff, 231
Hetero-Diels-Alder   7, 32, 48, 73, 122, 136, 154, 192, 194, 229, 284
Hofmann degradation   21
Hofmann elimination   302
Jacobsen hydrolytic kinetic resolution (HKR) 275, 293, 299
Jacobsen-Katsuki epoxydation   116ff, 177, 233
Juliá-Colonna epoxydation   177f
Kazlauskas rule   174

Asymmetric Synthesis – The Essentials.
Edited by Mathias Christmann and Stefan Bräse
Copyright © 2007 WILEY-VCH Verlag GmbH & Co. KGaA, Weinheim
ISBN: 978-3-527-31399-0

Keck allylation 262
Kinugasa reaction 72ff
Lewis acid 27, 29, 35, 41, 49, 106, 110ff, 132f, 137, 153ff, 228f, 263
Lewis base 29, 121ff, 161
Lindlar reduction 244
Mannich reaction 34, 48, 113, 202f
Meerwein-Ponndorf-Verley reduction 119, 123
Michael reaction 23, 48, 69, 73, 132, 191ff, 202, 243, 257, 258, 270
Mitsunobu reaction 242f
Monsanto catalyst 236, 267, 292
Mukaiyama aldol reaction 105ff, 132, 139, 163
Nitro-Michael reaction 203
Nitrosoaldol reaction 156
Norrish-Yang cyclization 168
Noyori reduction 236, 321ff
Payne rearrangement 252, 254
Pearlman's catalyst 258
Roche ester 278
Schiff base 34
Sharpless asymmetric aminohydroxylation (SAA) 234f
Sharpless asymmetric dihydroxylation (SAD) 234f, 275
Sharpless asymmetric epoxidation (SAE) 10, 116, 198, 232, 299
Sharpless-Katsuki epoxydation, see Sharpless asymmetric epoxidation
Shi epoxidation 118, 233
Soai reaction 181ff, 212ff
Somitomu cyclopropanation 293
Staudinger reduction 259
Staudinger β-lactam synthesis 187f
Strecker reaction 34
Suzuki coupling, 7, 219
Swern oxidation 128
Takasago 292f
Tishchenko reaction 142, 145
Ugi reaction 34
Weinreb amide 19
Weitz-Scheffer reaction 176
Williamson etherification 21
Wittg olefination 52, 105
Ziegler-Natta polymerization 126f
Zimmerman-Traxler transition state 101

List of Contributors

Varinder Aggarwal
School of Chemistry
University of Bristol
Bristol BS8 1TS
Great Britain

Thorsten Bach
Lehrstuhl für Organische Chemie
Department Chemie
Technische Universität München
Lichtenbergstraße 4
85748 Garching
Germany

Jan E. Bäckvall
Organisk kemi
Stockholms Universitet
Arrheniuslaboratoriet
106 91 Stockholm
Sweden

Albrecht Berkessel
Institut für Organische Chemie
Universität Köln
Greinstraße 4
50939 Köln
Germany

Wolfgang Bettray
Institut für Organische Chemie
RWTH Aachen
Landoltweg 1
52074 Aachen
Germany

Donna G. Blackmond
Department of Chemistry
Imperial College
London SW7 2AZ
Great Britain

Hans-Ulrich Blaser
Solvias AG
P.O. Box
4002 Basel
Schweiz

Siegfried Blechert
Chemistry Department
Technische Universität Berlin
Straße des 17. Juni
10623 Berlin
Germany

Carsten Bolm
Institut für Organische Chemie
der RWTH Aachen
Landoltweg 1
52056 Aachen
Germany

Stefan Bräse
Institut für Organische Chemie
Universität Karlsruhe (TH)
Fritz-Haber-Weg 6
76131 Karlsruhe
Germany

Asymmetric Synthesis – The Essentials.
Edited by Mathias Christmann and Stefan Bräse
Copyright © 2007 WILEY-VCH Verlag GmbH & Co. KGaA, Weinheim
ISBN: 978-3-527-31399-0

Karel M. J. Brands
Department of Process Research
Merck Research Laboratories
Rahway, NJ 07065
USA

Gerhard Bringmann
Institut für Organische Chemie
Am Hubland
97074 Würzburg
Germany

Paul G. Bulger
Department of Chemistry
The Scripps Research Institute
10550 North Torrey Pines Road
La Jolla, CA 92037
USA

Reinhard Brückner
Institut für Organische Chemie
und Biochemie
Albert-Ludwigs-Universität Freiburg
Albertstraße 21
79104 Freiburg
Germany

Stefan Dahmen
cynora GmbH
Kaiserstraße 100
52134 Herzogenrath
Germany

Armin de Meijere
Institut für Organische und
Biomolekulare Chemie
Georg-August-Universität Göttingen
Tammannstraße 2
37077 Göttingen
Germany

Samuel Danishefsky
Department of Chemistry
Columbia University
3000 Broadway
mail code 3106
New York, NY 10027
USA

Franklin A. Davis
Department of Chemistry
Beury Hall 201
1901 N. 13th Street
Philadelphia, PA 19122
USA

Klaus Ditrich
BASF Aktiengesellschaft
GVF/E – A030
Ludwigshafen
Germany

Dieter Enders
Institut für Organische Chemie
RWTH Aachen
Landoltweg 1
52074 Aachen
Germany

David A. Evans
Department of Chemistry
and Chemical Biology
Harvard University
Cambridge, MA 02138
Great Britain

Ben L. Feringa
Department of Organic and
Molecular Inorganic Chemistry
University of Groningen
Nijenborgh 4
9747 AG Groningen
The Netherlands

Craig Forsyth
Department of Chemistry
Smith Hall
207 Pleasant St SE
Minneapolis, MN 55455-0431
USA

Gregory Fu
Massachusetts Institute of Technology
77 Massachusetts Ave.
Cambridge, MA 02139-4307
USA

Alois Fürstner
Max-Planck-Institut
für Kohlenforschung
Kaiser-Wilhelm-Platz 1
45470 Mülheim an der Ruhr
Germany

Hans-Joachim Gais
Institut für Organische Chemie
RWTH Aachen
Landoltweg 1
52074 Aachen
Germany

Jean-Pierre Genêt
Laboratoire de Synthèse
Sélective Organique et
Produits Naturels
Ecole Nationale Supérieure
de Chimie de Paris
11 Rue Pierre et Marie Curie
75231 Paris Cedex 05
France

William A. Greenberg
Department of Chemistry
The Scripps Research Institute
10550 North Torrey Pines Road
La Jolla, CA 92037
USA

Tanja Gulder
Institut für Organische Chemie
Am Hubland
97074 Würzburg
Germany

Tobias A. M. Gulder
Institut für Organische Chemie
Am Hubland
97074 Würzburg
Germany

Jorma Hassfeld
Institut für Organische Chemie
Leibniz Universität Hannover
Schneiderberg 1b
30167 Hannover
Germany

Tamio Hayashi
Laboratory of Organic Chemistry
Department of Chemistry
Kyoto University
Oiwake-Cho Sakyo-Ku
Kyoto City 606-8502
Japan

Günter Helmchen
Organisch-Chemisches Institut
der Universität Heidelberg
Im Neuenheimer Feld 270
69120 Heidelberg
Germany

Reinhard W. Hoffmann
Fachbereich Chemie
Philipps-Universität Marburg
Hans-Meerwein-Straße
36032 Marburg
Germany

Nicole Holub
Chemistry Department
Technische Universität Berlin
Straße des 17. Juni
10623 Berlin
Germany

Dieter Hoppe
Organisch-Chemisches Institut
Westfälische Wilhelms-Universität Münster
Corrensstraße 40
48149 Münster
Germany

Herbert Hugl
Gemarkenweg 9
51467 Bergisch Gladbach
Germany

Yasushi Imada
Department of Chemistry
Graduate School of Engineering Science
Osaka University
Osaka
Japan

Andreas Job
Saltigo GmbHBL Pharma –
Process Development Building Q 18
51369 Leverkusen
Germany

Karl Anker Jørgensen
Department of Chemistry
Aarhus University
Langelandsgade 140
8000 Aarhus C
Denmark

Henri Kagan
Laboratoire de Synthése Asymétrique
ICMMO Bât 420
Université Paris Sud
91405 Orsay cedex
France

Markus Kalesse
Institut für Organische Chemie
Leibniz Universität Hannover
Schneiderberg 1b
30167 Hannover
Germany

Tsutomu Katsuki
Department of Chemistry
Faculty of Science
Graduate School
Kyushu University 33
Hakozaki, Higashi-ku
Fukuoka 812-8581
Japan

Tsuneomi Kawasaki
Department of Applied Chemistry
Tokyo University of Science
Kagurazaka, Shinjuku-ku,
Tokyo, 162-8601
Japan

Florian Kleinbeck
Departement Chemie und
Angewandte Biowissenschaften
ETH Hönggerberg
8093 Zürich
Switzerland

William S. Knowles
661 East Monroe Avenue
St. Louis, MO 63122
USA

Shoū Kobayashi
Graduate School of
Pharmaceutical Sciences
The University of Tokyo
Hongo, Bunkyo-ku
Tokyo 113-0033
Japan

Horst Kunz
Institut für Organische Chemie
Universiät Mainz
Duesbergweg 10–14
55128 Mainz
Germany

Oleg V. Larionov
Institut für Organische und
Biomolekulare Chemie
Georg-August-Universität Göttingen
Tammannstraße 2
37077 Göttingen
Germany

Torben Leßmann
Max-Planck Institute für
Molecular Physiology
University Dortmund
Otto-Hahn-Straße 11
44202 Dortmund
Germany

Steven V. Ley
Department of Chemistry
University of Cambridge
Cambridge, CB2 1 EW
Great Britain

Bo Liang
Brown Laboratories of Chemistry
Purdue University
560 Oval Drive
West Lafayette, IN 47907-2084
USA

Benjamin List
Max-Planck-Institut
für Kohlenforschung
Kaiser-Wilhelm-Platz 1
45470 Mülheim an der Ruhr
Germany

Fernando López
Department of Organic and
Molecular Inorganic Chemistry
University of Groningen
Nijenborgh 4
9747 AG Groningen
The Netherlands

Florian Lotz
Insitut für Organische und
Biomolekulare Chemie
Universität Göttingen
Tammannstraße 2
37077 Göttingen
Germany

Keiji Maruoka
Department of Chemistry
Graduate School of Science
Kyoto University
Kyoto 606-8502
Japan

Shigeki Matsunaga
Graduate School of
Pharmaceutical Sciences
The University of Tokyo
Hongo, Bunkyo-ku
Tokyo 113-0033
Japan

Kazuhiro Matsumoto
Department of Chemistry
Faculty of Science
Graduate School
Kyushu University 33
Hakozaki, Higashi-ku
Fukuoka 812-8581
Japan

Albert I. Meyers
Chemistry Department
Colorado State University
Ft. Collins, CO 80523
USA

Kenji Mori
The University of Tokyo
1-20-6-1309 Mukogaoka
Bunkyo-ku
Tokyo 113-0023
Japan

Kilian Muniz
Kekulé-Institut für
Organische Chemie
und Biochemie
Gerhard-Domagk-Straße 1
53121 Bonn
Germany

Shun-ichi Murahashi
Department of Applied Chemistry
Okayama University of Science
Ridai-cho 1-1
Okayama 700-0005
Japan

H. C. Ei-ichi Negishi
Brown Laboratories of Chemistry
Purdue University
560 Oval Drive
West Lafayette, IN 47907-2084
USA

Kyriacos C. Nicolaou
Department of Chemistry
The Scripps Research Institute
10550 North Torrey Pines Road
La Jolla, CA 92037
USA

Tibor Novak
Brown Laboratories of Chemistry
Purdue University
560 Oval Drive
West Lafayette, IN 47907-2084
USA

Ryoji Noyori
Department of Chemistry and
Research Center for Materials Science
Nagoya University
Chikusa
Nagoya 464-8602
Japan

Chikako Ogawa
Graduate School of
Pharmaceutical Sciences
The University of Tokyo
Hongo, Bunkyo-ku
Tokyo 113-0033
Japan

Takashi Ohshima
Department of Chemistry
Graduate School of Engineering
Science
Osaka University
Toyonaka
Osaka 560-8531
Japan

Takeshi Ohkuma
Department of Chemistry and
Research Center for Materials Science
Nagoya University
Chikusa
Nagoya 464-8602
Japan

Takashi Ooi
Department of Chemistry
Graduate School of Science
Kyoto University
Kyoto 606-8502
Japan

Ian Paterson
Department of Chemistry
Cambridge University
Cambridge, CB2 1EW
Great Britain

Andreas Pfaltz
Department of Chemistry
University of Basel
St. Johanns-Ring 19
4056 Basel
Switzerland

Philipp J. Pye
Department of Process Research
Merck Research Laboratories
Rahway, NJ 07065
USA

Viresh H. Rawal
Department of Chemistry
University of Chicago
5735 South Ellis Avenue
Chicago, IL 60637
USA

Manfred T. Reetz
Max-Planck-Institut
für Kohlenforschung
Kaiser-Wilhelm-Platz 1
45470 Mülheim an der Ruhr
Germany

Kai Rossen
Sanofi-Aventis Deutschland GmbH
Industriepark Höchst, G838
65926 Frankfurt am Main
Germany

Magnus Rüping
Institut für Organische Chemie
und Chemische Biochemie
Chemical Biology
Johann Wolfgang Goethe-University
Marie-Curie-Straße 11
60439 Frankfurt am Main
Germany

Christian A. Sandoval
Department of Chemistry and
Research Center for Materials Science
Nagoya University
Chikusa
Nagoya 464-8602
Japan

Masakatsu Shibasaki
Graduate School of
Pharmaceutical Sciences
The University of Tokyo
Hongo, Bunkyo-ku
Tokyo 113-0033
Japan

Kenso Soai
Department of Applied Chemistry
Tokyo University of Science
Kagurazaka, Shinjuku-ku,
Tokyo, 162-8601
Japan

Andreas Stolle
Saltigo GmbHBL Pharma –
Process Development
Building Q 18
51369 Leverkusen
Germany

Ze Tan
Brown Laboratories of Chemistry
Purdue University
560 Oval Drive
West Lafayette, IN 47907-2084
USA

Avinash N. Thadani
Department of Chemistry
University of Chicago
5735 South Ellis Avenue
Chicago, IL 60637
USA

Lutz F. Tietze
Insitut für Organische und
Biomolekulare Chemie
Universität Göttingen
Tammannstraße 2
37077 Göttingen
Germany

Herbert Waldmann
Max-Planck Institute für
Molecular Physiology
University Dortmund
Otto-Hahn-Straße 11
44202 Dortmund
Germany

Chi-Huey Wong
Department of Chemistry
The Scripps Research Institute
10550 North Torrey Pines Road
La Jolla, CA 92037
USA

Hisashi Yamamoto
Department of Chemistry
5735 S. Ellis Ave. (SCL 317)
Chicago, IL 60637
USA

Part I

Chiral Auxiliaries in Asymmetric Synthesis

David A. Evans, Harvard University, USA, Günter Helmchen, Organisch-Chemisches Institut der Universität Heidelberg, Germany, and Magnus Rüping, Johann Wolfgang Goethe-University Frankfurt, Germany

Background

The use of chiral auxiliaries in the synthesis of enantiomerically pure compounds has found wide application for a variety of reactions over the last three decades. Despite the extensive developments in this area by many academic and industrial research groups, new auxiliary controlled reactions continue to evolve frequently [1]. First objectives in this area have been to develop chiral enolate-derived reactions, wherein the chiral auxiliary (X_c) is both readily available and easily recovered after the desired bond construction has been achieved (Scheme 1).

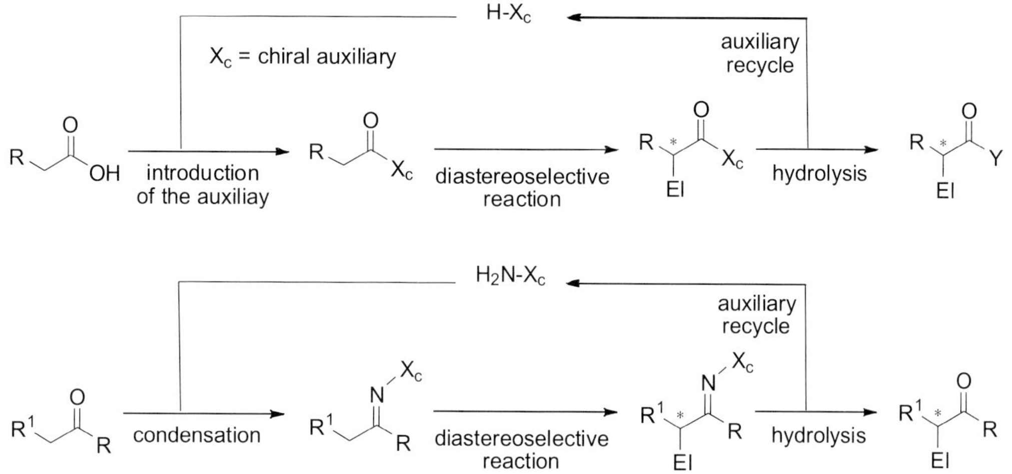

Scheme 1 Diastereoselective synthesis with chiral auxiliaries.

Generally, the major issues which have to be addressed in the development of diastereoselective transformations using chiral auxiliaries are threefold in nature. Subsequent to a facile introduction, the chiral auxiliary X_c must provide a strong predisposition for a highly selective enolization process; it must provide a strong

---

Asymmetric Synthesis – The Essentials.
Edited by Mathias Christmann and Stefan Bräse
Copyright © 2007 WILEY-VCH Verlag GmbH & Co. KGaA, Weinheim
ISBN: 978-3-527-31399-0

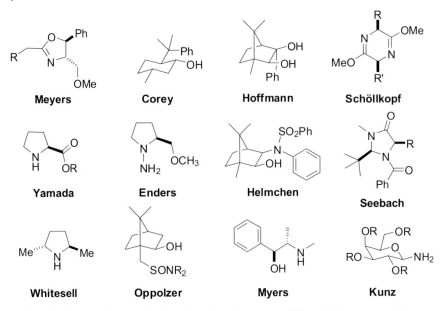

Figure 1 Selected chiral auxiliaries which have been successfully applied in asymmetric synthesis.

bias for enolate diastereoface selection in the new bond construction; and its non-destructive and mild cleavage must occur without racemization of the desired products. Today an arsenal of chiral auxiliaries is available meeting the above criteria in full or in part. Of the numerous chiral auxiliaries that have been developed over the past years some of the effectively applied auxiliaries are shown in Fig. 1. The majority of chiral auxiliaries are derived from inexpensive, chiral natural sources and most of the diastereoselective reactions reported proceed with high levels of diastereoselection. The most widely employed auxiliary controlled reactions are the asymmetric alkylations, aldol and Diels-Alder reactions.

Results

From the numerous auxiliary controlled reactions reported, a notable early example of an effective diastereoselective alkylation and Diels-Alder reaction has been developed by the Helmchen group, using the concave camphor-derived chiral auxiliaries **1** and **2** (Scheme 2) [2]. In this asymmetric alkylation procedure, a selective deprotonation leads to the corresponding E- or Z- ester enolate, which upon reaction with an alkyl halide and subsequent reduction results in enantiopure pure alcohols, valuable chiral building blocks and synthons for the synthesis of natural products. Remarkably, both diastereomers can be selectively obtained starting from the same chiral camphor derivative by simply changing the solvent.

One of the most utilized type of auxiliaries is the class of chiral oxazolidinones **1**, initially developed in the Evans group [3]. These chiral imides have been applied to a wide range of asymmetric transformations and the methodology devel-

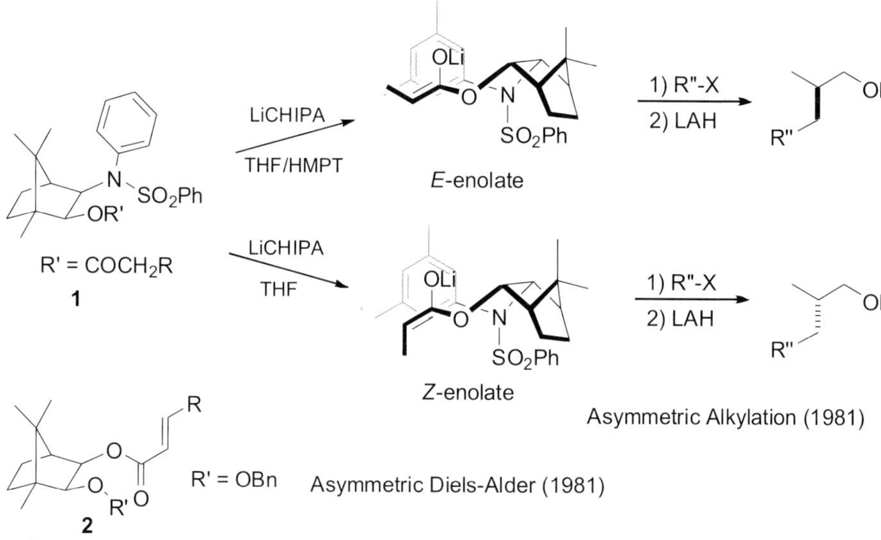

Scheme 2 Asymmetric alkylations and Diels-Alder reaction using Helmchen's camphor-derived auxiliaries.

oped has been most successful in the stereoselective construction of numerous chiral building blocks, as well as natural products, antibiotics and medicinally important compounds. Subsequent to the initial reports regarding oxazolidininone **1**, many structural variants (Fig. 2) have been developed which display different cleavage reactivity or complimentary diastereoselectivity compared to **1**.

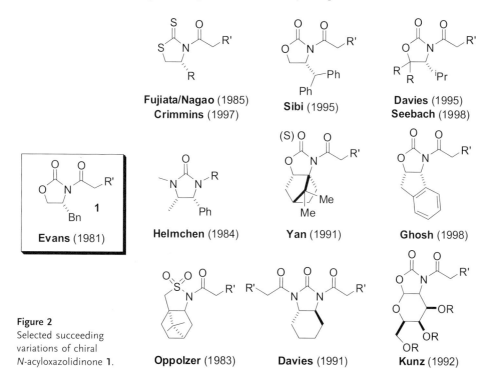

Figure 2 Selected succeeding variations of chiral N-acyloxazolidinone **1**.

Oxazolidinones of type **1**, which were initially developed for an efficient asymmetric C–C bond construction in the synthesis of several polyketide-derived natural products, have proved to be a gold standard and have continually been employed by the Evans group and numerous other groups over the last 20 years.

The first asymmetric reactions involving these chiral enolate synthons were the aldol and alkylation reactions. In these reactions selective enolization to form the Z-enolates ($Z:E > 100$) were achieved using either lithium and sodium amide bases or dibutylboryl trifluorosulfonate. Subsequent alkylation or aldol reaction of the corresponding metal enolates resulted in the products with highest levels of asymmetric induction (Scheme 3). Based on these seminal observations many other reactions employing chiral oxazolidinones have been reported over the years and the application will continue to be of great importance in the future [4].

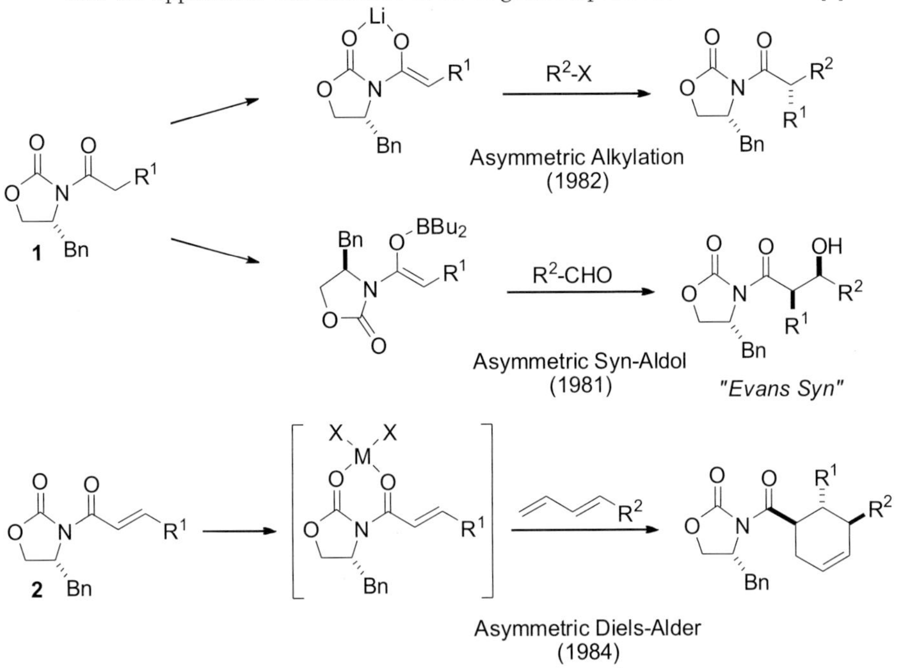

Scheme 3 Initial asymmetric alkylation, aldol and Diels-Alder reactions.

The original reports on the asymmetric aldol reactions mediated by boron or titanium resulted in the *syn* aldol product with very high diastereoselectivity. More recent studies by the Evans group have demonstrated an extension of the aldol process, which employs the same oxazolidinone **1** or the thiazolidine thione **2**, in the presence of catalytic amounts of magnesium salts, forming the *anti* aldol products, which were previously more difficult to access (Scheme 4) [5].

The significant cost effectiveness and facile scale-up of these magnesium halide catalyzed anti aldol reactions render them valuable methods for the preparation of various chiral building blocks and biologically important compounds, especially as all four diastereoisomers can be prepared from a single isomer of the auxiliary.

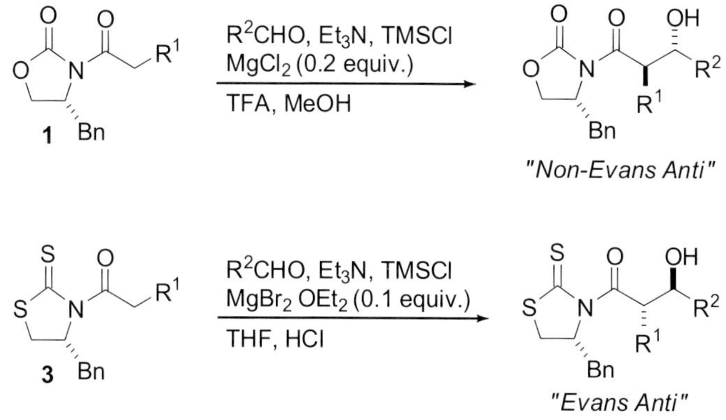

Scheme 4 Catalytic diastereoselective anti-aldol reactions.

The application of aldol reactions in natural product synthesis has recently been highlighted in the synthesis of FR-182877 by Evans, where all stereochemical relationships in the target structure were obtained from chiral oxazolidinone auxiliary controlled aldol reactions. Similar to the syntheses of himachalene [6] and phomoidride B [7], the asymmetric aldol reaction was the fundamental step for the construction of the key fragments of FR-182877, which were then united via a Suzuki coupling, followed by macrolactonization and oxidation. A subsequent Diels-Alder-Hetero-Diels-Alder reaction cascade culminated in the synthesis of hexacyclic FR-182877 (Scheme 5) [8].

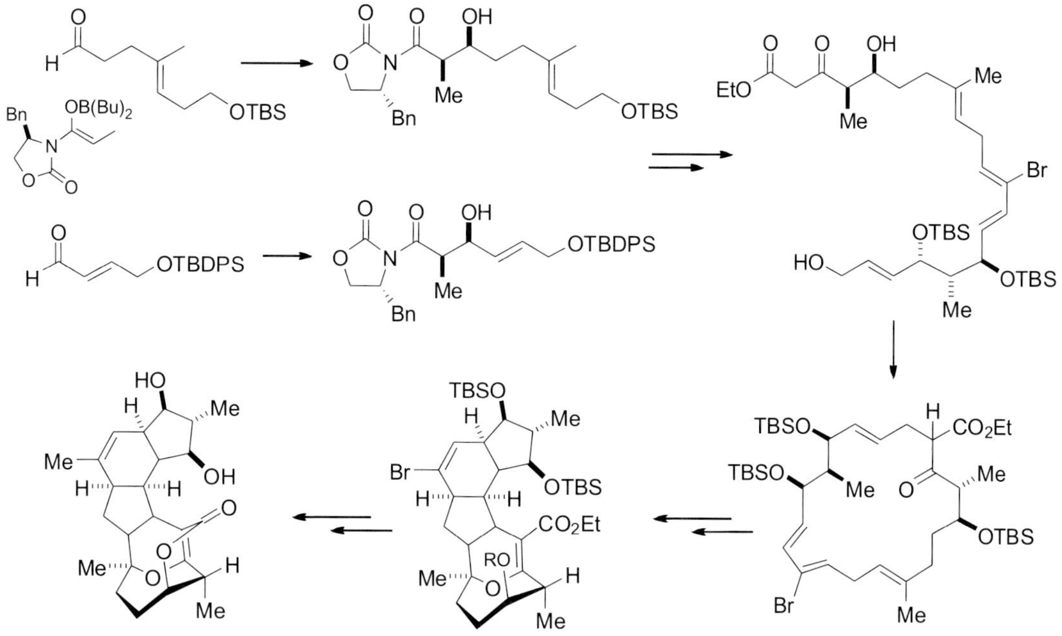

Scheme 5 Synthesis of FR-182877 using the auxiliary controlled aldol reactions and a Diels-Alder reaction cascade.

Summary

Asymmetric reactions employing chiral auxiliaries have experienced a remarkable progress over the past decades. Recent results from our groups, as well as many others, demonstrate that auxiliary-controlled reactions are still essential tools in the construction of complex molecular targets. The ready availability of the starting materials, the facile and versatile cleavage, as well as the applicability and reliability in a variety of stereoselective transformations, allows chiral auxiliaries to endure today as excellent synthetic intermediates in asymmetric synthesis.

CV of David A. Evans

David A. Evans was born in Washington D.C. in 1941. He received his A.B. degree from Oberlin College in 1963. He obtained his Ph.D. at the California Institute of Technology in 1967, where he worked under the direction of Professor Robert E. Ireland. In that year he joined the faculty at the University of California, Los Angeles. In 1973 he was promoted to the rank of Full Professor and shortly thereafter returned to Caltech where he remained until 1983. He then joined the Faculty at Harvard University and in 1990 he was appointed as the Abbott and James Lawrence Professor of Chemistry.

CV of Günter Helmchen

Günter Helmchen (b. 1940) is a Full Professor at the University of Heidelberg and director of the Institute of Organic Chemistry. He pursued undergraduate studies at the TH Hannover (Dipl.-Chem. 1965). His graduate work, completed in 1971, was carried out under the guidance of Professor V. Prelog at the ETH Zürich in the area of stereochemistry. He then joined the group of H. Muxfeldt for postdoctoral studies in the area of natural product synthesis and carried out a Habilitationsarbeit at the Technical University of Stuttgart (1975–1980). In 1980 he was appointed Professor C3 at the University of Würzburg. In 1985 he moved to his present position. His interest in catalysis dates back to ca. 1990. His scientific work has been recognized by a variety of scientific prizes and research awards, international lectureships and the invitation to join the advisory boards of scientific journals.

CV of Magnus Rüping

Magnus Rüping was born in Telgte, Germany, in 1972. He studied at the Technical University of Berlin, Trinity College Dublin and ETH Zürich. He obtained his Ph.D. in 2002 from ETH under the guidance of Professor Dieter Seebach. After carrying out postdoctoral work with Professor David A. Evans at Harvard University, he joined the Johann Wolfgang Goethe-University Frankfurt as Degussa Endowed Professor of Chemistry in fall 2004.

Selected References

1. (a) *Houben-Weyl, Methods in Organic Chemistry, Stereoselective Synthesis*, G. Helmchen, R.W. Hoffmann, J. Mulzer, E. Schaumann (Eds.), Thieme-Verlag, Stuttgart, **1995**; (b) *Compendium of Chiral Auxiliary Applications*, G. Roos (Ed.), Academic Press, New York, **2002**.
2. (a) R. Schmierer, G. Grotemeier, G. Helmchen, A. Selim, *Angew. Chem. Int. Ed. Engl.* **1981**, *20*, 207–208, Functional Groups at Concave Sites: Asymmetric Alkylation of Esters with Very High Stereoselectivity and Reversal of Configuration by Change of Solvent; (b) G. Helmchen, R. Schmierer, *Angew. Chem. Int. Ed. Engl.* **1981**, *20*, 205–207, Functional Groups at Concave Sites: Asymmetric Diels-Alder Synthesis with Almost Complete (Lewis-Acid Catalyzed) or High (Uncatalyzed) Stereoselectivity.
3. D. A. Evans. *Aldrichim. Acta* **1982**, *15*, 23–32, Studies in Asymmetric Synthesis. The Development of Practical Chiral Enolate Synthons.
4. D. J. Ager, I. Prakash, D. R. Schaad, *Aldrichim. Acta* **1997**, *30*, 3–12, Chiral Oxazolidinones in Asymmetric Synthesis.
5. (a) D. A. Evans, J. S. Tedrow, J. T. Shaw, C. W. Downey, *J. Am. Chem. Soc.* **2002**, *124*, 392–393, Diasteroselective Magnesium Halide Catalyzed Anti-Aldol Reactions of Chiral N-Acyloxazolidinones; (b) D. A. Evans, C. W. Downey, J. T. Shaw, J. S. Tedrow, *Organic Lett.* **2002**, *4*, 1127–1130, *Magnesium Halide-Catalyzed Anti-Aldol Reaction of Chiral N-Acylthiazoldinethiones*.
6. D. A. Evans, D. H. B. Ripin, J. S. Johnson, E. A. Shaughnessy, *Angew. Chem. Int. Engl.* **1997**, *36*, 2119–2121, *A New Strategy for Extending N-Acyl Imides as Chiral Auxiliaries for Aldol and Diels-Alder Reactions: Application to an Enantioselective Synthesis of α-Himachalene*.
7. N. Waizumi, T. Itoh, T. Fukuyama *J. Am. Chem. Soc.* **2000**, *122*, 7825–7826, Total Synthesis of (–)-CP-263,114 (Phomoidride B).
8. D. A. Evans, J. T. Starr *Angew. Chem. Int. Ed.* **2002**, *41*, 1787–1790, A Cascade Cycloaddition Strategy Leading to the Total Synthesis of (–)-FR182877.

Asymmetric Epoxidation of Pentadienols

Reinhard Brückner, University of Freiburg, Germany

Background

The asymmetric epoxidation of allylic alcohols – or Sharpless asymmetric epoxidation ("SAE") or Sharpless-Katsuki epoxidation – was a breakthrough in asymmetric synthesis [1]. Arguably it is one of the top ten transformations of organic chemistry [2]. Indeed, it became a Nobel Prize winning reaction [3]. SAEs of achiral primary allylic alcohols **1** lead to glycidols of controllable configuration **2** or *ent*-**2** (Scheme 1). These compounds can be carried on to an abundance of follow-up species by elaborating any or all of the *three* functionalities at C^1, C^2 or C^3 [4]. In contrast, SAEs of racemic secondary allylic alcohols *rac*-**3** affect one enantiomer of the substrate and enrich the other, i.e. *R*-**3** or *S*-**3**, accomplishing a kinetic resolution.

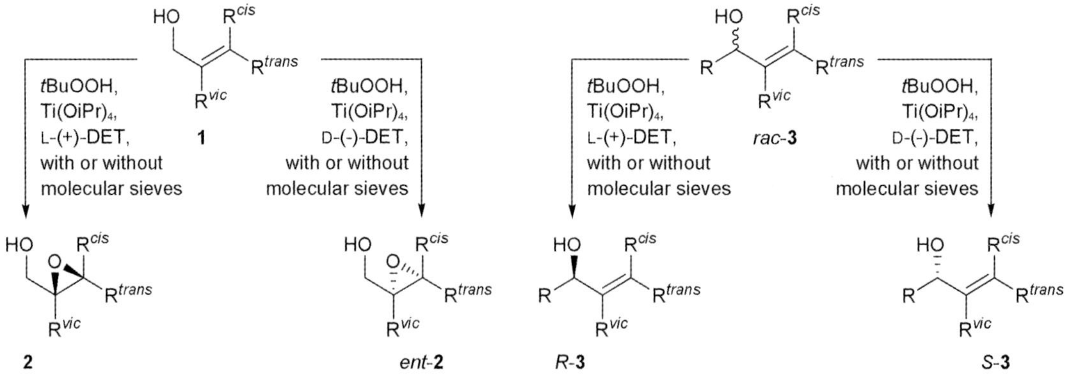

Scheme 1 SAEs of achiral primary allylic alcohols **1** or racemic secondary allylic alcohols *rac*-**3**.

Conceptually most intriguing are the desymmetrizing SAEs depicted in Scheme 2. Divinylcarbinol (**4**), a prochiral alcohol, and bis(allylic alcohol) **5**, a *meso* alcohol, provided epoxyalcohols **7** with > 99.7 % *ee* [5] and **8** with "> 99.99999 % *ee* expected" [6], respectively. These enantioselectivities distinctly surpass those found for achiral primary allylic alcohols. Interestingly, this out-

Asymmetric Synthesis – The Essentials.
Edited by Mathias Christmann and Stefan Bräse
Copyright © 2007 WILEY-VCH Verlag GmbH & Co. KGaA, Weinheim
ISBN: 978-3-527-31399-0

come could be predicted by Schreiber's insightful analysis [5a]. The tertiary *meso*-dialkenylcarbinol **6** was desymmetrized similarly, albeit only when Zr(OiPr)$_4$ was used and not Ti(OiPr)$_4$ [7].

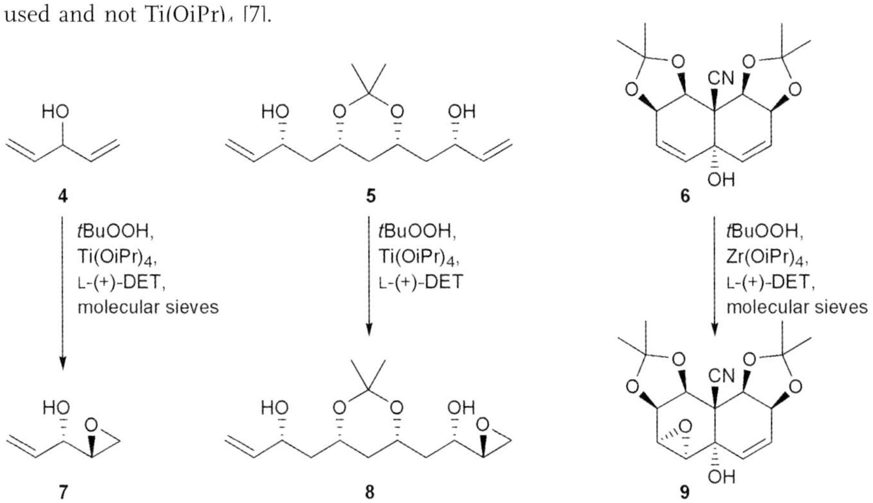

Scheme 2 Desymmetrization of achiral secondary alcohols **4** and **5** and tertiary allylic alcohol **6** by SAEs [Ti(OiPr)$_4$-mediated] and by a related epoxidation [Zr(OiPr)$_4$-mediated], respectively.

Objective: Making Building Blocks for the Synthesis of 1,3,5,7,...-Polyols from Epoxyalcohols

Polyol/polyene macrolide antibiotics contain extended stretches of unbranched 1,3,5,7,...polyols **12** (Scheme 3). The latter are neither "isotactically" nor "syndiotactically" configured but comprise, rather, random sequences of *syn*- and *anti*-configured 1,3-diol subunits. This feature suggests that a set of 1,3-diol building blocks **13** of all four conceivable configurations would be useful for constructing such polyols. We accessed two sets of such molecules **13** via SAEs: one from the conjugated dienol *trans*-**11** and one starting from the nonconjugated dienol *cis*-**15**. In both substrates, one C=C bond was oxidized and one preserved – initially, namely until its presence allowed follow-up transformations.

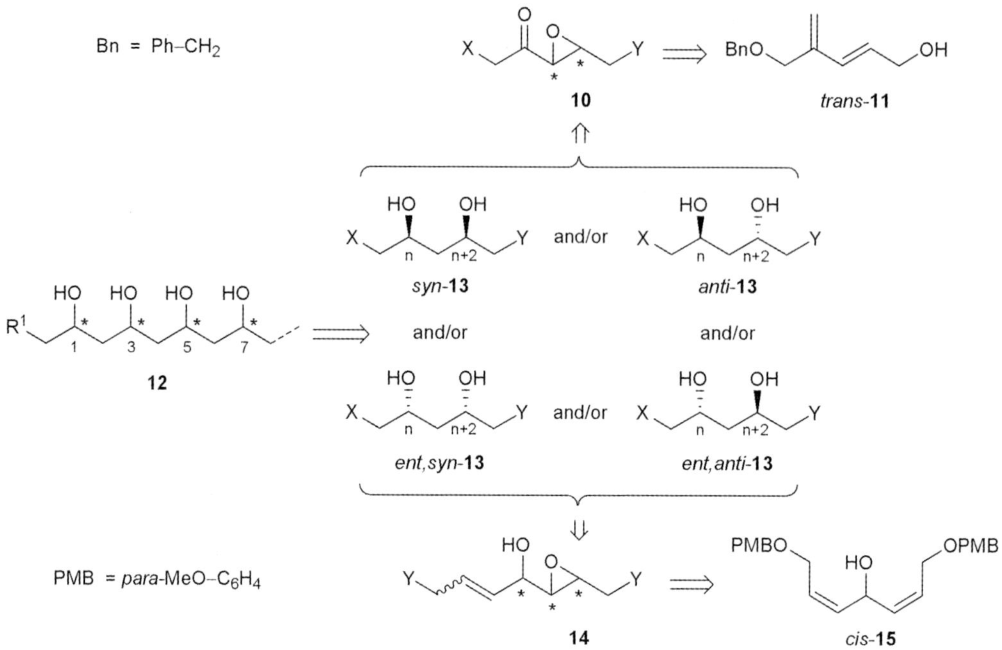

Scheme 3 Tracing back 1,3,5,7,...-polyols (12) via 1,3-diol building blocks (13) to epoxidized allylic alcohols.

Results

Asymmetric Epoxidation of a Conjugated Pentadienol [8]

SAE of pentadienol *trans*-11 functioned best (89 % yield/97.7 % ee) with near-stoichiometric [1a] rather than catalytic [1b] amounts both of L-(+)-diisopropyl tartrate and Ti(OiPr)$_4$ (Scheme 4). Epoxyalcohol 17 was silylated, the C=C bond ozonolyzed, and the resulting epoxyketone 18 reduced, either by chelation-controlled hydride addition (→19) or by electron transfer (→21). Renewed reduction followed by transacetalization yielded acetonides *anti*-20 and *syn*-20, respectively. Their protecting groups were selectively removable. Accordingly, these species are realizations of the 1,3-diol building blocks *syn*-, *anti*-, *ent,syn*-, and *ent,anti*-13.

Asymmetric Epoxidation of a Nonconjugated Pentadienol [9, 10]

The substrate of this approach to 1,3-diol motifs of variable stereostructure was the bis(*cis*-alkenyl)carbinol *cis*-15 (readily obtained from propargyl ether 22 and ethyl formiate; Scheme 5). SAE in the presence of molecular sieves [1b] and stoichiometric Ti(OiPr)$_4$/diisopropyl tartrate [1a] proceeded with 95–96 % ee. Since SAE of the analogous mono-PMB ether of *cis*-2-butene-1,4-diol gives only 85–88 % ee, the "Schreiber effect" is likely to have intervened. Whereas epoxyalcohols *anti*- and *ent,anti*-23 formed with only ∼75:25 ds, the epimers *syn*- and *ent,syn*-23 resulted diastereopure from the Zr(OiPr)$_4$-mediated AE of the same pentadienol *cis*-15 [7]. In these conditions, we raised the ee of epoxyalcohol *syn*-23 up to 99 %

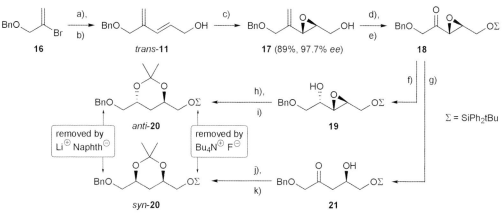

Scheme 4 a) $H_2C=CH-CO_2Me$, $Pd(OAc)_2$ (cat.), LiCl, Bu_4NCl, K_2CO_3, DMF; 57%. b) DIBAL, CH_2Cl_2; 85%. c) tertBuOOH, $Ti(OiPr)_4$ (56 mol%), L-(+)-diisopropyl tartrate (64 mol%), CH_2Cl_2, molecular sieves 4 Å. d) tertBuPh$_2$SiCl, imidazole, THF; 90%. e) O_3, CH_2Cl_2; PPh_3; 81%. f) $Zn(BH_4)_2$, toluene; 73%. g) Zn, Cp_2TiCl_2, 1,4-cylohexadiene; 60%. h) Same as (g); 67%. i) $Me_2C(OMe)_2$, camphor sulfonic acid (cat.), acetone; 79%. j) Et_2BOMe, MeOH, THF; $NaBH_4$; 73%. k) Same as (i); 85%. (Ref. [8].)

by allowing for some over-oxidation, and *proved* that this over-oxidation consumes most of the initially present minor enantiomer, furnishing bisepoxyalcohol *syn*, *syn*-**24** (Figure 1).

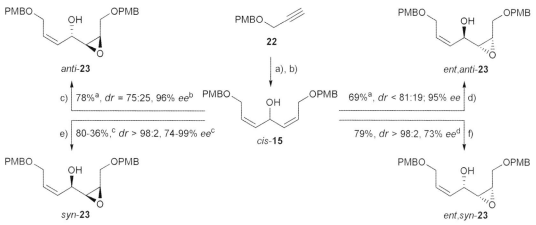

Scheme 5 a) nBuLi, THF; HCO_2Et; 85%. b) $ZnCl_2$ pre-treated with K, THF/MeOH; 75%. c) tertBuOOH, $Ti(OiPr)_4$ (1.0 equiv.), L-(+)-diisopropyl tartrate (1.1 equiv.), CH_2Cl_2, molecular sieves 4 Å. d) Same as (c) with D-(−)-diisopropyl tartrate (1.1 equiv.). (Ref. [9]) e) Same as (c) with $Zr(OiPr)_4$ (1.0 equiv.). f) Same as (e) with D-(−)-diisopropyl tartrate (1.1 equiv.). (Ref. [10].) [a]Taking into account 5–8% re-isolated *cis*-**15**. [b]Improvement of Ref. [10] vs. Ref. [9]. [c]Yield and *ee* were time-dependent (cf. Figure 1). [d]This conversion aimed for high yield rather than high *ee*.

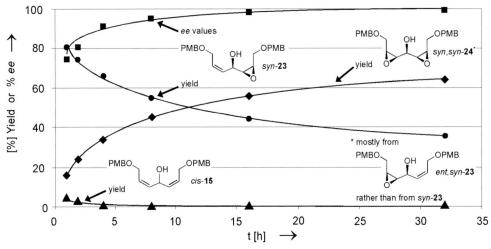

Figure 1 Time-resolved product analysis of the epoxidation of pentadienol cis-**15a** with tertBuOOH (2.0 equiv.), Zr(OiPr)$_4$ (1.0 equiv.), L-(+)-diisopropyl tartrate (1.1 equiv.), and molecular sieves (4 Å) in CH$_2$Cl$_2$ at −20 °C.

Scheme 6 depicts the extraction of stereodefined 1,3-diol building blocks from the pentadienol oxides **23** for a representative enantiomer from both the *anti*- and the *syn*-series: through reduction by Red-Al™ at 60 °C. The reagent effects two transformations in a single operation, namely regioselective opening of the epoxide and chemoselective cleavage of the allylic ether (by what we believe to be an OH-directed S$_N$2' reaction).

Scheme 6 a) Red-Al™, toluene; 83 % (analogous reduction of *ent,anti*-**23**: 95 %) (Ref. [9]). b) Same as (a); 51 % (Ref. [10]).

CV of Reinhard Brückner

Reinhard Brückner (born 1955) studied chemistry at the Universität München, acquiring his doctoral degree with Rolf Huisgen (1984). After post-doctoral studies with Paul A. Wender (Stanford University), he realized a habilitation under the auspices of Reinhard W. Hoffmann (Universität Marburg). Brückner has been a Professor of Organic Chemistry at the Universities of Würzburg (1990–92), Göttingen (1992–98), and Freiburg (since 1998) and a Visiting Professor at the Universities of Wisconsin (Madison), Santiago de Compostela (Spanien), Indiana (Bloomington), and Tokyo (Tokyo University). He received the Literature Prize of the Fonds der Chemischen Industrie for his textbook on organic reaction mechanisms and the Chemistry Prize of the Akademie der Wissenschaften Göttingen.

References

1. (a) T. Katsuki, K. B. Sharpless, *J. Am. Chem. Soc.* **1980**, *102*, 5974–5976. The First Practical Method for Asymmetric Epoxidation. (b) R. M. Hanson, K. B. Sharpless, *J. Org. Chem.* **1986**, *51*, 1922–1925. Procedure for the Catalytic Asymmetric Epoxidation of Allylic Alcohols in the Presence of Molecular Sieves.

2R. (a) T. Katsuki, V. S. Martín, *Org. React.* **1996**, *48*, 1–299. Asymmetric Epoxidation of Allylic Alcohols: The Katsuki-Sharpless Epoxidation Reaction. (b) R. A. Johnson, K. B. Sharpless, in *Catalytic Asymmetric Synthesis*, I. Ojima (Ed.), Wiley-VCH, Weinheim, 2nd edn., **2000**, pp. 231–286. Catalytic Asymmetric Epoxidation of Allylic Alcohols.

3. K. B. Sharpless, *Angew. Chem. Int. Ed.* **2002**, *41*, 2024–2032. The Search for New Chemical Reactivity.

4R. P. C. A. Pena, S. M. Roberts, *Curr. Org. Chem.* **2003**, *7*, 555–571. The Chemistry of Epoxy Alcohols.

5. (a) S. L. Schreiber, T. S. Schreiber, D. B Smith, *J. Am. Chem. Soc.* **1987**, *109*, 1525–1529. Reactions that Proceed with a Combination of Enantiotopic Group and Diastereotopic Face Selectivity can Deliver Products with Very High Enantiomeric Excess: Experimental Support of a Mathematical Model. (b) D. B. Smith, Z. Wang, S. L. Schreiber, *Tetrahedron* **1990**, *46*, 4793–4808. The Asymmetric Epoxidation of Divinyl Carbinols: Theory and Applications.

6. S. L. Schreiber, M. T. Goulet, G. Schulte, *J. Am. Chem. Soc.* **1987**, *109*, 4718–4720. Two-Directional Chain Synthesis: The Enantioselective Preparation of Syn-Skipped Polyol Chains from Meso Presursors.

7. A. C. Spivey, S. J. Woodhead, M. Weston, B. I. Andrews, *Angew. Chem. Int. Ed.* **2001**, *40*, 769–771. Enantioselective Desymmetrization of meso-Decalin Diallylic Alcohols by a New Zr-Based Sharpless AE Process: A Novel Approach to the Asymmetric Synthesis of Polyhydroxylated Celastraceae Sesquiterpene Cores.

8. S. Weigand, R. Brückner, *Synlett* **1997**, 225–228. Building Blocks for the Stereocontrolled Synthesis of 1,3-Diols of Various Configurations.

9. T. Berkenbusch, R. Brückner, *Synlett* **2003**, 1813–1816. Concise Synthesis of Optically Pure syn-1,3-Diols by Stereoselective Desymmetrization of a Divinylcarbinol.

10. R. Kramer, R. Brückner, *Synlett* **2006**, 33–38. Desymmetrizing Asymmetric Epoxidations of Bis(cis-Configured) Divinylcarbinols: Unusual syn-Selectivity Combined with ee-Enhancement through Kinetic Resolution.

Asymmetric Synthesis using Sulfur–Nitrogen Reagents

Franklin A. Davis, Temple University, Philadelphia, USA

Background

When we began exploring the chemistry of sulfur–nitrogen reagents such as sulfenamides ($RSNR_2$) in the late sixties there were no thoughts of asymmetric synthesis. The rationale for studying this underexplored class of compounds was the hope that new and perhaps useful chemistries would be discovered. In this context it was found that when a disulfide, silver nitrate, ammonia, and an aldehyde were mixed a new class of sulfur–nitrogen compounds resulted that we called sulfenimines (N-sulfenyl imines) [1R]. Sequential oxidation of sulfenimines with m-CPBA produced sulfinimines, sulfonimines, and N-sulfonyloxaziridines, respectively (Figure 1). At this time sulfinimines and N-sulfonyloxaziridines were unknown classes of compounds.

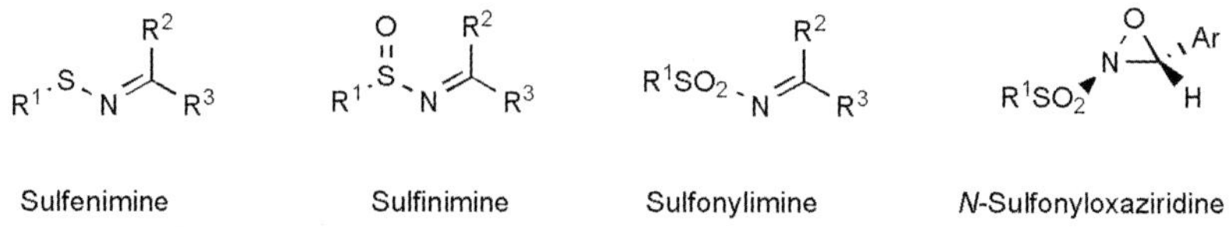

Sulfenimine   Sulfinimine   Sulfonylimine   N-Sulfonyloxaziridine

Figure 1 Sulfur–nitrogen reagents.

Results

Asymmetric Oxidations using N-Sulfonyloxaziridines

It was soon discovered that N-sulfonyloxaziridines were oxidizing reagents capable of selectively epoxidizing alkenes and oxidizing sulfides to sulfoxides, selenides to selenoxides, disulfides to thiosulfinates, thiols to sulfenic acids (RSOH), and amines to amine oxides [2R]. Because they are aprotic and neutral oxidizing agents they are among the very few reagents able to oxidize lithium and Grignard reagents to alcohols and enolates to α-hydroxy carbonyl compounds [2R, 3R, 4R].

The exploration of the asymmetric oxygen-transfer reactions of N-sulfonyloxaziridine required enantiopure examples and the (camphorsulfonyl)oxaziridines 1–4 have proven to be the most useful (Figure 2). They not only produce high ees, with predictable stereochemistry, but they are easily prepared, in both enantiomerically pure forms and on a large scale by biphasic oxidation of the corresponding (camphorsufonyl)imines [2R, 3R, 4R,5]. Oxaziridines (+)-1 and (+)-2 are commercially available.

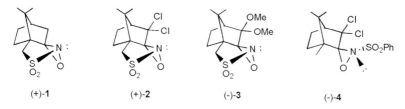

(+)-1   (+)-2   (-)-3   (-)-4

Figure 2 Enantiopure (camphoryl)sulfonyloxaziridines [3R].

Without question, the most widely used applications of N-sulfonyloxaziridines 1–3 is for the asymmetric oxidation of enolates to α-hydroxy carbonyl compounds, moieties found in many biologically active molecules [3R, 4R]. Because the absolute configuration of the oxaziridine three-member ring controls the stereochemistry of the product, either enantiomer is readily available through choice of the antipodal oxaziridine. Some examples of α-hydroxy carbonyl compounds and the oxaziridines used to produce them are given in Figure 3.

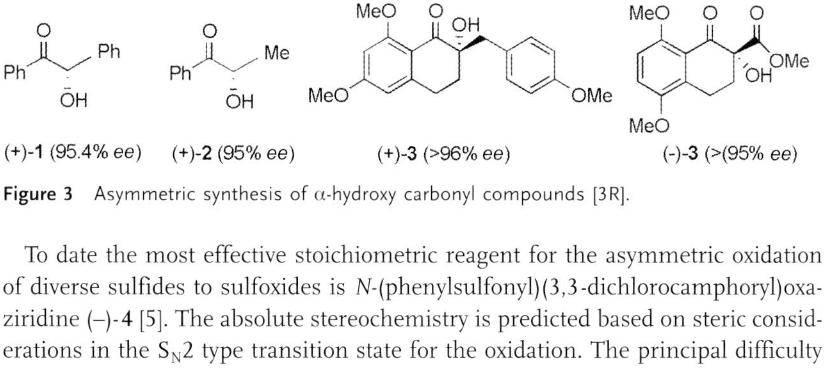

(+)-1 (95.4% ee)   (+)-2 (95% ee)   (+)-3 (>96% ee)   (-)-3 (>(95% ee)

Figure 3 Asymmetric synthesis of α-hydroxy carbonyl compounds [3R].

To date the most effective stoichiometric reagent for the asymmetric oxidation of diverse sulfides to sulfoxides is N-(phenylsulfonyl)(3,3-dichlorocamphoryl)oxaziridine (–)-4 [5]. The absolute stereochemistry is predicted based on steric considerations in the S_N2 type transition state for the oxidation. The principal difficulty in preparing selenoxides in high enantiomeric purity is their moisture sensitivity; i. e. the formation of achiral hydrates (ArSe(OH)$_2$Me). Because (–)-4 is aprotic and these oxidations can be conducted under anhydrous conditions, it was possible, for the first time, to obtain optically enriched selenoxides [6]. Examples of optically active (S)-sulfoxides and an (S)-selenoxide prepared using N-sulfonyloxaziridine (–)-4 are given in Figure 4.

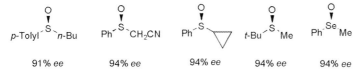

p-Tolyl-S-n-Bu   Ph-S-CH$_2$CN   Ph-S-△   t-Bu-S-Me   Ph-Se-Me

91% ee   94% ee   94% ee   94% ee   94% ee

Figure 4 Sulfoxides and a selenoxide synthesized asymmetrically using oxaziridine (-)-4.

Asymmetric Fluorination of Enolates with N-Fluoro-2,10-Camphorsultams

Fluorination of camphorsultam **5** with 10% F_2/N_2 gave (−)-N-fluoro-2,10,-camphorsultam (**6**) in 67% yield [7, 8R]. On treatment with this chiral electrophilic fluorinating reagent enolates afford enantiomerically enriched α-fluorocarbonyl compounds (Scheme 1).

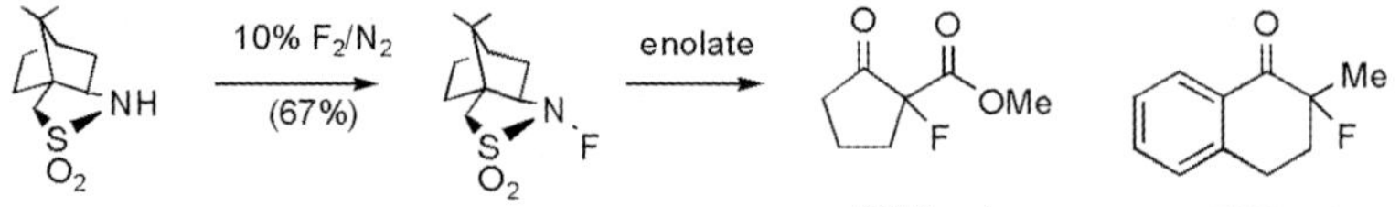

Scheme 1 Asymmetric fluorination of enolates with (−)-**6** [8R].

Asymmetric Synthesis of Amine Derivatives using Enantiopure Sulfinimines (N-Sulfinyl Imines)

Asymmetric oxidation of the sulfenimines corresponding to **7** gave the first examples of enantiomerically enriched sulfinimines derived from aldehydes [9R]. Although crystallization improved the ee to >95% the asymmetric synthesis of sulfinimines by this method proved to be impractical. However, it was soon learned that these chiral imine building blocks could be easily prepared, on a large scale, by condensation of (R)- or (S)-p-toluenesulfinamide (**8**) with an aldehyde or ketone in the presence of a Lewis acid dehydrating agent such as Ti(OEt)$_4$ (Scheme 2). Sulfinamide **8**, now commercially available, is prepared from menthyl p-toluenesulfinate (Andersen reagent) and LiHMDS.

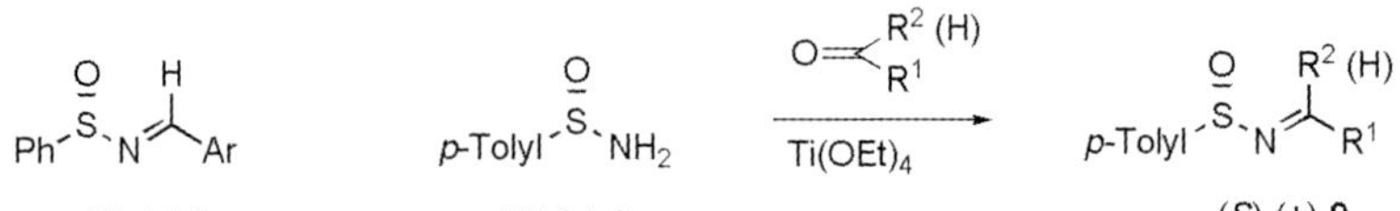

Scheme 2 Asymmetric synthesis of sulfinimines [9R].

Sulfinimines **9** offer a general solution to the problem of addition of organometallic reagents to the C=N double bond of chiral imines because the N-sulfinyl group activates the C=N bond for addition, is highly stereodirecting and is easily removed without epimerization of the amine product. In fact the most direct and reliable method for the asymmetric syntheses of diverse amines is the addition of an organometallic reagent to an enantiopure sulfinimine [9R]. Some examples of amine derivatives prepared via sulfinimine chemistry are given in Figure 5.

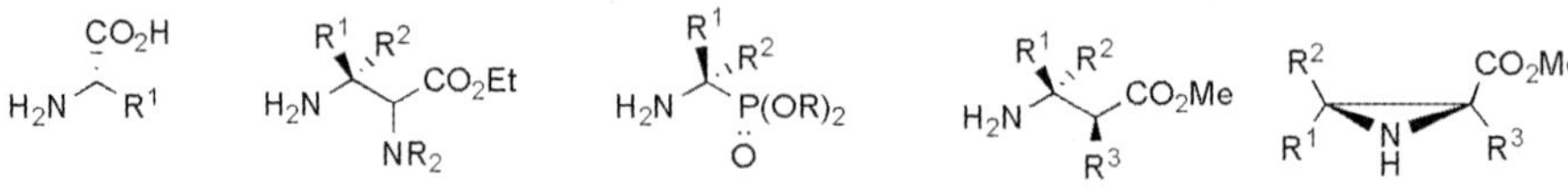

Figure 5 Asymmetric synthesis of amine derivatives from sulfinimines [9R].

Outlook

The current focus of our research is the design and synthesis of sulfinimine-derived chiral building blocks for the asymmetric synthesis of polyfunctionalized amine derivatives. We require these building blocks to be easily prepared in both enantiomerically pure forms and to provide efficient access to diverse classes of amines with a minimum of chemical manipulation and protecting group chemistry. Examples include δ-amino-β-ketoesters for pyrrolidine (proline) and piperidine (quinolizidine, indolizidine) synthesis; β-amino-β-ketoester enaminones for the synthesis of 2,4,5-trisubstituted piperidines; β-amino Weinreb amides for the syntheses of β-amino aldehydes and ketones; β-amino ketones for the synthesis of 2,3,4,5,6-pentasubstituted piperidines [10]; aziridine 2-carboxylates for the synthesis of α-amino acids and 2H-azirine 2-carboxylates; and aziridine 2-phosphonates for the synthesis of α-amino phosphonates and 2H-azirine-3-phosphonates, a new chiral iminodienophile (Scheme 5).

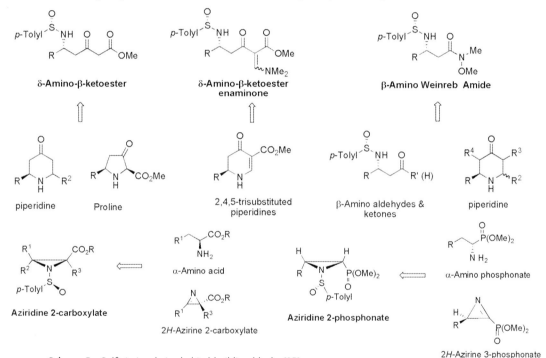

Scheme 5 Sulfinimine-derived chiral building blocks [9R].

CV of Franklin A. Davis

Franklin A. Davis was born in Des Moines, Iowa, USA. He received his BS degree from the University of Wisconsin in 1962 and his Ph.D. from Syracuse University in 1966 working with Donald C. Dittmer. After 2 years with the late Michael J. S. Dewar as a Welch Postdoctoral Fellow at the University of Texas he joined the faculty of Drexel University in Philadelphia where he was the George S. Sasin

Professor of Chemistry until 1995. At this time he moved across town to Temple University as Professor of Chemistry. Dr. Davis received Drexel University's Research Achievement Award in 1980, the Philadelphia American Chemical Society Award in 1992, a Fellowship from the Japan Society for the Promotion of Science in1992, Temple University Research Award in1999, and the Philadelphia Organic Chemist's Club Award in 2002. In 2006 he received the American Chemical Society Arthur C. Cope Scholar Award.

Selected Publications

1R. F. A. Davis, U. K. Nadir, *Org. Prep. Proced. Int.* **1979**, *11*, 33–51. Synthesis of Sulfenamide Derivatives. A Review.

2R. F. A. Davis, A. C. Sheppard, *Tetrahedron* **1989**, *45*, 5703–5742. Applications of Oxaziridines in Organic Synthesis.

3R. F. A. Davis, B.-C. Chen, *Chem. Rev.* **1992**, *92*, 919–934. Asymmetric Hydroxylation of Enolates with N-Sulfonyloxaziridines.

4R. B-C. Chen, P. Zhou, F. A. Davis, E. Ciganek, *Org. Reactions*, **2003**, *62*, 1–356. α-Hydroxylation of Enolates and Silyl Enol Ethers.

5. F. A. Davis, R. T. Reddy, W. Han, P. J. Carroll, *J. Am. Chem. Soc.* **1992**, *114*, 1428–1437. Chemistry of Oxaziridines 17. N-(Phenylsulfonyl)(3,3-dichlorocamphoryl)oxaziridine: A Highly Efficient Reagent for the Asymmetric Oxidation of Sulfides to Sufoxides.

6. F. A. Davis, R. T. Reddy, *J. Org. Chem.* **1992**, *57*, 2599–2606. Asymmetric Oxidation of Simple Selenides to Selenoxides in High Enantiopurity. Stereochemical Aspects of the Allyl Selenoxide/Allyl Selenenate Rearrangement.

7. F. A. Davis, P. Zhou, C. K. Murphy, G. Sundarabu, H. Qi, W. Han, R. M. Przeslawski, B.-C. Chen, P. J. Carroll, *J. Org. Chem.* **1998**, *63*, 2273–2280. Asymmetric Fluorination of Enolates with Nonracemic N-Fluoro-2,10,-Camphorsultams.

8R. F. A. Davis, P. V. N. Kasu, *Org. Prep. Proced. Int.* **1999**, *31*, 125–143. Synthesis of α-Fluoro Aldehydes and Ketones. A Review.

9R. P. Zhou, B.-C. Chen, F. A. Davis, *Tetrahedron* **2004**, *60*, 8003–8030. Recent Advances in Asymmetric Reactions using Sulfinimines (N-Sulfinyl Imines).

10. F. A. Davis, B. Yang, *J. Am. Chem. Soc.* **2005**, *127*, 8398–8407. Asymmetric Synthesis of α-Substituted β-Amino Ketones form Sulfinimines (N-Sulfinylimines). Synthesis of the Indolizidine Alkaloid (-)-223A.

Asymmetric Syntheses with SAMP-/RAMP-Hydrazones

Dieter Enders and Wolfgang Bettray, RWTH Aachen, Germany

Background

Regio-, diastereo- and enantioselective carbon–carbon and carbon–heteroatom bond formations in α position to the carbonyl group are among the most important procedures in organic synthesis. While problems with side reactions exist in classical enolate chemistry, the corresponding aza-enolates in the form of lithiated hydrazones give better yields and selectivities. Inspired by early results of Yamada et al. (1969) and Eder, Sauer, Wiechert (Schering AG, 1971) employing proline derived enamine chemistry in a stoichiometric and catalytic fashion, respectively, and based on postdoctoral work with Professor Corey at Harvard on lithiated dimethylhydrazones (DMHs, 1975) the time was ripe to develop a first practical method for the asymmetric α-alkylation of aldehydes and ketones. In parallel Meyers et al. followed similar ideas employing lithiated imines derived from an acyclic amino acid auxiliary.

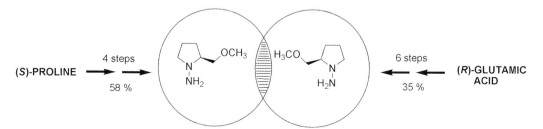

Scheme 1 (S)- or (R)-1-Amino-2-methoxymethylpyrrolidine.

The cyclic amino acid (S)-proline was converted in four steps by two LAH-reductions, an N-nitrosation and a Williamson etherification to (S)-1-amino-2-methoxymethylpyrrolidine (SAMP, 1975). The procedure can be carried out routinely on a 1-mol scale leading to the chiral auxiliary in up to 58% overall yield. If one wants to avoid the N-nitrosamine intermediates, an alternative six-step route has been developed with an N-amination via Hofmann degradation of an intermediate urea derivative as the key step [1]. Because (R)-proline is rather

Asymmetric Synthesis – The Essentials.
Edited by Mathias Christmann and Stefan Bräse
Copyright © 2007 WILEY-VCH Verlag GmbH & Co. KGaA, Weinheim
ISBN: 978-3-527-31399-0

expensive, the enantiomeric RAMP-auxiliary is usually prepared from the relatively cheap (R)-glutamic acid via pyroglutamic acid in six steps and 35 % overall yield (Scheme 1).

Thus, the combination of the useful DMH-methodology with an enantiopure cyclic amino acid derivative resulted in the now widely used SAMP-/RAMP-hydrazone methology [2R, 3R].

Strategy

Under classical conditions of electrophilic substitutions in the α-position to the carbonyl group of aldehydes and ketones the racemic carbonyl compounds are formed. In order to get an asymmetric induction and to make this process overall enantioselective, the aza-enolate strategy involves four steps, namely the formation of an enantiopure hydrazone (step 1), the metalation (step 2), the trapping of the chiral enolate equivalent with an electrophile (step 3) and the removal of the chiral auxiliary (step 4) (Scheme 2). For the final cleavage step numerous hydrolytic, oxidative, reductive and even photochemical protocols have been developed over the years and allow the racemization-free isolation of the carbonyl products [4R].

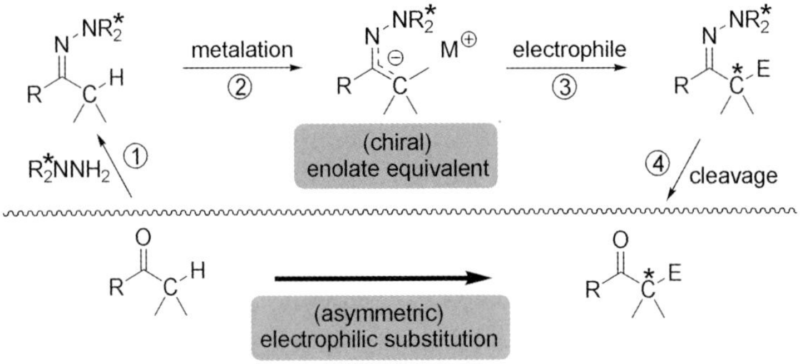

Scheme 2 Electrophilic substitutions α to the carbonyl group via metalated chiral hydrazones.

Thus, the easily available SAMP-hydrazones **1** are deprotonated, leading to the $E_{CC}Z_{CN}$-species of the monomeric azaenolate **2** with an intramolecular chelation of the lithium atom by the methoxy group (see the chapter by Meyers, page 37), which is located below the CCNN-plane. The electrophilic attack on this rigid intermediate proceeds under very high diastereofacial differentiation, resulting in the highly diastereomerically enriched hydrazones **3** (Scheme 3).

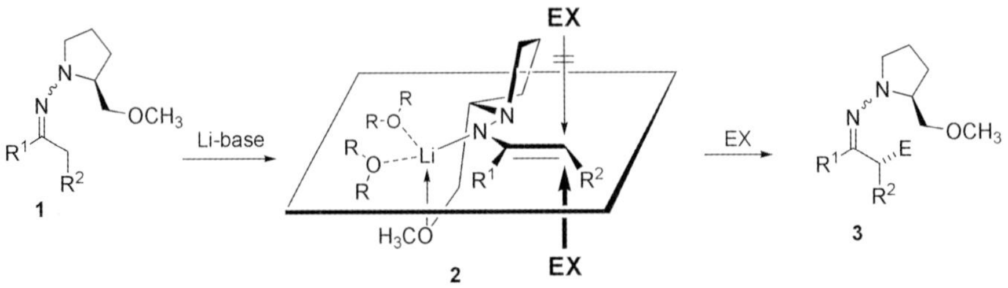

Scheme 3 S_E2'-front metalloretentive mechanism of the SAMP-/RAMP-hydrazone methodology.

Results

In asymmetric electrophilic substitution a great variety of electrophiles can be used, such as alkyl halides, Michael acceptors, carbonyl compounds, halide-substituted esters, oxiranes, aziridines and various hetero electrophiles. The methodology may be exemplified with the simple case of the propylation of diethyl-ketone, leading in one of the first asymmetric pheromone syntheses to the alarm pheromone of the leaf cutting ant *Atta texana* (**4**) with virtually complete asymmetric induction [1] (Scheme 4).

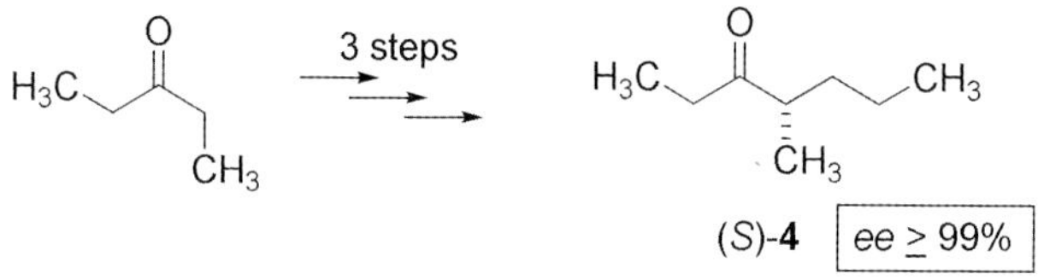

Scheme 4 Asymmetric α-alkylation of acyclic ketones.

In Nature dihydroxyacetone phosphate plays a pivotal role as a C_3 building block for which 2,2-dimethyl-1,3-dioxan-5-one (**5**) constitutes a versatile synthetic equivalent. In combination with our SAMP-/RAMP-hydrazone methodology it turned out to be broadly applicable and provides diastereo- and enantioselective access to various α,α'-disubstituted cyclic ketones **6** [5R] (Scheme 5).

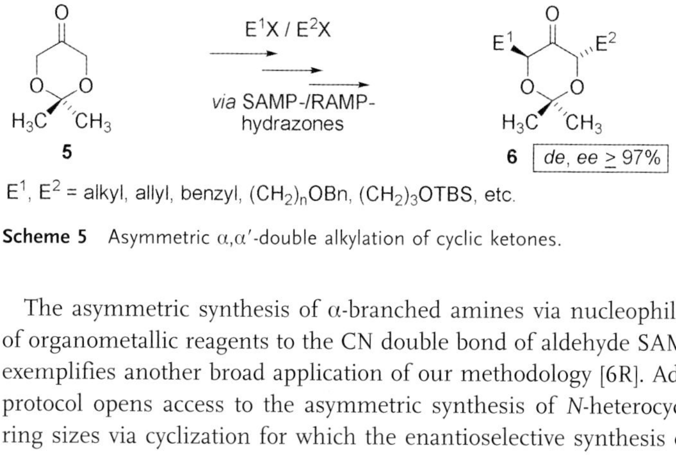

E^1, E^2 = alkyl, allyl, benzyl, $(CH_2)_nOBn$, $(CH_2)_3OTBS$, etc.

Scheme 5 Asymmetric α,α'-double alkylation of cyclic ketones.

The asymmetric synthesis of α-branched amines via nucleophilic 1,2-addition of organometallic reagents to the CN double bond of aldehyde SAMP-hydrazones exemplifies another broad application of our methodology [6R]. Additionally, this protocol opens access to the asymmetric synthesis of N-heterocycles of various ring sizes via cyclization for which the enantioselective synthesis of coniine, the alkaloid of the hemlock plant *Conium maculatum* is an example. Alternatively, the α-alkylation protocol allowed the asymmetric synthesis of stenusine, the propulsion fluid of the beetle *Stenus comma* [7R] (Scheme 6).

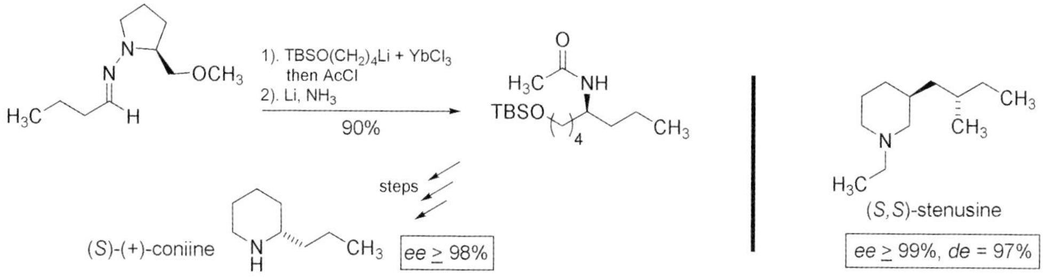

Scheme 6 Asymmetric synthesis of N-heterocycles.

In an asymmetric version of the Brehme reaction, formaldehyde SAMP-hydrazone (**7**) can be used as a synthetic equivalent of the formyl and cyanide anion in C–C bond formations under neutral conditions. For example, the 1,4-addition to nitroalkenes provides an efficient entry to the aldehydes **8** or nitriles **9** of high enantiomeric purity [8] (Scheme 7).

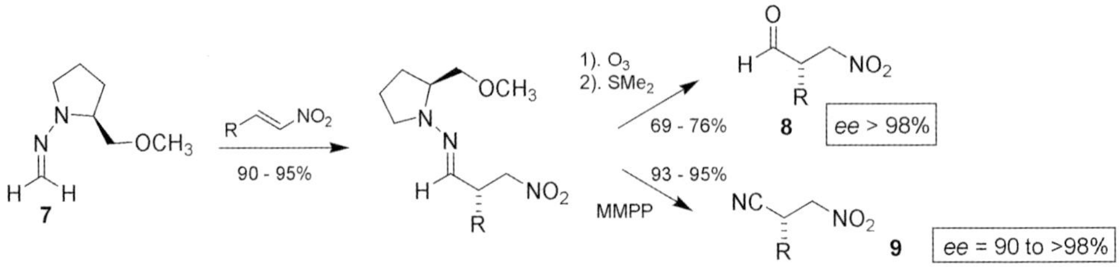

Scheme 7 Formaldehyde SAMP-hydrazone – a neutral chiral formyl anion and cyanide equivalent.

Application for the Synthesis of Natural Products
Over the years the SAMP-/RAMP-hydrazone methodology has turned out to be quite useful in the asymmetric synthesis of many natural and biologically active compounds, over 50 in our group alone [3R]. Typical examples where several hydrazone α-alkylations have been used as key steps are callystatin A [9], the banana weevil pheromone sordidin [10] and the cytotoxic attenol A [11] (Figure 1).

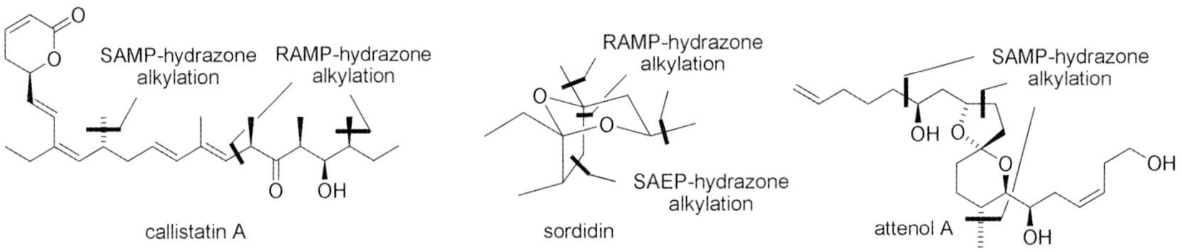

Figure 1 Typical natural products synthesized with the hydrazone methodology.

Other Asymmetric Syntheses from the Enders Group
Currently several other major projects in the area of asymmetric synthesis are being investigated, such as asymmetric nucleophilic acylations via α-aminonitriles [12], asymmetric syntheses of sulfonates and derivatives [13] and organocatalysis with nucleophilic carbenes [14] and in *de novo* carbohydrate synthesis [15].

Conclusions and Future Perspectives
Since the pioneering times of the mid 1970s, when the first practical and generally applicable methods in stoichiometric asymmetric synthesis were developed, such as the oxazoline method of Meyers (see his chapter in this book) and the SAMP-/RAMP-hydrazone method described here, we have seen an explosive growth. Thirty years later efficient methods for almost all important asymmetric C–C and C–heteroatom bond formations are now available. The next innovation

step, namely the development of catalytic versions of all these processes, is climbed rapidly. However, the very important α-alkylations of ketones and aldehydes are still waiting for a general catalytic solution.

CV of Dieter Enders

Dieter Enders was born in 1946 in Butzbach, Germany. He studied chemistry at the Justus Liebig University Gießen and received his Dr. rer. nat. in 1974 under the supervision of Professor D. Seebach. After postdoctoral studies at Harvard University with Professor E. J. Corey he returned to Gießen obtaining his Habilitation in 1979. In 1980 he moved to the University of Bonn as an associate professor and in 1985 to his present position as Professor of Organic Chemistry at the RWTH Aachen. He has been the recipient of many prizes among them the Leibniz Prize of the Deutsche Forschungsgemeinschaft (1993), the Yamada Prize (Japan, 1995) the Max Planck Research Award of the Max Planck Gesellschaft and the Alexander von Humboldt Research Foundation (2000) and the Emil Fischer Medal of the Gesellschaft Deutscher Chemiker (2002).

Selected Publications

1. D. Enders, P. Fey, H. Kipphardt, *Org. Synth.* **1987**, *65*, 173–182. *(S)-(–)-1-Amino-2-methoxymethylpyrrolidine (SAMP) and (R)-(+)-1-Amino-2-methoxymethylpyrrolidine (RAMP). Versatile Chiral Auxiliaries*; D. Enders, H. Kipphardt, P. Fey, *Org. Synth.* **1987**, *65*, 183–202. *Asymmetric Synthesis Using the SAMP-/RAMP-Hydrazone Method. (S)-(+)-4-Methyl-3-heptanone.*
2R. D. Enders, in *Asymmetric Synthesis* Vol. 3, J.D. Morrison (Ed.), Academic Press, Orlando, **1984**, pp. 275–339. *Alkylation of Chiral Hydrazones.*
3R. A. Job, C. F. Janeck, W. Bettray, R. Peters, D. Enders, *Tetrahedron* **2002**, *58*, 2253–2329. *The SAMP/RAMP-Hydrazone Methodology in Asymmetric Synthesis.*
4R. D. Enders, L. Wortmann, R. Peters, *Acc. Chem. Res.* **2000**, *33*, 157–169. *Recovery of Carbonyl Compounds from N,N-Dialkylhydrazones.*
5R. D. Enders, M. Voith, A. Lenzen, *Angew. Chem. Int. Ed. Engl.* **2005**, *44*, 1304–1325. *The Dihydroxyacetone Unit – a Versatile C_3-Building Block in Organic Synthesis.*
6R. D. Enders, U. Reinhold, *Tetrahedron: Asymmetry* **1997**, *8*, 1895–1946. *Asymmetric Synthesis of Amines by Nucleophilic 1,2-Addition of Organometallic Reagents to the CN-Double Bond.*
7R. D. Enders, M. Boudou, J. Gries in *New Methods for the Asymmetric Synthesis of Nitrogen Heterocycles*, J. L. Vicario, D. Badia, L. Carrillo (Eds.), Research Signpost, Trivandrum, India, **2005**, 1–31. *Asymmetric Syntheses of Four-, Five- and Six-Membered N-Heterocycles Employing the SAMP/RAMP-Hydrazone Methodology.*
8. D. Enders, M. Bolkenius, J. Vázquez, J.-M. Lassaletta, R. Fernández, *J. Prakt. Chem./Chem. Ztg.* **1998**, *340*, 281–285. *Formaldehyde SAMP-Hydrazone – a Neutral Chiral Formyl Anion and Cyanide Equivalent.*
9. J. L. Vicario, A. Job, M. Wolberg, M. Müller, D. Enders, *Chem. Eur. J.* **2002**, *8*, 4272–4284. *Asymmetric Total Synthesis of (–)-Callystatin A and (–)-20-epi-Callystatin A Employing Chemical and Biological Methods.*
10. D. Enders, I. Breuer, A. Nühring, *Eur. J. Org. Chem.* **2005**, 2677–2683. *First Asymmetric Synthesis of (+)-Sordidin and (–)-7-epi-Sordidin, Aggregation Pheromones of the Banana Weevil Cosmopolites sordidus.*
11. D. Enders, A. Lenzen, *Synlett* **2003**, 2185–2187. *Asymmetric Total Synthesis of Attenol A and B.*
12R. D. Enders, J. P. Shilvock, *Chem. Soc. Rev.* **2000**, *29*, 359–373. *Some Recent Applications of α-Amino Nitrile Chemistry.*

13. D. Enders, N. Vignola, O. M. Berner, W. Harnying, *Tetrahedron* **2005**, *61*, 3231–3243. *Efficient Asymmetric Synthesis of α-Alkylated Benzylic Methyl Sulfonates.*
14R. D. Enders, T. Balensiefer, *Acc. Chem. Res.* **2004**, *37*, 534–541. *Nucleophilic Carbenes in Asymmetric Organocatalysis.*
15. D. Enders, C. Grondal, *Angew. Chem. Int. Ed. Engl.* **2005**, *44*, 1210–1212. *Direct Organocatalytic De Novo Synthesis of Carbohydrates.*

Asymmetric Allylation Reactions

Reinhard W. Hoffmann, University of Marburg, Germany

Background

Asymmetric synthesis, the topic of this volume, depends on the energy difference between the competing transition states in a reaction leading to two stereoisomeric products. High stereoselectivity requires large energy differences ($\Delta\Delta G^{\ddagger} > 3$ kcal mol^{-1}) between these transition states. This energy difference is caused by differences in polar or steric interactions in the transition states between groups in the substrate and the reagent. Such interactions will be substantial and well defined in reactions that proceed via cyclic (ordered), compact transition states which are late on the reaction coordinate, i.e. when the forming bonds are short and close to their final bond length.

Results

Type (I) Allylmetallation of Aldehydes

With such considerations in mind we were attracted to allylmetallation reactions of the type I [1], i.e. those proceeding through cyclic transition states with a two point contact between the allylmetal and the aldehyde (Scheme 1), as opposed to the type II reactions with only a single point contact between the reactands in the transition state. In order to reach a two point contact, the metal in the allylmetal species must be Lewis-acidic to coordinate to the lone pair of the aldehyde carbonyl group. This led us to consider boron as the most promising metal of the allylmetal species, and indeed allylmetallation reactions of allylboranes and allylboronates were known at the outset of our investigations [2].

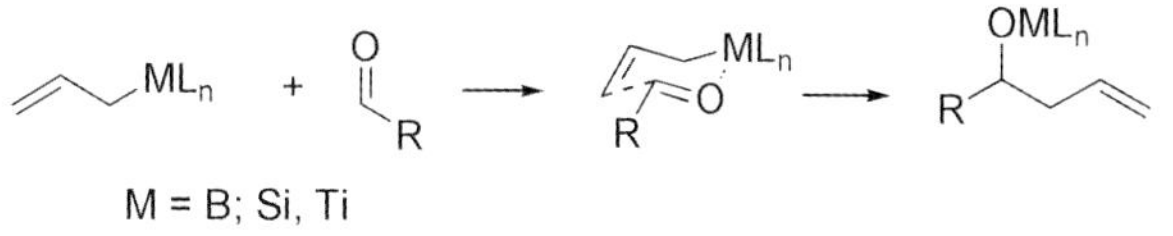

M = B; Si, Ti

Scheme 1 Cyclic transition state of type (I) allylmetallation.

The attractive feature of the allylboration is that chiral information can readily be coupled to the reaction via the alkoxyl moieties on boron. Luckily, our first experiments in this direction [3] led to levels of asymmetric induction that were spectacular at that time (Scheme 2).

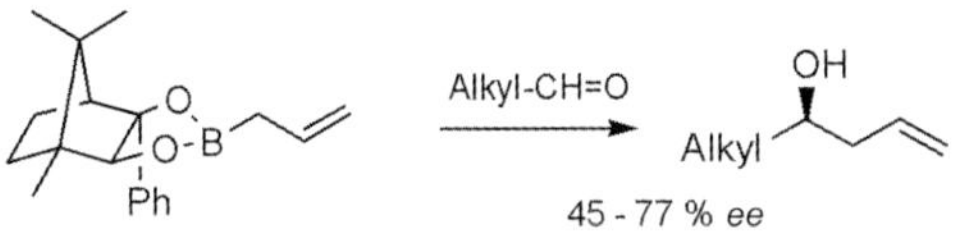

Scheme 2 First asymmetric allylboration reaction

This level of asymmetric induction, however, no longer meets the standards of today. Fortunately D. Hall revived our earlier results by the finding that under Sc(OTf)$_3$ catalysis lower reaction temperatures and higher *ee* values can be attained (Scheme 3) [4].

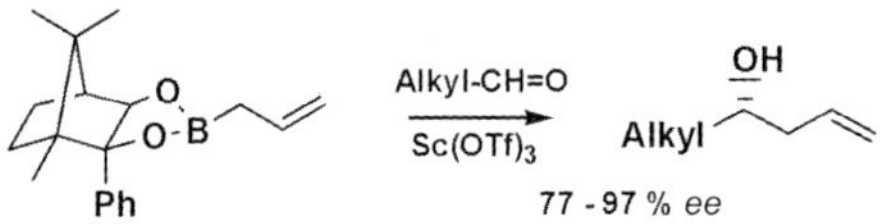

Scheme 3 Sc(OTf)$_3$-catalysis enhances the *ee*.

The asymmetric allylboration of aldehydes was rapidly expanded using allylboronates with other chiral auxiliaries [5]. Due to the good performance and the commercial availability of the chiral auxiliary the allylboration with allyl-diisopinocampheyl-*boranes* developed by H.C.Brown soon became the standard method (Scheme 4) [6].

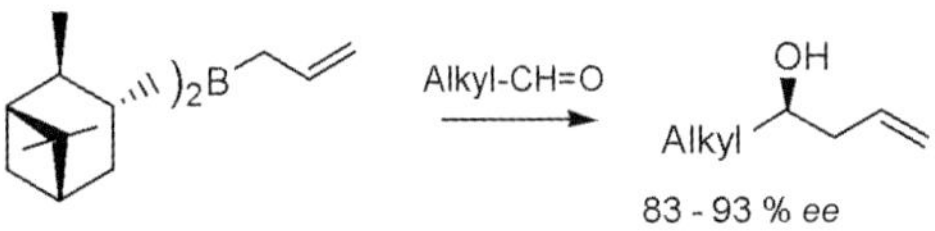

Scheme 4 H.C. Brown's allylboration with allyl-di-isopinocampheyl-borane.

It is the hallmark of type I allylmetallation reactions that with γ-substituted allylmetals the geometry of the double bond is translated with high fidelity into the relative configuration of the two new stereocenters in the product such that E → anti and Z → syn (Scheme 5) [5, 7].

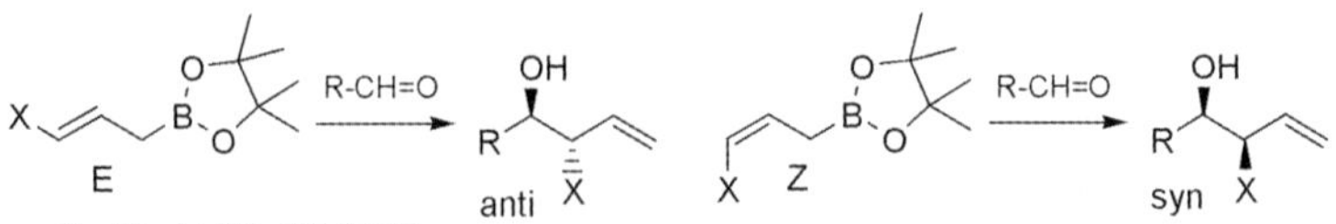

X = Me, Cl, RS, RO, AcNR

Scheme 5 Control of relative configuration at the two newly formed stereocenters.

But boron is not the only metal that allows type I allylmetallation reactions to proceed. The monopolic situation of the Brown allylboranes in natural product synthesis was challenged by the titanium reagents of Duthaler [8] and the recent move to allyl-chlorosilanes [9] may well mark the reagents of the future (Scheme 6).

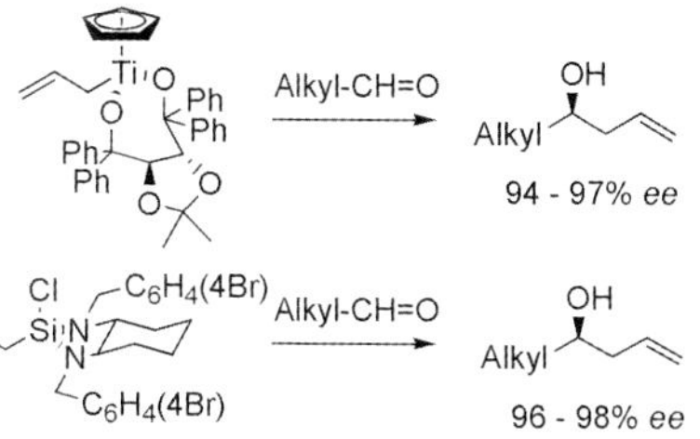

Scheme 6 Duthaler's and Leighton's allylmetallation reagents.

Finally, silicon offers the opportunity to do away with stoichiometric chiral auxiliaries by turning to Lewis-base asymmetric catalysis. It appears strange, at first sight, that coordination of a (chiral) Lewis base to tetracoordinate silicon enhances its Lewis acidity, but this is just the case on going to a pentacoordinate siliconate with its high tendency to form a hexacoordinate siliconate (Scheme 7) [10].

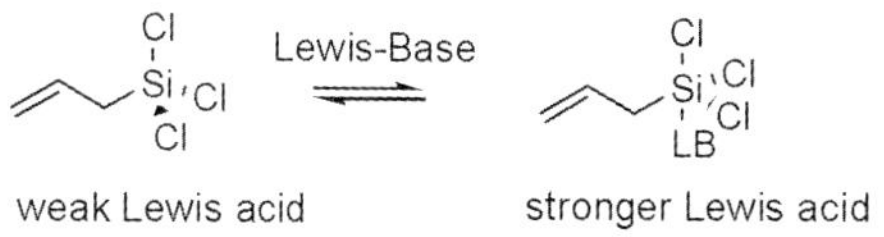

Scheme 7 Enhancement of Lewis acidity on silicon by coordination of a base.

This feature has been used to develop (chiral) Lewis-base catalysed asymmetric allylmetallation reactions (Scheme 8) [11], the full power of which still has to be established.

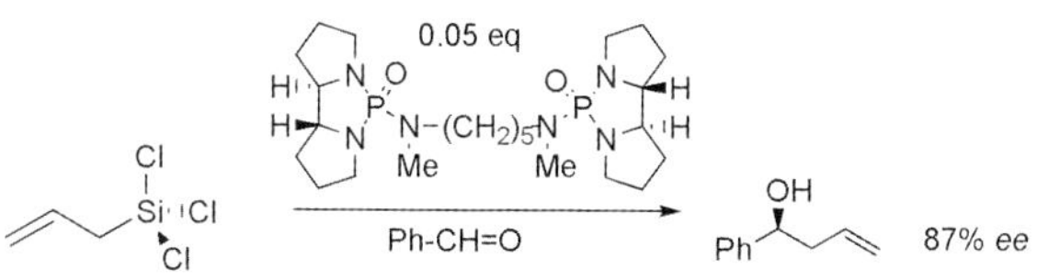

Scheme 8 Chiral Lewis-base catalysed asymmetric allylmetallation.

Second Generation Allylmetallation Reagents

The above-mentioned auxiliary-based asymmetric allylmetallation reagents of the first generation perform well on reaction with achiral aldehydes. Their limitations become manifest, sooner or later, on reaction with chiral aldehydes. No problems arise when the direction of the asymmetric induction of the allylmetallation reagent and of the aldehyde are consonant (matched pair), but when the reagent has to override the asymmetric induction from the aldehyde (mismatched case) the level of asymmetric induction of these first generation allylmetallation reagents may turn out to be insufficient. This situation induced us to develop second generation allylboronate reagents, in which the chiral information is not attached to the periphery of the transition state, but rather resides directly on the atoms that form the cyclic transition state. We thus focussed on allylboronates that have a stereogenic center in the allyl moiety directly adjacent to the boron atom.

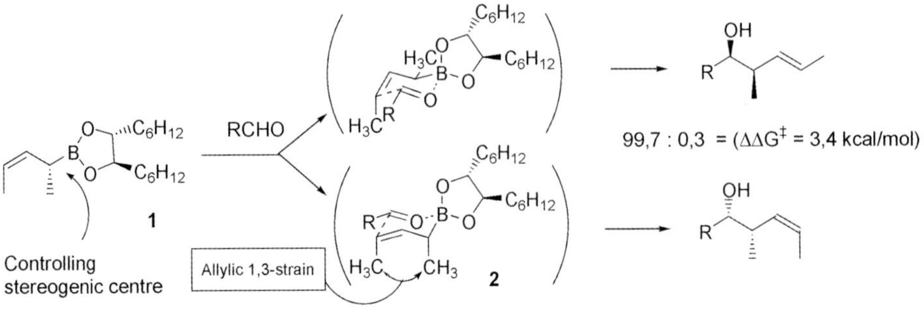

Scheme 9 Chiral alpha-substituted allylboronates, second generation allylboration reagents.

With these reagents the competing transition states that lead to the two stereoisomeric products differ in the arrangement (axial or equatorial) of the substituent at the stereogenic center (Scheme 9). The energy difference $\Delta\Delta G^{\ddagger}$ between the competing transition states is maximised (up to 3.4 kcal mol^{-1}) when using the reagents **1**, due to the destabilisation of transition state **2** by 1,3-allylic strain. Hence **1** is probably the allylmetallation reagent with the highest asymmetric induction available to override substrate-based asymmetric induction on reaction with chiral aldehydes [12].

The corresponding E-crotyl reagents with a chlorine or a methoxy group in the α-position to boron rely on differences of polar interactions in the competing transition states reaching $\Delta\Delta G^{\ddagger}$ values of up to 2.6 kcal mol^{-1} (Scheme 10) [13].

Thus, these second generation chiral allylboration reagents are available whenever, in reaction with chiral aldehydes, reagent control of diastereoselectivity has to be enforced [14].

CV of Reinhard W. Hoffmann

Professor Hoffmann studied chemistry from 1951 to 1958 at the University of Bonn, finishing with a doctorate under the guidance of Professor B. Helferich. Two years of postdoctoral studies at the Pennsylvania State University were followed by a second postdoctorate with Professor G. Wittig at the University of Heidelberg. There Professor Hoffmann started his independent research that led to his habilitation in 1964. Three years later he was appointed as Dozent at the Technische Hochschule Darmstadt. Since 1970 he has held the position of Professor of Organic Chemistry at the Universität Marburg (emeritus status since 2001). Professor Hoffmann had the pleasure of being visiting professor at the University of Wisconsin, the Universität Bern, the University of California at Berkeley, and Kyoto University. Over the years Professor Hoffmann's scientific interests changed from reactive intermediates (benzyne, nucleophilic carbenes) over electron rich alkenes, stereochemistry of 2,3-sigmatropic rearrangements, the topic of this

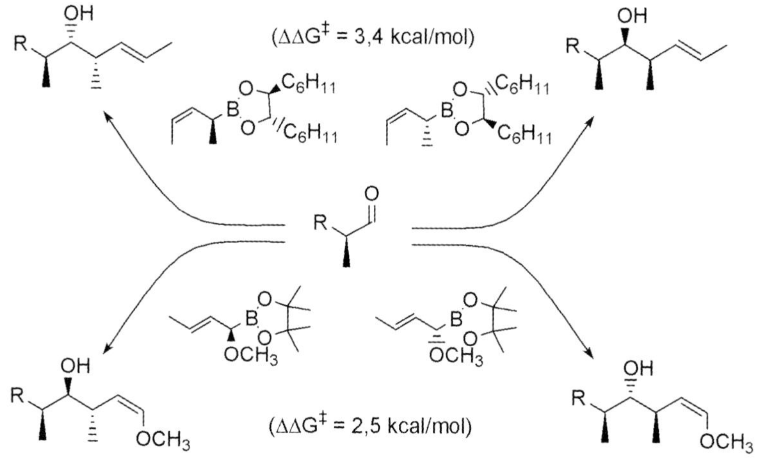

Scheme 10 Reagent control of diastereoselectivity using second generation allylboration reagents.

account: development of the allylboration reaction for the stereoselective synthesis of natural products, to conformation design.

References

1. S. E. Denmark, E. J. Weber, *Helv. Chim. Acta* **1983**, *66*, 1655–1660.
2. (a) E. Favre, M. Gaudemar, *Compt. Rend. Acad. Sci., Ser. C.* **1966**, *263*, 1543; (b) B. M. Mikhailov, *Organomet. Chem. Rev., Sect. A* **1972**, *8*, 1.
3. T. Herold, R. W. Hoffmann, *Angew. Chem.* **1978**, *90*, 822–823; *Angew. Chem., Int. Ed. Engl.* **1978**, *17*, 768–769.
4. H. Lachance, X. Lu, M. Gravel, D. G. Hall, *J. Am. Chem. Soc.* **2003**, *125*, 10160–10161.
5. W. R. Roush, in *Houben Weyl: Methods of Organic Chemistry*, G. Helmchen, R. W. Hoffmann, J. Mulzer, E. Schaumann (Eds.), Thieme, Stuttgart, vol. E21b, **1995**, pp. 1410–1486.
6. P. K. Jadhav, K. S. Bhat, P. T. Perumal, H. C. Brown, *J. Org. Chem.* **1986**, *51*, 432–439.
7. R. W. Hoffmann, H.-J. Zeiß, *J. Org. Chem.* **1981**, *46*, 1309–1314.
8. A. Hafner, R. O. Duthaler, R. Marti, G. Rihs, P. Rothe-Streit, F. Schwarzenbach, *J. Am. Chem. Soc.* **1992**, *114*, 2321–2336.
9. K. Kubota, J.L. Leighton, *Angew. Chem.* **2003**, *115*, 976–978; *Angew. Chem., Int. Ed. Engl.* **2003**, *42*, 946–948.
10. S. E. Denmark, D. M. Coe, N. E. Pratt, B. D. Griedel, *J. Org. Chem.* **1994**, *59*, 6161–6163.
11. S.E. Denmark, J. Fu, *J. Chem. Soc., Chem. Commun.* **2003**, 167–170.
12. (a) R. W. Hoffmann, K. Ditrich, G. Köster, R. Stürmer, *Chem. Ber.* **1989**, *122*, 1783–1789; (b) S. D. Rychnovsky, C. R. Thomas, *Org. Lett.* **2000**, *2*, 1217–1219.
13. R.W. Hoffmann, S. Dresely, *Chem. Ber.* **1989**, *122*, 903–909.
14. R. Stürmer, R. W. Hoffmann, *Chem. Ber.* **1994**, *127*, 2519–2526.

Carbohydrates as Chiral Auxiliaries

Horst Kunz, Universität Mainz, Germany

Introduction

As a consequence of the different pharmacological effects of enantiomers, there has been an increasing interest in stereoselective syntheses of biologically active compounds. Of the stereoselective C–C bond forming processes directed by chiral auxiliaries, the reactions of deprotonated hydrazones derived from L- and R-prolinol were an early milestone [1]. Although carbohydrates are inexpensive natural products containing a high density of stereogenic centers, they have been long ignored as stereoselective auxiliaries. For example, earlier experiences in aldol reactions with carbohydrate-derived components turned out disappointing in terms of selectivity and yield [2]. However, alkylation of carbohydrate ester enolates gave the branched products with good stereoselectivity [3], provided ketene formation could be prevented. While these enolate reactions confirmed the impression that carbohydrates are too complex for use as chiral auxiliaries, they showed promising stereodifferentiating effects in other types of reactions.

Results
Carbohydrate Auxiliaries in Cycloaddition Reactions

Dienol ethers of glucopyranose have been successfully subjected to hetero Diels-Alder reaction with glyoxylic esters to give the cycloadducts with high diastereoselectivity but low endo selectivity. On the other hand, glucosyloxy butadiene and alkyl acrylates in water formed the Diels-Alder adducts with complete endo selectivity, but modest diastereoselectivity [4].

Substantial improvement in stereoselective Diels-Alder reactions with carbohydrate auxiliaries was achieved by Lewis acid promotion [5]. Lewis acids enhance the reactivity of dienophiles. In addition, they arrange the reacting and shielding groups more rigidly for stereodifferentiation. The reactions of acrylates of dihydroglycals of D-glucose **1** and L-rhamnose **2** with cyclopentadiene, for example, show another advantage of carbohydrate auxiliaries (Scheme 1).

They are pseudoenantiomeric and alternatively form the enantiomeric carbobicyclic adducts **3** and **4** with excellent diastereoselectivity [6].

Asymmetric Synthesis – The Essentials.
Edited by Mathias Christmann and Stefan Bräse
Copyright © 2007 WILEY-VCH Verlag GmbH & Co. KGaA, Weinheim
ISBN: 978-3-527-31399-0

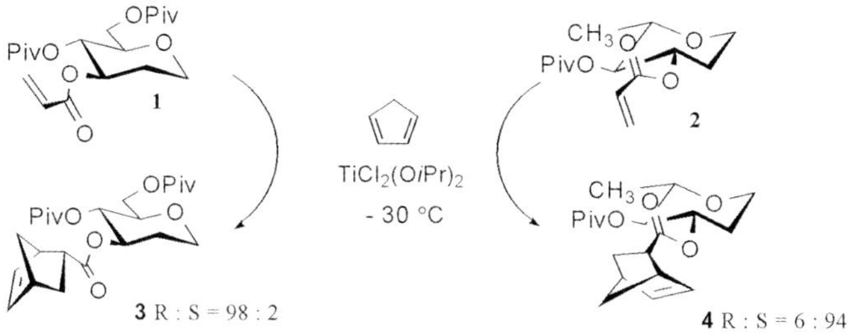

Scheme 1

Conjugate Addition Reactions Directed by Carbohydrate Auxiliaries

Carbohydrate-derived oxazolidinones are efficient tools in novel 1,4-addition reactions of organoaluminum compounds to α,β-unsaturated carboxylic derivatives like the N-acyl galactopyranosido-oxazolidinone **5** to give the branched products **6** (Scheme 2) [7].

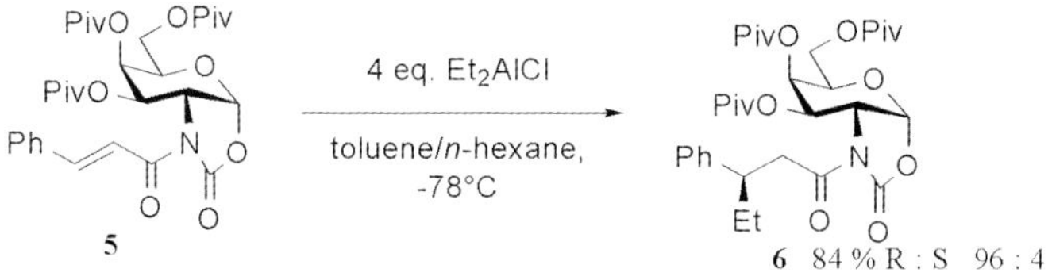

Scheme 2

The corresponding glucopyranose-derived compounds **7** gave addition products analogous to **6** but in slightly lower stereoselectivity. However, they proved more efficient in the domino 1,4 addition/α-chlorination reactions, as is shown for the formation of **8** (Scheme 3) [8].

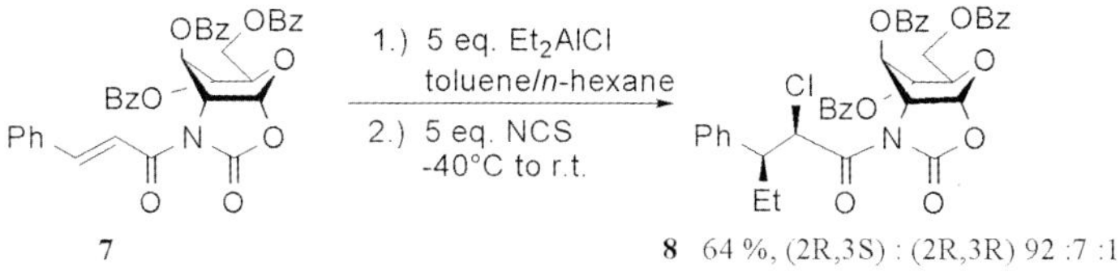

Scheme 3

An interesting feature of these addition reactions of organoaluminum chlorides to the N-acyl oxazolidinones concerns the different reactivity of dimethylaluminum chloride and higher diorganylaluminum analogues. In particular with N-acyl oxazolidinones containing the Evans auxiliary [9], dimethylaluminum chloride only reacted on irradiation [10]. D-Arabinopyranose-derived N-acyl oxazolidinones are useful for the stereoselective synthesis of β-branched carboxylic derivatives of opposite enantiomeric configuration [11].

Glycosylamines in Stereoselective Synthesis [12]

Gycosylamines **9** react with aldehydes to give Schiff bases. Addition of trimethylsilyl cyanide to **10** promoted by zinc chloride furnished α-amino nitriles **11** (Scheme 4) [13].

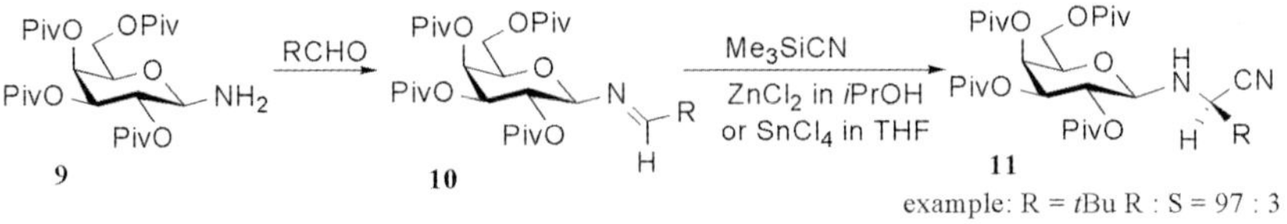

Scheme 4

The asymmetric induction in this stereoselective Strecker reaction can be reversed by changing the solvent to chloroform. The stereoselectivity exhibited by the carbohydrate can be rationalized by the exo-anomeric effect and the shielding coordination of zinc chloride to the imine and the pivaloyl group in the 2-position.

This stereodifferentiation is even more efficient in Ugi reactions. Galctosylamine **9** was treated in one pot with aldehydes, isocyanides and carboxylic acids in the presence of zinc chloride to furnish α-amino acid amides **12** in high yield and diastereoselectivity [14]. For the synthesis of the analogous L-amino acid amides *O*-pivaloylated D-arabinopyranosylamine **13** was used as the auxiliary [15]. After recrystallization or flash chromatography the pure diastereomers **12** or **14**, respectively, were obtained in high yield (Scheme 5).

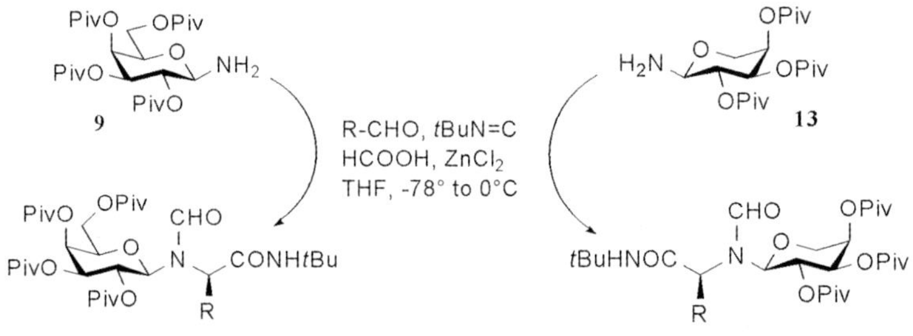

12 R = 4-Cl-Ph, 92 % pure, R : S = 97 : 3

14 R = 4-Cl-Ph, 92 % pure, R : S = 2 : 98

Scheme 5

Glycosyl imines also enabled the asymmetric synthesis of β-amino acids, for example, when **10** was subjected to a Mannich reaction with bissilyl ketene acetals. As a rule, one out of four possible diastereomers of the β-amino acids was formed in excellent selectivity [16].

Of the numerous addition reactions of glycosyl imines with various nucleophiles [12], the one-pot domino Mannich-Michael reactions are briefly outlined. Glycosyl imines **10** react with the Danishefsky diene promoted by zinc chloride to give *N*-glycosyl 5,6-dehydropiperidin-4-ones **15** with excellent diastereoselectivity [17]. Further conversion of the pure diastereomers by C=C bond reduction,

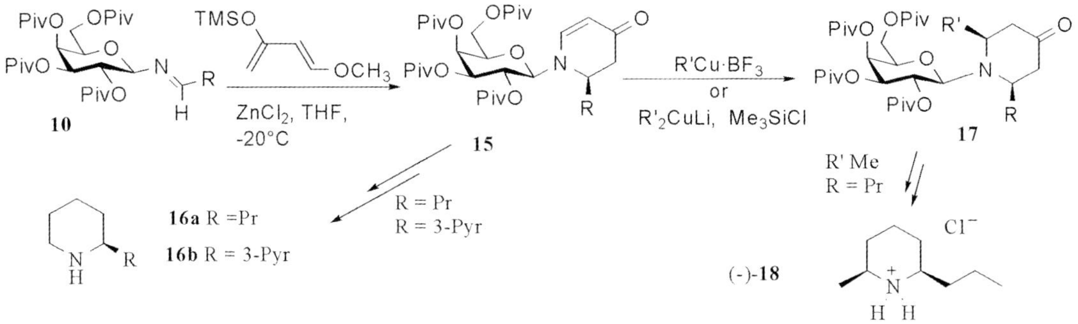

Scheme 6

dithiolane formation, desulfurization and release from the carbohydrate produced, for example, pure (R)-coniine **16a** and (S)-anabasine **16b** (Scheme 6).

Conjugate addition of organocuprates to enaminones **15** was only achieved after promotion by hard Lewis acids and formed 2,6-cis-disubstituted piperidinones **17** with high diastereoselectivity. Reduction and detachment from the carbohydrate gave (–)-dihydropinidine **18**, for example [17b]. Application of this reaction sequence using D-arabinopyranosylamine **13** furnished the opposite enantiomers of these alkaloids [18]. Interestingly, the enolates resulting after conjugate addition to the N-glycosyl dehydropiperidinones are protonated stereoselectively, as was demonstrated in the transformation of an N-galactosyl piperidinone via the octahydroquinolinone to 4a-epi-pumiliotoxin C [19].

These stereoselective syntheses of piperidines have also been performed in the solid phase in combinatorial approaches [20].

The outlined syntheses of piperidines are supplemented by a sequence of N-glycosylations of pyridones and subsequent conjugate addition reactions of Grignard compounds. This conversion of galactosyl-4-pyridone **19** afforded the product **20** which contains the 2-substituted piperidine derivative of opposite configuration compared to compounds **15** (Scheme 7) [21, 22].

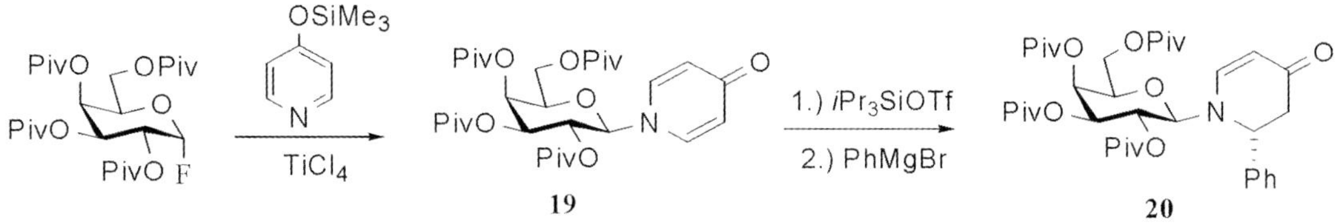

Scheme 7

Conclusions

The given examples illustrate that carbohydrates are powerful chiral auxiliaries in a number of asymmetric syntheses, if their multi-functionality is exploited for a suitable arrangement of reacting and shielding groups. This stereodiffentiation is even more amplified by chelation effects.

CV of Horst Kunz

Horst Kunz was born in Frankenhausen, Saxony, in 1940. He studied chemistry at the Humboldt-Universität Berlin and at the Universität Mainz. He received his Ph.D. in 1969 with Leopold Horner and completed his habilitation at Mainz in 1977. He was appointed Associate Professor in 1979 and Full Professor of Organic Chemistry at the Universität Mainz in 1988. He was awarded the Max Bergmann medal in 1992, the Emil Fischer medal in 2000 and the Adolf Windaus medal in 2001, and has been a member of the Saxony Academy of Sciences since 1998.

Selected Publications

1. D. Enders, H. Eichenauer, *Angew. Chem. Int. Ed. Engl.* **1976**, *15*, 549.
2. C. H. Heathcock, C. T. White, J. J. Morrison, D. van Deveer, *J. Org. Chem.* **1981**, *46*, 1296.
3. H. Kunz, J. Mohr, *Chem. Commun.* **1988**, 1315.
4. S. David, A. Lubineau, A. Thieffry, *Tetrahedron* **1978**, *34*, 299; A. Lubineau, Y. Queneau, *J. Org. Chem.* **1987**, *52*, 1001.
5. H. Kunz, B. Müller, D. Schanzenbach, *Angew. Chem. Int. Ed. Engl.* **1987**, *26*, 267.
6. W. Stähle, H. Kunz, *Synlett* **1991**, 260.
7. K. Rück, H. Kunz, *Synlett* **1992**, 343.
8. K. Rück-Braun, A. Stamm, S. Engel, H. Kunz, *J. Org. Chem.* **1997**, *62*, 967.
9. D. A. Evans, K. T. Chapman, J. Bisaha, *J. Am. Chem. Soc.* **1984**, *106*, 4261.
10. K. Rück, H. Kunz, *Angew. Chem. Int. Ed. Engl.* **1991**, *30*, 694.
11. S. Elzner, S. Maas, S. Engel, H. Kunz, *Synthesis* **2004**, 2153.
12. S. Knauer, B. Kranke, L. Krause, H. Kunz, *Curr. Org. Chem.* **2004**, *8*, 1739.
13. H. Kunz, W. Sager, *Angew. Chem. Int. Ed. Engl.* **1987**, *26*, 557.
14. H. Kunz, W. Pfrengle, *J. Am. Chem. Soc.* **1988**, *110*, 651.
15. H. Kunz, W. Pfrengle, W. Sager, *Tetrahedron Lett.* **1989**, *36*, 4109.
16. H. Kunz, A. Burgard, D. Schanzenbach, *Angew. Chem. Int. Ed. Engl.* **1997**, *36*, 386.
17. (a) H. Kunz, W. Pfrengle, *Angew. Chem. Int. Ed. Engl.* **1989**, *28*, 1067; for correct stereochemistry, see: (b) M. Weymann, W. Pfrengle, D. Schollmeyer, H. Kunz, *Synthesis* **1997**, 1151.
18. B. Kranke, D. Hebrault, M. Schultz-Kukula, H. Kunz, *Synlett* **2004**, 671.
19. M. Weymann, M. Schultz-Kukula, S. Knauer, H. Kunz, *Monatsh. Chem.* **2002**, *133*, 571.
20. G. Zech, H. Kunz, *Angew. Chem. Int. Ed.* **2003**, *42*, 787; *Chem. Eur. J.* **2004**, *10*, 4136.
21. M. Follmann, H. Kunz, *Synlett* **1998**, 989.
22. E. Klegraf, M. Follmann, D. Schollmeyer, H. Kunz, *Eur. J. Org. Chem.* **2004**, 3346.

The Use of Chiral Oxazolines in early C–C Bond Forming Reactions

Albert I. Meyers, Colorado State University, USA

Background

In the 1960s and early 1970s we were fascinated by the possibility that many heterocyclic systems were hiding useful chemistry in the chemical properties that they exhibited [1]. We found that cyclic iminoethers **1**, **3** were among these examples and they readily gave nonheterocyclic products **2**, **4** when appropriately treated (Scheme 1). Soon thereafter, in the early 1970s, we considered the chiral variant of these heterocycles and thus began our entry into asymmetric synthesis. We examined the oxazoline ring system **3**, which has been known for almost 100 years, and sought a route to make chiral derivatives. We had access to a chiral amino alcohol **5** from the Parke Davis laboratories and, as shown in Scheme 2, we prepared the chiral oxazoline **6**. This now allowed us to study analogous de-

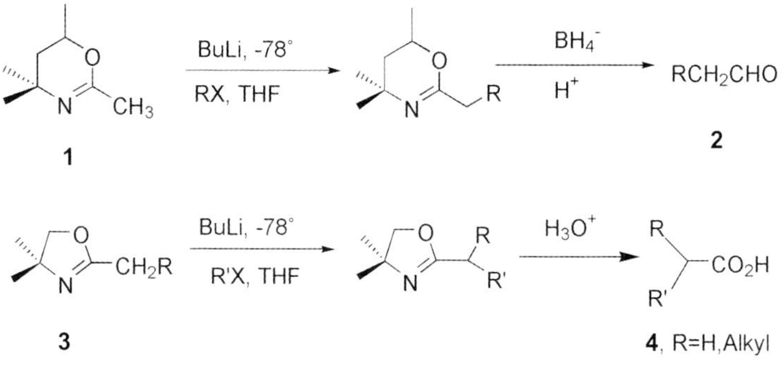

Scheme 1 Heterocycles as vehicles to non-heterocyclic compounds.

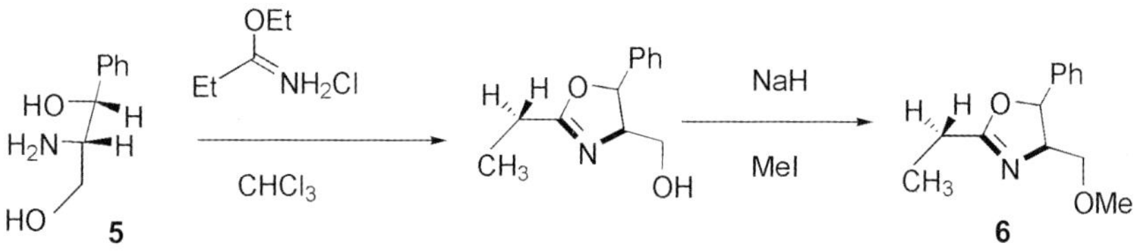

Scheme 2 First synthesis of a chiral oxazoline.

Asymmetric Synthesis – The Essentials.
Edited by Mathias Christmann and Stefan Bräse
Copyright © 2007 WILEY-VCH Verlag GmbH & Co. KGaA, Weinheim
ISBN: 978-3-527-31399-0

protonation and alkylation of the chiral variant which we had hoped would lead to chiral, nonracemic 2,2-substituted acetic acids **4** containing an asymmetric carbon.

Results

With **6** in hand, containing a prochiral α-carbon, we deprotonated and alkylated in a manner similar to that shown for oxazoline **3**. After hydrolysis to remove the oxazoline auxiliary, we obtained a series of 2,2-alkyl acetic acids **8** in 75–80% ees (Scheme 3) [2]. This was a significant increase over other carbanion alkylations in the literature at that time. We believed that the presence of the methoxy group in **6** was of critical importance to the relatively high ees observed and this was confirmed by repeating the deprotonation on a chiral oxazoline **9** which was devoid of the methoxy group. In this case the same 2,2 dialkyl acetic acids **8** were obtained but in only 15–18% ee. The rationale for the results obtained was based on the fact that two enolates **7A, 7B** were formed at the deprotonation step in a highly favored ratio, favoring the Z species (**7A**) by about 95:5 (determined by enriched ^{13}C NMR of the C–Me group) [3R]. The absolute configurations of the acids **8** were known in the literature and in order to account for this we felt that the alkylation–hydrolysis steps, **7A** to **S-8**, must have occurred from the α-face of the enolate, e.g., **10**. This was made possible by initial coordination of the halide lone pairs to the lithium cation. Due to the methoxy ligand we concluded that the Li cation most likely resides at the α-face, as shown in **10**.

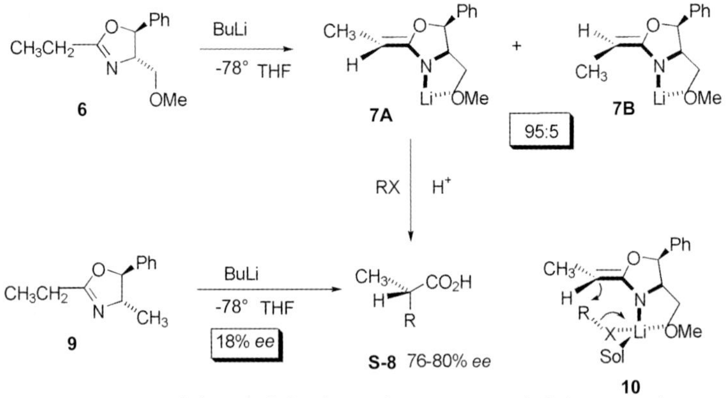

Scheme 3 Asymmetric alkylation of chiral oxazoline to (S)-2,2,-dialkylacetic acids.

We also showed in 1975 that the oxazoline methoxy ligand may take on a similar role in the nucleophilic conjugate addition of alkyl metallics (Mg, Li) to α,β-unsaturated chiral oxazolines (Scheme 4)[3R]. Therefore, addition of organolithium reagents to the unsaturated oxazoline **11** at –78 °C gave conjugate addition products in > 97% de, and after hydrolysis, produced the 3,3,-disubstituted propionic acid **12** in > 97% ee. To account for the absolute configuration of the known products, we again assumed that the initial complexes, **13A** and **13B**, were only able to deliver the organic moiety (R) when the properly aligned orbitals

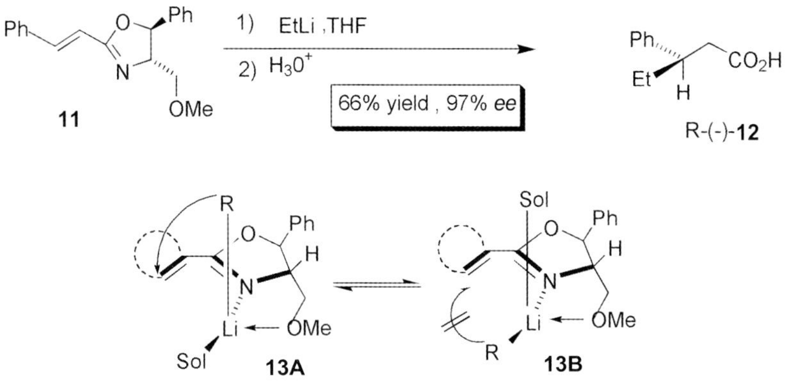

Scheme 4 Conjugate addition to α,β-unsaturated oxazolines.

(**13A**) were interacting with the π system. Since it is likely that both complexes **13A** and **13B** would be present in solution, but rapidly equilibrating via ligand exchange, only **13A** would add to the olefinic bond, since the orbitals in **13B** are improperly aligned [4a,b, 5].

CV of A.I. Meyers

Born in New York City, 1932. Studied at New York University for both BS (1954) and Ph. D degrees (1957, J. J. Ritter). Special NIH Fellow, Harvard University 1965–1966 (E. J. Corey). Faculty, Louisiana State University, 1958–1970; LSU Boyd Professor, 1968–1970; Professor, Wayne State University, 1970–1972; Colorado State University, 1972–present; University Distinguished Professor (1985), John K. Stille Professor (1991–2001); A. v-Humboldt senior scientist (Würzburg), 1984; ACS Award in Synthesis, 1985; A. C. Cope Scholar, 1986; National Academy Sciences, 1994; Yamada Prize (Japan), 1996; International Heterocyclic Award, 1997; Emeritus Professor, 2001.

Selected Publications

1. A. I. Meyers, *Heterocycles in Organic Synthesis*, Wiley-Interscience, New York, **1974**.
2R. A. I. Meyers, E. D. Mihelich, *New Synthetic Methods*, Vol. 5, Verlag Chemie, **1979**; *Angew. Chem. Int. Ed. Engl.* **1976**, 270. *The Synthetic Utility of 2-Oxazolines*.
3R. A. I. Meyers, K. A. Lutomski *Asymmetric Syntheses via Chiral Oxazolines* in *Asymmetric Synthesis*, Vol. 3, J. D. Morrison (Ed.), Academic Press, New York, **1984**; *Acc. Chem. Res.* **1978**, 11, 375. *Asymmetric Carbon-Carbon Bond Formation from Chiral Oxazolines*.
4R. (a) T. G. Gant, A. I. Meyers, *Tetrahedron Reports* in *Tetrahedron* **1994**, 50, 2297, *The Chemistry of Oxazolines* (1985–1994); (b) G. S. K. Wong, W. Wu, *2-Oxazolines* in *Chemistry of Heterocyclic Compounds, Oxazoles*, Vol. 60, Part B, D. Palmer (Ed.), J. Wiley and Sons, Hoboken, NJ, **2004**.
5. A. I. Meyers, *J. Org. Chem.*, 70, **2005**, 6137. *Chiral Oxazolines and Their Legacy in Asymmetric Carbon-Carbon-Bond Forming Reactions*.

Asymmetric Synthesis of Amines and Amino Acids from Amines

Shun-Ichi Murahashi, Okayama University of Science, Japan
Yasushi Imada, Osaka University, Japan

Background

Asymmetric synthesis of α-substituted amines and amino acids from the corresponding amines is extremely useful for the synthesis of biologically active nitrogen compounds such as β-lactam antibiotics. We explored a general method for substitution α to the nitrogen of the amines upon treatment with nucleophiles, because we discovered the environmentally benign catalytic oxidative transformation of secondary amines to nitrones in 1984. Reaction of nitrones with nucleophiles gives α-substituted N-hydroxylamines, which can be converted to the corresponding amines [1R]. This method is in contrast to the conventional method for substitution α to the nitrogen of amines, that is, protection of the N–H bond, deprotonation of the α C–H bond with strong base, subsequent reaction with electrophiles, and removal of the protection [2A]. In this reaction nucleophiles can be introduced at the α position of secondary amines in place of electrophiles.

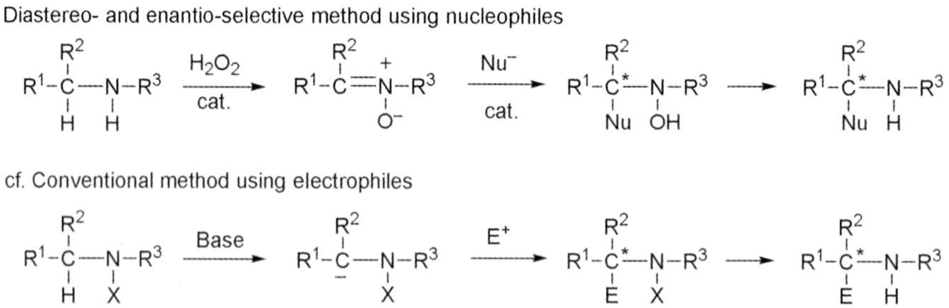

Scheme 1 Strategy for the synthesis of optically active α-substituted amines from secondary amines.

Asymmetric Synthesis – The Essentials.
Edited by Mathias Christmann and Stefan Bräse
Copyright © 2007 WILEY-VCH Verlag GmbH & Co. KGaA, Weinheim
ISBN: 978-3-527-31399-0

Strategy

As a strategy for the synthesis of optically active amines and amino acids from secondary amines, one can carry out diastereoselective and enantioselective reactions of nitrones, derived from secondary amines [3], with nucleophiles. For the diastereoselective method either addition of chiral nucleophiles to nitrones or addition of achiral nucleophiles to chiral nitrones can be carried out.

Results

Diastereoselective Reactions of Nitrones

The reaction of nitrones with chiral α-sulfinyl carbanions gives α-substituted N-hydroxylamines with high diastereoselectivity. Crystalline pure compounds are obtained in good to excellent yields by simple recrystallization [4].

Lewis acid-promoted reactions of nitrones with ketene silyl acetals give N-hydroxyl-β-amino acids highly selectively. Typically, zinc iodide-promoted reaction of (R)-(Z)-ketene silyl acetal **1** with nitrone **2** gave N-hydroxyl amino acid ester **3** (99% de). Catalytic hydrogenation, ring closing, and ruthenium-catalyzed oxidation with peracetic acid gave β-lactam **4**, which is an important precursor of β-methylcarbapenem (Scheme 2) [5].

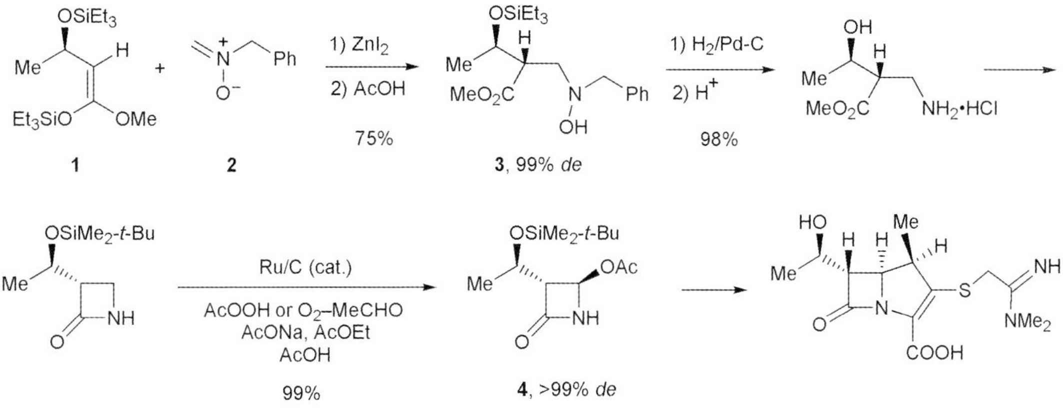

Scheme 2 Synthesis of a common precursor of β-lactam antibiotics.

In order to raise the reactivity of nitrones we generated oxyiminium ions upon treatment of nitrones with acid halides. With these intermediates α-substituted amino compounds can be obtained readily and selectively in comparison with nitrones. This method can be applied for synthesis of chiral β-amino acids and key synthetic intermediates of indolidine alkaloids. The reaction of the oxyiminium ion derived from **5** with chiral titanium enolate **6** gave **7** (96% de) (Scheme 3) [6].

The diastereoselective reaction of chiral nitrones with nucleophiles is also an alternative and practical method for the synthesis of chiral amines. Typically, the chiral pyrroline N-oxide, obtained regioselectively from the tungstate-catalyzed oxidation of 4-tert-butyldimethylsilyloxyprrolidine, is a highly useful intermediate

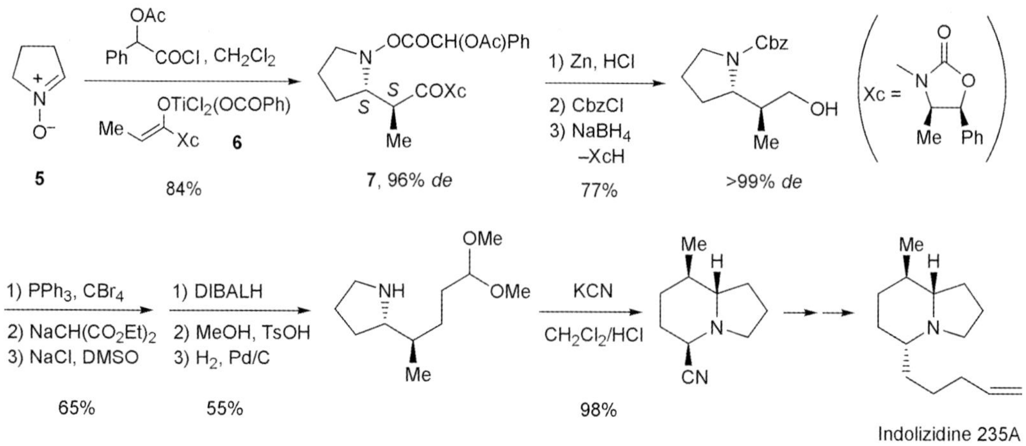

Scheme 3 Asymmetric synthesis of a key intermediate of indolizidine alkaloids.

for asymmetric synthesis of Geissman-Weiss lactone, which is a key intermediate for the synthesis of pyrrolidine alkaloids [7].

Enantioselective Reactions of Nitrones

Enantioselective reaction of nitrones with nucleophiles is the most attractive method for the synthesis of chiral amines and β-amino acids from secondary amines. Three typical examples: enantioselective ruthenium-catalyzed hydrosilylation, iridium-catalyzed hydrogenation, and titanium-catalyzed carbon–carbon bond formation reactions are described.

The enantioselective hydrosilylation of nitrones provides an efficient method for the synthesis of optically active N-hydroxylamines. Asymmetric hydrosilylation of nitrones with diphenylsilane in the presence of TolBINAP ruthenium complex in dioxane at 0 °C followed by hydrolysis gives the corresponding N-hydroxylamines in up to 91 % ee [8].

Enantioselective hydrogenation of nitrones is more attractive for the synthesis of optically active N-hydroxylamines and α-substituted amines. Hydrogenation of nitrones with homogeneous catalysts is extremely difficult; however, we succeeded in asymmetric hydrogenation, for the first time, using the iridium catalyst bearing chiral phosphine ligands. The hydrogenation of nitrones at 80 kg cm^{-2} of hydrogen pressure in the presence of the iridium catalyst prepared from [IrCl (cod)]$_2$, (S)-BINAP (1.1 equiv. of Ir), and n-Bu$_4$NBH$_4$ (1.0 equiv. of Ir) in situ gave the corresponding chiral N-hydroxylamines with up to 90 % ee (Scheme 4). Using (R)-BINAP as a ligand their enantiomers were obtained with the same enantioselectivity [9].

Importantly, asymmetric synthesis of β-amino acids can be carried out by enantioselective carbon–carbon bond formation by addition of carbon nucleophiles to nitrones. Thus, titanium–BINOL-catalyzed reaction of nitrones with enolates gives N-hydroxyl-β-amino acids enantioselectively. The enantioselectivity

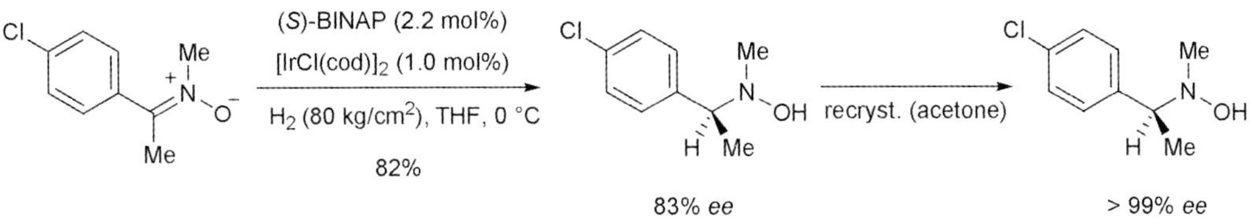

Scheme 4 Asymmetric hydrogenation of nitrones.

is strongly dependent on an additional ligand to the chiral titanium catalyst. When 4-*tert*-butylcatechol was used as an additional ligand, (S)-β-amino acid ester was obtained in 99% yield with 92% ee (Scheme 5) [10].

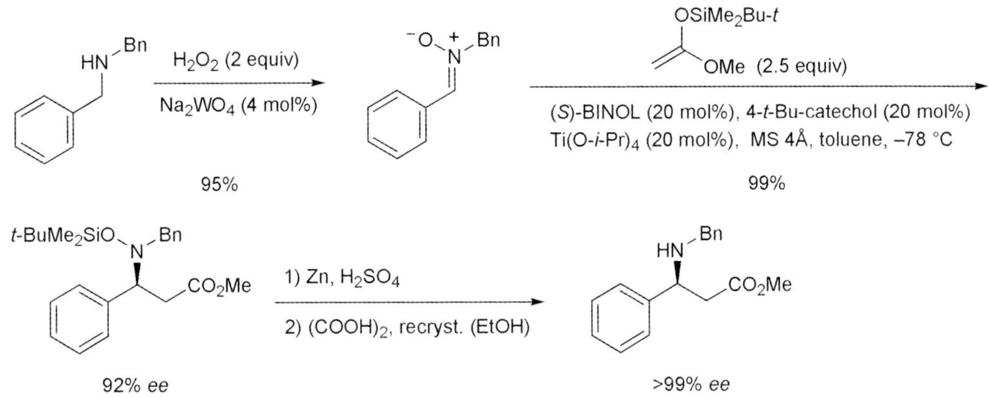

Scheme 5 Asymmetric synthesis of optically active β-aryl-β-amino acid derivatives.

The catalytic system can be applied to the synthesis of chiral isoquinoline derivatives. 1,2,3,4-Tetrahydroisoquinoline-2-acetic acid derivatives, which are the precursors of isoquinoline alkaloids, have been obtained in up to 90% ee.

Conclusions and Future Perspectives

Asymmetric synthesis of α-substituted amines and β-amino acids is performed starting from secondary amines via nitrones, which are readily obtained by simple environmentally benign catalytic oxidation of secondary amines on a large scale. Enantioselective reaction of nitrones with various nucleophiles will provide highly useful methods for the fine chemicals and pharmaceutical industries.

CV of Shun-Ichi Murahashi

Shun-Ichi Murahashi received his Ph. D. degree from Osaka University in 1967 under the guidance of Professor I. Moritani. He spent two years (1968–1970) as a postdoctoral fellow at Columbia University with Professor R. Breslow. He was appointed Assistant Professor in 1963 and promoted to Full Professor of the Department of Chemistry, Faculty of Engineering Science, Osaka University in 1979. Since 2001 he has been Emeritus Professor of Osaka University and Professor of Okayama University of Science. His awards include the Award for Young Chemists of the Chemical Society of Japan (1979), the Chemical Society of Japan Award (1997), and the Humboldt Award (2000). He served as the President of the Chemical Society of Japan in 2000. He has been interested in the exploitation of new catalytic reactions with transition metal catalysts and organic catalysts, which include catalytic reactions for green chemistry and biomimetic oxidation reactions.

Selected Publications

1R. S.-I. Murahashi, *Angew. Chem. Int. Ed. Engl.* **1995**, *34*, 2443–2465. Synthetic Aspects of Metal-Catalyzed Oxidations of Amines and Related Reactions.

2A. P. Beak, A. Basu, D. J. Gallagher, Y. S. Park, S. Thayumanavan, *Acc. Chem. Res.* **1996**, *29*, 552–560. Regioselective, Diastereoselective, and Enantioselective Lithiation–Substitution Sequences: Reaction Pathways and Synthetic Applications.

3. (a) H. Mitsui, S. Zenki, T. Shiota, S.-I. Murahashi, *J. Chem. Soc., Chem. Commun.* **1984**, 874–875. Tungstate Catalysed Oxidation of Secondary Amines with Hydrogen Peroxide. A Novel Transformation of Secondary Amines to Nitrones. (b) S.-I. Murahashi, T. Shiota, *Tetrahedron Lett.* **1987**, *28*, 2383–2386. Selenium Dioxide Catalyzed Oxidation of Secondary Amines with Hydrogen Peroxide. Simple Synthesis of Nitrones from Secondary Amines. (c) S.-I. Murahashi, H. Mitsui, T. Shiota, T. Tsuda, S. Watanabe, *J. Org. Chem.* **1990**, *55*, 1736–1744. Tungstate-Catalyzed Oxidation of Secondary Amines to Nitrones. α-Substitution of Secondary Amines via Nitrones. (d) S.-I. Murahashi, T. Shiota, Y. Imada, *Org. Syntheses* **1992**, *70*, 265–271. Oxidation of Secondary Amines to Nitrones. 6-Methyl-2,3,4,5-tetrahydropyridine N-Oxide.

4. S.-I. Murahashi, J. Sun, T. Tsuda, *Tetrahedron Lett.* **1993**, *34*, 2645–2648. The Reaction of Nitrones with (R)-(+)-Methyl p-Tolyl Sulfoxide Anion; Asymmetric Synthesis of Optically Active Secondary Amines.

5. H. Ohtake, Y. Imada, S.-I. Murahashi, *J. Org. Chem.* **1999**, *64*, 3790–3791. Highly Diastereoselective Addition of a Chiral Ketene Silyl Acetal to Nitrones: Asymmetric Synthesis of β-Amino Acids and Key Intermediates of β-Lactam Antibiotics.

6. T. Kawakami, H. Ohtake, H. Arakawa, T. Okachi, Y. Imada, S.-I. Murahashi, *Org. Lett.* **1999**, *1*, 107–110. Asymmetric Synthesis of β-Amino Acids by Addition of Chiral Enolates to N-Acyloxyiminium Ions and Application for Synthesis of Optically Active 5-Substituted 8-Methylindolizidines.

7. S.-I. Murahashi, H. Ohtake, Y. Imada, *Tetrahedron Lett.* **1998**, *39*, 2765–2766. Synthesis of (R)-and (S)-3-(tert-Butyldimethylsilyloxy)-1-Pyrroline N-Oxides–Chiral Nitrones for Synthesis of Biologically Active Pyrrolidine Derivative, Geissman-Weiss Lactone.

8. S.-I. Murahashi, S. Watanabe, T. Shiota, *J. Chem. Soc., Chem. Commun.* **1994**, 725–726. Asymmetric Hydrosilylation of Nitrones using Ru(II) Phosphine Complex Catalysts; Syntheses of Optically Active N,N-Disubstituted Hydroxylamines and Secondary Amines.

9. S.-I. Murahashi, T. Tsuji, S. Ito, *Chem. Commun.* **2000**, 409–410. Synthesis of Optically Active N-Hydroxylamines by Asymmetric Hydrogenation of Nitrones with Iridium Catalysts.

10. S.-I. Murahashi, Y. Imada, T. Kawakami, K. Harada, Y. Yonemushi, N. Tomita, *J. Am. Chem. Soc.* **2002**, *124*, 2888–2889. Enantioselective Addition of Ketene Silyl Acetals to Nitrones Catalyzed by Chiral Titanium Complexes. Synthesis of Optically Active β-Amino Acids.

Part II

Metal-catalyzed Asymmetric Synthesis

Masakatsu Shibasaki and Shigeki Matsunaga, The University of Tokyo, Japan

Introduction

Asymmetric syntheses are important in the pharmaceutical, agrochemical, fragrance, and flavor industries. For example, chirality-control in pharmaceuticals can be significant in terms of safety; often only one of the enantiomers of a pharmaceutical has desirable biological activity, while the other enantiomer is often inactive or even, in the worst case, has adverse effects. Thus, it is highly desirable to produce medicines in enantiomerically pure form.

Asymmetric catalysis is an ideal method for synthesizing optically active compounds. A tiny amount of chiral promoter produces natural and unnatural chiral compounds in large quantity. The chirality multiplication efficiency of asymmetric synthesis can be evaluated with a value of [{(major enantiomer of product) − (minor enantiomer of product)}/chiral source]. In principle, the chirality multiplication efficiency of asymmetric catalysis can be infinite. In that respect, asymmetric catalysis is different from intra- and inter-molecular chirality transfer reactions, in which there is no chirality multiplication. Thus, tremendous effort has been devoted to developing new asymmetric catalysis reactions and to improving the enantioselectivity and reaction efficiency of existing asymmetric catalysis reactions [1]. To realize maximum chiral multiplication, catalysts should strictly differentiate enantiotopic groups or faces of prochiral molecules. For that purpose, homogeneous metal complexes are some of the most powerful and general tools available. Indeed, metal-mediated asymmetric catalysis has been intensively studied over the last four decades since the first report of homogeneous Cu-catalyzed asymmetric cyclopropanation [2]. Although, at first, enantioselectivity was only modest, notable advances have been achieved during the last three decades. Rh-phosphine complex catalyzed asymmetric hydrogenation of dehydro amino acids [3–5] and Ti-tartrate complex catalyzed asymmetric epoxidation of allylic alcohols [6] were early milestones in metal-catalyzed asymmetric catalysis, realizing high enantioselectivity and broad substrate scope. These reactions demonstrated the high practicality of asymmetric catalysis. Nowadays, the list of asymmetric reactions that realize excellent enantiomeric excess ($> 98\%$ *ee*) has grown to include hydrogenation, dihydroxylation of C–C double bonds, epoxidation,

Asymmetric Synthesis – The Essentials.
Edited by Mathias Christmann and Stefan Bräse
Copyright © 2007 WILEY-VCH Verlag GmbH & Co. KGaA, Weinheim
ISBN: 978-3-527-31399-0

hydroformylation, organometallic addition to aldehydes, allylic alkylation, Diels-Alder reaction, hetero-Diels-Alder reaction, ene-reaction, aldol reaction, Mannich reaction, Michael reaction, conjugate addition of organometallic reagents, and many others [1]. There have also been notable advances in kinetic resolution, including dynamic kinetic resolution. Many catalysts with spectacular selectivity factors have appeared during the last decade. Numerous new chiral metal complexes are reported every year, and the number of new asymmetric reactions is still increasing. In Part 2 some representative examples of metal-catalyzed asymmetric syntheses are introduced.

Design of Asymmetric Metal Catalysis
In metal-catalyzed asymmetric catalysis, high selectivity and reactivity can be realized by selecting the proper catalyst, substrate, and reaction conditions. Catalytic activity is typically determined by the central metal used, while reactivity and enantioselectivity are more finely tuned by chiral organic ligands coordinated to the central metal. A delicate balance of electronic and steric factors determines the efficiency of a chiral metal catalyst. The proper choice of chiral ligands is crucial to realizing efficient intermolecular chirality transfer from catalyst to substrate. Ligands with central chirality, axial chirality and/or planar chirality have proven to be effective. The diverse catalytic activities of metallic species combined with the virtually infinite availability of chiral organic ligands have so far made possible a broad spectrum of asymmetric reactions. Given the countless possible combinations of metal and chiral ligands, future prospects in this area seem bright.

On the basis of the tremendous accumulated knowledge of organometallic compounds brought to us through mechanistic study, rational design and optimization of catalysts is possible. For well-precedented reactions, rational design can be considered appropriate. On the other hand, for unprecedented asymmetric reactions, the screening of suitable metals, chiral ligands, and reaction conditions is still required during early development. The synthesis of chiral ligands, which must themselves be optically pure, can be laborious. Thus, easily modifiable and readily tunable chiral ligands are desirable. Combinatorial approaches to the development of libraries of chiral ligands have recently been reported [7R]. During the screening of library members and reaction conditions, rapid reliable methods for enantioselectivity determination are essential. After initial screening, mechanistic studies take on a key role for further optimization. A good mechanistic understanding of a reaction provides clues for rationally optimizing the metal, chiral ligands, and reaction conditions. Unfortunately, it is difficult to know the precise transition state of an asymmetric reaction at present, even though rapid improvements in computational methods are enabling better and better transition state models to be constructed. Currently, both screening and rational optimization through mechanistic studies are required to achieve highly enantioselective metal-catalyzed asymmetric methods. Either random screening or rational design alone seldom lead to superior chiral catalysts.

When considering basic catalyst design, one does not need to be restricted to mono-metallic complexes. Using bimetallic and polymetallic chiral complexes, bifunctional and multifunctional asymmetric catalysis can be achieved [8, 9R]. Under the right conditions, all of the metals involved in a polymetallic complex can work synergistically to improve reactivity and selectivity in an asymmetric reaction. Proven bifunctional asymmetric catalysts simultaneously control the orientation of a nucleophile and an electrophile while activating them both. High enantio-induction can be achieved through dual activation in an organized chiral environment. Bifunctional asymmetric catalysis, using either metal complexes or organocatalysts, has been one of the most intensively studied fields of catalytic asymmetric methodology during the last decade. Chiral metal complexes exhibiting various modes of bifunctional catalysis have been reported. Some of these modes include Lewis acid–Brønsted base catalysis [10R, 11R], Brønsted acid-assisted-Lewis acid catalysis [12R], Lewis acid–Lewis acid catalysis [12], and Lewis acid–Lewis base catalysis [13]. Among successful bifunctional asymmetric catalysts, both intramolecular and intermolecular cooperativities have been realized. In the intramolecular version, two functionalities, such as a Lewis acid and a Brønsted base, are incorporated in a single chiral metal complex. In intermolecular bifunctional catalysis, two different chiral metal complexes function synergistically in the transition state. Typically, one metal activates a nucleophile, while the other activates an electrophile in the reaction.

Future Prospects for Metal-catalyzed Asymmetric Synthesis

Despite spectacular recent achievements in asymmetric catalysis, there are problems to be tackled in future research. The construction of chiral tetrasubstituted carbon stereocenters, especially quaternary stereocenters, is one of the most important and difficult tasks in asymmetric catalysis. The number of asymmetric catalytic methods applicable in this regard is small, and the enantioselectivity of such reactions is still in need of improvement. Catalytic asymmetric hydrogenation, one of the most sophisticated and practical asymmetric catalyses, cannot afford quaternary carbon centers. Therefore, the development of new powerful and practical catalytic asymmetric carbon–carbon bond-forming reactions is of great importance. At the same time, atom-economy [14] in asymmetric reactions must be considered; to be useful, newly developed asymmetric catalysis needs to be environmentally benign.

Currently, most asymmetric catalysts promote only one specific transformation. Sequential asymmetric catalysis, in which one asymmetric catalyst promotes more than one mechanistically distinct asymmetric reaction in one pot, is also desirable. Although there are several recent examples of sequential asymmetric catalysis [15R], the scope of the reaction is still limited. More powerful and sophisticated catalyst design is required in the future research. Ideally, enantiomerically enriched complex molecules should be synthesized rapidly through the sequential asymmetric catalysis.

CV of Masakatsu Shibasaki

Masakatsu Shibasaki received his PhD from the University of Tokyo in 1974 under the direction of the late Professor Shun-ichi Yamada before doing postdoctoral studies with Professor E. J. Corey at Harvard University. In 1977 he returned to Japan and joined Teikyo University as an Associate Professor. In 1983 he moved to Sagami Chemical Research Center as a group leader, and in 1986 took up a Professorship at Hokkaido University, before returning to the University of Tokyo as a Professor in 1991. He has received the Pharmaceutical Society of Japan Award for Young scientists (1981), Inoue Prize for Science (1994), Fluka Prize (Reagent of the Year, 1996), the Elsevier Award for Inventiveness in Organic Chemistry (1998), the Pharmaceutical Society of Japan Award (1999), the Molecular Chirality Award (1999), the Naito Foundation Research Prize for 2001 (2002), ACS Award (Arthur C. Cope Senior Scholar Award) (2002), the National Prize of Purple Ribbon (2003), the Toray Science Award (2004), and the Japan Academy Prize (2005).

CV of Shigeki Matsunaga

Shigeki Matsunaga received his PhD, with a thesis on the development of a novel chiral ligand linked-BINOL, from the University of Tokyo under the direction of Professor M. Shibasaki. He started his academic career in 2001 as an Assistant Professor in Professor Shibasaki's group at the University of Tokyo. He is the recipient of the 2001 Yamanouchi Award for Synthetic Organic Chemistry, Japan. His current research interests lie in the development and mechanistic studies of new catalytic reactions, including asymmetric catalysis.

Selected Publications

1. I. Ojima (Ed.) *Catalytic Asymmetric Synthesis*, 2nd edn., Wiley, New York, **2000**.
2. H. Nozaki, S. Moriuti, H. Takaya, R. Noyori, *Tetrahedron Lett.* **1966**, 5239–5244. *Asymmetric Induction in Carbenoid Reaction by Means of a Dissymmetric Copper Chelate.*
3. S. Knowles, M. J. Sabacky, *J. Chem. Soc., Chem. Commun.* **1968**, 1445–1446. *Catalytic Asymmetric Hydrogenation Employing a Soluble, Optically Active, Rhodium Complex.* See also contribution by Knowles in Part 5.
4. T. P. Dang, H. B. Kagan *J. Chem. Soc., Chem. Commun.* **1971**, 481–481. *The Asymmetric Synthesis of Hydratropic Acid and Amino Acids by Homogeneous Catalytic Hydrogenation.*
5. R. Noyori, *Asymmetric Catalysis in Organic Synthesis*, Wiley, New York, **1994**.
6. T. Katsuki, K. B. Sharpless, *J. Am. Chem. Soc.* **1980**, *102*, 5974–5976. *The 1st Practical Method for Asymmetric Epoxidation.*
7R. K. D. Shimizu, M. L. Snapper, A. H. Hoveyda, in *Comprehensive Asymmetric Catalysis, Supplement 1*, E. N. Jacobsen, A. Pfaltz, H. Yamamoto (Eds.) Springer, Berlin, **2004**, pp.171–187. *Combinatorial Approaches.*
8. M. Shibasaki, Y. Yamamoto (Eds.), *Multimetallic Catalysts in Organic Synthesis*, Wiley - VCH, Weinheim, **2004**.
9R. J.-A. Ma, D. Cahard, *Angew. Chem. Int. Ed. Engl.* **2004**, *43*, 4566–4583. *Towards Perfect Catalytic Asymmetric Synthesis: Dual Activation of the Electrophile and the Nucleophile.*
10R. M. Shibasaki, H. Sasai, T. Arai, *Angew. Chem., Int. Ed. Engl.* **1997**, *36*, 1236–1256. *Asymmetric Catalysis with Heterobimetallic Compounds.*

11R. M. Shibasaki, N. Yoshikawa, *Chem. Rev.* **2002**, *102*, 2187–2209. *Lanthanide Complexes in Multifunctional Asymmetric Catalysis.*

12R. H. Yamamoto, K. Futatsugi, *Angew. Chem. Int. Ed. Engl.* **2005**, *44*, 1924–1942. *"Designer Acids": Combined Acid Catalysis for Asymmetric Synthesis.*

13. M. Kanai, N. Kato, E. Ichikawa, M. Shibasaki, *Synlett* **2005**, 1491–1508. *Power of Cooperativity: Lewis Acid-Lewis Base Bifunctional Asymmetric Catalysis.*

14. B. M. Trost, *Science* **1991**, *254*, 1471–1477. *The Atom Economy – A Search for Synthetic Efficiency.*

15R. A. Ajamian, J. L. Gleason, *Angew. Chem. Int. Ed. Engl.* **2004**, *43*, 3754–3760. *Two Birds with one Metallic Stone: Single-Pot Catalysis of Fundamentally Different Transformations.*

Catalytic Asymmetric Sulfur Ylide Mediated Epoxidation of Carbonyl Compounds

Varinder Aggarwal, University of Bristol, United Kingdom

Background

Epoxides are important functional groups in synthesis and can be prepared from carbonyl compounds in two ways: by Wittig olefination (which controls relative stereochemistry) followed by enantioselective oxidation of the prochiral double bond, or by enantioselective alkylidenation using either an ylide, carbene or Darzens reagent (Scheme 1). Whilst there are many examples of the two-step process involving a Wittig reaction followed by asymmetric oxidation, the one-step sulfur ylide route had not been widely used. We felt that if we could render the process both catalytic and asymmetric, the advantages of the new one-step route could begin to compete with the more traditional oxidative approach.

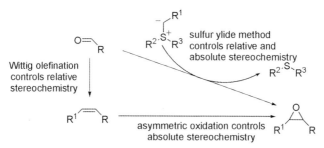

Scheme 1 Synthesis of epoxides from carbonyl compounds: competing methods.

The reaction of a sulfonium ylide with an aldehyde initially forms betaine intermediates that undergo subsequent ring closure to furnish an epoxide with regeneration of the sulfide. In order to render the process catalytic (with respect to the sulfide), the sulfide must be converted back into the ylide under suitably mild conditions. This is particularly important when chiral sulfides are employed. Two general methods exist for converting a sulfide into a sulfur ylide: alkylation with a suitable electrophile followed by deprotonation of the resulting sulfonium salt, or reaction with a diazo compound in the presence of a metal catalyst [1].

Asymmetric Synthesis – The Essentials.
Edited by Mathias Christmann and Stefan Bräse
Copyright © 2007 WILEY-VCH Verlag GmbH & Co. KGaA, Weinheim
ISBN: 978-3-527-31399-0

The method involving reaction of a chiral sulfide with an alkyl halide and base in the presence of an aldehyde has been extensively explored by Furukawa, Metzner, and Goodman but is generally limited to the synthesis of stilbene oxide derivatives [1A]. This is because reactive halides are required for reasonable rates in the alkylation step, selective deprotonation of the resulting salt should result, and the aldehydes should not undergo significant side reactions (e.g. aldol reactions). Nevertheless, this method has delivered high levels of enantioselectivity for this class of epoxides.

As indicated above, an alternative method for the generation of sulfur ylides is by reaction of a sulfide with a diazo compound in the presence of a metal catalyst. One of the principle advantages of this method is that the reaction is conducted under neutral conditions allowing base sensitive substrates to be employed. Furthermore, the intermediate metal carbenes should be more reactive than the alkyl halides and should therefore facilitate ylide formation, even with less reactive sulfides.

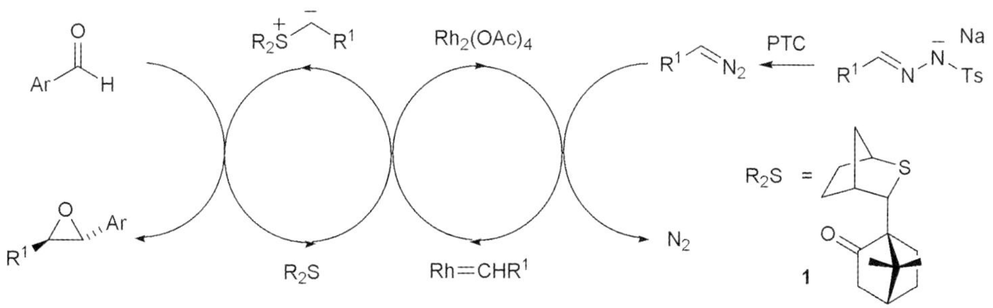

Scheme 2 Catalytic asymmetric epoxidation of aldehydes using tosylhydrazone salts.

Results

The catalytic cycle shown in Scheme 2 was developed [2]. This involves decomposition of the diazo compound in the presence of a transition metal complex (0.5–1 mol % loading) to yield a metallocarbene which is then transferred to a sulfide (5–20 mol % loading), forming a sulfur ylide, which finally undergoes reaction with an aldehyde to give the desired epoxide, returning the sulfide to the cycle to make it available for further catalysis. The diazo compound is itself generated *in situ* from the corresponding tosylhydrazone salt, thus providing a user-friendly process that has been conducted on relatively large scale (50 mmol). The best substrates were found to be aromatic hydrazone salts, unsaturated hydrazones giving lower yields. Most aldehydes and activated ketones were effective, although hindered substrates and those containing basic groups (e.g. pyridines) often gave reduced yields [2].

The use of the chiral bicyclic sulfide **1**, which could be prepared from camphor sulfonyl chloride in four steps with 48 % overall yield, was highly effective in our epoxidation process, giving good yields and high enantio- and diastereo-selectivities in most cases (> 87 % *ee*) [2, 3].

The model for the origin of enantioselectivity is shown in Scheme 3 [4]. We believe that a single ylide diastereomer is formed **2** and, as a result of electronic repulsions, this can exist in either of the two conformations **2A** and **2B** where the lone pairs of electrons on carbon and sulfur are orthogonal. Once a single diastereomeric sulfonium ylide is formed, enantioselectivity is governed by three main factors: (i) ylide conformation, (ii) facial selectivity of the ylide reaction, and (iii) the degree of reversibility in betaine formation.

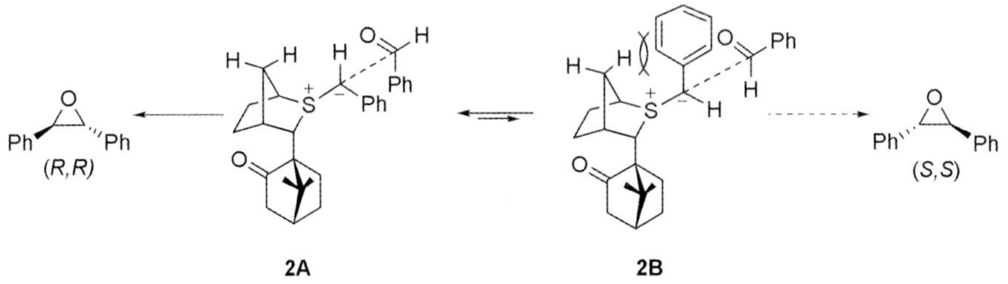

Scheme 3 Rationale for enantioselectivity.

The asymmetric ylide-mediated epoxidation process has also been extended to the preparation of cyclopropanes and aziridines and again high enantio- and diastereo-selectivity has been achieved [5]. Furthermore, the cyclopropane methodology has been applied to the preparation of cyclopropylamino acids [5], while the aziridination methodology has been applied to the synthesis of the side chain of taxol (Scheme 3).

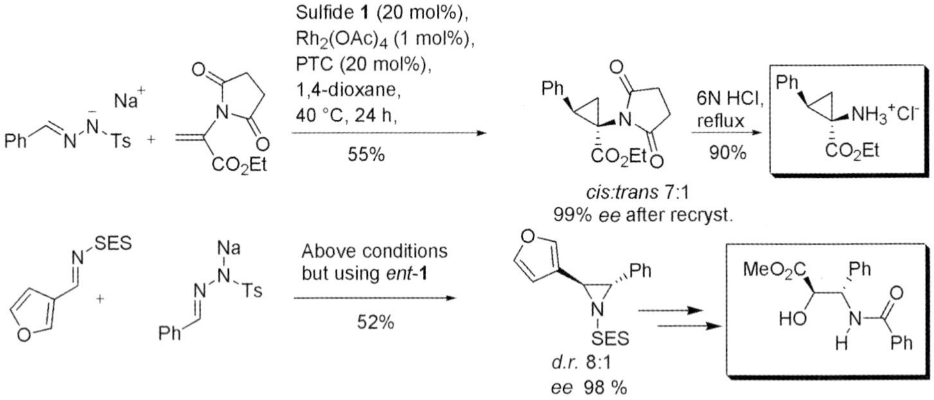

Scheme 4 Application of catalytic asymmetric cyclopropanation and aziridination.

Although the catalytic process has quite broad scope, like most catalytic processes it also has its limitations (although it is less limited than the alkylation/deprotonation approach discussed above). Thus, as an alternative, we considered the use of a stoichiometric process, ideally involving efficient recovery of the sulfide. This has proved to be very successful and every substrate that proved problematic in the catalytic process has worked well in the stoichiometric process [6]. This methodology has been applied to the synthesis of the anti-inflammatory agent, CDP-840 (Scheme 5) [6]. Chiral sulfonium salt **3** was reacted with 4-pyridine carboxaldehyde using the phosphazene (EtP$_2$) base and the desired epoxide **4** was obtained as a 7:3 mixture of diastereoisomers, with almost complete control of enantioselectivity (> 98% ee for each diastereoisomer). The sulfide itself was recovered in near quantitative yield. A mixture of cis and trans epoxides was acceptable because they were both obtained with the same, high enantioselectivity at C2 and following ring opening and removal of the hydroxy group CDP-840 was obtained in enantiomerically pure form.

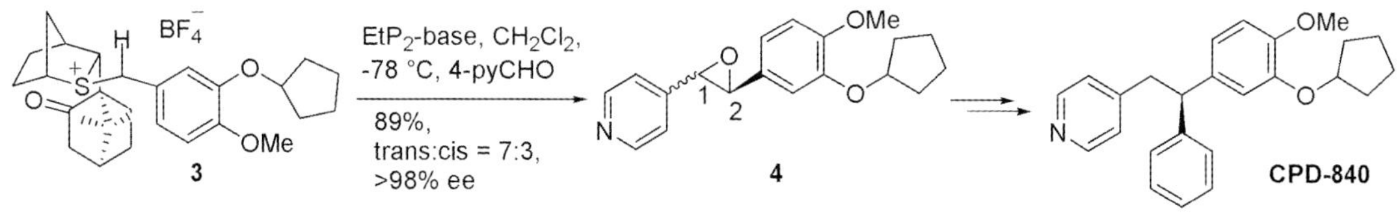

Scheme 5 Application of stoichiometric sulfur ylide epoxidation.

Other Reactions of Sulfur Ylides from the Aggarwal Group

The reactions of aryl stabilized sulfonium ylides with trialkyl/triarylboranes have recently been studied and found to be highly enantioselective in all cases (Scheme 6).

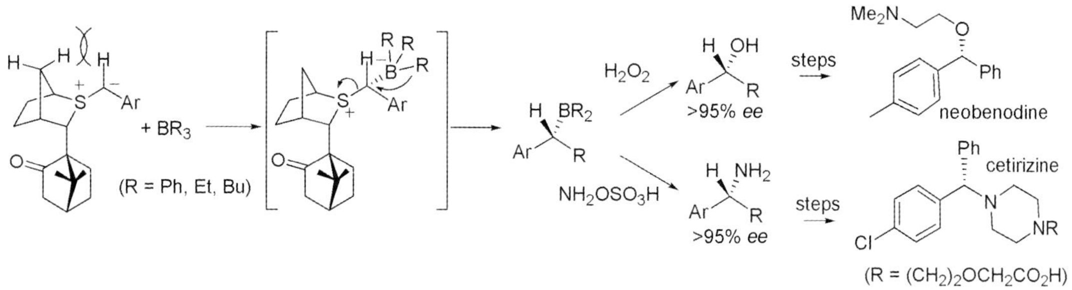

Scheme 6 Reactions of sulfur ylides with organoboranes.

The ylide-borane reaction has been applied to short syntheses of the anti-inflammatory agents neobenodine and cetirizine [7].

Conclusions and Future Perspectives

We have developed a catalytic asymmetric sulfur ylide mediated epoxidation process that shows fairly broad scope. In parallel, a stoichiometric process has also been introduced to address the limitations of the catalytic process. The combined catalytic and stoichiometric processes allow access to a very broad range of epoxides including glycidic amides and α,β-unsaturated epoxides, aziridines and cyclopropanes with control of both relative and absolute stereochemistry in many instances. This broad substrate scope now allows the sulfur ylide disconnection to be applied with confidence in ever more complex total syntheses and studies in this area are ongoing. Reactions of sulfur ylides with organoboranes provide a new route to chiral organoboranes with high enantiomeric excess. This reaction shows considerable promise for asymmetric organic synthesis.

CV of Varinder Aggarwal

Varinder K. Aggarwal was born in Kalianpur in North India in 1961 and emigrated to the United Kingdom in 1963. He received his B.A. (1983) and Ph.D. (1986) from Cambridge University, the latter under the guidance of Dr Stuart Warren. He was subsequently awarded a Harkness fellowship to carry out postdoctoral work with Professor Gilbert Stork at Columbia University, NY (1986–1988). He returned to a lectureship at Bath University and in 1991 moved to Sheffield University, where in 1997, he was promoted to Professor of Organic Chemistry. In 2000, he moved to the University of Bristol to take up the Chair of Synthetic Chemistry. He is the recipient of the AstraZeneca Award, Pfizer Award, GlaxoWelcome Award, Novartis Lectureship, RSC Hickinbottom Fellowship, Nuffield Fellowship, RSC Corday Morgan Medal and the Liebigs Lectureship, RSC Green Chemistry Award and RSC Organic Reaction Mechanisms Award.

Selected Publications

1A. V. K. Aggarwal, C. L. Winn, *Acc. Chem. Res.* **2004**, 611–620. Catalytic Asymmetric Sulfur Ylide Mediated Synthesis of Epoxidation of Carbonyl Compounds: Scope, Selectivity and Applications in Synthesis.

2. V. K. Aggarwal, E. Alonso, I. Bae, G. Hynd, K. M. Lydon, M. J. Palmer, M. Patel, M. Porcelloni, J. Richardson, R. Stenson, J. R. Studley, J.L. Vasse, C. L. Winn, *J. Am. Chem. Soc.* **2003**, 125, 10926–10940. A New Protocol for the In Situ Generation of Aromatic, Heteroaromatic, and Unsaturated Diazo Compounds and Its Application in Catalytic and Asymmetric Epoxidation of Carbonyl Compounds. Extensive Studies To Map Out Scope and Limitations, and Rationalization of Diastereo- and Enantioselectivities.

3. V. K. Aggarwal, E. Alonso, G. Hynd, K. M. Lydon, M.J. Palmer, M. Porcelloni,, J. R. Studley, *Angew. Chem. Int. Ed.* **2001**, 40, 1430–1433. Catalytic Asymmetric Synthesis of Epoxides from Aldehydes Using Sulfur Ylides with In Situ Generation of Diazocompounds.

4. V. K. Aggarwal, J. Richardson, *Chem. Commun.* **2003**, 21, 2644–2651. The Complexity of Catalysis: Origins of Enantio- and Diastereocontrol in Sulfur Ylide Mediated Epoxidation Reactions.

5. V. K. Aggarwal, E. Alonso, G. Fang, M. Ferrara, G. Hynd, M. Porcelloni, *Angew. Chem. Int. Ed. Engl.* **2001**, 40, 1433–1436. Application of Chiral Sulfides to Catalytic Asymmetric Aziridination and Cyclopropanation with In Situ Generation of the Diazocompound.

6. V. K. Aggarwal, I. Bae, H.-Y. Lee, D. T. Williams, *Angew. Chem. Int. Ed. Engl.* **2003**, 42, 3274–3278. Sulfur Ylide Mediated Synthesis of Highly Functionalised and Trisubstituted Epoxides with High Enantioselectivity. Application to the Synthesis of CDP 840.

7. V. K. Aggarwal, G. Yu Fang, A. Schmidt, *J. Am. Chem. Soc.* **2005**, 127, 1642–1643. Synthesis and Applications of Chiral Organoboranes Generated from Sulfur Ylides.

Asymmetric Baeyer-Villiger Reactions

Carsten Bolm, RWTH Aachen University, Germany

Background

In 1899 a new reaction was discovered when, in their laboratories at the Academy of Sciences in Munich, the German chemists Baeyer and Villiger treated the cyclic ketone menthone with Caro's reagent, peroxomonosulfuric acid. A ring-expanded product resulted, which had incorporated one oxygen atom stemming from the inorganic oxidant (Scheme 1). During the following century several well-designed studies elucidated mechanistic details and demonstrated the pronounced synthetic value of the "Baeyer-Villiger reaction" yielding lactones and esters from cyclic and acyclic ketones, respectively [1R].

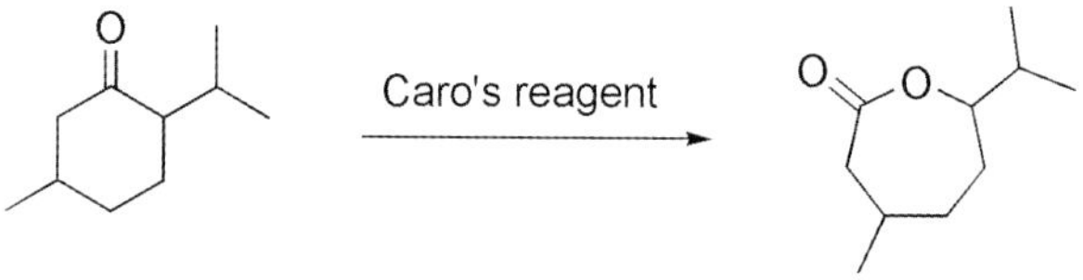

Scheme 1 The first Baeyer-Villiger reaction published in 1899.

Among the most important characteristics of the Baeyer-Villiger reaction are: (i) A wide range of substrates can be converted, and most functional groups are tolerated. (ii) Generally, the oxygen insertion is highly regioselective, and the carbonyl substituent, which is able to better stabilize a developing positive charge, migrates with high preference. Thus, in most cases the regioselectivity is predictable and unsymmetrically substituted carbonyl compounds can selectively be converted into single isomeric products. (iii) When chiral ketones with stereogenic centers α to the carbonyl group undergo this type of oxidative ring expansion, the stereochemistry is retained, and the migration occurs with retention of configuration.

Interestingly, all three of the forementioned attributes of the Baeyer-Villiger reaction were already relevant in the first example (Scheme 1) described by the discoverers in 1899. Even more so, in that publication Baeyer and Villiger made mechanistic proposals, which are still discussed today, including the intermediacy of a dioxirane, which they depicted in one of their schemes.

Other stereochemical issues of the Baeyer-Villiger reaction resulting from asymmetric versions starting from prochiral substrates or kinetic resolutions involving racemic ketones have first been addressed by enzyme catalysis. Initially, major reports in this area were published by Walsh and Schwab, and recently significant progress has been achieved by Furstoss, Roberts and others.

Strategy
Early on it was shown that various oxidizing agents such as peracids, hydrogen peroxide and alkyl peroxides could be applied in Baeyer-Villiger reactions, and that their structure affected the regiochemistry of the oxygen insertion. Furthermore, acids and bases were known to catalyze the ketone to ester/lactone conversion. On that basis in the early 90s we initiated a study of the catalyzing effect of metal salts in *aerobic* Baeyer-Villiger reactions, and in 1993 we published our first communication on this topic. Already at this stage we wondered about enantioselective versions with chiral catalysts, but it took an additional year until we found an appropriate catalyst/reagent/substrate combination, which allowed us to report on the *first metal-catalyzed asymmetric Baeyer-Villiger reaction* giving an optically active lactone from a racemic starting material (Scheme 2) [2].

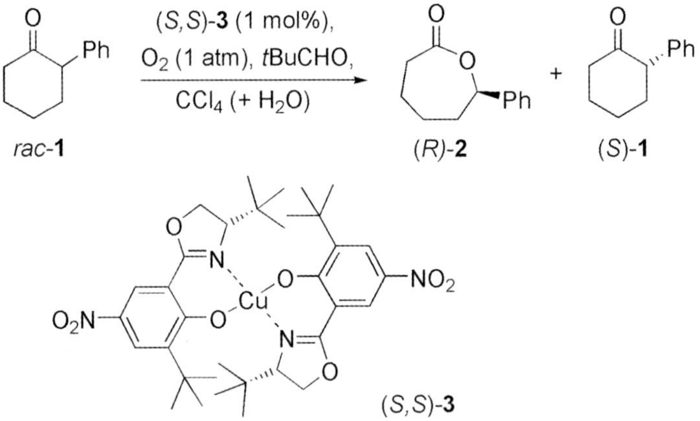

Scheme 2 First asymmetric Baeyer-Villiger reaction catalyzed by a chiral metal complex.

Although with this initial copper-based catalyst system both yield and maximal enantioselectivity (69% ee) were rather limited, the demonstration of the reaction potential was regarded as a conceptual breakthrough. Interestingly, at the same time Strukul presented his first asymmetric Baeyer-Villiger catalysis involving a chiral platinum/BINAP catalyst, which converted ketones into optically active lactones with hydrogen peroxide as oxidant.

Results

After these early demonstrations of metal-catalyzed asymmetric Baeyer-Villiger reaction by Strukul and us, a number of groups all around the world began to work on this topic, and various metal systems were found to be applicable, yielding optically active lactones with, in some cases, > 90% ee [1R]. For example, Kanger et al. in Estonia introduced a titanium-based (Sharpless epoxidation type) catalyst, which converted racemic bicyclic ketones into lactones with up to 75% ee. Furthermore, Katsuki and coworkers in Japan developed highly enantioselective cobalt and platinum catalysts yielding lactones with up > 99% ee.

Based on the concept that an efficient Baeyer-Villiger reaction requires a Lewis-acidic metal with an appropriate chiral ligand environment, we continued our studies by investigating the effect of various metal/ligand combinations on the oxidative conversions of chiral and prochiral substrates such as **4**, **7**, and **9**, respectively (Scheme 3).

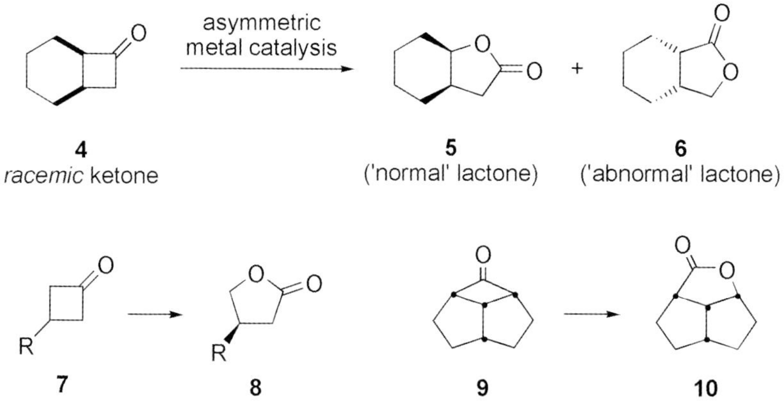

Scheme 3 Metal-catalyzed Baeyer-Villiger reactions with chiral and prochiral substrates.

After inital applications of copper complex **3**, which, for example, revealed that metal-catalyzed Baeyer-Villiger reactions of bicyclic (racemic) ketones such as **4** proceeded (in analogy to some of their enzymatic counterparts) in an *enantiodivergent* manner giving 'abnormal' lactone **6** with very high ee [3] and that tricyclic ketone **9** led to the highest enantioselectivities in conversions of prochiral ketones, we began to focus our attention on the use of other metal catalysts, which were based on zirconium, magnesium, and aluminum [1R]. In particular, the latter metal proved most versatile, and with BINOL derivatives such as VANOL (**11**) as ligands and alkyl hydroperoxides as oxidants even simple prochiral substrates such as **7** (with R = aryl or alkyl substituents) were converted into lactones **8** with up to 84% ee [4]. An example illustrating this catalyst system, which for such a simple ketone is one of the best known to date, is shown in Scheme 4.

Due to the fact that a detailed mechanistic understanding of catalyzed Baeyer-Villiger reactions is still lacking, it is difficult to predict catalyst performances and stereochemical results. With the hope of shining more light into the darkness of

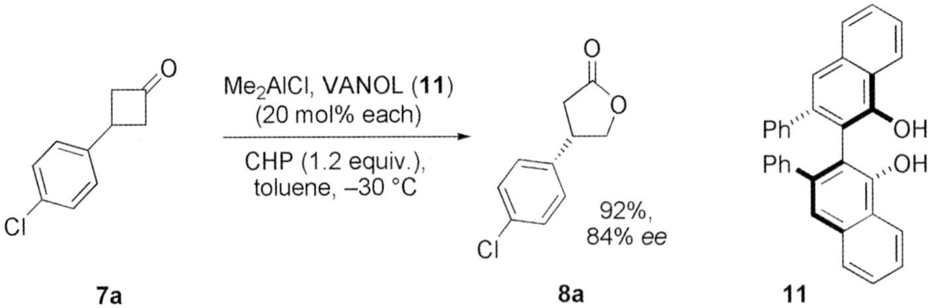

Scheme 4 Asymmetric Baeyer-Villiger reaction with an Al/VANOL catalyst.

this more than 100 years old oxygen insertion reaction, studies on this topic are currently in progress in my laboratories.

Other Research Topics from the Bolm Group
The study of catalyzed Baeyer-Villiger reactions is only one of our projects related to oxidative transformations of organic substrates. Others focus on enantioselective sulfide oxidations, which give rise to sulfoxides with high enantiomeric excesses. There, vanadium and iron catalysts have been developed, which utilize hydrogen peroxide under mild reaction conditions [5]. In subsequent reactions the sulfoxides are then used as intermediates for sulfoximines, which serve as building-blocks of pseudopeptides [6] and ligands for metal complexes being capable of catalyzing asymmetric C–C bond-forming reactions [7]. The latter reaction type has also been examined in the context of catalyzed aryl transfer reactions onto aldehydes, which yield diaryl methanols with excellent enantioselectivities [8]. Furthermore, organocatalytic asymmetric anhydride openings have been studied [9], and novel amino and hydroxy acid derivatives have been prepared [10].

CV of Carsten Bolm
Carsten Bolm was born in Braunschweig, Germany in 1960. He studied chemistry at the TU Braunschweig (Germany) and at the University of Wisconsin, Madison (USA). In 1987 he obtained his doctorate with Professor Reetz in Marburg (Germany). After postdoctoral training with Professor Sharpless at MIT, Cambridge (USA), Carsten Bolm worked in Basel (Switzerland) with Professor Giese to obtain his habilitation. In 1993 he became Professor of Organic Chemistry at the University of Marburg (Germany), and since 1996 he has held a chair of Organic Chemistry at the RWTH Aachen University (Germany). He has held visiting professorships at the universities in Madison, Wisconsin (USA), Paris (France), Florence (Italy), Milan (Italy), and Namur (Belgium). His awards include the Heinz-Maier-Leibnitz prize, the ADUC-Jahrespreis for habilitands, the annual prize for Chemistry of the Akademie der Wissenschaften zu Göttingen, the Otto-Klung prize, the Otto-Bayer award, and a fellowship of the Japan Society for the Promotion of Science.

Selected Publications

1R. C. Bolm, C. Palazzi, O. Beckmann, in *Transition Metals For Organic Chemistry: Building Blocks and Fine Chemicals*, Vol. 2 (M. Beller, C. Bolm (Eds.), Wiley-VCH, Weinheim, 2nd edn., **2004**, pp. 267–274. *Metal-catalyzed Baeyer-Villiger reactions.*

2. C. Bolm, G. Schlingloff, K. Weickhardt, *Angew. Chem.* **1994**, *106*, 1944–1946; *Angew. Chem. Int. Ed. Engl.* **1994**, *33*, 1848–1849. *Optically active lactones by Baeyer-Villiger analogous oxidation with molecular oxygen.*

3. C. Bolm, G. Schlingloff, *J. Chem. Soc., Chem. Commun.* **1995**, 1247–1248. *Metal-catalyzed enantiospecific aerobic oxidations of cyclobutanones.*

4. C. Bolm, J.-C. Frison, Y. Zhang, W. D. Wulff, *Synlett* **2004**, 1619–1621. *Vaulted Biaryls: Efficient ligands for the aluminum-catalyzed asymmetric Baeyer-Villiger reaction.*

5. J. Legros, C. Bolm, *Chem. Eur. J.* **2005**, *11*, 1086–1092. *Investigations on the iron-catalyzed asymmetric sulfide oxidation.*

6. C. Bolm, D. Müller, C. Dalhoff, C. P. R. Hackenberger, E. Weinhold, *Bioorg. Med. Chem. Lett.* **2003**, *13*, 3207–3211. *The Stability of pseudopeptides bearing sulfoximines as chiral backbone modifying element towards proteinase K.*

7. H. Okamura, C. Bolm, *Chem. Lett.* **2004**, *33*, 482–487. *Sulfoximines: Synthesis and catalytic applications.*

8. S. Özçubukçu, F. Schmidt, C. Bolm, *Org. Lett.* **2005**, *7*, 1407–1410. *Organosilanols as catalysts in asymmetric aryl transfer reactions.*

9. C. Bolm, I. Schiffers, I. Atoridisei, C. P. R. Hackenberger, *Tetrahedron: Asymmetry* **2003**, *14*, 3455–3467. *The alkaloid-mediated desymmetrization of meso-anhydrides by nucleophilic ring opening with benzyl alcohol and its application in the synthesis of highly enantiomerically enriched β-amino acids.*

10. C. Bolm, S. Saladin, A. Claßen, A. Kasyan, E. Veri, G. Raabe, *Synlett* **2005**, 461–464. *Enantiopure α-silyl-substituted α-hydroxyacetic acids using O–H insertion methodology and boron-based asymmetric reductions.*

Planar Chiral Ligands Based on [2.2]Paracyclophanes
Stefan Bräse, University of Karlsruhe (TH), Germany

Background

Planar chirality plays a pivotal role in modern ligand systems. In particular, the success of ferrocenyl ligands has not to date been matched by any other planar chiral ligands. Both metallocene and metalarene-based ligand backbones exhibit a common feature, they only become planar chiral upon addition of (at least) two substituents on one ring. [2.2]Paracyclophanes, however, only require one substituent (Figure 1) [1A].

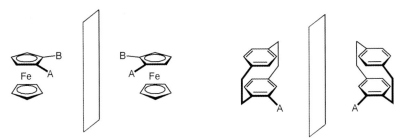

Figure 1 Planar chirality in ferrocenyl and paracyclophanyl systems.

Since the initial reports of Reich and Cram, the field of [2.2]paracyclophane chemistry has expanded considerably. The chemical behavior of [2.2]paracyclophanes is now well understood and their comparison with "rocks" is justified because they are usually extremely stable compounds. However, [2.2]paracyclophanes are sometimes very unreactive in a desired reaction. All in all, they can impose a reasonable amount of resistance against synthetic efforts and therefore present an interesting challenge to dedicated chemists.

When we first entered this field of [2.2]paracyclophane ligand synthesis, successful applications were relatively rare. The most prominent example was the PHANEPHOS ligand developed by Rossen and Pye, which was applied in asymmetric hydrogenation reactions. The PHANEPHOS ligand exhibits the well-established C_2 symmetry that is present in various flourishing systems (BINOL, Salen ligand, bisoxazolines). Through the applications of other ligand systems such as

Asymmetric Synthesis – The Essentials.
Edited by Mathias Christmann and Stefan Bräse
Copyright © 2007 WILEY-VCH Verlag GmbH & Co. KGaA, Weinheim
ISBN: 978-3-527-31399-0

the oxazoline phosphanes developed by Helmchen, Pfaltz, and Williams, it was confirmed further that, in certain cases, other types of symmetry can lead successfully to high degrees of enantioselectivity.

The use of planar chiral and central chiral ligands based on paracyclophane systems still remained a relatively unexplored frontier, with the notable exception of the reactions examined by the groups of Rozenberg and Berkessel.

Strategy

Central intermediates in our strategy were the known *ortho*-acylated hydroxy[2.2] paracyclophanes **1** (R^1 = H), **2** (R^1 = alkyl) and **3** (R^1 = aryl). These are then condensed with amines to give imines **4**, ketimines **5** and **6**, or reduced to amino alcohols **7–9**, respectively (Figure 2). The ligand structure is therefore vastly variable. Steric factors such as flexibility in the backbone and side chains as well as electronic factors (e.g. sp^2 versus sp^3 configuration of the N-donors) can be easily modulated. The introduction of central chirality via chiral amine side chains or different donor atoms is also possible. The interaction of planar and central chirality, usually referred to as *chiral cooperativity*, can thus be studied in a ligand system which has planar as well as central chiral elements.

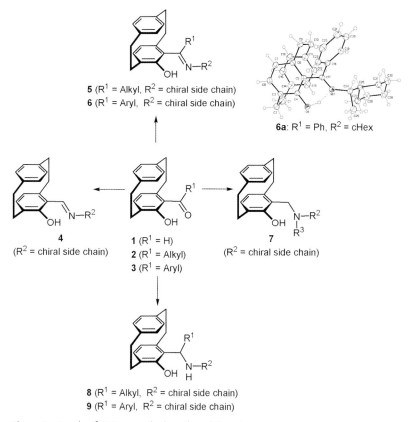

Figure 2 Family of [2.2]paracyclophane-based ligands.

One of the three important intermediates in our strategy was the 5-formyl-4-hydroxy[2.2]paracyclophane (FHPC, **1**), which can be regarded as a planar-chiral analog of salicyl aldehyde. This compound was first prepared by Rozenberg, Belokon et al. and its development was followed by various reports on alternative routes to the racemic as well as the enantiomerically pure compound.

Results

Aliphatic and aromatic aldehydes **10** react readily with aryl, alkyl and alkenyl zinc reagents [2–4] in the presence of [2.2]paracyclophane ligands like 69 to give chiral secondary alcohols **11** in good to excellent enantiomeric excess (up to 98% *ee*, up to quantitative yield) (Scheme 1). In particular, aliphatic aldehydes are superior substrates for this ligand class.

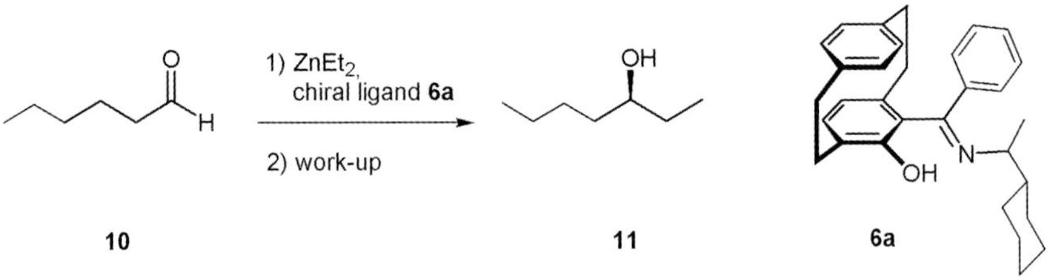

Scheme 1 Diethyl zinc addition to hexanal (**10**). Conditions: hexanal (0.5 mmol), ligand (5%), toluene (1 ml), diethyl zinc (1.0 ml, 1.0 M in hexane, 2 equiv.), 0°C, 16 h under argon.

A special feature of this class of ligands is its ability to catalyze the 1,4-addition of zinc reagents to α,β-unsaturated carbonyl compounds **12** in the absence of copper ions [5] (Scheme 2).

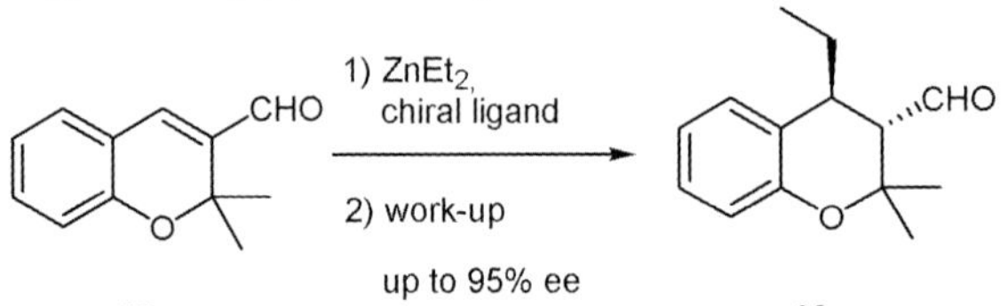

Scheme 2 Diethyl zinc addition to aldehyde **12**. Conditions: aldehyde **12** (0.5 mmol), ligand (5%), toluene (1 ml), diethyl zinc (1.0 ml, 1.0 M in hexane, 2 equiv.), 0°C, 16 h under argon.

The reaction of aliphatic and notably aromatic zinc reagents with imines prepared *in situ* from sulfonyl amines **14** led to the formation of α-chiral amines **16** in excellent enantiomeric excess (Scheme 3). This addition works with both alkyl and aryl zinc reagents. While carbamates or acetamides are less suitable substrates, formamides achieve high enantiomeric excess. Diminishing solubility results in strong temperature dependence. The removal of the formyl group proceeds without racemization [6, 7].

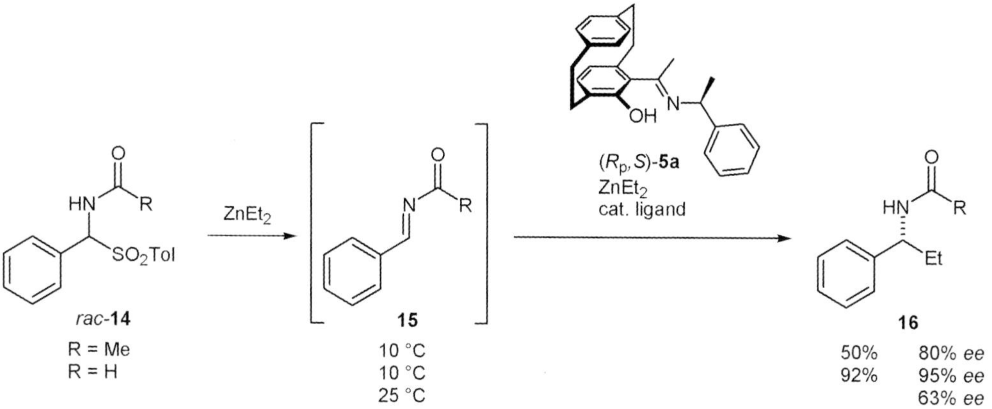

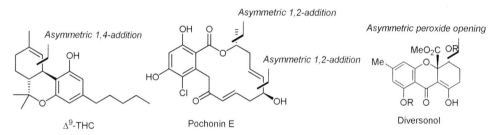

Scheme 3 Asymmetric synthesis of α-branched amines **16**.

Application for the Synthesis of Natural Products

Currently, we are using these strategies for the synthesis of natural products such as tetrahydrocannabinol, pochonin E, and diversonol (Figure 3).

Figure 3 Targets for natural product syntheses.

Other Asymmetric Syntheses from the Bräse Group

In addition to the described systems, we used desymmetrization reactions [8] and organo-catalysis [9, 10] for the construction of complex molecules.

Conclusions and Future Perspectives

Over the last five years, we have designed, synthesized, and applied new ligands for asymmetric 1,2- and 1,4-addition reactions. Suitable ligands were found for the addition of alkyl, aryl and alkenyl zinc reagents aromatic to α-branched and unbranched aliphatic aldehydes and imines. Although some substrates such as ketones and α,β-cyclic unsaturated carbonyl compounds remain a challenge, we believe that this system provides an excellent entry into various classes of chiral intermediates. Application of these synthesized complex molecules is our current pursuit in our laboratories.

CV of Stefan Bräse

Stefan Bräse was born in Kiel, Germany in 1967. He studied at the Universities of Göttingen, Bangor (UK) and Marseille. He received his Ph.D. in 1995 after working with Armin de Meijere in Göttingen. After post-doctoral appointments at Uppsala University (Jan E. Bäckvall) and The Scripps Research Institute (K. C. Nicolaou) as DAAD fellow, he began his independent research career at the RWTH Aachen in 1997 (associated to Dieter Enders). In June 2001, he finished his Habilitation and moved to the University of Bonn as a Professor of Chemistry. Since 2003, he has held a chair at the University of Karlsruhe (TH). He was the recipient of the OrChem prize of the Gesellschaft Deutscher Chemiker (2000) and Eli Lilly Lecturer (2001).

Selected Publications

1A. S. Bräse, S. Dahmen, S. Höfener, F. Lauterwasser, M. Kreis, R. E. Ziegert, *Synlett* **2004**, 2647–2669. *Planar and Central Chiral [2.2]Paracyclophanes as Powerful Catalysts for Asymmetric 1,2 and Conjugate Addition Reactions.*

2. S. Dahmen, S. Bräse, *Chem. Commun.* **2002**, 26–27. *Planar and central chiral [2.2]paracyclophane-based N,O-ligands as highly active catalysts in the diethylzinc addition to aldehydes.*

3. S. Höfener, F. Lauterwasser, S. Bräse, *Adv. Synth. Catal.* **2004**, 364, 755–759. *Second-generation paracyclophane imine ligands for the dialkyl zinc addition to aldehydes. Optimization of the branched side chain leads to improvement for aliphatic aldehydes.*

4. S. Dahmen, S. Bräse, *Org. Lett.* **2001**, 3, 4119–4122. *[2.2]Paracyclophane-based N,O-Ligands in Alkenylzinc Additions to Aldehydes.*

5. S. Bräse, S. Höfener, *Angew. Chem. Int. Ed.* **2005**, 44, 7879–7881. *The Copper-free Asymmetric Conjugate Addition of Zinc Reagents to α,β-unsaturated Ketones and Aldehydes.*

6. S. Dahmen, S. Bräse, *J. Am. Chem. Soc.* **2002**, 124, 5941–5942. *The Asymmetric Dialkylzinc Addition to Imines Catalyzed by [2.2]Paracyclophane-based N,O-Ligands.*

7. N. Hermanns, S. Dahmen, C. Bolm, S. Bräse, *Angew. Chem.* **2002**, 114, 3844–3846; *Angew. Chem. Int. Ed.* **2002**, 40, 3692–3694. *Asymmetric, Catalytic Phenyl Transfer to Imines: Highly Enantioselective Synthesis of Diarylmethylamines.*

8. S. Bräse, *Synlett* **1999**, 1654–1656. *Synthesis of Bis(enolnonaflates) and their 4-exo-trig-Cyclizations by Intramolecular Heck Reactions.*

9. H. Vogt, S. Vanderheiden, S. Bräse, *Chem. Commun.* **2003**, 2448–2449. *The proline-catalysed asymmetric amination of α,α-disubstituted aldehydes – Synthesis of configurationally stable enantioenriched α-aminoaldehydes.*

10. H. Vogt, T. Baumann, S. Bräse, *Eur. J. Org. Chem*, in press. *Asymmetric Synthesis of α-Amino acids by an Organo-Catalytic 1,3-Dipolar-Addition Reaction.*

Asymmetric Syntheses of 3-(*trans*-2-Nitrocyclopropyl)alanine and 3-(*trans*-2-Aminocyclopropyl)alanine

Armin de Meijere and Oleg V. Larionov, Georg-August-Universität Göttingen, Germany

Background

Nature makes use of cyclopropyl groups even in amino acids, and most of the over two dozen known naturally occurring amino acids containing a cyclopropyl group as well as most of the cyclopropyl analogs of natural amino acids, play a pivotal role in the observed biological activities of compounds containing such constituents.

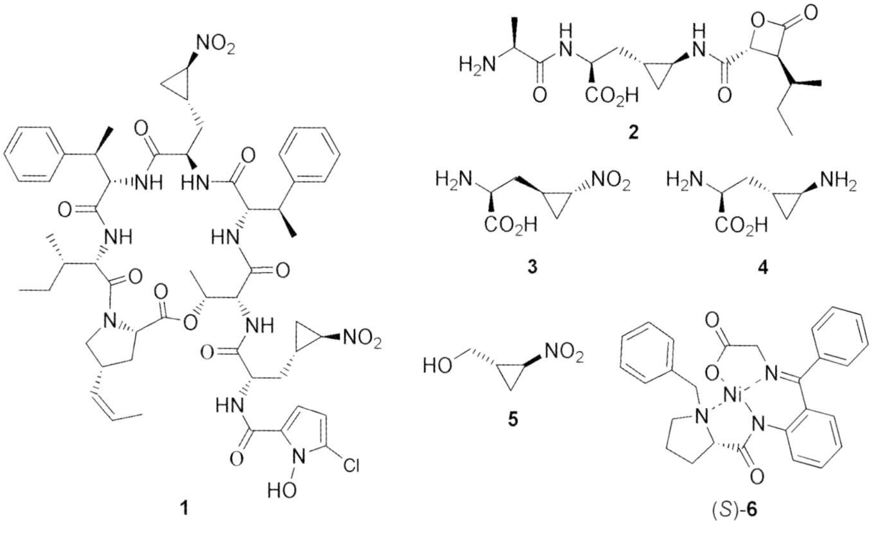

Figure 1 Hormaomycin **1**, Belactosin A **2**, two unusual cyclopropyl-group containing amino acids **3**, **4**, as well as the starting material **5** and Belokon's chiral glycine equivalent (S)-**6** for the asymmetric syntheses of **3** and **4**.

Among the natural products with cyclopropane motifs, the peptidolactone Hormaomycin **1**, which influences the secondary metabolite production of certain bacteria, and the recently isolated Belactosin A **2**, which itself, as well as analogs and mimics thereof, exhibits remarkable proteasome inhibitory activity, are espe-

Asymmetric Synthesis – The Essentials.
Edited by Mathias Christmann and Stefan Bräse
Copyright © 2007 WILEY-VCH Verlag GmbH & Co. KGaA, Weinheim
ISBN: 978-3-527-31399-0

cially intriguing, as they contain the previously unknown 3-(trans-2-nitrocyclopropyl)alanine [(NcP)Ala] (3) and 3-(trans-2-aminocyclopropyl)alanine (4) [(AcP)Ala] as key constituents (Figure 1) [1].

Hormaomycin 1 incorporates the (2S,1'R,2'R)- and (2R,1'R,2'R)-diastereomers of 3-(trans-2-nitrocyclopropyl)alanine (3). On the other hand, the 3-(trans-2-aminocyclopropyl)alanine (4) moiety in Belactosin A (2) possesses the (2S,1'R,2'S)-configuration. Hence, there was an apparent need for a convergent and productive access to at least three diastereomers of the amino acid 3, which in turn would serve as a precursor to the enantiomerically pure amino acid 4. (trans-2-Nitrocyclopropyl)methanol 5, which had already been used in previous syntheses of (2RS,1'R,2'S)-3 and (2S,1'RS,2'RS)-3, as well as several other cyclopropyl-containing amino acids [1–4], was chosen as an appropriate precursor for all four diastereomers of 3 and 4.

Strategy and Results

To begin with, a new productive synthesis of racemic (trans-2-nitrocyclopropyl)-methanol (5) was developed. Towards that, tert-butyl 2,3-dibromopropionate rac-7 was transformed into tert-butyl trans-2-nitrocyclopropanecarboxylate rac-8 in a sequence of alkylation with nitromethane and base-induced diastereoselective ring closure. This ester was reduced to (trans-2-nitrocyclopropyl)methanol (rac-5), which in turn was converted to the iodide rac-9 (Scheme 1).

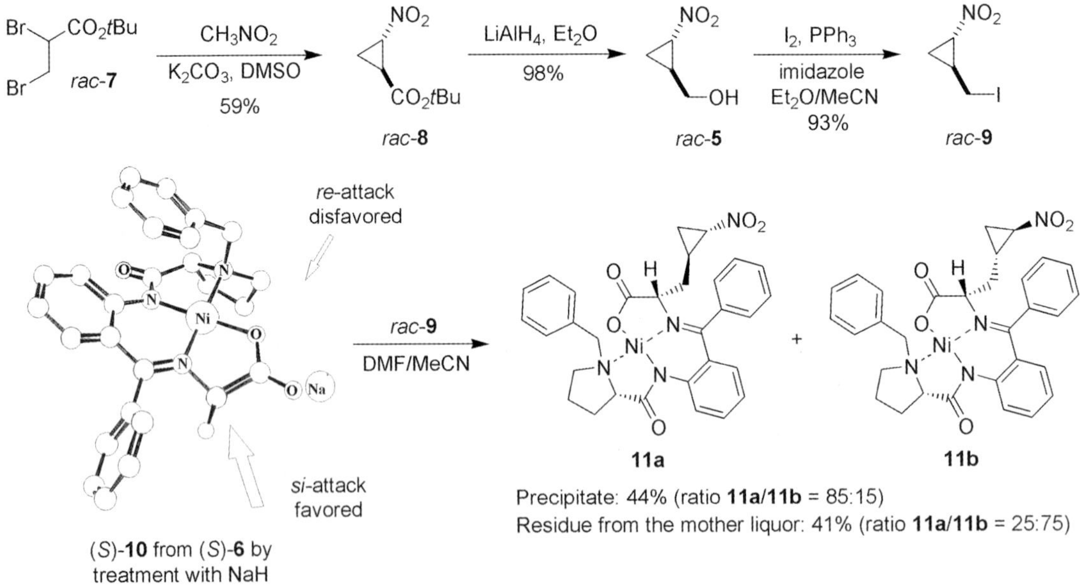

Scheme 1 Synthesis of the (trans-2-nitrocyclopropyl)methyl iodide rac-9, and diastereoselective alkylation of the enolate of Belokon's chiral glycine equivalent (S)-6 with subsequent separation of diastereomers.

Among many different synthetic approaches to enantiomerically pure α-amino acids, the asymmetrically induced C-alkylations of Belokon's chiral nickel(II) complexes, derived from 2-[(N-benzylprolyl)amino]benzophenone [BPB] and simple amino acids, for example glycine and alanine, have proved to be an efficient route to more than 100 enantiomerically pure proteinogenic and non-proteinogenic α-amino acids. The glycyl moiety in Belokon's chiral glycine equivalent (S)-**6** becomes a sufficiently strong CH-acid to be deprotonated in the presence of NaH, NaOH, KOH or K_2CO_3 to provide the enolate, which can be alkylated by alkyl halides or Michael acceptors with a high degree of diastereofacial control. This is ensured by the rigid framework, the benzyl group on the prolyl moiety and the phenyl substituents of the imine component, which efficiently control the direction of the electrophilic attack (Scheme 1) [5R].

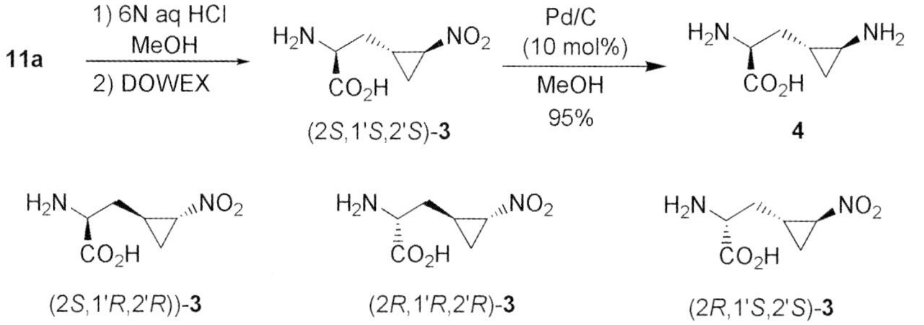

Scheme 2 Liberation of (NcP)Ala **3** from the corresponding nickel complexes **11** and reduction of **3** to **4**.

The racemic iodide *rac*-**9** was employed in the alkylation of the enolate (S)-**10** (Scheme 1). After completion of the reaction, the formed precipitate, isolated in 44% yield, was composed of **11a** and **11b** in a ratio of 85:15. On the other hand, a mixture of the same **11a** and **11b**, (41% yield), but in a ratio of 25:75, was secured by evaporation of the mother liquor. Recrystallization of the fraction enriched in **11a** afforded **11a** with a diastereomeric excess of 95–98%, which was subsequently decomposed to give the amino acid (2S,1′S,2′S)-**3** (Scheme 2). The diastereomer (2S,1′R,2′R)-**3** with an enantiomeric purity of over 96% can be analogously obtained from the fraction enriched in **11b**, by first liberation and then threefold recrystallization of the crude amino acid. The same overall protocol was also applied with (R)-**6**, to give the diastereomers (2R,1′S,2′S)-**3** and (2R,1′R,2′R)-**3**. Thus, all four diastereomers of 3-(*trans*-2-nitrocyclopropyl)-alanine **3** can be efficiently prepared from racemic *trans*-2-nitrocyclopropylmethyl iodide (*rac*-**9**), employing one chiral auxiliary to establish three stereogenic centers [6].

Quite unexpectedly, the hydrogenative reduction of the nitro group, both in the free amino acid **3** and in its hydrochloride, under typical conditions with Pd/C in water to a large extent led to cyclopropane-ring cleavage. Gratifyingly, when methanol was used instead of water, the undesired ring opening was completely suppressed, allowing a highly selective reduction of **3** to **4** (Scheme 2) [7].

Outlook

The appropriate diastereomers of 3-(*trans*-2-nitrocyclopropyl)alanine **3**, prepared according to the developed procedure, were successfully employed in the total syntheses of Hormaomycin **1** [8–10] and several analogs, as well as Belactosin A [11].

CV of Armin de Meijere

Armin de Meijere, born 1939 in Homberg (Niederrhein), Germany, studied chemistry at the Universities of Freiburg and Göttingen, where he obtained his doctorate in 1966 (W. Lüttke). Following postdoctoral training at Yale University (K. B. Wiberg) he finished his "Habilitation" in 1971 in Göttingen. He became Full Professor at the University of Hamburg in 1977, and returned to Göttingen as Professor of Organic Chemistry in 1989. He has been a visiting scientist at many institutions around the world. His current research interests include the development of new metal-catalyzed cascade reactions towards complex organic skeletons, new small-ring building blocks and their application in the synthesis of natural and non-natural compounds, new highly strained polycyclic compounds with interesting properties, as well as the development of new methodology based on metal-mediated and -catalyzed transformations. His scientific achievements have been published in more than 630 original papers, review articles and book chapters. Among other honors he received the French Alexander-von-Humboldt-Gay-Lussac Award in 1996, the appointment as an Honorary Professor of St. Petersburg State University in 1997, and the Adolf-von-Baeyer Medal of the Gesellschaft Deutscher Chemiker in 2005.

CV of Oleg V. Larionov

O. V. Larionov, born in 1979 in Shevchenko (Aktau), Kazakhstan, graduated from the Higher Chemical College of the Russian Academy of Sciences in 2002 under the supervision of Prof. Yu. N. Belokon (Moscow). He obtained his doctorate in 2005 (A de Meijere) at the University of Göttingen, and he is currently pursuing postdoctoral work at the Max-Planck-Institut für Kohleforschung (A. Fürstner).

Selected Publications

1. J. Zindel, A. Zeeck, W. A. König, A. de Meijere, *Tetrahedron Lett.* **1993**, *34*, 1917–1920. *Synthesis of 3-(trans-2′-Nitrocyclopropyl)alanine – an Unusual Natural Amino Acid.*

2. J. Zindel, A. de Meijere, *Synthesis* **1994**, 190–194. *A Short and Efficient Diastereoselective Synthesis of 2′-Substituted 2-Cyclopropylglycines.*

3. J. Zindel, A. de Meijere, *J. Org. Chem.* **1995**, *60*, 2968–2973. *Synthesis of 3-(trans-2′-Nitrocyclopropyl)alanine, a Constituent of the Natural Peptide Lactone Hormaomycin.*

4. M. Brandl, S. I. Kozhushkov, K. Loscha, O. V. Kokoreva, D. S. Yufit, J. A. K. Howard, A. de Meijere, *Synlett* **2000**, *12*, 1741–1744. *Synthesis of trans-(2-Aminocyclopropyl)alanine – a Key Constituent of the Novel Antitumor Antibiotic Belactosin A.*

5R. Yu. N. Belokon, *Janssen Chim. Acta* **1992**, *10* (2), 4–12. *(S)-2-[(N-benzylprolyl)amino]benzophenone (BPB) – a Reagent for the Synthesis of Optically Pure α-Amino Acids.*

6. O. V. Larionov, T. F. Savel'eva, K. A. Kochetkov, N. S. Ikonnikov, S. I. Kozhushkov, D. S. Yufit, J. A. K. Howard, V. N. Khrustalev, Yu. N. Belokon, A. de Meijere, *Eur. J. Org. Chem.* **2003**, 869–877. *Productive Asymmetric Synthesis of All Four Diastereomers of 3-(trans-2-nitrocyclopropyl)alanine from Glycine with (S)- or (R)-2-[(N-Benzylprolyl)amino]benzophenone as a Reusable Chiral Auxiliary.*

7. O. V. Larionov, S. I. Kozhushkov, M. Brandl, A. de Meijere, *Mendeleev Commun.* **2003**, *5*, 199–200. *Rational Synthesis of All Four Stereoisomers of 3-(trans-2-Aminocyclopropyl)alanine.*

8. B. D. Zlatopolsky, K. Loscha, P. Alvermann, S. I. Kozhushkov, S. V. Nikolaev, A. Zeeck, A. de Meijere, *Chem. Eur. J.* **2004**, *10*, 4708–4717. *Final Elucidation of the Absolute Configuration of the Signal Metabolite Hormaomycin.*

9. B. D. Zlatopolsky, A. de Meijere, *Chem. Eur. J.* **2004**, *10*, 4718–4727. *First Total Synthesis of Hormaomycin, a Naturally Occurring Depsipeptide with Interesting Biological Activities.*

10. U. M. Reinscheid, B. D. Zlatopolski, A. Zeeck, C. Griesinger, A. de Meijere, *Chem. Eur. J.* **2005**, *11*, 2929–2945. *The Structure of Hormaomycin and an Aza-Analog in Solution – Synthesis and Biological Activities of New Hormaomycin Analogs.*

11. O. V. Larionov, A. de Meijere, *Org. Lett.* **2004**, *6*, 2153–2156. *Total Syntheses of Belactosin A, Belactosin C and its Homo-Analogue.*

Copper-Bis(oxazoline) Catalyzed Synthesis of β-Lactams – Enantioselective Reaction of Alkynes with Nitrones

David A. Evans, Harvard University, USA,
Florian Kleinbeck, ETH Zürich, Switzerland,
and Magnus Rüping, Johann Wolfgang Goethe-University Frankfurt, Germany

Background

β-Lactams are frequently found in natural and unnatural products with biological activity. They find application as antibiotics and protease inhibitors and are useful reagents in organic synthesis. Consequently, their stereoselective synthesis has attracted great interest. Most successful approaches in the enantioselective synthesis of β-lactams are dependent on chiral auxiliary based methods [1] whereas only a few catalytic enantioselective reactions have been reported. Among those the copper catalyzed reaction of alkynes with nitrones, the Kinugasa reaction, has provided a particularly attractive route to functionalized β-lactams (Scheme 1) [2]. In this transformation a copper acetylide is formed by the reaction of the Cu catalyst with the terminal alkyne followed by a [3+2] dipolar cycloaddition with the nitrone. The β-lactam is then obtained by ring opening fragmentation to the ketene, recyclization and protonation.

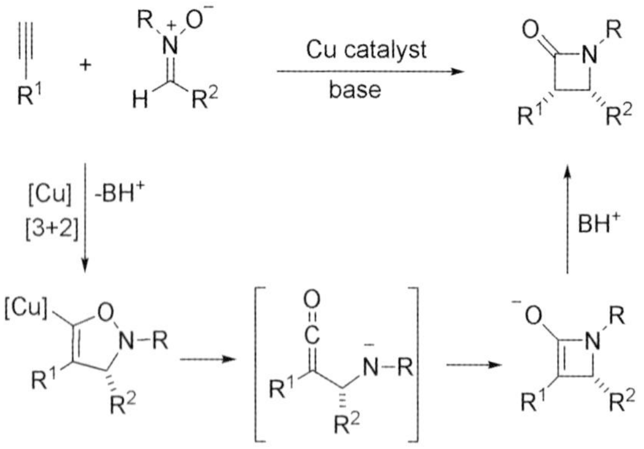

Scheme 1 Mechanism of the copper-catalyzed Kinugasa reaction.

Inspired by Kinugasa, the first asymmetric reaction using catalytic amounts of copper (I) iodide and bisoxazoline **A** (Figure 1) was reported by Miura and coworkers [3]. The β-lactam was obtained in moderate enantio- and diastereoselectivity in the single reaction of phenylacetylene with N-α-diphenylnitrone. Subsequently, Fu and coworkers have reported a catalytic enantioselective Kinugasa reaction using copper(I) chloride and a C_2-symmetric planar chiral bis(azaferrocene) ligand **B** (Figure 1)[4]. The reaction products were obtained in good to excellent enantio- and diastereoselectivities when conducted at low temperatures in the presence of Cy_2NMe. Most recently, chiral Cu(II)-tris(oxazoline) complexes **C** (Figure 1) were examined by Tang and coworkers in the enantioselective coupling of alkynes with nitrones, giving β-lactams in good enantio- and diastereoselectivities, but lower yields [5].

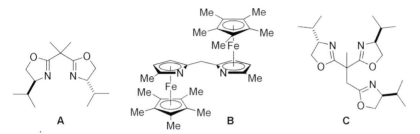

Figure 1 Chiral ligands used in the Kinugasa reaction.

Results

Recent reports from this laboratory and others have focused on the development of bis(oxazoline)-Cu(II) complexes as chiral Lewis acids for a variety of highly enantioselective transformations [6], such as Diels-Alder and hetero Diels-Alder reactions [7-8], cycloadditions [9], ene reactions [10], aldol reactions [11], Michael reactions [12], and amination reactions [13].

Intrigued by the Kinugasa reaction, this mild and practicable approach to the β-lactams, we speculated that sterically more demanding Cu-bisoxazoline catalysts could lead to better diastereo- and enantioselection. Therefore, we decided to test various box ligands in combination with copper salts in the Kinugasa reaction.

1a R^1 = Me   **1f** R^1 = -$CH_2C_6F_5$
1b R^1 = Et   **1g** R^1 = -$CH_2(2,6\text{-}Cl_2C_6H_3)$
1c R^1 = H   **1h** R^1 = -CH_2(pyridyl)
1d R^1 = -$(CH_2)_2$-   **1i** R^1 = -$CH_2(4\text{-}iPr\text{-}C_6H_4)$
1e R^1 = Bn   **1j** R^1 = -$CH_2(4\text{-}tBu\text{-}C_6H_4)$

From a survey of different ligands, as well as copper salts, solvents, and bases we found that the residues R^1 of the bis(oxazoline) skeleton **1** had the greatest impact on the reaction selectivities. While Cu catalysts prepared from the known indabox ligands **1a–1e** yielded only moderate enantioselectivities we decided to prepare the new ligands **1f–1j** and applied those in the Kinugasa reaction. From this comparison the [(**1j**)Cu]Br$_2$ complex proved to be the catalyst of choice providing the β-lactam **4a** in 96% *ee* with 95:5 *cis*:*trans* diastereoselection (Scheme 2).

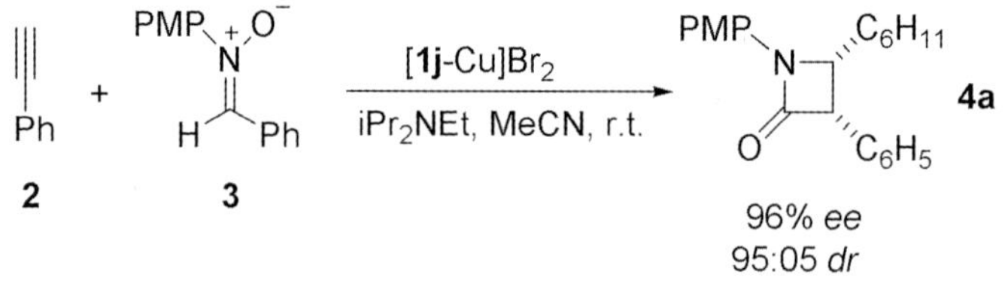

Scheme 2 Copper(II)-indabox catalyzed Kinugasa reaction.

Using the optimized conditions, we explored the scope of the [(1j)Cu]Br$_2$ catalyzed Kinugasa reaction. Table 1 shows examples of the copper bisoxazoline catalyzed reaction. All reactions were carried out using 10 mol % of catalyst in acetonitrile at ambient temperature. In general β-lactams generated from both aryl- and alkyl- substituted nitrones and alkynes exhibit high cis-diastereoselectivities and good to excellent enantiomeric excess.

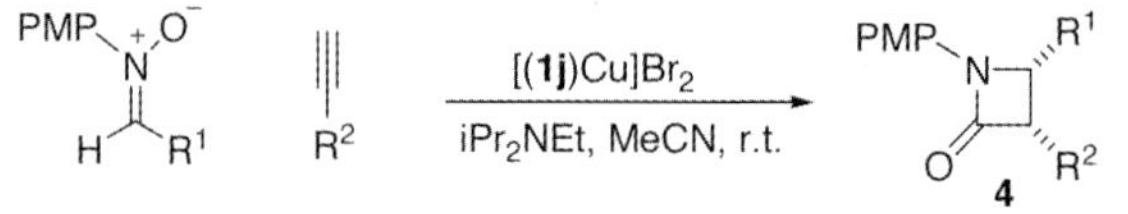

product	R^1	R^2	cis/trans	%yield (cis)	%ee (cis)
4a	C$_6$H$_{11}$	C$_6$H$_5$	95:05	64	96
4b	C$_6$H$_{11}$	4-(CF$_3$)C$_6$H$_4$	95:05	68	97
4c	C$_6$H$_{11}$	4-(OMe)C$_6$H$_4$	97:03	71	98
4d	C$_6$H$_{11}$	4-(CH$_3$)C$_6$H$_4$	91:09	70	98
4e	C$_6$H$_{11}$	1-cyclohexenyl	90:10	53	92
4f	CH(CH$_3$)$_2$	4-(CF$_3$)C$_6$H$_4$	94:06	60	98
4g	CH(CH$_3$)$_2$	4-(OMe)C$_6$H$_4$	90:10	69	98
4h	CH(CH$_3$)$_2$	4-(CH$_3$)C$_6$H$_4$	90:10	72	97
4i	CH(CH$_3$)$_2$	1-cyclohexenyl	90:10	41	91
4j	4-(CO$_2$Me)C$_6$H$_4$	C$_6$H$_5$	97:03	68	90
4k	4-(CO$_2$Me)C$_6$H$_4$	4-(OMe)C$_6$H$_4$	96:04	62	90
4l	4-(CO$_2$Me)C$_6$H$_4$	4-(CH$_3$)C$_6$H$_4$	97:03	65	88
4n	4-(OMe)C$_6$H$_4$	C$_6$H$_5$	98:02	68	86
4o	4-(OMe)C$_6$H$_4$	4-(OMe)C$_6$H$_4$	93:07	43	87
4p	4-(Br)C$_6$H$_4$	C$_6$H$_5$	95:05	72	86
4q	4-(Br)C$_6$H$_4$	4-(OMe)C$_6$H$_4$	94:06	69	89

Table 1 Examples of the copper bisoxazoline-catalyzed reactions performed at room temperature with 10 mol % catalyst loading for two days.

Additionally, one of the obtained β-lactams, cis-(3S,4R)-4b was epimerized to trans-4b by treatment with DBU in CH$_2$Cl$_2$, establishing that the copper bisoxazoline catalyzed process effectively provides stereoselective access to all possible isomers.

The absolute configuration of the obtained β-lactams can be explained by a proposed model based on the X-ray crystal structure of complex [(1j)Cu]Br$_2$ (Figure 2).

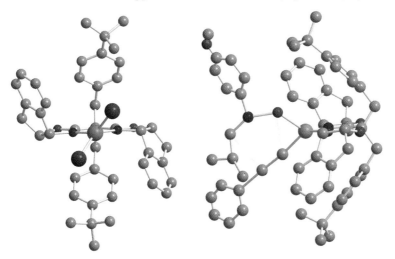

The observed stereochemistry of the β-lactam may be rationalized by assuming that in the transition state the alkyne and nitrone are bound to the copper catalyst and that the nitrone is approached from the *Si* face (Figure 2, right).

Figure 2 X-ray crystal structure of catalyst [(**1j**)Cu]Br$_2$ (left) and proposed model of a transition structure (right).

Summary

In summary, we have developed a highly diastereo- and enantioselective reaction of alkynes with nitrones, catalyzed by chiral copper-bis(oxazoline) complexes. This process provides a general and practical route to versatile β-lactams. The mild reaction conditions, functional group tolerance, and availability of alkynes and nitrones render it an attractive approach to various β-lactams. In addition to the reports of numerous other transformations, this study demonstrates once again that the structural flexibility of the readily accessible bisoxazoline ligands provides highly efficient Cu-catalysts that in many reactions are superior in performance to other catalyst systems.

CV of David A. Evans

David A. Evans was born in Washington D.C. in 1941. He received his A.B. degree from Oberlin College in 1963. He obtained his Ph.D. at the California Institute of Technology in 1967, where he worked under the direction of Professor Robert E. Ireland. In that year he joined the faculty at the University of California, Los Angeles. In 1973 he was promoted to the rank of Full Professor and shortly thereafter returned to Caltech where he remained until 1983. He then joined the Faculty at Harvard University and in 1990 he was appointed as the Abbott and James Lawrence Professor of Chemistry.

CV of Magnus Rüping

Magnus Rüping was born in Telgte, Germany, in 1972. He studied at the Technical University of Berlin, Trinity College Dublin and ETH Zürich. He obtained his Ph.D. in 2002 from ETH under the guidance of Professor Dieter Seebach. After carrying out postdoctoral work with Professor David A. Evans at Harvard University, he joined the Johann Wolfgang Goethe-University Frankfurt as Degussa Endowed Professor of Chemistry in fall 2004.

CV of Florian Kleinbeck

Florian Kleinbeck was born in Stuttgart, Germany, in 1978. He studied at the Technical University of München and ETH Zürich and completed his diploma thesis with Professor David A. Evans. In 2004 he joined Professor Erick M. Carreiras group for his Ph. D. studies.

Selected Publications

1. (a) D. A. Evans, E. B. Sjogren, *Tetrahedron Lett.* **1985**, *26*, 3783–3786. The Asymmetric Synthesis of β-Lactam Antibiotics – I. Application of Chiral Oxazolidones in the Staudinger Reaction. (b) D. A. Evans, E. B. Sjogren, *Tetrahedron Lett.* **1985**, *26*, 3787–3790. The Asymmetric Synthesis of β-Lactam Antibiotics -II. The First Enantioselective Synthesis of the Carbacephalosporin Nucleus. Recent reviews: (c) M. Liu, M. P. Sibi, *Tetrahedron* **2002**, *58*, 7991–8035. Recent Advances in the Stereoselective Synthesis of β-Amino Acids. (d) G. S. Singh, *Tetrahedron* **2003**, *59*, 7631–7649. Recent Progress in the Synthesis and Chemistry of Azetidinones.
2. M. Kinugasa, S. Hashimoto, *Chem. Commun.* **1972**, 466–467. The Reactions of Copper(I) Phenylacetylide with Nitrones.
3. M. Miura, M. Enna, K. Okuro, M. Nomura, *J. Org. Chem.* **1995**, *60*, 4999–5004. Copper Catalyzed Reaction of Terminal Alkynes with Nitrones. Selective Synthesis of 1-Aza-1-buten-3-yne and 2-Azetidinone Derivatives.
4. (a) M. M. C. Lo, G. C. Fu *J. Am. Chem. Soc.* **2002**, *124*, 4572–4573. Cu(I)/Bis(azaferrocene)-Catalyzed Enantioselective Synthesis of β-Lactams via Couplings of Alkynes with Nitrones. (b) R. Shintani, G. C. Fu, *Angew. Chem. Int. Ed. Engl.* **2003**, *42*, 4082–4085. Catalytic Enantioselective Synthesis of β-Lactams: Intramolecular Kinugasa Reactions and Interception of an Intermediate in the Reaction Cascade.
5. M. C. Ye, J. Zhou, Z. Z. Huang, Y. Tang, *Chem. Commun.* **2003**, 2554–2555. Chiral Tris(oxazoline)/Cu(II) Catalyzed Coupling of Terminal Alkynes and Nitrones.
6. (a) D. A. Evans, T. Rovis, J. S. Johnson, *Pure Appl. Chem.* **1999**, *71*, 1407–1415. Chiral Copper(II) Complexes as Lewis Acids for Catalyzed Cycloaddition, Carbonyl Addition, and Conjugate Addition Reactions. (b) J. S. Johnson, D. A. Evans, *Acc. Chem. Res.* **2000**, *33*, 325–335. Chiral Bi-

s(oxazoline) Copper(II) Complexes: Versatile Catalysts for Enantioselective Cycloaddition, Aldol, Michael, and Carbonyl Ene Reactions.

7. D. A. Evans, S. J. Miller, T. Lectka, *J. Am. Chem. Soc.* **1993**, *115*, 6460–6461. Bis(oxazoline) copper(II) complexes as chiral catalysts for the enantioselective Diels Alder reaction.

8. D. A. Evans, J. S. Johnson, *J. Am. Chem. Soc.* **1998**, *120*, 4895–4896. Catalytic Enantio-selective Hetero Diels-Alder Reactions of α,β-Unsaturated Acyl Phosphonates with Enol Ethers.

9. D. A. Evans, J. M. Janey, *Org. Lett.* **2001**, *3*, 2125–2128. C_2-Symmetric Cu(II) Complexes as Chiral Lewis Acids. Catalytic, Enantioselective Cycloadditions of Silyl Ketenes.

10. (a) D. A. Evans, S. W. Tregay, C. S. Burgey, N. A. Paras, T. Vojkovsky, *J. Am. Chem. Soc.* **2000**, *122*, 7936–7943. C_2-Symmetric Copper(II) Complexes as Chiral Lewis Acids. Catalytic Enantioselective Carbonyl-Ene Reactions with Glyoxylate and Pyruvate Esters. (b) D. A. Evans, C. S. Burgey, N. A. Paras, T. Vojkovsky, S. W. Tregay, *J. Am. Chem. Soc.* **1998**, *120*, 5824–5825. C_2- Symmetric Copper(II) Complexes as Chiral Lewis Acids. Enantioselective Catalysis of the Glyoxylate-Ene Reaction.

11. (a) D. A. Evans, D. Seidel, M. Rueping, H. W. Lam, J. T. Shaw, C. W. Downey, *J. Am. Chem. Soc.* **2003**, *125*, 12692–12693. A New Copper Acetate-Bis(oxazoline)-Catalyzed, Enantio-selective Henry Reaction. (b) D. A. Evans, M. C. Kozlowski, C. S. Burgey, D. W. C. MacMillan, *J. Am. Chem. Soc.* **1997**, *119*, 7893–7894. C_2-Symmetric Copper(II) Complexes as Chiral Lewis Acids. Catalytic Enantioselective Aldol Additions of Enolsilanes to Pyruvate Esters. (c) D. A. Evans, J. A. Murry, M. C. Kozlowski, *J. Am. Chem. Soc.* **1996**, *118*, 5814–5815. C_2-Symmetric Copper(II) Complexes as Chiral Lewis Acids. Catalytic Enantioselective Aldol Additions of Silylketene Acetals to (Benzyloxy)acetaldehyde.

12. D. A. Evans, K. A. Scheidt, J. N. Johnston, M. C. Willis, *J. Am. Chem. Soc.* **2001**, *123*, 4480–4491. Enantioselective and Diastereoselective Mukaiyama-Michael Reactions Catalyzed by Bis(oxazoline) Copper(II) Complexes. D. A. Evans, M. C. Willis, J. N. Johnston, *Org. Lett.* **1999**, *1*, 865–868. Catalytic Enantioselective Michael Additions to Unsaturated Ester Derivatives Using Chiral Copper(II) Lewis Acid Complexes.

13. D. A. Evans, D. S. Johnson, *Org. Lett.* **1999**, *1*, 595–598. Catalytic Enantioselective Amination of Enolsilanes Using C_2-Symmetric Copper(II) Complexes as Chiral Lewis Acids.

Catalytic Asymmetric Conjugate Addition Reactions of Organometallic Reagents

Fernando López and Ben L. Feringa, University of Groningen (RUG), The Netherlands

Background

The conjugate addition (CA) of organometallic reagents to α,β-unsaturated compounds is one of the most widely used methods for C–C bond formation. Because of the versatility of the products, an enormous effort has been devoted over the last three decades to the development of asymmetric versions of this key synthetic transformation. The first successful approaches were based on the copper mediated CA of organolithium and Grignard reagents to α,β-unsaturated systems covalently modified with chiral auxiliaries [1R]. Other strategies made use of organocopper compounds with chiral nontransferable groups, such as chiral alkoxycuprates and amidocuprates, leading in some cases to excellent levels of asymmetric induction (up to 97% *ee*) [2R]. The use of organolithium reagents in the presence of stoichiometric amounts of a chiral ligand such as (–)-sparteine was also explored, providing high enantioselectivities in the CAs to α,β-unsaturated *N*-cyclohexylimines and hindered esters (up to 99% *ee*) [2R]. Despite these important achievements, catalytic rather than stoichiometric processes are required in order to provide truly efficient synthetic methods.

The First Catalytic Enantioselective Approaches

It was not until the late 80s that the feasibility of a catalytic ($\leq$ 10 mol% chiral catalyst) and enantioselective CA was demonstrated. In 1988, Lippard reported the first enantioselective CA of a Grignard reagent to an enone, using catalytic amounts of a copper amide complex **1** [3]. Following this seminal work, a variety of catalytic systems, based on *e.g.* copper thiolates **2–5** [4], and monophosphine ligands **6**, were introduced for the CA of Grignard reagents [5]. However, the scope remained limited and *ee*s rarely reached the 90% level (Figure 1).

The difficulties encountered at this stage in the development of an effective enantioselective method (i.e. the fast uncatalyzed reaction and the high sensitivity towards various reaction parameters) encouraged the synthetic community to explore less reactive organometallics, such as organozinc, aluminum, or boron reagents, in combination with different metal sources (Cu, Rh, Ni, Co, etc).

Asymmetric Synthesis – The Essentials.
Edited by Mathias Christmann and Stefan Bräse
Copyright © 2007 WILEY-VCH Verlag GmbH & Co. KGaA, Weinheim
ISBN: 978-3-527-31399-0

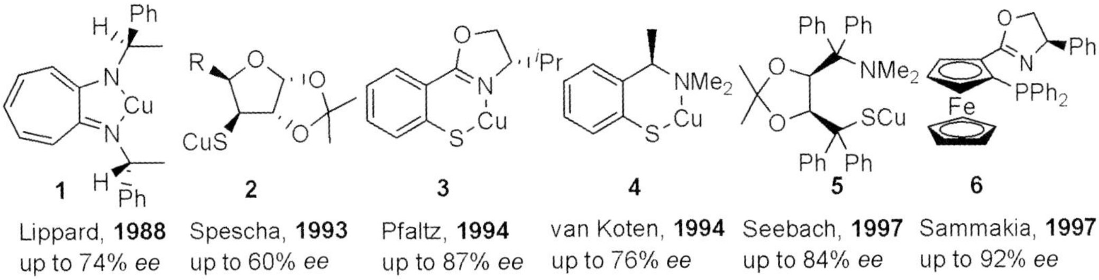
1 Lippard, 1988 up to 74% ee
2 Spescha, 1993 up to 60% ee
3 Pfaltz, 1994 up to 87% ee
4 van Koten, 1994 up to 76% ee
5 Seebach, 1997 up to 84% ee
6 Sammakia, 1997 up to 92% ee

Figure 1 Selected catalytic systems developed for the CA of Grignard reagents to enones.

Results

Cu-catalyzed Enantioselective CA of Organozinc Reagents

In 1997, our group reported the first highly enantioselective Cu-catalyzed CA of dialkylzinc reagents to enones using a novel class of monodentate phosphoramidite ligands [6]. Thus, binaphthol-based phosphoramidite **8** and $Cu(OTf)_2$, provided enantioselectivities up to 98% in the CA to cycloenones **7** (Scheme 1, (eq. 1)). Notably, a tandem catalytic and enantioselective 1,4-addition–aldol reaction of dialkylzinc reagents to cyclopentenones, in the presence of aldehydes, was also developed. The potential of this three-component protocol was demonstrated in the short total synthesis of prostaglandin E_1 methyl ester (Scheme 1, (eq. 2)) [7].

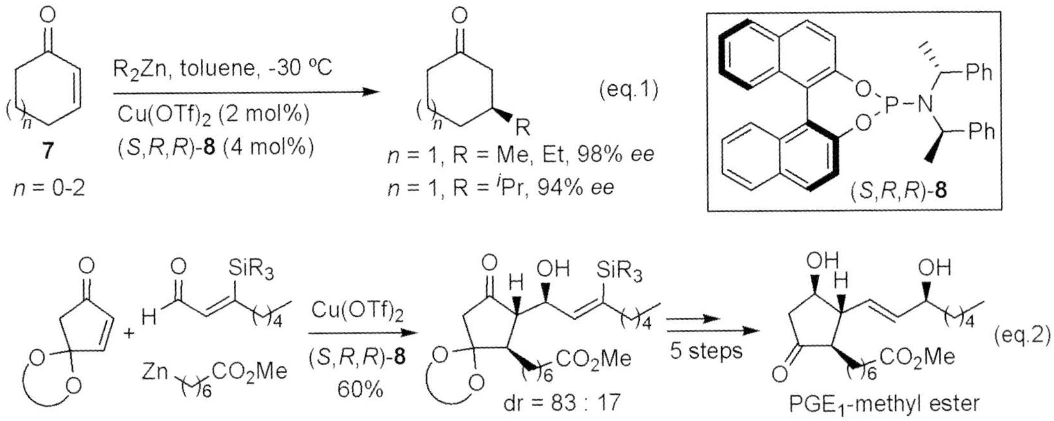
R2Zn, toluene, -30 °C
Cu(OTf)2 (2 mol%)
(S,R,R)-8 (4 mol%)
7
n = 0-2
(eq.1)
n = 1, R = Me, Et, 98% ee
n = 1, R = iPr, 94% ee
(S,R,R)-8
Cu(OTf)2
(S,R,R)-8
60%
dr = 83 : 17
5 steps
(eq.2)
PGE1-methyl ester

Scheme 1 Cu-catalyzed enantioselective CA of dialkylzinc reagents to enones. Application to the synthesis of PGE_1.

This discovery has stimulated numerous modifications to our original catalytic system as well as the introduction of a range of new phosphorus-based catalysts. For instance, it is remarkable that a dynamic atropisomeric biphenyl unit (in **9**) can play the role of the chiral binaphthol, providing also high *ee*s for some CAs. Hoveyda's peptidic phosphine ligands **10**, are among the most successful

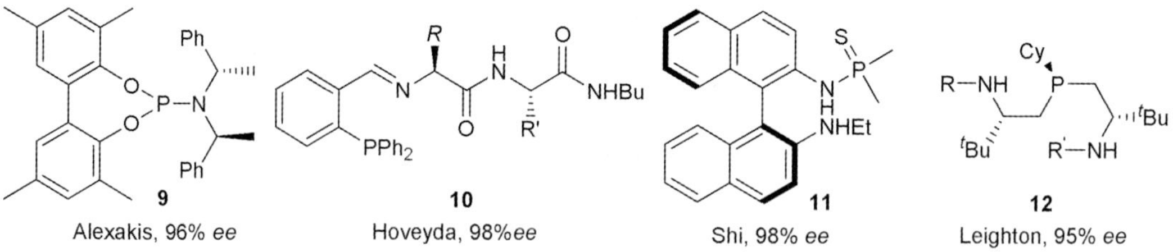

Figure 2 Some recent successful ligands for the Cu-catalyzed CA of dialkylzinc reagents to unsaturated systems.

systems, providing high *ee*s for various types of unsaturated substrates (Figure 2) [8]. Several reviews summarize the impressive progress in this field [9R, 10R, 11R].

Phosphoramidites **8** and **9**, as well as phosphines such as **10**, are also excellent ligands for the Cu-catalyzed CA of dialkylzinc reagents to nitroalkenes and *ee*s over 90% for a variety of substrates have been reported [11R]. On the other hand, for the introduction of aryl and vinyl groups, the Rh-catalyzed CA of boron reagents developed by Miyaura, Hayashi, and our group, among others, is still the method of choice [12R].

Highly Enantioselective Cu-catalyzed CA of Grignard Reagents

Notwithstanding the major success achieved in the enantioselective CA of dialkylzinc reagents, we anticipate that there are significant advantages in the use of common Grignard reagents, including their ready availability, the transfer of all the alkyl groups of the organometallic compound and the higher reactivity of the magnesium enolates formed. These features, and the lack of synthetically useful catalytic CAs of RMgX reagents to enones and enoates, encouraged us to realize this asymmetric transformation. Such a catalyst should also be able to avoid the competing fast noncatalyzed 1,2-addition of the organomagnesium reagent to the carbonyl group.

We reported in 2004, the first Cu-catalyzed asymmetric CA of Grignard reagents to cyclic enones with *ee*s up to 96% (Scheme 2). The unprecedented level of stereocontrol was achieved with CuCl or CuBr·SMe$_2$, simple alkylmagnesium bromides and commercially available ferrocenyl diphosphines as chiral ligands (i.e. Taniaphos and Josiphos) [13].

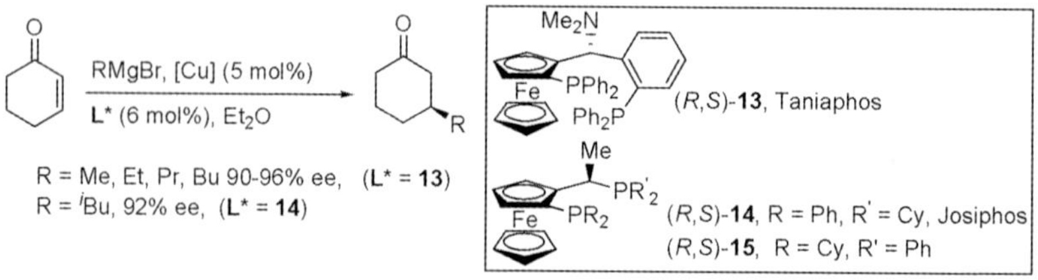

Scheme 2 Cu-catalyzed enantioselective CA of Grignard reagents to cyclic enones ([Cu] = CuCl or CuBr·SMe$_2$).

The Josiphos ligands **14** and **15**, in combination with CuBr·SMe$_2$, were the diphosphines of choice for the enantioselective CA of Grignard reagents to acyclic aliphatic enones (*ees* up to 97% and regioselectivities up to 98%) [14], and particularly significant, to α,β-unsaturated esters (*ees* and regioselectivities up to 99%) [15], for which no other previously reported combinations of catalysts and alkylorganometallic reagents had been successful (Scheme 3).

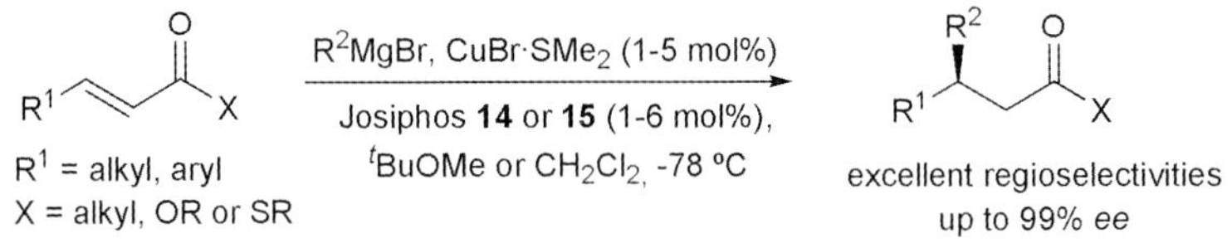

Scheme 3 Cu-catalyzed enantioselective CA of Grignard reagents to acyclic enones and unsaturated acid derivatives.

Interestingly, the introduction of a methyl group to an α,β-unsaturated ester (R^2=Me, X=OR), providing an essential propionate building block for natural product synthesis, proceeded with low conversions, reflecting the lower reactivity of MeMgBr. It was better accomplished by using, instead of an ester, the more reactive but equally readily accessible α,β-unsaturated thioester. The application of this methodology to a diastereo- and enantioselective iterative route to both *syn*- and *anti*-1,3-dimethyl arrays and deoxypropionate chains was recently reported by our group, and the versatility of the method was further illustrated in the synthesis of (–)-Lardolure, a multi-methyl-branched insect natural pheromone (Scheme 4) [16].

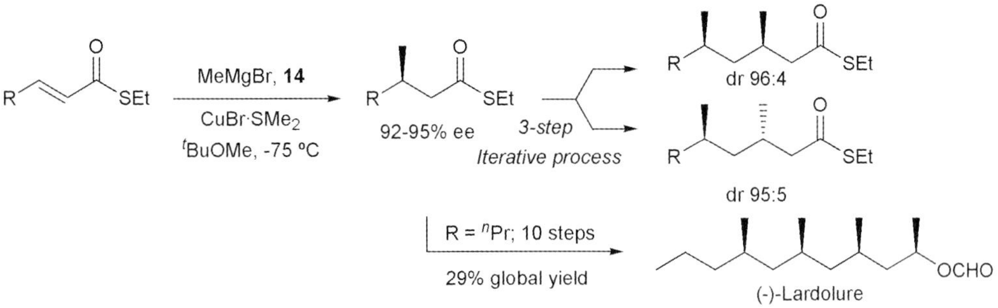

Scheme 4 An iterative catalytic route to enantiopure deoxypropionate subunits and (–)-lardolure.

Conclusion and Future Perspectives

In 1997 we developed the first highly enantioselective Cu-catalyzed CA of organozinc reagents to α,β-unsaturated systems by using binaphthol-based phosphoramidites. Many applications, modifications as well as new efficient catalysts for these processes, have been reported since then by our group and many others. A highly enantioselective CA of Grignard reagents, however, remained unrealized until 2004, when we introduced commercially available Josiphos and Taniaphos ligands for this transformation. In view of the success so far obtained, there is no doubt that in the coming years new catalysts will be reported for these accessible organometallic reagents, keeping them in a predominant position in this field. On the other hand, the challenge to develop catalysts with similar activities in related C–C bond formation using organolithium reagents remains as strong as ever.

CV of Fernando López

Fernando López was born in 1975, he studied chemistry at the University of Santiago de Compostela (Spain) where he obtained his degree in 1998. During his PhD, under the supervision of Professor José L. Mascareñas, he carried out two predoctoral stays at the ETH-Zentrum in Zürich, (with Prof. E. M. Carreira), and at the University of Yale (with Prof. John F. Hartwig). After receiving his PhD in 2003, he was awarded a Marie Curie Intraeuropean Postdoctoral fellowship, and joined the group of Professor Ben L. Feringa at the University of Groningen, where his current work is centered on enantioselective catalysis. Recently he was awarded a Ramon y Cajal contract from the Spanish Ministry of Education and Science.

CV of Ben L. Feringa

Ben L. Feringa received his PhD degree from the University of Groningen in 1978 with Professor Hans Wynberg. He was a research scientist with Royal Dutch Shell, both at the Shell Research Center in Amsterdam and at the Shell Biosciences Laboratories in Sittingbourne, UK, from 1978 to 1984. He joined the University of Groningen in 1984 as a lecturer and was appointed Professor at the same University in 1988. He was the recipient of a JSPS fellowship, the 1997 Pino gold medal of the Italian Chemical Society, the joint 2003 Koerber European Science Award, the 2003 Guthikonda Award (Columbia University), and the 2004 Dauben Lectureship (Berkeley). In 2004 he was elected foreign honorary member of the American Academy of Arts and Science. He has recently received the Spinoza award (the highest scientific distinction in the Netherlands) from the Netherlands Organization of Scientific Research and the 2005 Prelog gold medal. Professor Feringa is Jacobus van't Hoff Professor of Molecular Sciences and the Scientific Editor of the RSC journal Organic & Biomolecular Chemistry.

Selected Publications

1R. B. E. Rossiter, N. M. Swingle, *Chem. Rev.* **1992**, *92*, 771–806. *Asymmetric Conjugate Addition.*

2R. K. Tomioka, Y. Nagaoka, in *Comprehensive Asymmetric Catalysis*; E. N. Jacobsen, A. Pfaltz, H. Yamamoto (Eds.), Springer-Verlag, New York, 1999, Vol. 3, pp. 1105–1120. *Conjugate Addition of Organometallic Reagents.*

3. G. M. Villacorta, C. P. Rao, S. J. Lippard, *J. Am. Chem. Soc.* **1988**, *110*, 3175–3182. *Synthesis and Reactivity of Binuclear Tropocoronand and Related Organocopper(I) Complexes. Catalytic Enantioselective Conjugate Addition of Grignard Reagents to 2-Cyclohexen-1-one.*

4. Q-L. Zhou, A. Pfaltz, *Tetrahedron* **1994**, *50*, 4467-4478, *Chiral Mercaptoaryl-oxazolines as Ligands in Enantioselective*

Copper-catalyzed 1,4-additions of Grignard Reagents to Enones, and references therein.
5. E. L. Stangeland, T. Sammakia, *Tetrahedron* **1997**, *53*, 16503–16510. *New Chiral Ligands for the Asymmetric Copper Catalyzed Conjugate Addition of Grignard Reagents to Enones.*
6. B. L. Feringa, M. Pineschi, L. A. Arnold, R. Imbos, A. H. M. de Vries, *Angew. Chem. Int. Ed. Engl.* **1997**, *36*, 2620–2623. *Highly Enantioselective Catalytic Conjugate Addition and Tandem Conjugate Addition-Aldol Reactions of Organozinc Reagents.*
7. L. A. Arnold, R. Naasz, A. J. Minnaard, B. L. Feringa, *J. Am. Chem. Soc.* **2001**, *123*, 5841–5842. *Catalytic Enantioselective Synthesis of Prostaglandin E_1 Methyl Ester Using a Tandem 1,4-Addition-Aldol Reaction to a Cyclopenten-3,5-dione Monoacetal.*
8. A. H. Hoveyda, A. W. Hird, M. A. Kacprzynski, *Chem. Commun.* **2004**, 1779–1785. *Small Peptides as Ligands for Catalytic Asymmetric Alkylations of Olefins. Rational Design of Catalysts or of Searches that Lead to Them?* and references therein.
9R. B. L. Feringa, R. Naasz, R. Imbos, L. A. Arnold, in *Modern Organocopper Chemistry*, Krause, N. (Ed.), Wiley-VCH, Weinheim, 2002, pp. 224–258. *Copper-catalyzed Enantioselective Conjugate Addition Reactions of Organozinc Reagents.*
10R. N. Krause, A. Hoffmann-Röder, *Synthesis*, **2001**, 171–196. *Recent Advances in Catalytic Enantioselective Michael Additions.*
11R. K. Tomioka, in *Comprehensive Asymmetric Catalysis. Supplement 2*; E. N. Jacobsen, A. Pfaltz, H. Yamamoto (Eds.) Springer-Verlag, New York, 2004; Vol. 31.1, pp. 109–124. *Conjugate Addition of Organometallic Reagents to Activated Olefins.*
12R. T. Hayashi, K. Yamasaki, *Chem. Rev.* **2003**, *103*, 2829–2844. *Rhodium-Catalyzed Asymmetric 1,4-Addition and Its Related Asymmetric Reactions.*
13. B. L. Feringa, R. Badorrey, D. Peña, S. R. Harutyunyan, A. J. Minnaard, *Proc. Natl. Acad. Sci. U.S.A.* **2004**, *101*, 5834–5838. *Copper-catalyzed Asymmetric Conjugate Addition of Grignard Reagents to Cyclic Enones.*
14. F. López, S. R. Harutyunyan, A. J. Minnaard, B. L. Feringa, *J. Am. Chem. Soc.* **2004**, *126*, 12784–12785. *Copper-catalyzed Enantioselective Conjugate Addition of Grignard Reagents to Acyclic Enones.*
15. F. López, S. R. Harutyunyan, A. Meetsma, A. J. Minnaard, B. L. Feringa, *Angew. Chem., Int. Ed. Engl.* **2005**, *44*, 2752–2756. *Copper-catalyzed Enantioselective Conjugate Addition of Grignard Reagents to α,β-Unsaturated Esters.*
16. R. Des Mazery, M. Pullez, F. López, S. R. Harutyunyan, A. J. Minnaard, B. L. Feringa, *J. Am. Chem. Soc.* **2005**, *127*, 9966–9967. *An Iterative Catalytic Route to Enantiopure Deoxypropionate Subunits: Asymmetric Conjugate Addition of Grignard Reagents to α,β-Unsaturated Thioesters.*

Catalytic Asymmetric Synthesis of Allylic Alcohols via Dynamic Kinetic Resolution

Hans-Joachim Gais, RWTH Aachen, Germany

Introduction

Catalytic dynamic kinetic resolution, that is the complete conversion of a racemic substrate to one of the enantiomers of a product without intermediate separation of materials, is an interesting alternative to catalytic asymmetric synthesis starting from a prochiral substrate [1, 2]. Chiral allylic alcohols are important intermediates in organic synthesis. Several catalytic methods for their asymmetric synthesis from prochiral compounds are available, including enantioselective reduction (Scheme 1) [3]. A catalytic dynamic kinetic resolution of a number of allylic acetates has been accomplished by the combination of a lipase catalyzed kinetic resolution with a Pd(0)/L-catalyzed racemization of the remaining enantiomer [4]. This method, however, requires the application of two different catalysts. An interesting alternative is the Pd(0)/L*-catalyzed reaction of racemic allylic carbonates with carboxylates yielding allylic esters [5]. However, this method has been applied only to symmetrical cyclic substrates and requires monitoring of the reaction in order to avoid racemization of the product which is also a substrate for the Pd(0)/L*-catalyst [6].

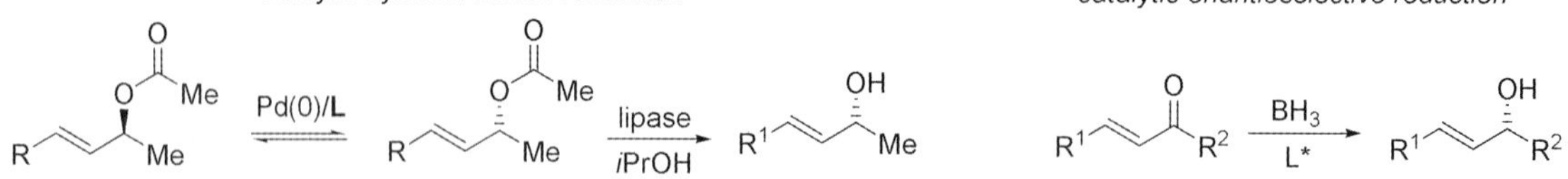

Scheme 1 Catalytic asymmetric synthesis of allylic alcohols from racemic and prochiral substrates.

Strategy

During a study of the Pd(0)-catalyzed allylic alkylation of thioacetate with the racemic allylic carbonate *rac*-**1** and Pd(0)/**L1*** [7] in CH_2Cl_2/water [8] we observed the formation of a mixture of the allylic alcohol **2** and the allylic thioester **3** (Scheme 2). Surprisingly, alcohol **2** was enantiomerically enriched and had the same absolute configuration as thioester **3**. Obviously, a competing Pd(0)/**L1***-

Asymmetric Synthesis – The Essentials.
Edited by Mathias Christmann and Stefan Bräse
Copyright © 2007 WILEY-VCH Verlag GmbH & Co. KGaA, Weinheim
ISBN: 978-3-527-31399-0

catalyzed allylic substitution with an O-nucleophile had taken place finally leading to the formation of alcohol **2**. This observation suggested the possibility of a Pd(0)/L*-catalyzed asymmetric synthesis of allylic alcohols by a simple treatment of the corresponding racemic carbonates with a chiral Pd(0)/L* catalyst and water.

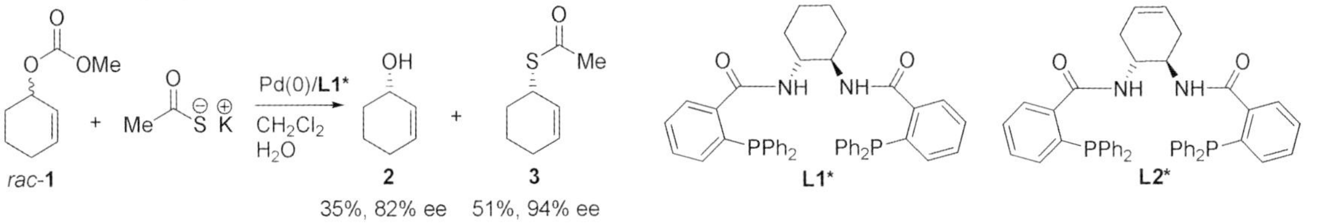

Scheme 2 Pd(0)/L*-catalyzed allylic alkylation of thioacetate in the presence of water.

Results

Reaction of the cyclic carbonate *rac*-**1** with 2–6 mol% of Pd(0)/L* and water in CH_2Cl_2 gave the allylic alcohol **2** in 71–94% yield and 97–99% ee (Scheme 3) [9]. A similar treatment of the bicyclic carbonate *rac*-**4** in presence of $KHCO_3$ afforded indenol **5** in 88% yield with 97% ee [10]. Indenol **5** and cyclopentenol **2** ($n = 0$) are building blocks for the synthesis of a HIV-protease inhibitor and a chiral catalyst for asymmetric Diels-Alder reactions, respectively. An application of this protocol to the acyclic carbonates *rac*-**6** furnished the acyclic allylic alcohols **7** in high yields with high *ee*-values. Similar results were obtained in the reaction of carbonates *rac*-**1** and *rac*-**6** by using the Pd(0)/L2*-catalyst containing the unsaturated bisphosphane **L2*** as ligand [11]. Interestingly, the use of $KHCO_3$ as additive in the above transformations resulted in higher reaction rates. Acetates can also serve as substrates in combination with external hydrogen carbonate. This has been applied to the synthesis of cyclopentenol **2** ($n = 0$) of 99% ee (71% yield) from racemic cyclopentenol acetate. Pd(0)/L1* and $KHCO_3$ in CH_2Cl_2/water.

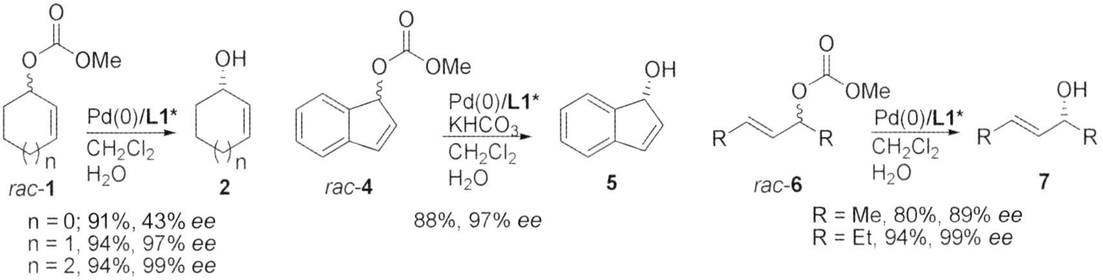

Scheme 3 Asymmetric synthesis of symmetrical cyclic and acyclic allylic alcohols.

A mechanistic scheme for the Pd(0)/L1*-catalyzed dynamic kinetic resolution of allylic carbonates leading to the corresponding allylic alcohols is proposed in Figure 2. Both enantiomers **1** and *ent*-**1** react with the chiral Pd(0) catalyst with formation of the π-allyl-Pd(II) complex **8** containing the methyl carbonate ion. Because of the symmetrical carbon skeleton of the substrates and the C_2-symmetry of ligand **L1***, only one π-allyl-Pd(II) complex is formed. Then water causes an irreversible hydrolysis of the methyl carbonate ion to the hydrogen carbonate ion with formation of complex **9**. The hydrogen carbonate ion in turn reacts with the π-allyl-Pd(II) cation as the O-nucleophile to give the hydrogen carbonate **10** which irreversibly decomposes with formation of alcohol **2** and CO_2. This renders the allylic carbonate to allylic alcohol transformation practically irreversible. Control experiments showed that neither the hydroxy ion nor water act as an O-nucleophile under the conditions used [9].

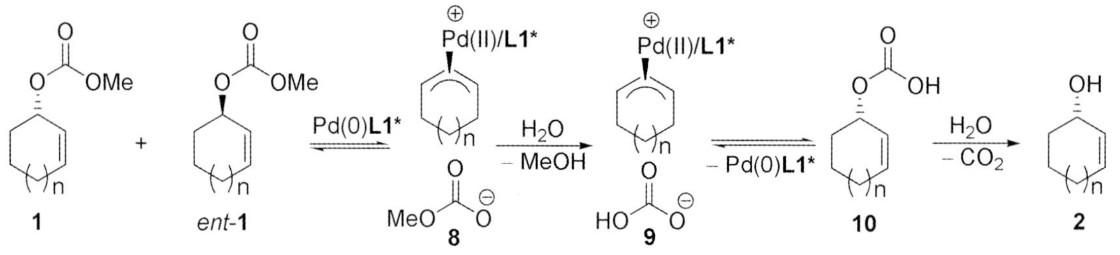

Figure 4 Mechanistic scheme for the Pd(0)/L1*-catalyzed reaction of allylic carbonates with water.

Having obtained these results, it was of interest to see whether racemic unsymmetrical allylic carbonates are also capable of undergoing an efficient catalytic transformation to the corresponding allylic alcohols with Pd(0)/**L1*** and water. Here the situation is more complex. Both enantiomers of the substrate react with the catalyst with formation of two diastereomeric π-allyl-Pd(II) complexes (Scheme 5). Only if the following prerequisites are met can the racemic substrate be completely converted to the chiral alcohol with high efficiency. First, the reactivity of the complexes must be different. Second, a fast interconversion of both complexes or a racemization of the substrate has to occur. Third, the chiral ligand must induce a high stereoselectivity. Fourth, the substituents have to provide for a high regioselectivity.

Gratifyingly, treatment of the unsymmetrical racemic carbonate *rac*-**11** (EWG = Ph) with the catalyst, $KHCO_3$ and water in CH_2Cl_2 gave alcohol **12** (EWG = Ph) in high yield and enantioselectivity (Scheme 6). Synthetically more relevant are the results obtained in the case of the functionalized racemic carbonate *rac*-**11** (EWG = CO_2Et, SO_2Ph). A similar treatment of the carbonate afforded alcohol **12** (EWG = CO_2Et, SO_2Ph) in high yield and enantioselectivity [10].

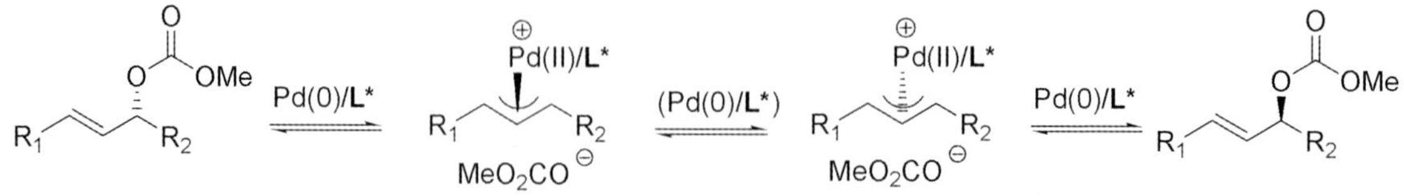

Scheme 5 Reaction of a Pd(0)/L* catalyst with unsymmetrical allylic carbonates.

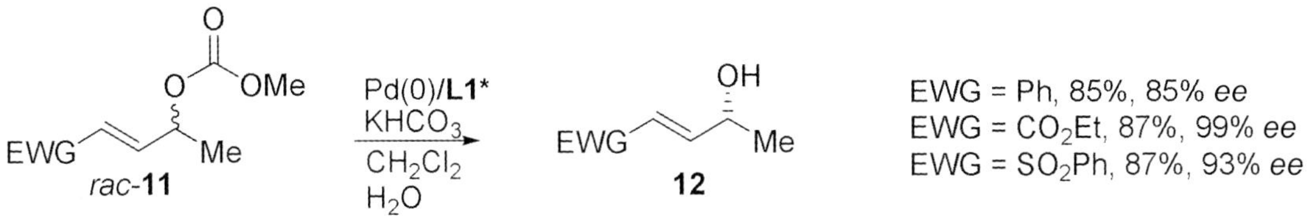

Scheme 6 Asymmetric synthesis of unsymmetrical allylic alcohols.

EWG = Ph, 85%, 85% ee
EWG = CO$_2$Et, 87%, 99% ee
EWG = SO$_2$Ph, 87%, 93% ee

Application to the Synthesis of Natural Products

Application of the Pd-catalyzed dynamic kinetic resolution to the *meso*-configured biscarbonate **13** gave the allylic alcohol **14** in 87% yield with 96% ee (Scheme 7) [10]. Alcohol **14** was converted via the silyl ether **15** and alcohol **16** to ketone **17**, the enantiomer of which is an important building block for the synthesis of prostaglandins [12].

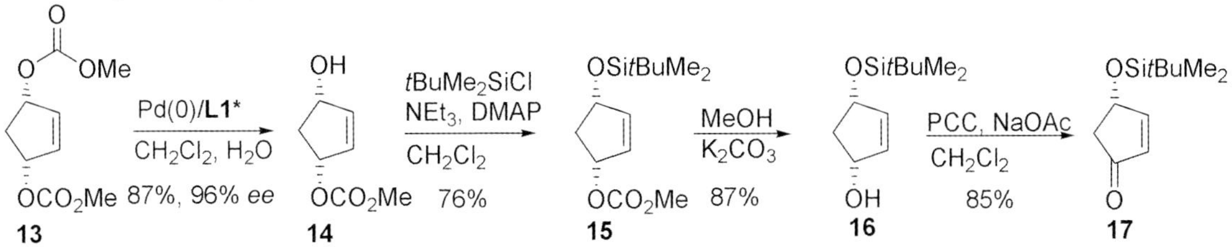

Scheme 7 Asymmetric synthesis of a prostaglandin building block.

Other Asymmetric Syntheses from the Gais Group

Currently other projects are being followed including the synthesis of allylic sulfur compounds [13, 14], α-amino acids [15], β-amino acids [16] and prostacyclin analogs [17], and a study of the structure and dynamics of (allylsulfoximine)titanium complexes [18, 19].

Conclusion and Future Perspectives

The high enantioselectivity of the Pd-catalyzed asymmetric synthesis of unsymmetrical allylic alcohols points to the operation of an efficient dynamic kinetic resolution involving either an interconversion of the diastereomeric π-allyl-Pd(II) complexes by an *anti*-substitution of the complexes by Pd(0)/L1* or a racemization of the carbonate by a Pd-centered *syn*-substitution of the π-allyl-Pd(II) complexes by the anion. Future investigation will be directed towards the synthesis of further synthetically useful allylic alcohols and elucidation of the factors determining the selectivity.

CV of Hans-Joachim Gais

Hans-Joachim Gais was born in 1942 in Darmstadt, Germany. He studied chemistry at the University of Technology in Darmstadt and received his Dipl.-Ing. in 1969 and his Dr. Ing. with Professor K. Hafner in 1973. After postdoctoral studies with Professor R. B. Woodward at Harvard University in 1974 and 1975, he returned to Darmstadt and finished his Habilitation in 1981. In 1986 he moved to the University of Freiburg as Professor of Organic Chemistry. Since 1991 he has been Professor of Organic Chemistry and holds a chair of Organic Chemistry at the Rheinisch-Westfälische Technische Hochschule (RWTH) Aachen. He was the recipient of the Carl-Duisberg-Gedächtnis Preis der GDCh (1986).

Selected Publications

1. E. Vedejs, M. Jure, *Angew. Chem. Int. Ed. Engl.* **2005**, *44*, 3974–4001. Efficiency in nonenzymatic kinetic resolution.
2. H.-J. Gais, F. Theil, in *Enzyme Catalysis in Organic Synthesis*, K. Drauz, H. Waldmann, (Eds.), Wiley-VCH: Weinheim, 2002, Vol. 2, pp. 335–578. Hydrolysis and formation of C–O bonds.
3. E. J. Corey, C. J. Helal, *Angew. Chem., Int. Ed. Engl.* **1998**, *37*, 1986–2012. Reduction of carbonyl compounds with chiral oxazaborolidine catalysts: A new paradigm for enantioselective catalysis and a powerful new synthetic method.
4. M.-J. Kim, Y. Ahn, J. Park, *Curr. Opin. Biotechnol.* **2002**, *13*, 578–587. Dynamic kinetic resolution and asymmetric transformation by enzymes coupled with metal catalysis.
5. B. M. Trost, M. G. Organ, *J. Am. Chem. Soc.* **1994**, *116*, 10320–10321. Deracemization of cyclic allyl esters.
6. For a futher method, see: S. Akai, K. Tanimoto, Y. Kanao, M. Egi, T. Yamamoto, Y. Kita, *Angew. Chem.* **2006**, *118*, 2654–2657. A dynamic kinetic resolution of allyl alcohols by the combination of lipases and $[VO(OSiPh_3)_3]$.
7. B. M. Trost, C. Lee in *Catalytic Asymmetric Synthesis* (Ed.: I. Ojima), Wiley-VCH, New York, 2000, pp. 592–650. Asymmetric allylic alkylation.
8. B. J. Lüssem, H.-J. Gais, *J. Org. Chem.* **2004**, *69*, 4041–4052. Palladium-catalyzed enantioselective alkylation of thiocarboxylate ions: Asymmetric synthesis of allylic thioesters and memory effect/dynamic kinetic resolution of allylic esters.
9. B. J. Lüssem, H.-J. Gais, *J. Am. Chem. Soc.* **2003**, *125*, 6066–6067. Palladium-catalyzed deracemization of allylic carbonates in water with formation of allylic alcohols: Hydrogen carbonate ion as nucleophile in the palladium-catalyzed allylic substitution and kinetic resolution.
10. H.-J. Gais, O. Bondarev, R. Hetzer, *Tetrahedron Lett.* **2005**, *46*, 6279–6283. Palladium-catalyzed asymmetric synthesis of allylic alcohols from unsymmetrical and symmetrical racemic allylic carbonates featuring C–O-bond formation and dynamic kinetic resolution.
11. W. Oppolzer, D. L. Kuo, M. W. Hutzinger, R. Leger, J.-O. Durand, C. Leslie, *Tetrahedron Lett.* **1997**, *38*, 6213–6216. Studies towards asymmetric catalyzed metallo-ene reactions.
12. R. Noyori, M. Suzuki, *Angew. Chem.* **1984**, *96*, 854–882. Prostaglandin syntheses by three-component coupling.
13. T. Jagusch, H.-J. Gais, O. Bondarev, *J. Org. Chem.* **2004**, *69*, 2731–2736. Palladium-catalyzed enantioselective 1,3-rearrangement of racemic allylic sulfinates: Asymmetric synthesis of allylic sulfones and kinetic resolution of an allylic sulfinate.
14. H.-J. Gais, T. Jagusch, N. Spalthoff, F. Gerhards, M. Frank, G. Raabe, *Chem. Eur. J.* **2003**, *9*, 4202–4221. Highly selective palladium catalyzed kinetic resolution and enantioselective substitution of racemic allylic carbonates with sulfur nucleophiles: asymmetric synthesis of allylic sulfide, allylic sulfones and allylic alcohols.
15. S. Koep, H.-J. Gais, G. Raabe, *J. Am. Chem. Soc.* **2003**, *125*, 13243–13251. Asymmetric synthesis of unsaturated, fused bicyclic proline analogues through amino alkylation of cyclic bis(allylsulfoximine)titanium complexes and migratory cyclization of δ-amino alkenyl aminosulfoxonium salts.
16. H.-J. Gais, R. Loo, D. Roder, P. Das, G. Raabe, *Eur. J. Org. Chem.* **2003**, 1500–1526. Asymmetric synthesis of protected β-substituted and β,β-disubstituted β-amino acids bearing branched hydroxyalkyl side

chains and protected 1,3-amino alcohols with three contiguous stereogenic centers from allylic sulfoximines and aldehydes.

17. M. Lerm, H.-J. Gais, K. Cheng, C. Vermeeren, *J. Am. Chem. Soc.* **2003**, *125*, 965–9667. Asymmetric synthesis of the highly potent anti-metastatic prostacyclin analogue cicaprost and its isomer isocicaprost.

18. H.-J. Gais, P. R. Bruns, G. Raabe, R. Hainz, M. Schleusner, J. Runsink, G. S. Babu, *J. Am. Chem. Soc.* **2005**, *127*, 6617–6631. Dynamic behavior of chiral sulfonimidoyl-substituted allyl and alkyl (dimethylamino)titanium(IV) complexes: Metallotropic shift, reversible β-hydride elimination/reinsertion, and ab initio calculations of allyl and alkyl aminosulfoxonium ylides.

19. H.-J. Gais, R. Hainz, H. Müller, P. R. Bruns, N. Giesen, G. Raabe, J. Runsink, S. Nienstadt, J. Decker, M. Schleusner, J. Hachtel, R. Loo, C.-W. Woo, P. Das, *Eur. J. Org. Chem.* **2000**, 3973–4009. N-Methyl-sulfonimidoyl-substituted (2-alkenyl)titanium complexes: Application to the synthesis of β- and δ-sulfonimidoyl-substituted chiral homoallylic alcohols, X-ray crystal structure analysis, and fluxional behavior.

Asymmetric Cross-Coupling Reactions

Tamio Hayashi, Kyoto University, Japan

Background

Nickel and palladium complexes catalyze the reaction of organometallic reagents (R–m) with alkenyl or aryl halides and related compounds (R'–X) to give cross-coupling products (R–R'), which provides one of the most useful synthetic means for making a carbon–carbon bond (Figure 1). Because the new carbon–carbon bond is usually formed on the sp^2 carbon center, the creation of chiral carbon centers or chiral molecules by the catalytic cross-coupling is not always easy. For asymmetric synthesis by this cross-coupling process, special systems have been designed. One is the reaction of secondary alkyl Grignard reagents where a kinetic resolution of the racemic reagents is expected and the other is the asymmetric synthesis of axially chiral molecules such as biaryls [1].

$$\text{R-m} + \text{R'-X} \xrightarrow{\text{[M] (catalyst)}} \text{R-R'} + \text{mX}$$

M = Ni, Pd
m = Mg, Li, Zn, Al, Zr, Sn, B, Si, Cu, In, Bi, Ti, etc.
R' = aryl, alkenyl
X = Cl, Br, I, OSO_2CF_3, $OPO(OR)_2$, etc.

Figure 1 Nickel- or palladium-catalyzed cross-coupling of organometallic reagents with organic halides.

Results

Asymmetric Cross-coupling of Secondary Alkyl Grignard Reagents

Asymmetric synthesis by the catalytic cross-coupling reaction has been most extensively studied with secondary alkyl Grignard reagents. The asymmetric cross-coupling with chiral catalysts allows transformation of a racemic mixture of the secondary alkyl Grignard reagent into an optically active product by a kinetic resolution of the Grignard reagent. Since the secondary alkyl Grignard reagents usually undergo racemization at a rate comparable to the cross-coupling, enantiomerically enriched coupling product is formed, even if the conversion of the Grignard reagent is quanti farin (Figure 2).

Asymmetric Synthesis – The Essentials.
Edited by Mathias Christmann and Stefan Bräse
Copyright © 2007 WILEY-VCH Verlag GmbH & Co. KGaA, Weinheim
ISBN: 978-3-527-31399-0

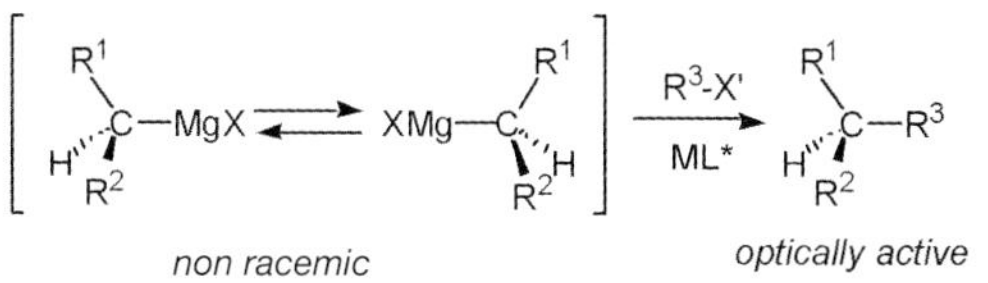

$$\left[\begin{array}{c}R^1\\\text{C—MgX}\\H\quad R^2\end{array}\rightleftarrows\begin{array}{c}R^1\\\text{XMg—C}\\R^2\quad H\end{array}\right]\xrightarrow[ML^*]{R^3\text{-X'}}\begin{array}{c}R^1\\\text{C—}R^3\\H\quad R^2\end{array}$$

non racemic   optically active

Figure 2 Asymmetric cross-coupling of secondary alkyl Grignard reagents.

Initial studies on the asymmetric cross-coupling of secondary alkyl Grignard reagents were reported in 1973 by Consiglio and in 1974 by Kumada by use of a nickel catalyst coordinated with (−)-diop ligand, but the enantiomeric purities of the coupling products were rather low (13–17% ee). After these reports, asymmetric cross-coupling of the secondary alkyl Grignard reagents was attempted using various kinds of optically active phosphine ligands. The reaction most extensively studied so far is that of 1-phenylethylmagnesium chloride (**1**) with vinyl bromide (**2**) forming 3-phenyl-1-butene (**3**) (Figure 3). We found that the ferrocenylphosphines containing (dialkylamino)alkyl group on the side chain are effective for the cross-coupling of **1** catalyzed by nickel or palladium complexes [2, 3]. Ferrocenylmonophosphine, (*S*)-(*R*)-PPFA and -bisphosphine, (*S*)-(*R*)-BPPFA gave the coupling product **3** with 68% ee and 65% ee, respectively. The presence of the (dialkylamino)alkyl side chain is of primary importance for high selectivity and the enantioselectivity is strongly affected by the structure of the dialkylamino group. Use of a C_2-symmetric ferrocenylbisphosphine ligand and an organozinc reagent in place of the Grignard reagent **1** increased the enantioselectivity up to 93% ee [4].

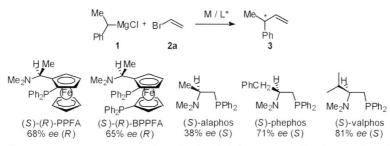

Figure 3 Asymmetric cross-coupling of 1-phenylethylmagnesium chloride (**1**) with vinyl bromide (**2a**).

Based on the high efficiency of the (dialkylamino)alkyl side chain on the ferrocenylphosphines, a series of β-(dialkylamino)alkylphosphines were prepared and used for the cross-coupling. Those substituted with sterically bulky alkyl group at the chiral carbon center are more effective than the ferrocenylphosphine ligands. For example, (*S*)-valphos, which was prepared starting with valine, gave the product **3** with 81% ee [5].

The asymmetric cross-coupling was successfully applied to the synthesis of optically active allylsilanes [6, 7] (Figure 4). The reaction of α-(trimethylsilyl)benzyl-

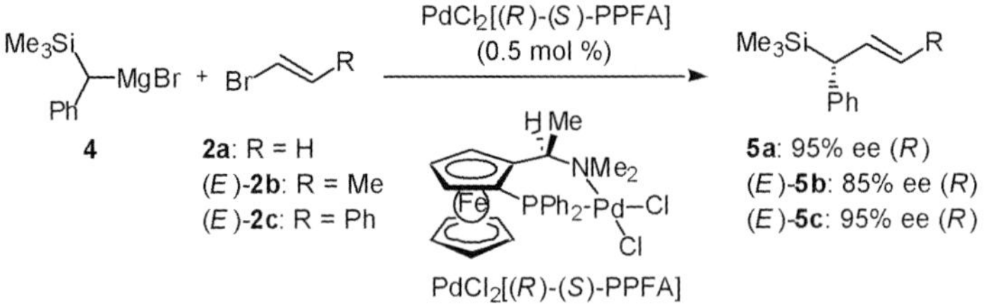

Figure 4 Asymmetric cross-coupling of α-(trimethylsilyl)benzylmagnesium bromide (**4**) giving allylsilanes **5**.

magnesium bromide (**4**) with vinyl bromide (**2a**), (*E*)-bromopropene ((*E*)-**2b**), and (*E*)-bromostyrene ((*E*)-**2c**) in the presence of 0.5 mol % of a palladium complex coordinated with chiral ferrocenylphosphine, (*R*)-(*S*)-PPFA, gave the corresponding (*R*)-allylsilanes (**5a–c**) with 95%, 85%, and 95% ee, respectively, which are substituted with phenyl group at the chiral carbon center bonded to the silicon atom and are very difficult to obtain by other methods. These allylsilanes were used for the S_E'-reactions forming optically active homoallyl alcohols and π-allyl-palladium complexes.

Asymmetric Cross-Coupling Forming Axially Chiral Biaryls

Preparation of axially chiral binaphthyls is one of the most exciting applications of the catalytic asymmetric cross-coupling reaction to organic synthesis. The reaction of 2-methyl-1-naphthylmagnesium bromide (**6**) with 1-bromo-2-methyl-naphthalene (**7a**) forming 2,2′-dimethyl-1,1′-binaphthyl (**8a**) has been examined with nickel catalysts coordinated with several chiral phosphine ligands. Use of ferrocenylphosphine ligand (*S*)-(*R*)-**PPF-OMe**, which is a chiral monophosphine ligand containing a methoxy group on the side chain, dramatically increased the selectivity to produce a high yield of (*R*)-**8a** with 95% ee (Figure 5) [8]. High enantioselectivity was also attained in the reaction of **6** with 1-bromonaphthalene (**7b**) which gave (*R*)-2-methyl-1,1′-binaphthyl (**8b**) with 83% ee.

Axially chiral biaryl molecules were also prepared by the enantioposition-selective asymmetric cross-coupling [9] (Figure 6). Reaction of achiral ditriflate **9** with 2 equiv of phenylmagnesium bromide in the presence of lithium bromide and 5 mol % of PdCl$_2$[(*S*)-phephos] at −30 °C for 48 h gave 87% yield of monophenylation product (*S*)-**10** which is 93% ee and 13% yield of diphenylation product **11**.

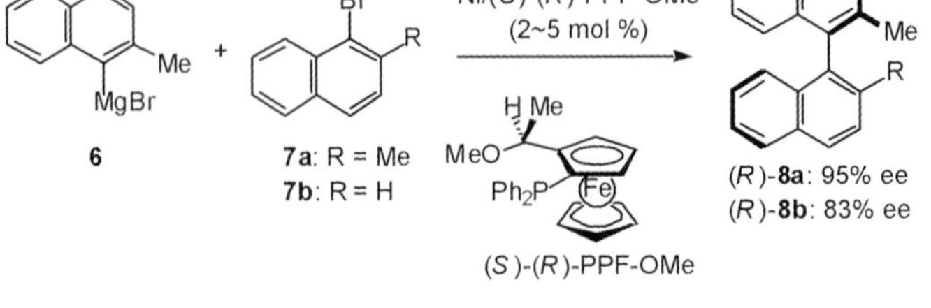

Figure 5 Asymmetric cross-coupling of 2-methyl-1-naphthylmagnesium bromide (**6**) giving 1,1′-binaphthyls.

The enantiomeric purity of the monophenylation product (S)-**10** is dependent on the yield of diphenylation product **11**. A kinetic resolution is demonstrated to take place at the second cross-coupling, forming **11**. The minor isomer at the first cross-coupling, that is (R)-**10**, is consumed five times faster than the major isomer (S)-**10** at the second cross-coupling, which causes an increase in enantiomeric purity of (S)-**10** as the amount of **11** increases.

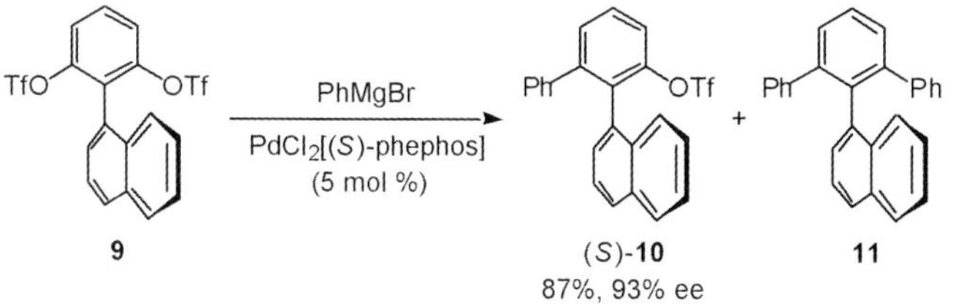

Figure 6 Asymmetric synthesis of axially chiral biaryls by enantioposition selective cross-coupling.

Recently we have reported another type of catalytic asymmetric cross-coupling giving axially chiral biaryls [10]. The ring-opening cross-coupling of dinaphtho[2,1-*b*:1′,2′-*d*]thiophene (**12**) with aryl Grignard reagents proceeded with high enantioselectivity in THF at 20 °C in the presence of 3 mol % of a nickel catalyst generated from Ni(cod)$_2$ and a chiral oxazoline-phosphine ligand to give high yields of axially chiral 2-mercapto-2′-aryl-1,1′-binaphthyls (**13**), whose enantiomeric excesses are over 93 % (Figure 7). The mercapto group in the chiral binaphthyl was converted into iodo, boryl, and phosphino groups without racemization.

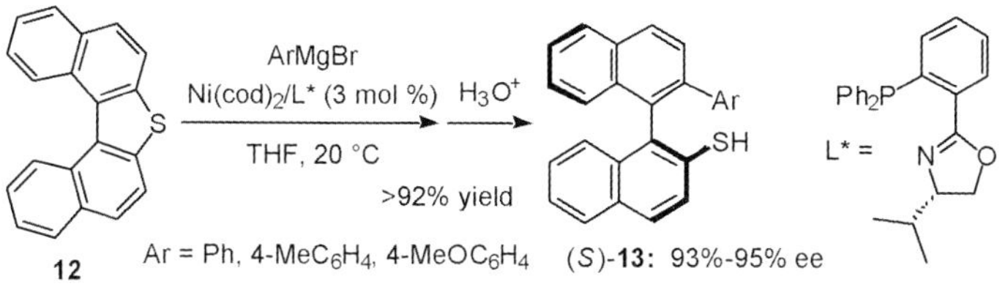

Figure 7 Asymmetric cross-coupling of dinaphthothiophene giving axially chiral 1,1′-binaphthyls.

Conclusions and Future Perspectives

Catalytic asymmetric cross-coupling is in a unique position in the field of asymmetric synthesis. It has opened a new and convenient route to enantiomerically enriched molecules which cannot be readily obtained by other methods. Typical examples are the synthesis of allylsilanes and axially chiral biaryls. Hopefully, asymmetric cross-coupling will find new applications in the synthesis of various types of chiral molecules.

CV of Tamio Hayashi

Tamio Hayashi was born in Gifu, Japan, in 1948. He graduated from Kyoto University in 1970. He received his Ph.D. in 1975 from Kyoto University under the direction of Professor Makoto Kumada, and was appointed as a Research Associate in the Faculty of Engineering, Kyoto University. He spent the year 1976–1977 as a postdoctoral fellow at Colorado State University with Professor Louis S. Hegedus. He was promoted to Full Professor in 1989 in the Catalysis Research Center, Hokkaido University. Since 1994, he has been Full Professor in the Faculty of Science, Kyoto University. His awards include the Award for Young Chemists of the Society of Synthetic Organic Chemistry, Japan in 1983, IBM (Japan) Prize in 1991, and The Chemical Society of Japan Award in 2003.

Selected Publications

1. T. Hayashi, *J. Oraganomet. Chem.* **2002**, *653*, 41–45. Catalytic Asymmetric Cross-Coupling.
2. T. Hayashi, M. Tajika, K. Tamao, M. Kumada, *J. Am. Chem. Soc.* **1976**, *98*, 3718–3719. High Stereoselectivity in Asymmetric Grignard Cross-Coupling Catalyzed by Nickel Complexes of Chiral (Aminoalkylferrocenyl)phosphines.
3. T. Hayashi, M. Konishi, M. Fukushima, T. Mise, M. Kagotani, M. Tajika, M. Kumada, *J. Am. Chem. Soc.* **1982**, *104*, 180–186. Asymmetric Synthesis Catalyzed by Chiral Ferrocenylphosphine-Transition Metal Complexes. 2. Nickel- and Palladium-Catalyzed Asymmetric Grignard Cross-Coupling.
4. T. Hayashi, M. Konishi, M. Fukushima, K. Kanehira, T. Hioki, M. Kumada, *J. Org. Chem.* **1983**, *48*, 2195–2202. Chiral (β-Aminoalkyl)phosphines. Highly Efficient Phosphine Ligands for Catalytic Asymmetric Grignard Cross-Coupling.
5. T. Hayashi, A. Yamamoto, M. Hojo, Y. Ito, *J. Chem. Soc., Chem. Commun.* **1989**, 495–496. A New Chiral Ferrocenylphosphine Ligand with C_2 Symmetry: Preparation and Use for Palladium-Catalysed Asymmetric Cross-Coupling.
6. T. Hayashi, M. Konishi, H. Ito, M. Kumada, *J. Am. Chem. Soc.* **1982**, *104*, 4962–4963. Optically Active Allylsilanes. 1. Preparation by Palladium-Catalyzed Asymmetric Grignard Cross-Coupling and Anti Stereochemistry in Electrophilic Substitution Reactions.
7. T. Hayashi, M. Konishi, Y. Okamoto, K. Kabeta, M. Kumada, *J. Org. Chem.* **1986**, *51*, 3772–3781. Asymmetric Synthesis Catalyzed by Chiral Ferrocenylphosphine-Transition-Metal Complexes. 3. Preparation of Optically Active Allylsilanes by Palladium-Catalyzed Asymmetric Grignard Cross-Coupling.
8. T. Hayashi, K. Hayashizaki, T. Kiyoi, Y. Ito, *J. Am. Chem. Soc.* **1988**, *110*, 8153–8156. Asymmetric Synthesis Catalyzed by Chiral Ferrocenylphosphine-Transition-Metal Complexes. 6. Practical Asymmetric Synthesis of 1,1'-Binaphthyls via Asymmetric Cross-Coupling with a Chiral (Alkoxyalkylferrocenyl)monophosphine/Nickel Catalyst.
9. T. Hayashi, S. Niizuma, T. Kamikawa, N. Suzuki, Y. Uozumi, *J. Am. Chem. Soc.* **1995**, *117*, 9101–9102. Catalytic Asymmetric Synthesis of Axially Chiral Biaryls by Palladium-Catalyzed Enantioposition-Selective Cross-Coupling.
10. T. Shimada, Y.-H. Cho, T. Hayashi, *J. Am. Chem. Soc.* **2002**, *124*, 13396–13397. Nickel-Catalyzed Asymmetric Grignard Cross-Coupling of Dinaphthothiophene Giving Axially Chiral 1,1'-Binaphthyls.

Asymmetric Allylic Substitutions

Günter Helmchen, Universität Heidelberg, Germany

Background

Transition metal catalyzed allylic substitution (Scheme 1) is a useful reaction in organic synthesis [1R]. The selectivity of this reaction is a function of many factors, for example the metal ion, ligands, nucleophile and substituents of the allyl system.

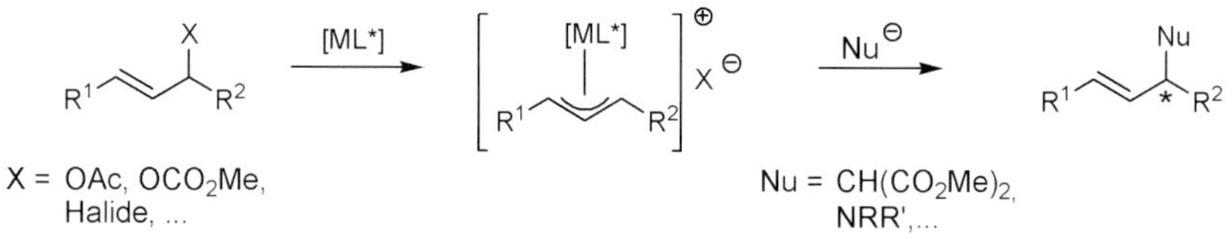

X = OAc, OCO$_2$Me, Halide, ...

Nu = CH(CO$_2$Me)$_2$, NRR',...

Scheme 1 Allylic substitution via π-allyl complexes.

As a rule, reactions via unsymmetrically substituted substrates ($R^1 \neq R^2$) proceed with conservation of enantiomeric purity, because isomerization processes of intermediary π-allyl complexes are slow. For this reason, substrates of either the type $R^1 = R^2$ or $R^2 = H$ (Scheme 2) are mainly used. The first case, i.e. symmetrically substituted substrates, is almost exclusively the domain of Pd catalysis. Numerous studies over a period of ca. 30 years have led to chiral ligands that quite generally allow products with > 90 % enantiomeric excess (% *ee*) to be prepared. Mechanistic aspects are *grosso modo* understood.

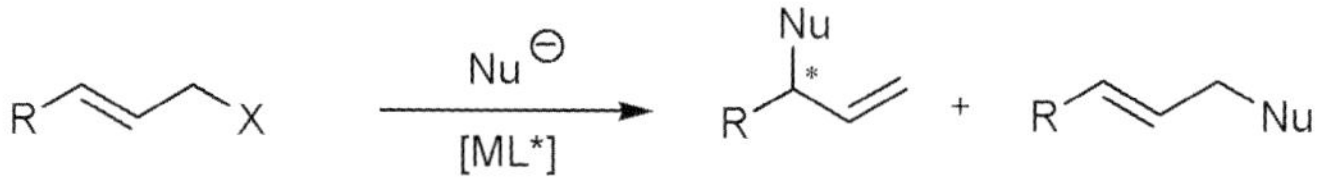

Scheme 2 Allylic substitution with monosubstituted allyl derivatives.

Preparation of the seemingly simple substrates of the type $R^1 = R^2$ is often quite cumbersome. For that reason, reactions with monosubstituted allyl derivatives according to Scheme 2 are of interest. With the exception of a few examples, these substrates are not suitable for use with Pd catalysts as they give rise to

Asymmetric Synthesis – The Essentials.
Edited by Mathias Christmann and Stefan Bräse
Copyright © 2007 WILEY-VCH Verlag GmbH & Co. KGaA, Weinheim
ISBN: 978-3-527-31399-0

achiral, linear substitution products. Remarkably, complexes of most of the other transition metals, e.g. Ir, Mo, and Ru, preferentially give rise to the branched, chiral products. While Mo- and Ru-complexes furnished good results in special cases, Ir catalysts are generally applicable to reactions with C- as well as N- and O-nucleophiles.

Results
Pd-Catalyzed Allylic Substitutions

In 1993, phosphinooxazoline ligands **L1** (Figure 1) were reported almost simultaneously by Pfaltz, Williams and our group [2R]. One of the reactions probed with these then new ligands was the Pd-catalyzed allylic substitution. Remarkably high enantiomeric excess was obtained with acyclic substrates. However, enantioselectivity with cyclic substrates was disappointingly low. Based on mechanistic studies, the ligands of type **L2** were devised and gave rise to very high enantiomeric excess [3]. The X-ray crystal structure of the (η^3-cyclohexenyl)Pd complex of **L2**, R = tert-Bu, (Figure 2) gives a clear illustration of the conformational rigidity of the complex, which in solution displays only one of two possible π-allyl isomers. The configurations of the substitution products can be rationalized by the assumptions that the solid state structure corresponds to the reactive conformer and that attack of the nucleophile at the allylic carbon *trans* to phosphorus is preferred.

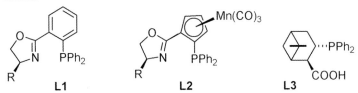

Figure 1 Ligands for Pd-catalyzed allylic substitutions.

Led by the idea of devising a P,O-chelate ligand, the ligand **L3** was prepared and gave rise to extremely high *ee*, often above 99%, with cyclic compounds [4]. It was soon found that **L3** coordinates in a monodentate manner, however, a reliable mechanistic model was elusive until very recently.

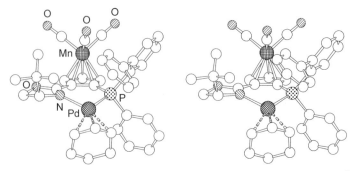

Figure 2 Stereoview of the crystal structure of the cation of the complex [(η^3-C$_6$H$_9$)L2Pd]PF$_6$. Hydrogen atoms are omitted.

Ir-Catalyzed Allylic Substitutions

Ir complexes as catalysts for allylic substitutions with monosubstituted allyl derivatives have been developed only recently (Scheme 3).

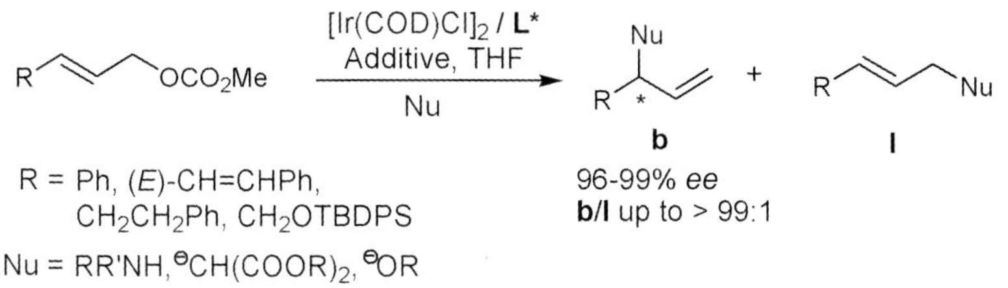

R = Ph, (E)-CH=CHPh, CH$_2$CH$_2$Ph, CH$_2$OTBDPS
Nu = RR'NH, $^\ominus$CH(COOR)$_2$, $^\ominus$OR

96-99% ee
b/l up to > 99:1

Scheme 3 Ir-catalyzed allylic substitutions.

So far phosphorus amidites, introduced by Alexakis and Feringa, are the most effective ligands for these reactions (Figure 3). High levels of regio- and enantioselectivity [5, 6] were obtained with catalysts prepared by transforming complexes [Ir(cod)ClL4] into P,C-chelate complexes **L5** by base-promoted C–H activation [7, 8].

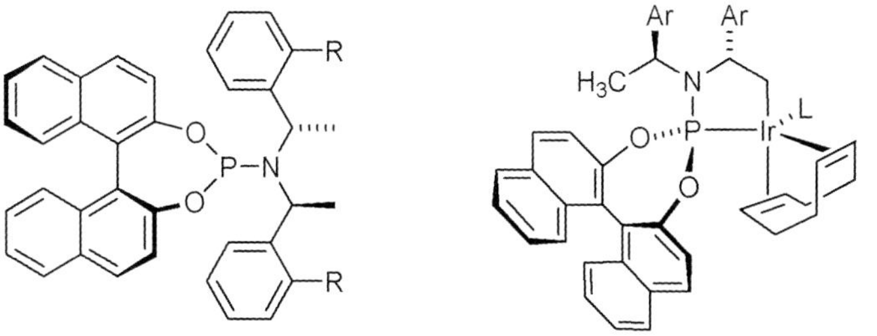

(S,S,aS)-**L4**   **L5**

Figure 3 Ligands used in Ir-catalyzed allylic substitutions.

Application for the Synthesis of Biologically Active Compounds

In contrast to methods that are biomimetic, such as the aldol reaction, allylic substitution is not an obvious element of a synthetic strategy. Nevertheless, there are numerous opportunities, as illustrated by examples from Trost's and our laboratory.

We have prepared a variety of natural products containing five-membered rings [Figure 4]. For example, a precursor of archaea cell wall lipids with ten chirality centers was constructed by combining six chiral building blocks with the help of Cu- or Ni-catalyzed cross-coupling reactions. The building block incorporating the five-membered ring was prepared via Pd-catalyzed allylic substitution, the methylated centers were generated by Rh-catalyzed catalytic hydrogenation. Remarkably, almost all C–C connections in this synthesis could be based on transition metal catalyzed reactions [9].

Ir-catalyzed allylic substitution has found only a few applications so far. It is particularly useful in combination with ring closing metathesis (RCM). Using this

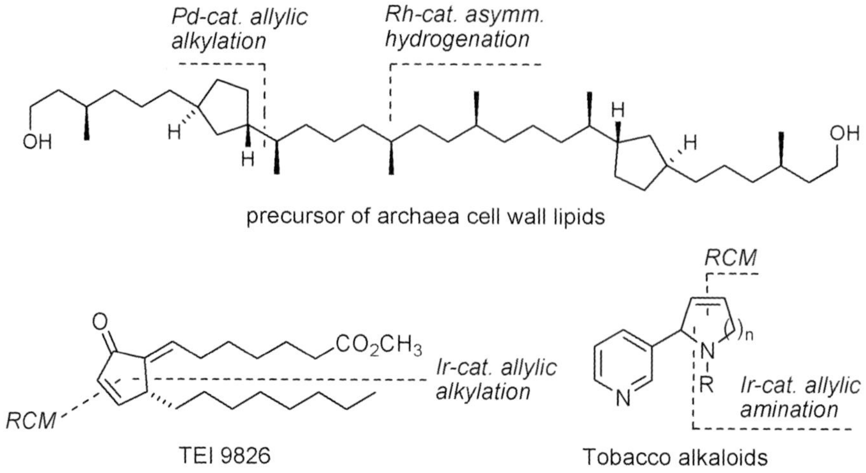

Figure 4 Targets pursued successfully with the help of Pd- and Ir-catalyzed allylic alkylation and amination.

strategy, we have accomplished remarkably short syntheses of the prostaglandin analog TEI 9826 and of a variety of tobacco alkaloids and their analogs [10].

Other Asymmetric Syntheses from the Helmchen Group
In addition to allylic substitutions, we have investigated asymmetric Diels-Alder reactions [11], catalytic hydrogenations [12], conjugate additions [13] and catalytic P–C cross-coupling reactions [14]. Most of the methods were applied in the construction of biologically active compounds.

Conclusions and Future Perspectives
Over the last ca. 12 years, catalysts for Pd-catalyzed allylic substitution have matured to the degree that almost any symmetrically substituted cyclic or acyclic substrate can be alkylated or aminated with a high degree of enantioselectivity. Areas that remain to be improved are reactions with non-stabilized carbanions and enolates. Furthermore, catalytic efficiency needs to be increased. Ir catalysis seems to be broadly applicable to substitutions with monosubstituted allylic derivatives. This juvenile field offers many possibilities for further development and applications.

CV of Günter Helmchen
Günter Helmchen (b. 1940) is a Full Professor at the University of Heidelberg and director of the Institute of Organic Chemistry. He pursued undergraduate studies at the TH Hannover (Dipl.-Chem. 1965). His graduate work, completed in 1971, was carried out under the guidance of Professor V. Prelog at the ETH Zürich in the area of stereochemistry. He then joined the group of H. Muxfeldt for postdoctoral studies in the area of natural product synthesis and carried out a Habilita-

tionsarbeit at the Technical University of Stuttgart (1975–1980). In 1980 he was appointed Professor C3 at the University of Würzburg. In 1985 he moved to his present position. His interest in catalysis dates back to ca. 1990. His scientific work has been recognized by a variety of scientific prizes and research awards, international lectureships and the invitation to join the advisory boards of scientific journals.

Selected Publications

1R. B. M. Trost, C. Lee, *Catalytic Asymmetric Synthesis*, 2nd edn., I. Ojima (Ed.), Wiley-VCH, New York, **2000**, pp. 593–649. *Asymmetric Allylic Alkylation Reactions*.

2R. G. Helmchen, A. Pfaltz, *Acc. Chem. Res.* **2000**, *33*, 336–345. *Phosphinooxazolines - A New Class of Versatile, Modular P,N-Ligands for Asymmetric Catalysis*.

3. S. Kudis, G. Helmchen, *Angew. Chem.* **1998**, *110*, 3210–3212; *Angew. Chem. Int. Ed. Engl.* **1998**, *37*, 3047–3050. *Enantioselective Allylic Substitution of Cyclic Substrates by Catalysis with Palladium Complexes of P,N-Chelate Ligands with a Cymantrene Unit*.

4. G. Knühl, P. Sennhenn, G. Helmchen, *J. Chem. Soc., Chem. Commun.* **1995**, 1845–1846. *New Chiral β-Phosphinocarboxylic Acids and their Application in Palladium-Catalyzed Asymmetric Allylic Alkylations*.

5. G. Lipowsky, N. Miller, G. Helmchen, *Angew. Chem.* **2004**, *116*, 4695–4698; *Angew. Chem. Int. Ed. Engl.* **2004**, *43*, 4995–4997. *Regio- and Enantioselective Iridium-Catalyzed Allylic Alkylation with in situ Aktivated P,C-Chelate Complexes*.

6. C. Welter, A. Dahnz, B. Brunner, S. Streiff, P. Dübon, G. Helmchen, *Org. Lett.* **2005**, *7*, 1239–1242. *Highly Enantioselective Syntheses of Heterocycles via Intramolecular Ir-Catalyzed Allylic Amination and Etherification*.

7. B. Bartels, C. García-Yebra, F. Rominger, G. Helmchen, *Eur. J. Inorg. Chem.* **2002**, 2569–2586. *Iridium-Catalyzed Allylic Substitution: Stereochemical Aspects and Isolation of Ir^{III}-Complexes Related to the Catalytic Cycle*.

8. C. A. Kiener, C. Chu, C. Incarvito, J. F. Hartwig, *J. Am. Chem. Soc.* **2003**, *125*, 14272–14273. *Identification of an Activated Catalyst in the Ir-Catalyzed Allylic Amination and Etherification*.

9. E. Montenegro, B. Gabler, G. Paradies, M. Seemann, G. Helmchen, *Angew. Chem.* **2003**, *115*, 2521–2523, *Angew. Chem. Int. Ed. Engl.* **2003**, *42*, 2419–2421. *Determination of the Configuration of Archaea Membrane Lipids Containing Cyclopentane Rings via Total Synthesis*.

10. C. Welter, M. Moreno, S. Streiff, G. Helmchen, *Org. Biol. Chem.* **2005**, *3*, 3266–3268. *Enantioselective Synthesis of (+)(R)- and (–)(S)-Nicotine based on Ir-catalysed Allylic Amination*.

11. I. Sagasser, G. Helmchen, *Tetrahedron Lett.* **1998**, 261–264. *(Phosphinooxazoline)copper(II) Complexes as Chiral Catalysts for Enantioselective Diels-Alder Reactions*.

12. M. Ostermeier, B. Brunner, Ch. Korff, G. Helmchen, *Eur. J. Org. Chem.* **2003**, 3453–3459. *Highly Enantioselective Rhodium-Catalyzed Hydrogenation of Itaconic Acid 4-Methyl Ester – A Convenient Access of Enantiomerically Pure Isoprenoid Building Blocks*.

13. J.-M. Becht, O. Meyer, G. Helmchen, *Synthesis* **2003**, 2805–2810. *Enantioselective Syntheses of (–)(R)-Rolipram, (–)(R)-Baclofen and Other GABA Analogues via Rhodium-Catalyzed Conjugate Addition of Arylboronic Acids*.

14. Ch. Korff, G. Helmchen, *Chem. Commun.* **2004**, 530–531. *Preparation of Chiral Triarylphosphines by Pd-catalysed Asymmetric P-C Cross Coupling*.

Asymmetric Homoaldol Reactions

Dieter Hoppe, Westfälische Wilhelms-Universität Münster, Germany

Background

The homoaldol reaction leads to γ-hydroxy carbonyl compounds **3** which are homologous to β-aldols. In principle, these originate from the addition of β-enolates **2** onto aldehydes **1**. Synthons **2** lack mesomeric stabilization and are usually not accessible by deprotonation reactions. Among some other approaches, α-metallated 2-alkenyl N,N-dialkylcarbamates **5** offer the most general solution (Scheme 1) [1R]. The carbamate moiety facilitates the deprotonation step by binding the lithium base onto the substrate **4**. In the five-membered chelate complex **5**, the lithium cation is held tightly at the α-position. Exchange of the lithium cation by YTiX$_3$ (Y, X = OiPr; Y = Cl, X = OiPr; Y = Cl, X = NEt$_2$) and subsequent addition of aldehydes (or ketones) reliably leads to δ-hydroxy-1-alkenyl N,N-dialkylcarbamates **6**, which are enol esters of **3**. Unlike enol acetates, compounds **6** are stable to acidic and alkaline reaction conditions, but are easily methanolized in the presence of a strong acid and catalytic amounts of Hg^{2+}.

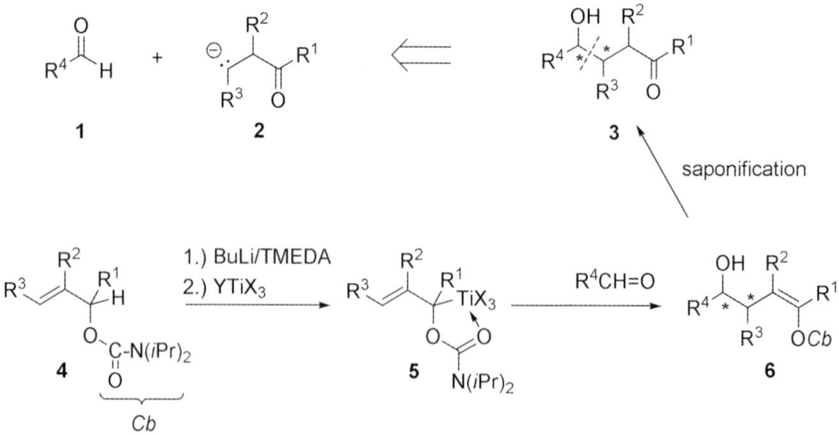

Scheme 1 Homoaldol reaction with metallated 2-alkenyl carbamates.

Asymmetric Synthesis – The Essentials.
Edited by Mathias Christmann and Stefan Bräse
Copyright © 2007 WILEY-VCH Verlag GmbH & Co. KGaA, Weinheim
ISBN: 978-3-527-31399-0

Applying the pure (E)-configured 2-alkenyl carbamate **7**, the reaction sequence affords highly diastereoselectively (Z)-anti diastereomers **9**, since the covalently bound cation enforces the reaction through a six-membered Zimmerman–Traxler transition state from the 1-endo-conformation of the metallated intermediate and a pseudo-equatorial position of the incoming residue of the aldehyde **1** (Scheme 2 [1R, 2].

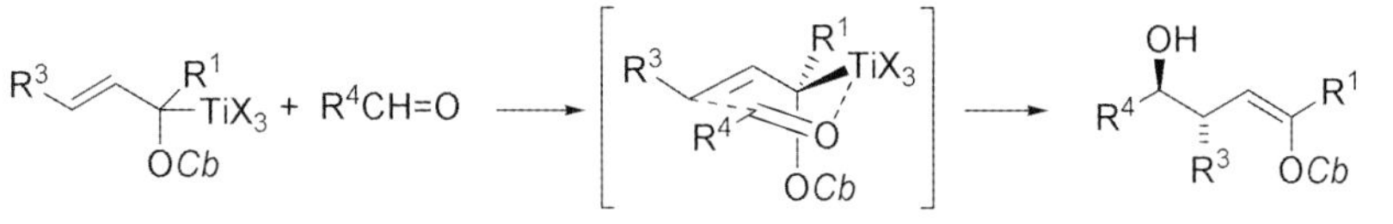

Scheme 2 anti-Diastereoselective homoaldol reaction.

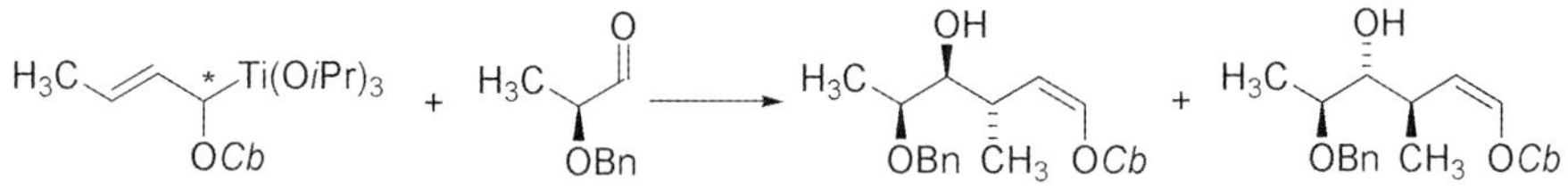

Since the addition of the racemic titanate rac-**10** onto (S)-O-benzyl-lactaldehyde **11** afforded the optically pure adducts **12** and **13** in essentially equal amounts, we concluded that both titanium intermediates (R)- and (S)-**10** take their independent reaction pathways and do not interconvert (Scheme 2) [3].

Results
Enantioselective Additions via Stereospecific Deprotonation

The above led to the conclusion that, if the allyltitanates **10** or **7** were available in enantioenriched form, enantioenriched homoaldol adducts of type **9**, **12** or **13** could be formed by efficient chirality transfer via transition states of type **8**. Indeed, the lithium-TMEDA carbanions of type **15**, derived from optically active secondary allyl carbamates of type **14**, turned out to be configurationally stable at −78 °C and to undergo lithium–titanium exchange with clean stereoinversion; the titanium intermediates **16** add to aldehydes forming homoaldol adducts **17** with complete chirality transfer (Scheme 3) [4].

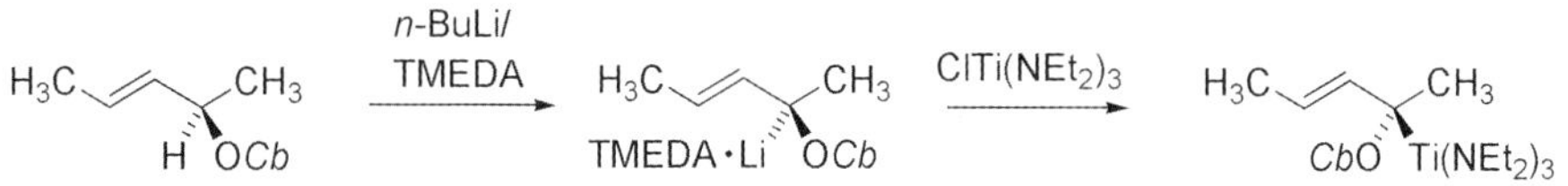

Scheme 3 Stereospecific deprotonation of a secondary allyl carbamate and subsequent homoaldol reaction.

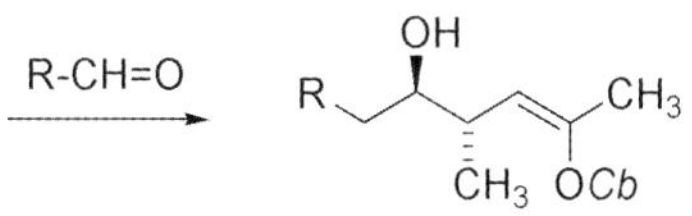

Enantioselective Additions via (−)-Sparteine-mediated α-Deprotonations

As we found in 1989–1990, the complexes formed from the alkaloid (−)-sparteine (**19**) and *n*- or *sec*-butyllithium have a high tendency to remove the *pro*-*S*-proton in primary allyl and alkenyl carbamates [5R, 6R, 7R]. Unfortunately, the *trans*-crotyl complex (*S*)-**20** epimerizes in part to (*R*)-**20** even at −78 °C (in pentane or toluene), but under certain conditions, a selective crystallization of (*S*)-**20** takes place, which shifts the equilibrium of diastereomers essentially completely towards (*S*)-**20**. After transmetallation with Ti(O*i*Pr)$_4$ from the solid, the titanate (*R*)-**10** is produced with complete stereoinversion. Its addition to achiral aldehydes leads to (*Z*)-*anti*-homoaldol adducts with up to 92 % ee (Scheme 4).

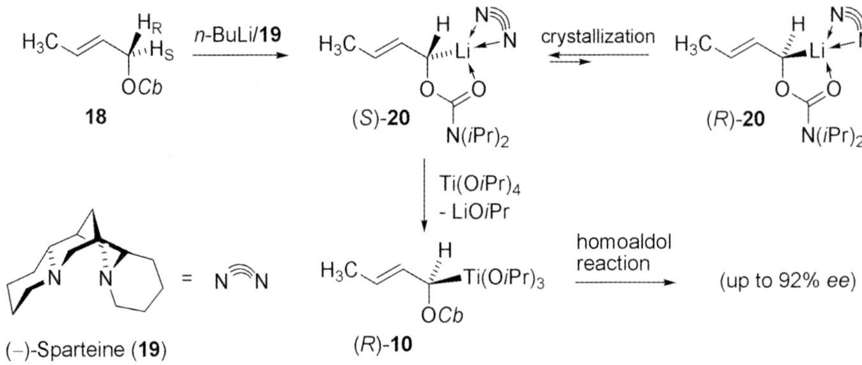

Scheme 4 (−)-Sparteine-mediated lithiation of (*E*)-but-2-enyl *N*,*N*-diisopropyl carbamate.

Although the mechanism is still unclear, the configurational stability of the lithium-sparteine complexes is enhanced by a higher substitution degree in the allyl residue. The lithium compound, obtained by (−)-sparteine-mediated deprotonation from **21**, is configurationally stable below −70 °C and leads, via the titanium intermediate (*R*)-**22**, to homoaldol adducts **23** with high ee. Methanolysis, followed by *Grieco* oxidation furnishes *cis*-fused bicyclic γ-lactones **24** (Scheme 5) [8].

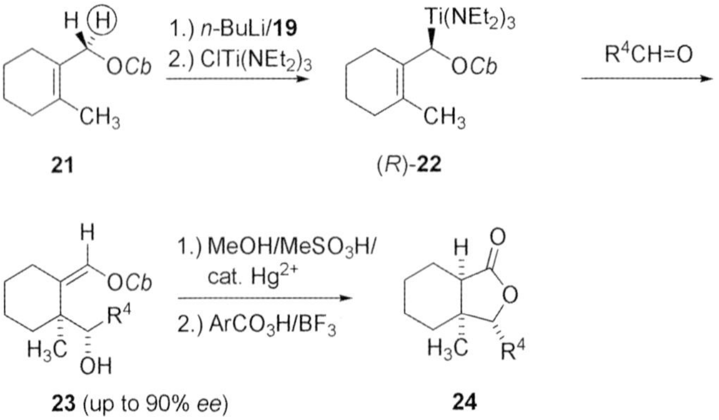

Scheme 5 Synthesis of bicyclic γ-lactones through asymmetric homoaldol reaction.

The stereochemical course of the (–)-sparteine-mediated lithiation of alkyl and allyl carbamates has been rationalized by E.-U. Würthwein by quantum-chemical calculations [9].

Enantioselective γ-Deprotonations

A surprisingly simple approach to chiral ketone homoenolate equivalents was recently found; it consists in the γ-deprotonation of achiral (Z)-enol carbamates **25** (A = aryl, Me$_3$Si), which bear a slightly electron-withdrawing residue A, by n-butyllithium/(–)-sparteine (**19**). The selective abstraction of the γ-pro-R-H is expected to proceed through a nine-membered transition state formed from the ternary associate **26** to deliver the configurationally stable α-chelate complex **27**, which undergoes the homoaldol reaction by formation of **28** [10]. A couple of further electrophiles have been used [10]. Intermediates of type **27**, which had been previously produced by stereospecific deprotonation of enantioenriched precursors or via kinetic resolution by (–)-sparteine-mediated lithiation, can now be accessed from simple achiral precursors (Scheme 6).

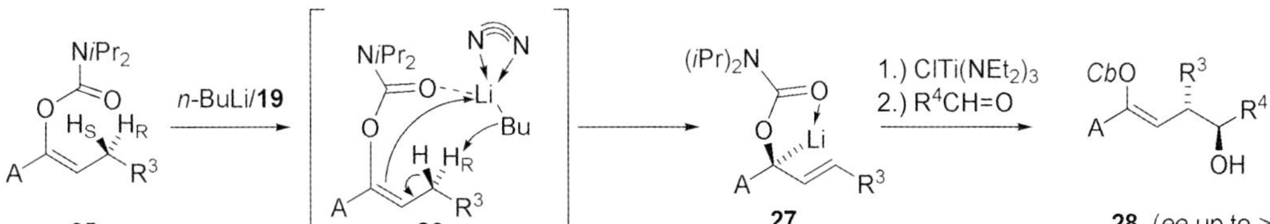

Scheme 6 Homoaldol reaction via enantioselective γ-deprotonation.

Synthetic Applications

Besides the solvolysis of the formed δ-hydroxyenol carbamates to produce the corresponding aldehydes, ketones, and carboxylic acid esters, many further highly diastereoselective transformations have been developed. The limited space does not allow their presentation here nor the elegant methods for homoaldol reactions of Beak, Helmchen, Thomas, and Ahlbrecht. The reader is asked to consult the reviews [5R, 6R, 7R].

CV of Dieter Hoppe

Dieter Hoppe was born in Berlin, Germany, in 1941. He studied at the University of Göttingen. He received his PhD in 1970 after working with Ulrich Schöllkopf in Göttingen, where he also began his independent work; the Habilitation was finished in 1977. From 1977 to 1978 he spent a year as a Research Scholar with R. B. Woodward at Harvard University (Cambridge, Massachussetts). In 1985 he moved to the University of Kiel as a Full Professor and has held since 1992 a chair at the University of Münster. He was awarded the Otto-Bayer-Preis (1993), the Max-Planck-Forschungspreis (1999) and the Adolf-von-Baeyer-Denkmünze of the Gesellschaft Deutscher Chemiker (2001).

Selected Publications

1R. D. Hoppe, *Angew. Chem.* **1984**, *96*, 930–946; *Angew. Chem. Int. Ed. Engl.* **1984**, *23*, 932–947. *The Homoaldol Reaction – How to Solve Problems of Regio- and Stereoselectivity.*

2. R. Hanko, D. Hoppe, *Angew. Chem.* **1982**, *94*, 378–379; *Angew. Chem. Int. Ed. Engl.* **1982**, *21*, 372–373. *Highly Diastereoselective and Regioselective Homoaldol Reactions by means of (1-Oxyallyl)titanium Derivatives.*

3. D. Hoppe, G. Tarara, M. Wilcken, *Synthesis* **1989**, 83–88. *Enantiomerically Pure 2-(Carbamoyloxy)oxiranes and Their Utility in the Synthesis of D- and L-3,6-Dideoxy-3-C-methylhexofuranosides via the Homoaldol Reactions.*

4. D. Hoppe, T. Krämer, J.-R. Schwark, O. Zschage, *Pure Appl. Chem.* **1990**, *62*, 1999–2006. *Metallated 2-Alkenyl Carbamates: Chiral Homoenolate Reagents for Asymmetric Synthesis.*

5R. D. Hoppe, T. Hense, *Angew. Chem.* **1997**, *109*, 2376–2410; *Angew. Chem. Int. Ed. Engl.* **1997**, *36*, 2282–2316. *Enantioselective Synthesis with Lithium/(–)-Sparteine Carbanion-Pairs.*

6R. D. Hoppe, F. Marr, M. Brüggemann, in D. M. Hodgson (Ed.), *Topics in Organometallic Chemistry; Organolithiums in Enantioselective Synthesis*, Vol. 5, pp. 61–138, **2003**, Springer, Berlin. *Enantioselective Synthesis by Lithiation Adjacent to Oxygen and Electrophile Incorporation.*

7R. D. Hoppe, G. Christoph, in Z. Rappoport, I. Marek (Eds.), *The Chemistry of Organolithium Compounds*, Wiley, Chichester **2004**, pp. 1058–1164. *Asymmetric Deprotonation with Alkyllithium/ (–)-Sparteine.*

8. M. Özlügedik, J. Kristensen, J. Reuber, R. Fröhlich, D. Hoppe, *Synthesis* **2004**, 2303–2316. *Stereoselective Synthesis of Highly Substituted Bicyclic γ-Lactones Using Homoaldol Addition of 1-(1-Cycloalkenyl)methyl Carbamates.*

9. E.-U. Würthwein, D. Hoppe, *J. Org. Chem.* **2005**, *70*, 4443–4451. *Enantioselective Deprotonation of O-Alkyl and O-Alk-2-enyl Carbamates in the Presence of (–)-Sparteine and (–)-α-Isosparteine. A Theoretical Study.*

10. M. Seppi, R. Kalkofen, J. Reupohl, R. Fröhlich, D. Hoppe, *Angew. Chem. Int. Ed. Engl.* **2004**, *43*, 1423–1427; *Angew. Chem.* **2004**, *116*, 1447–1451. *Highly Enantiomerically Enriched Ketone Homoenolate Reagents Prepared by (–)-Sparteine-Mediated Deprotonation of Achiral 1-Alkenyl Carbamates.*

Asymmetric Vinylogous Mukaiyama Aldol Reaction

Markus Kalesse and Jorma Hassfeld, University of Hannover, Germany

Introduction

The stereoselective assembly of polyketides plays a major role in natural product synthesis and biological chemistry. These compounds serve not only as lead structures but also as molecular probes in order to identify new biological targets and to unravel cellular processes.

Results

We approached this field during our synthesis of ratjadone [1] for which a rapid assembly of the tetrahydropyrane moiety (**5**) [2] had to be developed (Scheme 1). Following the proposed biosynthetic route, we envisioned a ring-closing epoxide opening as the key step in this synthesis. In our retrosynthetic analysis, the desired epoxide **4** could be generated from the corresponding α,β-unsaturated ester **3**.

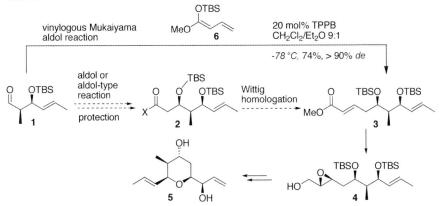

Scheme 1 Synthesis towards ratjadone using the vinylogous Mukaiyama aldol reaction.

At this point, standard sequences such as aldol reactions followed by functional group transformations and Wittig reaction would furnish compound **3**. Alternatively, a vinylogous Mukaiyama aldol reaction (VMAR) could provide the same in-

Asymmetric Synthesis – The Essentials.
Edited by Mathias Christmann and Stefan Bräse
Copyright © 2007 WILEY-VCH Verlag GmbH & Co. KGaA, Weinheim
ISBN: 978-3-527-31399-0

termediate in only one step. Based on literature precedence, BF$_3$-etherate was initially used as the Lewis acid to provide compound **3** in good yield and 3:1 selectivity in favor of the Felkin product. We rationalized that the observed moderate selectivity might be explained by loss of steric hindrance in **6** compared to classic Mukaiyama aldol reactions due to the vinylogous situation. In order to compensate for this drawback, we used tris(pentafluorophenyl)borane (TPPB) as a bulky substitute for BF$_3$-etherate. The reaction performed under the conditions employed for BF$_3$ gave selectivities of greater than 19:1 for the Felkin product. Interestingly, we also observed simultaneous TBS protection of the newly generated alcohol, which saves one additional step in the synthesis. More detailed investigations of the reaction mechanism showed that the Lewis acid liberated a positively charged silicon species that could also promote the vinylogous Mukaiyama aldol reaction.

In cases were a δ-methyl substituted ketene acetal (**8**) was used, the observed *syn/anti* selectivity dropped to 4:1. Here, the presence of an additional silicon scavenger (isopropyl alcohol) was necessary in order to suppress the undesired "silicon catalysis" pathway (Figure 1). The VMAR catalysed by TPPB exclusively produced the *syn*-Felkin product, probably due to tighter binding of the Lewis acid resulting in increased steric interactions in the boron-mediated reaction [3] (Scheme 2).

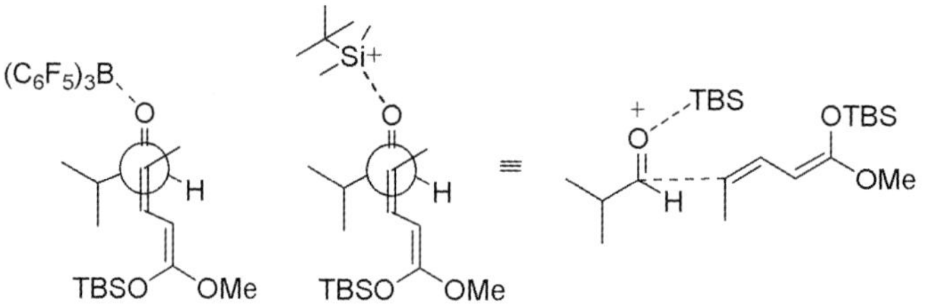

Figure 1 Alternative catalytic pathways – TPPB vs. positively charged silicon species.

This procedure now allows the rapid assembly of polyketide structures with ketene acetal **8** as a synthon for one propionate and one acetate moiety. The utility of this protocol has been exemplified for **10** as a building block for oleandolide [4] (Scheme 2).

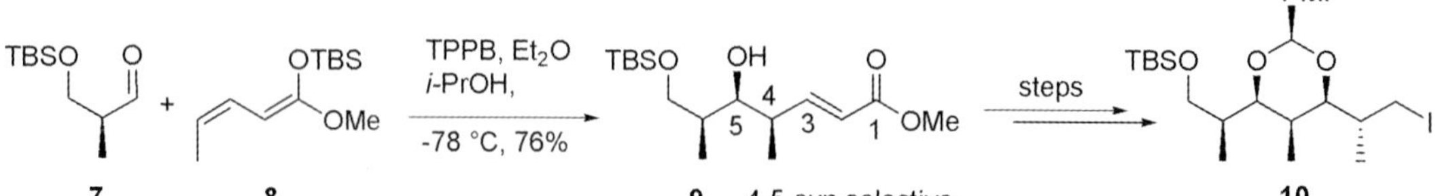

Scheme 2 VMAR employed in the synthesis of oleandolide with isopropyl alcohol as silicon scavenger.

When we changed the supplier of TPPB, reactions that we knew to give almost quantitative yields failed completely, and decomposition of the starting material was observed. This prompted us to investigate the role of the Lewis acid one more time and it turned out that we had used a partially hydrated TPPB species in previous experiments instead of the pure Lewis acid that turned out to be inferior for the VMAR. Therefore, the TPPB-monohydrate was generated prior to reactions by dissolving TPPB in pentane and adding one equivalent of water. The water adduct precipitated and was isolated by filtration. This TPPB-monohydrate was then used in vinylogous Mukaiyama aldol reactions for the rapid construction of advanced intermediates (12) in complex natural product synthesis, which has been demonstrated in the construction of tedanolide [5–7] (Scheme 3).

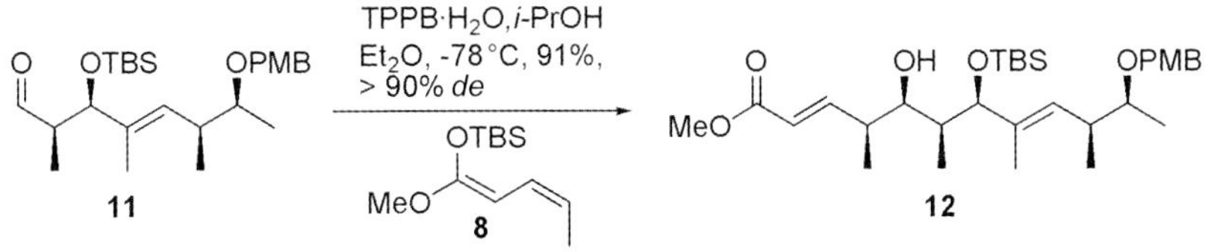

Scheme 3 Synthesis of the ketone segment of tedanolide.

Another example of a truly catalytic VMAR is given in the synthesis of the C19–C26 segment of amphidinolide. Glycerol aldehyde was reacted with ketene acetal 8 and only 2.5 mol% of TPPB to produce 60% of the unprotected Felkin product [8]. Interestingly, aldehydes with an α-oxygen substitution do not form vinylogous aldol products under silicon catalysis. In this case, only catalytic amounts of the Lewis acid were necessary (Scheme 4).

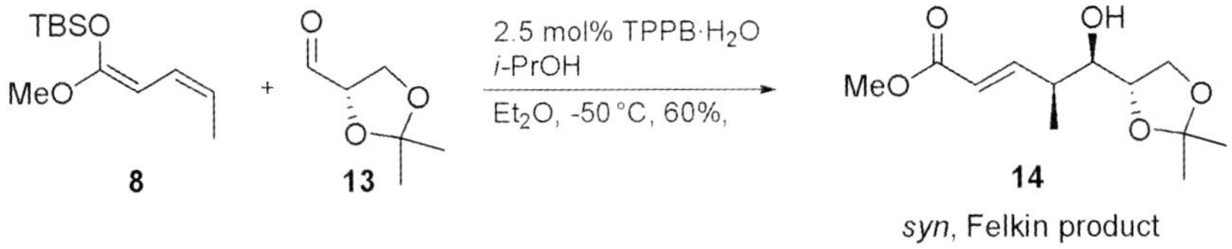

Scheme 4 Synthesis of the C19–C26 segment of amphidinolide.

The mechanistic investigation of the silicon-mediated reaction pathway indicated that a positively charged silicon species acted as Lewis acid in order to activate the aldehyde. A related intermediate was proposed by Jung and coworkers for a TBSOTf catalyzed semipinacol rearrangement (Scheme 5). We therefore subjected alcohol epoxides, together with the appropriate silyl ketene acetals, to TBSOTf treatment to obtain the rearranged VMAR products (Scheme 6). In this context, we found that the γ,δ-*E*-geometry of the ketene acetal translated into a 4,5-*anti*-relationship in the aldol product. This opens up access to a greater variety of aldol products [9].

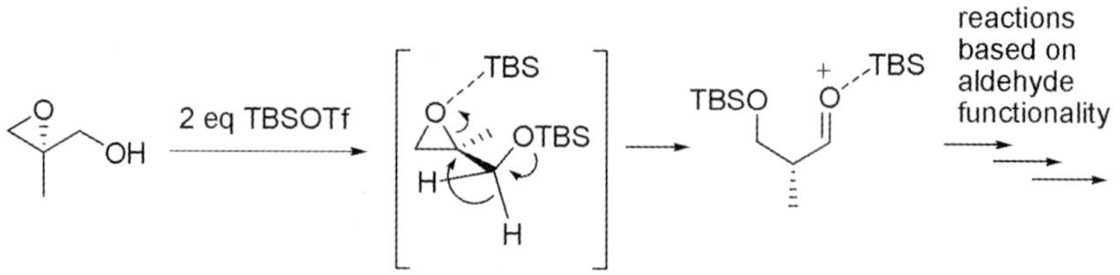

Scheme 5 Proposed mechanism for the TBSOTf-catalyzed epoxide-opening as proposed by Jung et al.

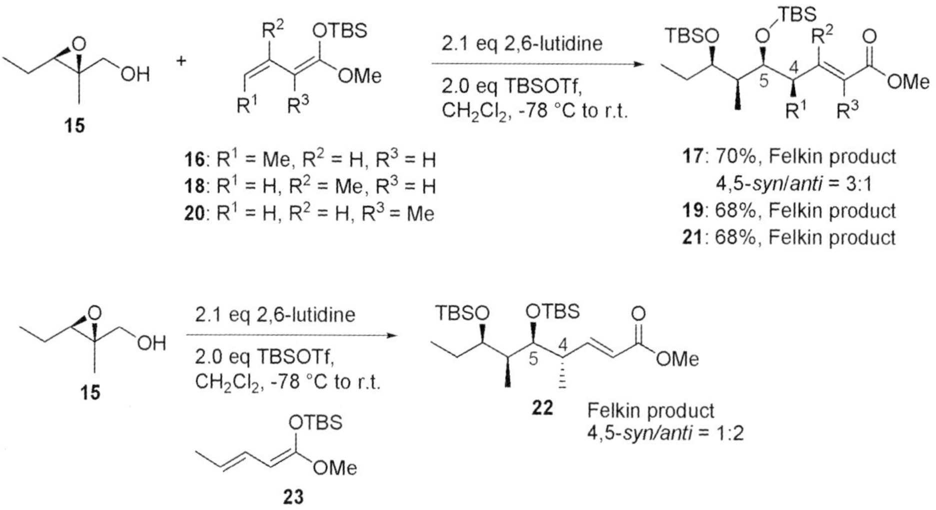

16: $R^1 = Me, R^2 = H, R^3 = H$
18: $R^1 = H, R^2 = Me, R^3 = H$
20: $R^1 = H, R^2 = H, R^3 = Me$

17: 70%, Felkin product
4,5-*syn/anti* = 3:1
19: 68%, Felkin product
21: 68%, Felkin product

22 Felkin product
4,5-*syn/anti* = 1:2

Scheme 6 Tandem VMAR/semi-pinacol rearrangement.

Conclusions and Further Perspectives

The vinylogous Mukaiyama aldol reaction allows the efficient construction of polyketide frameworks using ketene acetals as the synthon for two acetate or propionate subunits. It is noteworthy that, especially, saturated aliphatic aldehydes are good substrates for this reaction, exhibiting excellent yields and selectivities. Another great advantage of the VMAR protocol is the resulting α,β-unsaturated ester, which serves as a formidable starting point for a great variety of established subsequent transformations towards the endgame of natural product synthesis.

One open task that still remains is an enantioselective catalytic protocol that would allow one-step access to enantiomerically pure building blocks and reagent-controlled transformations. Even though first reports of a catalytic asymmetric VMAR are given by the Paterson group with Ti/BINOL as the catalytic system, these protocols fail with γ-substituted silyl ketene acetals. Nevertheless, the rapid access to advanced intermediates and the flexibility given for further transformations make this reaction one of the pivotal transformations in organic chemistry [10R].

CV of Markus Kalesse

Markus Kalesse, born in 1961, studied chemistry at the University of Hannover where he graduated in 1989. In 1991, he obtained his Doctorate in Chemistry (*summa cum laude*) with a dissertation in the laboratory of Professor Dieter Schinzer. From 1991 to 1992 he stayed as a post-doctoral researcher in the groups of Steven D. Burke and Laura L. Kiessling at the University of Wisconsin (Madison, USA). He returned to Hannover as a Habilitand with Prof. Eckehard Winterfeldt and obtained the *venia legendi* in 1997. He took the position of a Visiting Professor at the Universities of Wisconsin (1989) and Kiel (2000). He rejected an offer as a Full Professor from the University of Oslo (Norway) and accepted a C3 position at the Freie Universität Berlin in 2001. In 2003 he moved to the University of Hannover as a C4 professor in Organic Chemistry. He became Novartis Lecturer (2004/2005) and was appointed Director and Head of Medicinal Chemistry at the GBF (2005). His research interests lie in the field of natural product synthesis and biological chemistry.

CV of Jorma Hassfeld

Jorma Hassfeld, born in 1975, studied chemistry at the University of Hannover. In the course of his studies, he spent one year in the group of Professor Paul A. Wender at Stanford University, USA for undergraduate research work in 1998 to 1999. After finishing his doctoral studies under the supervision of Professor Markus Kalesse in 2005, he spent one year in the group of Dirk Menche as a post-doctoral researcher at the GBF in Braunschweig. In 2006, he moved to process research at Schering AG, Berlin.

Selected Publications

1. M. Christmann, U. Bhatt, M. Quitschalle, E. Claus, M. Kalesse, *Angew. Chem.* **2000**, *112*, 4535–4538; *Angew. Chem. Int. Ed.* **2000**, *39*, 4364–4366. *Total Synthesis of (+)-Ratjadone.*
2. M. Christmann, M. Kalesse, *Tetrahedron Lett.* **2001**, *42*, 1269–1271. *Vinylogous Mukaiyama Aldol Reactions with Triarylboranes.*
3. J. Hassfeld, M. Christmann, M. Kalesse, *Org. Lett.* **2001**, 3561–3564. *Rapid Access to Polyketide Scaffolds via Vinylogous Mukaiyama Aldol Reactions.*
4. J. Hassfeld, M. Kalesse, *Tetrahedron Lett.* **2002**, 5093–5095. *Advances in the Vinylogous Mukaiyama Aldol Reaction and its Application to the Synthesis of the C1-C7 Subunit of Oleandolide.*
5. J. Hassfeld, M. Kalesse, *Synlett* **2002**, 2007–2010. *Synthesis of the C1-C11 Segment of Tedanolide via Vinylogous Mukaiyama Aldol Reaction.*
6. M. Kalesse, J. Hassfeld, U. Eggert, *Synthesis* **2005**, 1183–1199. *Synthesis of the C1-C17 Macrolactone of Tedanolide.*
7. G. Ehrlich, M. Kalesse, *Synlett* **2005**, 655–657. *Synthesis of the C13-C23 Segment of Tedanolide.*
8. F. Liesener, M. Kalesse, *Synlett* **2005**, 2236–2238. *Synthesis of the C19-C26 Segment of Amphidinolide.*
9. N. Rahn, M. Kalesse, *Synlett* **2005**, 863–865. *A One-Pot Non-Aldol-Aldol Vinylogous Mukaiyama Aldol Tandem Sequence for the Rapid Construction of Polyketide Frameworks.*
10R. M. Kalesse, *Top. Curr. Chem.* **2004**, *244*, 43–76. *Vinylogous Aldol Reactions and their Applications in the Syntheses of Natural Products.*

Chiral Lewis Acid Catalysis in Aqueous Media

Shū Kobayashi and Chikako Ogawa, The University of Tokyo, Japan

Background

Organic reactions in aqueous media are of current interest because water is a key solvent for attaining environmentally benign chemical synthesis. Although several interesting reactions with unique reactivity and selectivity have been developed in water or water/organic solvent systems, asymmetric catalysis in such media is difficult because many chiral catalysts are not stable in the presence of water. In particular, chiral Lewis acid catalysis in aqueous media is extremely difficult because most chiral Lewis acids decompose rapidly in the presence of water. To address this issue, we have investigated Lewis acid catalysis in water and discovered water-compatible Lewis acids. Furthermore, catalytic asymmetric reactions using these Lewis acids and chiral ligands have been developed.

Results
Water-compatible Lewis Acid

We first found that rare earth triflates ($Sc(OTf)_3$, $Y(OTf)_3$, and $Ln(OTf)_3$) were water-compatible Lewis acids. They work as efficient Lewis acid catalysts in aqueous solvents. We then searched for other Lewis acids which could be used in aqueous solvents and determined the criteria for water-compatible Lewis acids. We screened Group 1–15 metal chlorides, perchlorates, and triflates in the aldol reaction of benzaldehyde with silyl enol ether **1** in water–THF (1/9) Eq. (1). This screening revealed that not only Sc(III), Y(III), and Ln(III) but also Fe(II), Cu(II), Zn(II), Cd(II), and Pb(II) worked as Lewis acids in this medium to afford the desired aldol adduct in high yields.

$$PhCHO + \underset{\text{1 (1.5 equiv)}}{\overset{OSiMe_3}{\diagup\!\!\!\diagdown Ph}} \xrightarrow[\substack{H_2O-THF \\ (1/9) \\ rt, 12\ h}]{MX_n\ (20\ mol\ \%)} \underset{Ph}{\overset{OH\ O}{\diagup\!\!\!\diagdown}} Ph \quad \text{(Eq. 1)}$$

From these results we noticed a correlation between the catalytic activity of the metal cations and two kinds of constants for the metal cations: hydrolysis con-

Asymmetric Synthesis – The Essentials.
Edited by Mathias Christmann and Stefan Bräse
Copyright © 2007 WILEY-VCH Verlag GmbH & Co. KGaA, Weinheim
ISBN: 978-3-527-31399-0

stants (K_h) and exchange rate constants for substitution of inner-sphere water ligands (water exchange rate constants (WERC)). The active metal compounds were found to have pK_h values in the range of about 4 (4.3 for Sc(III)) to 10 (10.08 for Cd(II)) and WERC values greater than 3.2×10^6 M^{-1} s^{-1}. Cations are generally difficult to be hydrolyzed when their pK_h values are large. When the pK_h values are less than 4, cations are easily hydrolyzed to produce certain amounts of protons. Under these conditions, the silyl enol ether decomposes rapidly. On the other hand, when the pK_h values are more than 10, the Lewis acidities of the cations are too low to catalyze the aldol reaction. Large WERC values may be necessary to secure fast exchange between hydrating water molecules and an aldehyde which must coordinate to the metal cation to be activated. "Borderline" elements such as Mn(II), Ag(I), and In(III), whose pK_h and WERC values are close to the criteria limits, gave the aldol adduct in modest yields. Although the precise activity as Lewis acids in aqueous media cannot be quantitatively predicted by pK_h and WERC values, these results have demonstrated the possibility of using several promising metal compounds such as Cu(II) and Pb(II) salts as water-compatible Lewis acid catalysts.

Asymmetric Aldol Reactions

Rare earth triflate [RE(OTf)$_3$]-catalyzed asymmetric aldol reactions in aqueous media have been realized for the first time using a chiral 18-crown-6-type ligand (**2**) having two pyridine rings. In the reaction of benzaldehyde with silicon enolate **1** in water/ethanol (1/9), RE(OTf)$_3$ having large metals such as La, Ce, Pr, and Nd gave the aldol adduct with high diastereo- and enantioselectivities, showing that size-fitting between the crown ether and the metal cations is an important factor to attain high selectivity. According to ^{1}H NMR studies, the crown ether was found to have strong binding ability to the La cation. A study on the reaction profiles of the Pr(OTf)$_3$-catalyzed aldol reaction indicates that the crown ether does not significantly reduce the activity of Pr(OTf)$_3$. This retention of the activity in the presence of the crown ether is the key to realizing the asymmetric induction in the present RE(OTf)$_3$-catalyzed aldol reactions. The X-ray structure of [Pr(NO$_3$)$_2$·**2**]$_3$[Pr(NO$_3$)$_6$] shows that all of the methyl groups at the asymmetric carbon atoms of the crown ether are located in the axial positions (Scheme 1(a)). This conformation of the crown ring should be crucial in creating an effective chiral environment around the Pr cation. From X-ray studies of several other RE(NO$_3$)$_3$–crown ether complexes, it was found that they had similar structures regardless of the RE cations and the crown ethers used. Accordingly, the binding ability of the crown ether with the RE cation and the catalytic activity of the complex are important for attaining high selectivity in the asymmetric aldol reaction. Various aromatic and α,β-unsaturated aldehydes and silicon enolates derived from ketones and a thioester can be employed in the catalytic asymmetric aldol reactions using Pr(OTf)$_3$ and **2**, to provide the aldol adducts in good to high yields and stereoselectivities.

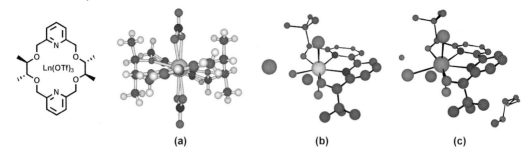

Scheme 1 (a) Chiral Pr, (b) chiral Sc, and (c) chiral Bi complexes.

Asymmetric Hydroxymethylation Reactions

Formaldehyde is one of the most important C1 electrophiles in organic synthesis. Although hydroxymethylation of enolates with formaldehyde provides an efficient method to introduce a C1 functional group at the α-position of carbonyl groups, there have been few successful examples of catalytic asymmetric hydroxymethylation. As for the source of formaldehyde, use of a commercial aqueous solution of formaldehyde is the most convenient because tedious and harmful procedures to generate formaldehyde monomer from formaldehyde oligomers such as paraformaldehyde and trioxane can be avoided. We found that a novel chiral scandium complex realized highly enantioselective, catalytic hydroxymethylation of silicon enolates with an aqueous formaldehyde solution (Scheme 2). Furthermore, X-ray crystal structural analysis of a ScBr$_3$–3 complex shows that the complex adopts a pentagonal bipyramidal structure in which the hydroxy groups of **3** coordinate to Sc^{3+} in a tetradentate manner (Scheme 1(b)). Formation of this type of structure may be a key to obtaining high enantioselectivity. In addition, on considering the absolute configurations of the hydroxymethylated products, it is clear that formaldehyde tends to react with the same face of the silicon enolates regardless of the substituents at the α-position.

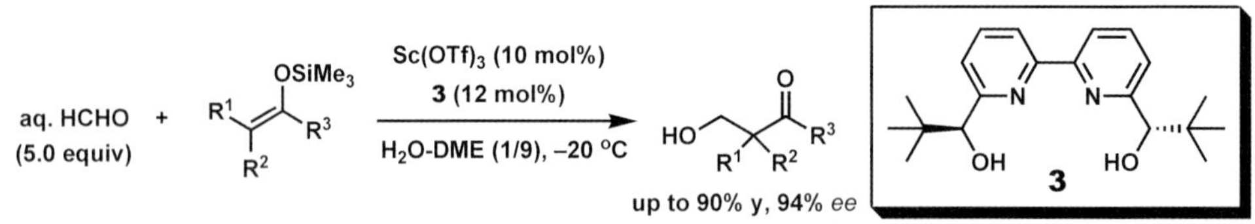

Scheme 2 Catalytic asymmetric hydroxymethylation using chiral Sc.

As an extension of this work, we tested other metal salts and chiral bipyridine **3** in the reaction of silicon enolate **1** with an aqueous formaldehyde solution, and found that Bi(OTf)$_3$ gave promising selectivity. This result was unexpected because (i) the ionic diameter of bismuth (2.34 Å for 8-coordination) is much larger than that of scandium (1.74 Å for 8-coordination) and (ii) Bi(OTf)$_3$ is known to be hydrolyzed. Indeed, only a trace amount of the hydroxymethylated adduct was obtained using Bi(OTf)$_3$ without the chiral bipyridine. In this case, silicon enolate **1** was rapidly decomposed by TfOH which is easily generated from Bi(OTf)$_3$ in water. On the other hand, decomposition of silicon enolate **1** was slow and the desired hydroxymethylation proceeded in the presence of Bi(OTf)$_3$ and **3**. Obviously, Bi(OTf)$_3$ was stabilized by chiral bipyridine **3** in water. The yield and the selectivity were improved when 1 mol% of Bi(OTf)$_3$, 3 mol% of **3**, and 5 mol% of bipyridine were used. This is the first example of highly enantioselective reactions using a chiral bismuth catalyst in aqueous media. It is noted that Sc and Bi with very different ionic diameters gave similar complexes with **3**. In addition to these characteristic points of this reaction, this result provides a new entry to "water-compatible Lewis acids." For a long time, Lewis acids were believed to be hydrolyzed rapidly in the presence of water. Contrary to this, we found that rare earth and other metal complexes were water-compatible. Further, we have added Bi(OTf)$_3$-**3** complex as a water-compatible Lewis acid. Bi(OTf)$_3$ is unstable in the presence of water but is stabilized by the basic ligand. There are many strong Lewis acids unstable in water; however, these Lewis acids would be available in water when combined with basic ligands. In particular, the use of chiral basic ligands leading to new types of water-compatible chiral Lewis acids would enable a wide range of asymmetric catalysis to be carried out in aqueous media.

Asymmetric Mannich-type Reactions
Asymmetric Mannich reactions provide useful routes for the synthesis of optically active β-amino ketones and esters, which are versatile chiral building blocks for the preparation of many nitrogen-containing biologically important compounds. In the past few years, various enantioselective Mannich reactions have been developed. Among them, catalytic enantioselective additions of silicon enolates to imines have been elaborated into one of the most powerful and efficient asymmetric Mannich-type reactions, because silicon enolates can be prepared regio- and stereoselectively from various carbonyl compounds. We reported the first catalytic asymmetric Mannich-type reactions of an α-hydrazono ester with silicon enolates in H$_2$O/THF by using a combination of a stoichiometric amount of ZnF$_2$ and a catalytic amount of a chiral diamine and trifluoromethanesulfonic acid (TfOH). In these reactions, both Zn^{2+} and fluoride anion were necessary to achieve high yields and high enantioselectivities, suggesting a double activation mechanism where Zn^{2+} activates the α-hydrazono ester and the fluoride anion simultaneously activates the silicon enolate. When chiral diamine ligands bearing methoxy substituted aromatic rings were employed, the reactions in aqueous THF were markedly accelerated. Furthermore, the use of these diamines facilitated the asymmetric Mannich-type reactions in water without any organic co-solvents. It is noteworthy that either *syn*- or *anti*-adducts were stereospecifically obtained from (*E*)- or (*Z*)-silicon enolates, respectively, in the present Mannich-type reaction. Interestingly, these reactions proceeded smoothly only in the presence of water. On the basis of several experimental results, the reaction mechanism is likely to be a fluoride-catalyzed one in which ZnF$_2$–chiral diamine complex is regenerated from Me$_3$SiF formed as the reaction progresses.

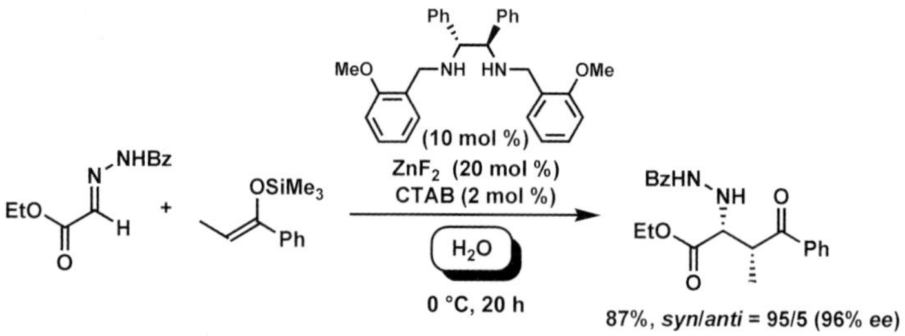

Scheme 3 Catalytic asymmetric Mannich-type reaction.

CV of Shū Kobayashi
Shū Kobayashi was born in Tokyo in 1959. He received his Ph. D. in 1988 under the direction of Professor T. Mukaiyama (The University of Tokyo). After spending eleven years at the Science University of Tokyo, he moved to the Graduate School of Pharmaceutical Sciences, the University of Tokyo, in 1998. His research interests include development of new synthetic methods and novel catalysts, organic reactions in water, solid-phase synthesis, total synthesis of biologically interesting compounds, and organometallic chemistry. He has received the Springer Award in Organometallic Chemistry (1997), the IBM Science Award (2001), the Organic Reactions Lecturer Award (2002), the Mitsui Chemical Catalysis Science Award (2005), the JSPS Prize (2005), and the Arthur C. Cope Scholar Award (2005).

CV of Dr. Chikako Ogawa
Chikako Ogawa received her Ph. D. in 2004 under the direction of Professor Shū Kobayashi (The University of Tokyo). After spending as a JSPS postdoctoral fellow, she joined into an ERATO program, JST at the University of Tokyo as a group leader. Her research interests include development of asymmetric reactions, polymer chemistry, and medicinal sciences.

Selected Publications

1. S. Kobayashi, T. Ogino, H. Shimizu, S. Ishikawa, T. Hamada, K. Manabe, *Org. Lett.* **2005** *7*, 4729–4731. *Bismuth Triflate-Chiral Bipyridine Complexes as Water-Compatible Chiral Lewis Acids.*
2. T. Hamada, K. Manabe, S. Kobayashi, *Chem. Eur. J.* **2006** *12*, 1205–1205. *Catalytic Asymmetric Mannich-type Reactions in Aqueous Media Activated by ZnF_2–Chiral Diamine.*
3. S. Ishikawa, T. Hamada, K. Manabe, S. Kobayashi, *J. Am. Chem. Soc.* **2004**, *126*, 12236–12237. *Catalytic Asymmetric Hydroxymethylation of Silicon Enolates Using an Aqueous Solution of Formaldehyde with a Chiral Scandium Complex.*
4. T. Hamada, K. Manabe, S. Kobayashi, *J. Am. Chem. Soc.* **2004**, *126*, 7768–7769. *Enantio- and Diastereoselective, Stereospecific Mannich-Type Reaction in Water.*
5. T. Hamada, K. Manabe, S. Ishikawa, S. Nagayama, M. Shiro, S. Kobayashi, *J. Am. Chem. Soc.* **2003**, *125*, 2989–2996. *Catalytic Asymmetric Aldol Reactions in Aqueous Media Using Chiral Bis-pyridino-18-crown-6–Rare Earth Metal Triflate Complexes.*

6. K. Manabe, S. Iimura, X.-M. Sun, S. Kobayashi, *J. Am. Chem. Soc.* **2002**, *124*, 11971–11978. *Dehydration Reactions in Water. Brønsted Acid– Surfactant-Combined Catalyst for Ester, Ether, Thioether, and Dithioacetal Formation in Water.*
7. S. Kobayashi, T. Hamada, K. Manabe, *J. Am. Chem. Soc.* **2002**, *124*, 5640–5641. *The Catalytic Asymmetric Mannich-type Reactions in Aqueous Media.*
8. Y. Mori, K. Manabe, S. Kobayashi, *Angew. Chem. Int. Ed. Engl.* **2001**, *40*, 2815–2818. *Catalytic Use of a Boron Source for Boron Enolate-Mediated Stereoselective Aldol Reactions in Water.*
9. K. Manabe, S. Kobayashi, *Chem. Eur. J.* **2002**, *8*, 4094–4101. *Catalytic Asymmetric Carbon–Carbon Bond-Forming Reactions in Aqueous Media.*
10. S. Kobayashi, K. Manabe, *Acc. Chem. Res.* **2002**, *35*, 209–217. *Development of Novel Lewis Acid Catalysts for Selective Organic Reactions in Aqueous Media.*

Asymmetric Epoxidation of Non-activated Olefins

Kazuhiro Matsumoto and Tsutomu Katsuki, Kyushu University, Japan

Introduction

Epoxides can be converted stereospecifically into a wide variety of useful compounds. Thus, asymmetric epoxidation of olefins is of tremendous importance for organic synthesis. Indeed, the introduction of the first practical asymmetric epoxidation by Katsuki and Sharpless changed the synthetic methods used for optically active compounds, though the substrates of the epoxidation are limited to allylic alcohols (Scheme 1) [1, 2R]: since then, most chiral compounds have been synthesized in an enantioselective manner. Asymmetric epoxidation of non-activated and unfunctionalized olefins subsequently became an important topic in organic synthesis. In this chapter we will present a brief overview of the development of asymmetric epoxidation of this class of olefins, focusing on our own efforts.

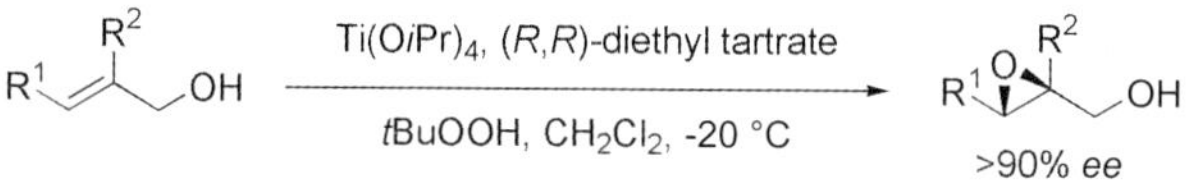

$$R^1\underset{}{\overset{R^2}{\diagup\!\!\!\diagdown}}OH \xrightarrow[\text{tBuOOH, CH}_2\text{Cl}_2,\ -20\ °\text{C}]{\text{Ti(O}i\text{Pr)}_4,\ (R,R)\text{-diethyl tartrate}} R^1\underset{}{\overset{R^2\ \ O}{\diagup\!\!\!\diagdown}}OH$$
$$>90\%\ ee$$

Scheme 1 The first practical asymmetric epoxidation.

Results

Manganese(salen) Catalyst

In 1986, Kochi et al. reported that manganese(salen) complexes catalyze the epoxidation of non-activated olefins efficiently. Coincidentally, Fujita et al. reported that an optically active vanadium(salen) complex catalyzes sulfoxidation, albeit with modest enantioselectivity. These seminal studies inspired the quest for asymmetric epoxidation of unfunctionalized olefins.

In 1990, Jacobsen et al. and our group independently reported enantioselective epoxidation using chiral Mn(salen) complexes (**1** and **2**) as catalyst (Figure 1) [3R]. Complex **2** is the first example of the C_2-symmetric salen catalyst consisting of two chiral subunits [4]. Combination of this synthetic strategy and module construction (diamine and salicylaldehyde subunits) of diastereomeric metallosalen complexes has realized an exceptionally wide scope of the catalytic performance

Asymmetric Synthesis – The Essentials.
Edited by Mathias Christmann and Stefan Bräse
Copyright © 2007 WILEY-VCH Verlag GmbH & Co. KGaA, Weinheim
ISBN: 978-3-527-31399-0

of M(salen)s [5R]. High enantioselectivity (ca. 80–95 % ee) has been achieved in the epoxidation of conjugated cis- and tri-substituted olefins with modified complexes of this type.

Subsequently, we were able to achieve remarkable improvement in enantio-

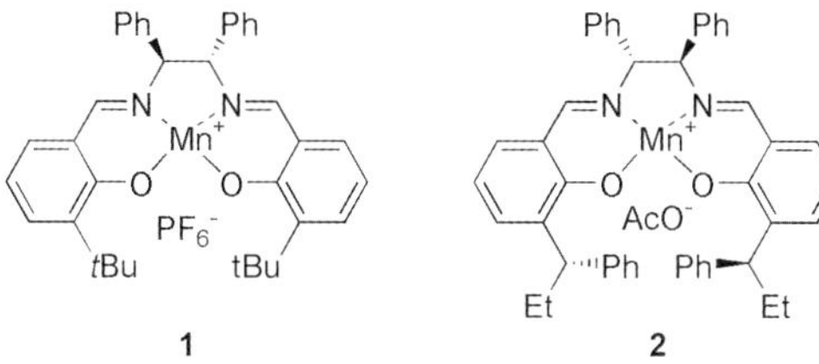

Figure 1 Initial chiral catalysts for enantioselective epoxidation reactions.

selectivity (90–>99 % ee) by introducing a new complex 3 that bears 2-phenyl-naphthyl substituents at C3 and C3' (Scheme 2) [6]. The substituents direct their 2-phenyl group toward the metal center and control the orientation of the incoming olefin, resulting in enhancement of enantioselectivity. Furthermore, we disclosed that the conformation of the salen ligand plays an important role in asymmetric induction by the Mn(salen) complex, together with steric and electronic repulsion between the salen ligand and olefin [5R]. In agreement with this proposal, it was shown, for the first time, that an achiral Mn(salen) complex could be used as the catalyst for asymmetric epoxidation if its conformation was suitably regulated by a chiral additive [5R].

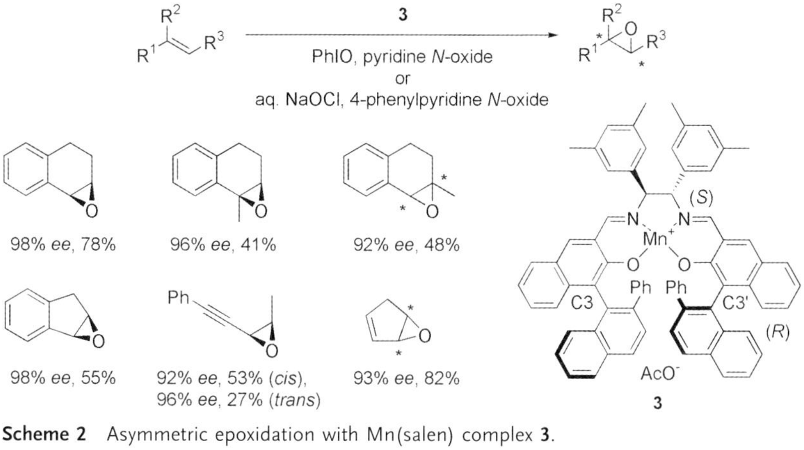

Scheme 2 Asymmetric epoxidation with Mn(salen) complex **3**.

Ruthenium(salen) Catalyst

Mn(salen) complexes are efficient catalysts for epoxidation of conjugated olefins; however, good substrates are limited to cis- and tri-substituted olefins and the epoxidation is not stereospecific (Scheme 2). On the other hand, some ruthenium-catalyzed asymmetric epoxidation reactions are known to be stereospecific, albeit with moderate selectivity. Thus, we synthesized (nitrosyl)ruthenium(salen) complex 4 and used it as catalyst [7]. It is noteworthy that complex 4 is coordinatively saturated and catalytically inactive. Fortunately, Mihara found that irradiation of visible light promotes the dissociation of the apical nitrosyl ligand and endows 4 with catalytic activity [8]. The epoxidation with 4 is stereospecific and in general shows good to high enantioselectivity, regardless of the olefinic substitution pattern: mono- to tetra-substituted olefins can be good substrates for this epoxidation (Scheme 3) [8]. To our best knowledge, complex 4 is the first molecular catalyst activated by photoirradiation.

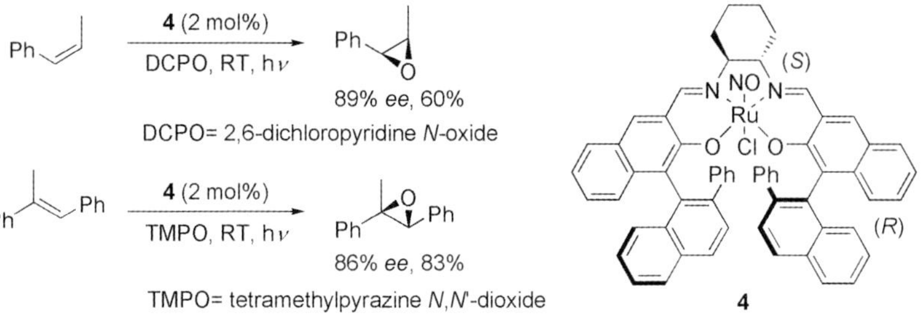

Scheme 3 Asymmetric epoxidations with Ru(salen) complex 4.

Titanium(salalen) Catalyst

From a practical point of view, metallosalen-catalyzed epoxidations still have two problems: (i) the substrates are limited to conjugated olefins, and (ii) the stoichiometric oxidants used for these oxidations are poorly atom-efficient and produce undesired by-product(s). To avoid the second problem, the use of aqueous hydrogen peroxide, that is readily available and easy to handle, is highly recommended, because its oxygen content is as high as 47% and it generates environmentally acceptable water as the sole by-product during the reaction. Thus, significant efforts have been devoted to the development of epoxidation using hydrogen peroxide. Recently, Shi et al. have reported that a fructose-derived chiral ketone serves as the catalyst for highly enantioselective epoxidation using hydrogen peroxide in acetonitrile; however, the turnover number of the catalyst is modest.

We have recently reported that di-μ-oxo titanium(salen) complex 5 is an excellent catalyst for enantioselective sulfoxidation using urea·hydrogen peroxide (UHP) as oxidant [9]. A monomeric peroxo cis-β-Ti(salen) species 6 has been proven, by ^{1}H NMR analysis, to be the active species for this sulfoxidation. Unfortunately, no epoxidation proceeds under the same conditions (Scheme 4).

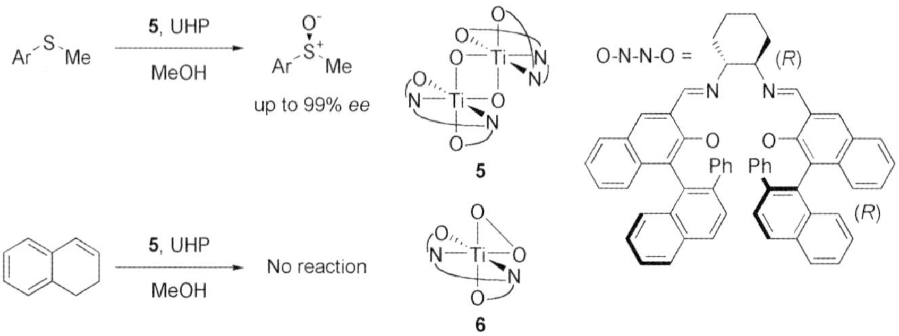

Scheme 4 Asymmetric sulfoxidation catalyzed by di-μ-oxo Ti(salen) complex 5.

The above results prompted us to synthesize a titanium complex that can generate an intermediary-activated peroxo species without diminishing the high asymmetry-inducing ability of complex **5**. Our effort led to a new di-µ-oxo titanium(salalen) complex **7** [salalen: a hybrid salen/salan tetradentate (ONNO) ligand] that efficiently catalyzes epoxidation using aqueous hydrogen peroxide as oxidant (Scheme 5) [10]. Treatment of a monomeric Ti(salen)Cl$_2$ complex with water gives the corresponding di-µ-oxo Ti(salen) complex, but we found that the Ti(salen)(OiPr)$_2$ complex underwent intramolecular Meerwein–Ponndorf–Verley reduction to give a Ti(salalen)(OiPr) complex which was converted to the di-µ-oxo complex **7** on treatment with water. Epoxidation of various olefins with complex **7** proceeds with high to excellent enantioselectivity at room temperature. The selectivity is very high and no by-product formation is observed under the optimized conditions.

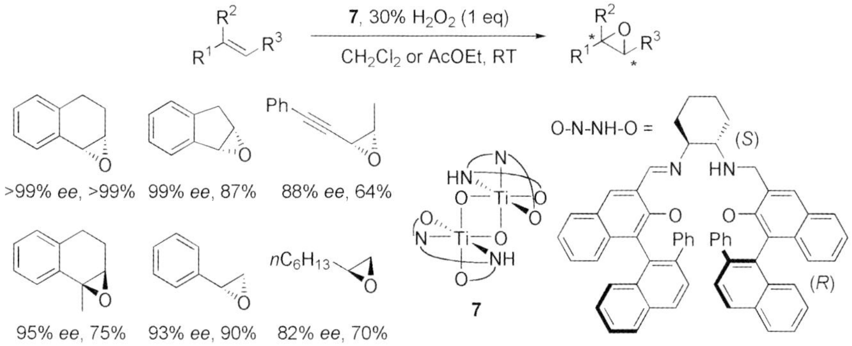

Scheme 5 Asymmetric epoxidation using hydrogen peroxide with titanium(salalen) complex **7**.

Thus, use of an almost stoichiometric amount of hydrogen peroxide is sufficient for this epoxidation and the turnover number of **7** amounted to 4600 in the epoxidation of 1,2-dihydronaphthalene. It is noteworthy that this epoxidation is stereospecific and a poorly reactive terminal and nonconjugated olefin, 1-octene, can be used as the substrate. Although the reaction mechanism is unclear at present, we have speculated that hydrogen bond formation between the amino proton of the salalen ligand and the peroxo oxygen atom participates in the activation of hydrogen peroxide. However, we are not sure whether the active peroxo species is monomeric or dimeric.

Conclusion

Thirty years ago, achievement of high enantioselectivity was a goal in synthetic chemistry. However, advancement of chemistry during the last 30 years shows us a new horizon of synthetic chemistry, namely ecologically benign, atom-efficient and sustainable synthesis. Our effort has been devoted to realizing this new goal in asymmetric epoxidation. We hope that a new route to a more ideal epoxidation extending beyond the horizon will be opened by an understanding of the catalysis of the Ti(salalen) complex.

CV of Tsutomu Katsuki

Tsutomu Katsuki was born in Saga, Japan, in 1946. He studied at Kyushu University, graduating in 1969, and received a doctoral degree in 1976 from the same university under the supervision of Professor M. Yamaguchi. He was a research associate at Kyushu University from 1971 to 1987. For two years from 1979, he was a post-doctoral fellow with Professor K. B. Sharpless at Stanford University and Massachusetts Institute of Technology, at which time he and Professor Sharpless published their paper on the asymmetric epoxidation of allylic alcohols. He has been a full Professor at Kyushu University since 1988. His current research interests are focused on asymmetric catalysis with organo-transition-metal complexes. His work has been recognized by the Synthetic Organic Chemistry Award, Japan, the Molecular Chirality Award, Japan, and the Chemical Society of Japan Award.

CV of Kazuhiro Matsumoto

Kazuhiro Matsumoto was born in Hiroshima, Japan, in 1980. He received his B.S (in 2003) and M.S. (in 2005) in chemistry from Kyushu University. He is now a doctoral student under the supervision of Professor T. Katsuki, and a Research Fellow of the Japan Society for the Promotion of Science from 2005.

Selected Publications

1. T. Katsuki, K. B. Sharpless, *J. Am. Chem. Soc.* **1980**, *102*, 5974–5976. The First Practical Method for Asymmetric Epoxidation.
2R. T. Katsuki, V. S. Martin, *Org. React.* **1996**, *48*, 1–299. Asymmetric Epoxidation of Allylic Alcohols: The Katsuki-Sharpless Epoxidation Reaction.
3R. T. Katsuki, *Coord. Chem. Rev.* **1995**, *140*, 189–214. Catalytic Asymmetric Oxidation Using Optically Active (Salen)manganese(III) Complexes as Catalysts.
4. R. Irie, K. Noda, Y. Ito, N. Matsumoto, T. Katsuki, *Tetrahedron Lett.* **1990**, *31*, 7345–7348. Catalytic Asymmetric Epoxidation of Unfunctionalized Olefins.
5R. T. Katsuki, *Synlett* **2003**, 281–297. Some Recent Advancements in Metallosalen Chemistry.
6. H. Sasaki, R. Irie, T. Hamada, K. Suzuki, T. Katsuki, *Tetrahedron* **1994**, *50*, 11827–11838. Rational Design of Mn-Salen Catalyst (2): Highly Enantioselective Epoxidation of Conjugated cis-Olefins.
7. T. Takeda, R. Irie, Y. Shinoda, T. Katsuki, *Synlett* **1999**, 1157–1159. Ru-Salen Catalyzed Asymmetric Epoxidation: Photoactivation of Catalytic Activity.
8. K. Nakata, T. Takeda, J. Mihara, T. Hamada, R. Irie, T. Katsuki, *Chem. Eur. J.* **2001**, *7*, 3776–3782. Asymmetric Epoxidation with a Photoactivated [Ru(salen)] Complex.
9. B. Saito, T. Katsuki, *Tetrahedron Lett.* **2001**, *42*, 3873–3876. Ti(salen)-catalyzed Enantioselective Sulfoxidation Using Hydrogen Peroxide as a Terminal Oxidant.
10. K. Matsumoto, Y. Sawada, B. Saito, K. Sakai, T. Katsuki, *Angew. Chem. Int. Ed.* **2005**, *44*, 4935–4939. Construction of Novel Pseudo Heterochiral- and Homochiral Di-μ-oxo Ti(Schiff-base) Dimers and Enantioselective Epoxidation Using Aqueous Hydrogen Peroxide.

Chiral Carbonyl Lewis Acid Complexes in Asymmetric Syntheses

Keiji Maruoka and Takashi Ooi, Kyoto University, Japan

Background

In 1995, we initiated a research project directed toward the design of bidentate Lewis acids possessing two metal centers aligned in the same direction with an ideal distance for capturing both of the carbonyl lone pairs (**A**). By choosing 1,8-biphenylenediol derivatives as a requisite spacer to fulfill the structural requirements and organoaluminums as an ideal main group element, in view of their high affinity toward an oxygen atom, we developed bis(organoaluminum) reagent **1** (Scheme 1) and succeeded in extracting its characteristic reactivity and selectivity in various useful synthetic transformations [1]. On the basis of this con-

Scheme 1

ceptual and methodological advance, a series of new chiral bis-titanium(IV) complexes have been devised and successfully applied to catalytic asymmetric syntheses, featuring an effective approach for obtaining high catalytic activity and enantioselectivity, as showcased in this section. Other research efforts on inventing unique carbon–carbon bond-forming reactions using chiral organoaluminum Lewis acids are also included.

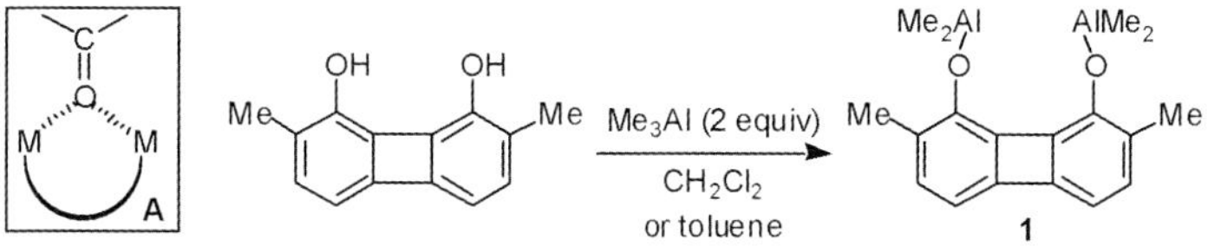

Results

A chiral titanium(IV) complex of type **2** has been designed and utilized as a practical catalyst for enabling precise enantiofacial discrimination of aldehyde carbonyls in the asymmetric allylation with allyltributyltin [2]. The crucial importance of two titanium centers is emphasized by the comparison with mono-titanium(IV) complex **3** (Scheme 2).

Asymmetric Synthesis – The Essentials.
Edited by Mathias Christmann and Stefan Bräse
Copyright © 2007 WILEY-VCH Verlag GmbH & Co. KGaA, Weinheim
ISBN: 978-3-527-31399-0

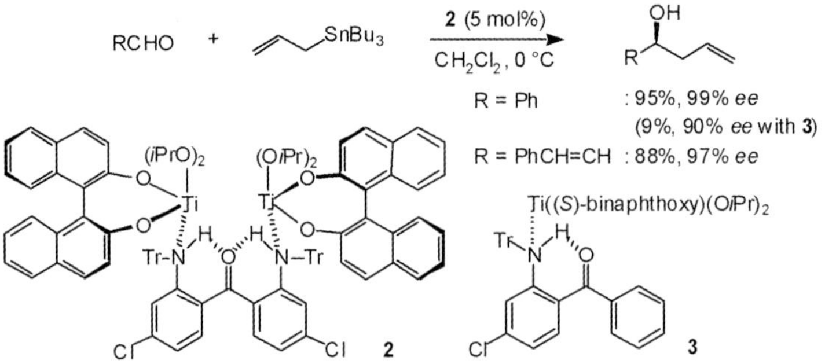

Scheme 2

$$RCHO + \text{allyl-SnBu}_3 \xrightarrow[\text{CH}_2\text{Cl}_2, 0\,°C]{\textbf{2 (5 mol\%)}} R\!\!-\!\!\text{CH(OH)CH}_2\text{CH=CH}_2$$

R = Ph : 95%, 99% ee
(9%, 90% ee with **3**)
R = PhCH=CH : 88%, 97% ee

3: Ti((S)-binaphthoxy)(OiPr)$_2$

This type of catalyst is applicable to asymmetric hetero-Diels-Alder reaction of aldehydes and Danishefsky's diene, and aldol reaction of aldehydes with ketene silyl thioacetal with high levels of enantioselectivities [3] (Scheme 3).

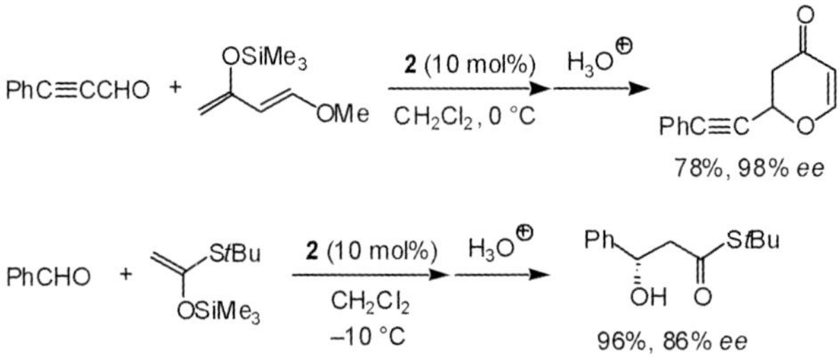

Scheme 3

$$\text{PhC≡CCHO} + \text{diene(OSiMe}_3\text{, OMe)} \xrightarrow[\text{CH}_2\text{Cl}_2, 0\,°C]{\textbf{2 (10 mol\%)}, \text{H}_3\text{O}^\oplus} \text{PhC≡C–pyranone}$$

78%, 98% ee

$$\text{PhCHO} + \text{CH}_2\text{=C(StBu)(OSiMe}_3\text{)} \xrightarrow[\substack{\text{CH}_2\text{Cl}_2 \\ -10\,°C}]{\textbf{2 (10 mol\%)}, \text{H}_3\text{O}^\oplus} \text{Ph-CH(OH)-CH}_2\text{-C(O)StBu}$$

96%, 86% ee

Interestingly, 4,6-bis(tritylamino)dibenzofuran also serves as an effective spacer, and bis-titanium(IV) complex **4** exhibits high catalytic and chiral efficiency in the asymmetric allylation of aldehydes and aryl ketones [4].

We have further demonstrated the potential utility of the M–O–M unit for the design of chiral Lewis acid catalysts by the preparation of a new, homochiral bis-titanium(IV) oxide (S,S)-**5** and its application to catalytic, highly enantioselective allylation, methallylation and propargylation of aldehydes [5, 6] (Scheme 4).

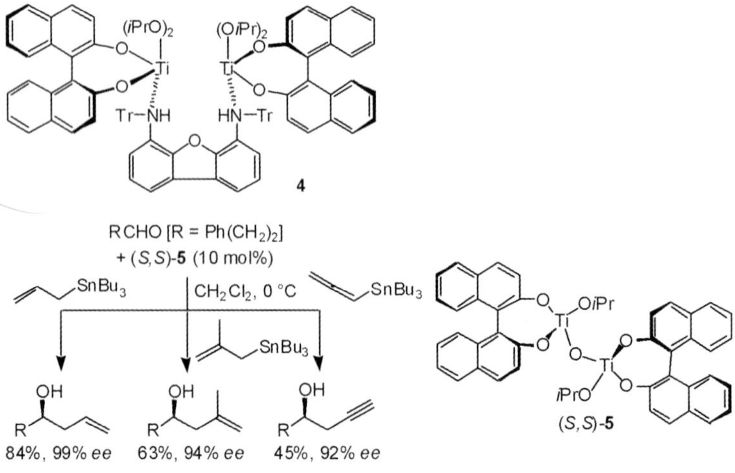

Scheme 4

RCHO [R = Ph(CH$_2$)$_2$] + (S,S)-**5** (10 mol%), CH$_2$Cl$_2$, 0 °C

with allyl-SnBu$_3$: 84%, 99% ee
with methallyl-SnBu$_3$: 63%, 94% ee
with propargyl-SnBu$_3$: 45%, 92% ee

This approach based on the utilization of the μ-oxo-type chiral bis-titanium(IV) oxide (S,S)-5 can be successfully extended to the asymmetric 1,3-dipolar cycloaddition reaction between various nitrones and acrolein, which gives rise to the corresponding isooxazolines with high to excellent enantioselectivities (Scheme 5) [7].

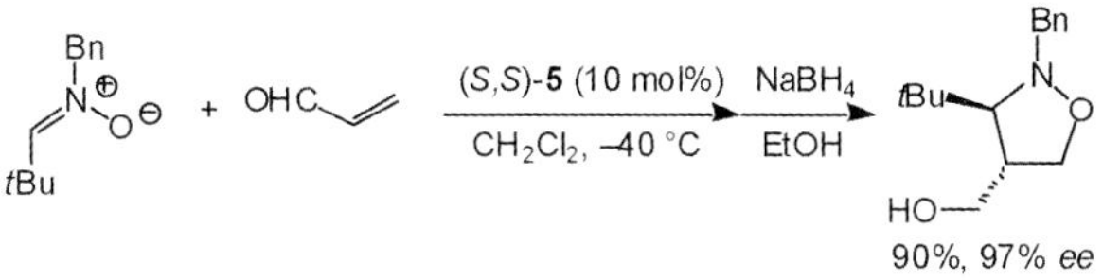

Scheme 5

On the other hand, prompted by the fact that 1 is capable of strongly activating substrates possessing the ether moiety, we designed chiral bis(organoaluminum) Lewis acid (S,S)-6 and evaluated its ability to activate the ether functionality in the asymmetric Claisen rearrangement of ally vinyl ethers (Scheme 6) [8].

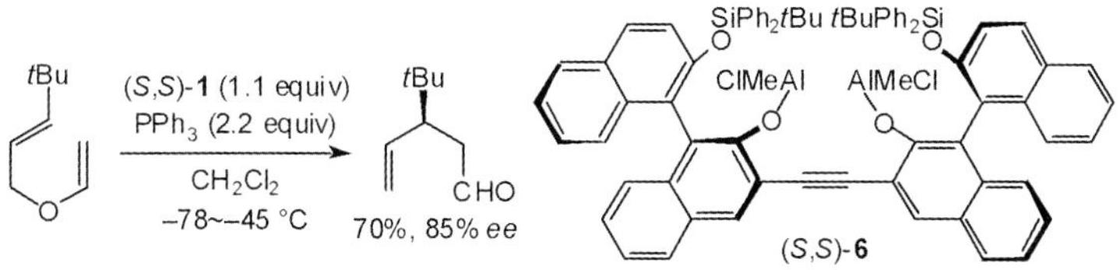

Scheme 6

Based on our discovery of the Meerwein–Ponndorf–Verley alkynylation of aldehydes, stereochemical control of this new carbonyl alkylation has been achieved by introducing an axially chiral aluminum alkoxide 7 having conformational flexibility (Scheme 7) [9].

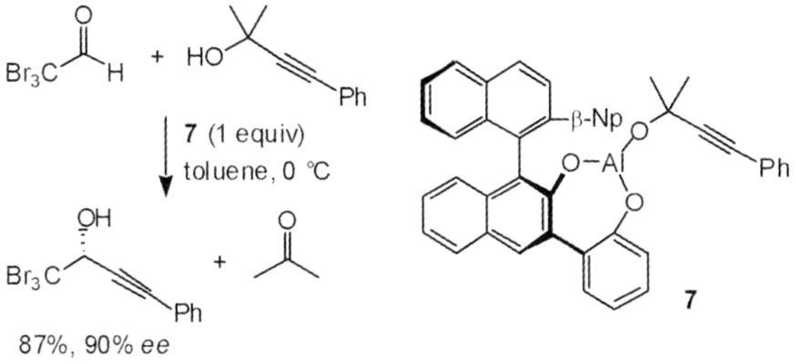

Scheme 7

We have also accomplished a new asymmetric skeletal rearrangement of symmetrically α,α-disubstituted α-amino aldehydes by the use of chiral organoaluminum Lewis acid (S)-8 as an ideal promoter, providing a facile access to optically active α-hydroxy ketones (Scheme 8) [10].

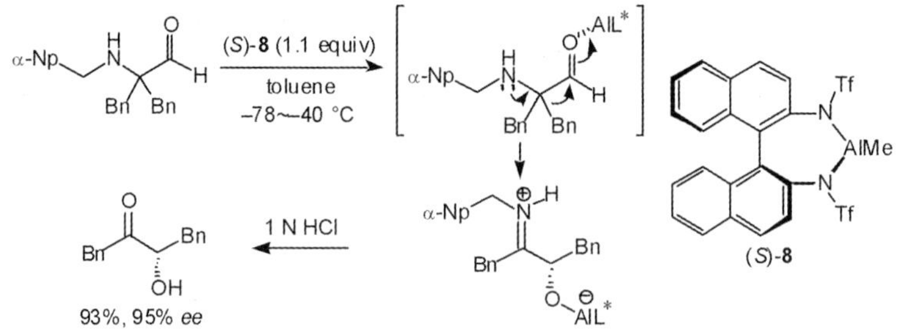

Scheme 8

Conclusions and Future Perspectives

During the past decade, we have designed several types of chiral Lewis acids bearing two metal centers within a molecule, and applied them to various synthetically useful asymmetric carbon–carbon bond-forming reactions, uncovering their potential utility in organic synthesis. This study originates from our fundamental interest in the coordination mode of Lewis acid and its chemical consequences. We believe that even deeper insight into the nature of Lewis acid–base complexation should benefit the future development of more reactive and selective chiral Lewis acid catalysts.

CV of Keiji Maruoka

Keiji Maruoka was born in 1953 in Mie, Japan. He graduated from Kyoto University (1976) and received his Ph.D. (1980) from the University of Hawaii (Thesis Director: Prof. H. Yamamoto). He became an Assistant Professor of Nagoya University (1980) and was promoted to lecturer (1985) and Associate Professor (1990) there. He moved to Hokkaido University as a full Professor (1995–2001), and currently is a Professor of Chemistry in Kyoto University. He was awarded the Japan Chemical Society Award for Young Chemist (1985), the Inoue Prize for Science (2000), the Ichimura Prize for Science (2001), and the Japan Synthetic Organic Chemistry Award (2003). He was an Associate Editor of *Chemistry Letters*, and is a member of the International Advisory Editorial Board of *Organic & Biomolecular Chemistry*. He has a wide range of research interests in synthetic organic chemistry. His current research interests include bidentate Lewis acids in organic synthesis, molecular recognition with bowl-shaped molecules, and practical asymmetric synthesis with chiral C_2-symmetric spiro-type phase-transfer catalysts.

CV of Takashi Ooi

Takashi Ooi was born in 1965 in Nagoya, Japan. He received his Ph.D. (1994) from Nagoya University under the direction of Professor Hisashi Yamamoto. He was granted a Fellowship of the Japan Society for the Promotion of Sciences (JSPS) for Japanese Junior Scientists (1992–1995), during which time he joined the group of Professor Julius Rebek, Jr. at MIT as a postdoctoral fellow (1994–1995). He was appointed as an Assistant Professor at Hokkaido University in 1995 and then promoted to a Lecturer (1998). In 2001 he moved to Kyoto University and currently is an Associate Professor of Chemistry. He was awarded the Japan Chemical Society Award for Young Chemists (1999). His current research interests are focused on the development of new and useful synthetic methodologies by designing main group metal and organic catalysts including chiral C_2-symmetric ammonium salts.

Selected Publications

1. T. Ooi, M. Takahashi, M. Yamada, E. Tayama, K. Omoto, K. Maruoka, *J. Am. Chem. Soc.* **2004**, *126*, 1150–1160. *(2,7-Disubstituted-1,8-biphenylenedioxy)bis(dimethylaluminum) as Bidentate Organoaluminum Lewis Acids: Elucidation and Synthetic Utility of the Double Electrophilic Activation Phenomenon.*
2. S. Kii, K. Maruoka, *Tetrahedron Lett.* **2001**, *42*, 1935–1939. *Practical Approach for Catalytic Asymmetric Allylation of Aldehydes with a Chiral Bidentate Titanium(IV) Complex.*
3. S. Kii, T. Hashimoto, K. Maruoka, *Synlett* **2002**, 931–932. *Catalytic, Enantioselective Hetero-Diels-Alder Reaction with Novel, Chiral Bis-Titanium(IV) Catalyst.*
4. H. Hanawa, S. Kii, K. Maruoka, *Adv. Synth. Catal.* **2001**, *343*, 57–60. *New Chiral Bis-Titanium(IV) Catalyst with Dibenzofuran Spacer for Catalytic Asymmetric Allylation of Aldehydes and Aryl Ketones.*
5. H. Hanawa, T. Hashimoto, K. Maruoka, *J. Am. Chem. Soc.* **2003**, *125*, 1708–1709. *Bis(((S)-binaphthoxy)(isopropoxy)titanium) Oxide as a μ-Oxo-Type Chiral Lewis Acid: Application to Catalytic Asymmetric Allylation of Aldehydes.*
6. H. Hanawa, D. Uraguchi, S. Konishi, T. Hashimoto, K. Maruoka, *Chem. Eur. J.* **2003**, *9*, 4405–4413. *Catalytic Asymmetric Allylation of Aldehydes and Related Reactions with Bis(((S)-binaphthoxy)(isopropoxy)titanium) Oxide as a μ-Oxo-Type Chiral Lewis Acid.*
7. T. Kano, T. Hashimoto, K. Maruoka, *J. Am. Chem. Soc.* **2005**, *127*, 11926–11927. *Asymmetric 1,3-Dipolar Cycloaddition Reaction of Nitrones and Acrolein with a Bis-Titanium Catalyst as Chiral Lewis Acid.*
8. E. Tayama, A. Saito, T. Ooi, K. Maruoka, *Tetrahedron* **2002**, *58*, 8307–8312. *Activation of ether functionality of allyl vinyl ethers by chiral bis(organoaluminum) Lewis acids: application to asymmetric Claisen rearrangement.*
9. T. Ooi, T. Miura, K. Ohmatsu, A. Saito, K. Maruoka, *Org. Biomol. Chem.* **2004**, *2*, 3312–3319. *Meewein-Ponndorf-Verley alkynylation of aldehydes: essential modification of aluminum alkoxides for rate acceleration and asymmetric synthesis.*
10. T. Ooi, A. Saito, K. Maruoka, *J. Am. Chem. Soc.* **2003**, *125*, 3220–3221. *Asymmetric Skeletal Rearrangement of Symmetrically α,α-Disubstituted α-Amino Aldehydes: A New Entry to Optically Active α-Hydroxy Ketones.*

ZACA Reaction: Zr-Catalyzed Asymmetric Carboalumination of Alkenes [1-10]

Ei-ichi Negishi, Bo Liang, Tibor Novak, and Ze Tan, Purdue University, USA

Background and Discovery

In 1995, the Zr-catalyzed asymmetric carboalumination of alkenes with Me$_3$Al was discovered [1]. Initial attempts to extend the scope of the reaction so as to achieve ethyl- and higher alkyl-alumination led to the formation of aluminacyclopentane derivatives in high yields but in a disappointingly low enantioselectivity of < 40% ee [2]. When the solvents were changed from hexanes to chlorinated hydrocarbons, such as CH$_2$Cl$_2$, and ClCH$_2$CH$_2$Cl, the course of the reaction completely changed from cyclic to acyclic, and the desired 2-alkyl-1-alkanols were obtained in good yields and in 90–95% ee [2] (Scheme 1).

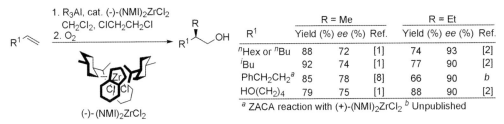

R	R = Me			R = Et		
	Yield (%)	ee (%)	Ref.	Yield (%)	ee (%)	Ref.
^{n}Hex or ^{n}Bu	88	72	[1]	74	93	[2]
^{i}Bu	92	74	[1]	77	90	[2]
PhCH$_2$CH$_2$^{a}	85	78	[8]	66	90	b
HO(CH$_2$)$_4$	79	75	[1]	88	90	[2]

^{a} ZACA reaction with (+)-(NMI)$_2$ZrCl$_2$ ^{b} Unpublished

Scheme 1 ZACA reaction with (NMI)$_2$ZrCl$_2$ as a catalyst.

The ZACA reaction represents a prototypical example of enantioselective carbon–carbon bond-forming reactions of alkenes of one-point binding, requiring no other hetero- or carbofunctional groups. It is catalytic in both Zr and chiral auxiliaries. Although there are some promising catalytic reactions for asymmetric carbon–carbon bond formation, the great majority of the widely used asymmetric carbon–carbon bond-forming reactions known today for the synthesis of deoxypolypropionates and related chiral compounds are stoichiometric. It is also instructive to compare the ZACA reaction with well-known catalytic enantioselective addition reactions of alkenes represented by Noyori reduction and Sharpless oxidation. These reactions involve C–H or C–O bond formation, often requiring geometrically defined allylic alcohols, i.e., substrates of two-point binding [13]. The relationship between the ZACA reaction and the Kaminsky version of the Ziegler-Natta alkene polymerization is intriguing. The ZACA reaction involves a controlled single-stage carboalumination with methylalanes and alkylalanes containing one to three RCH_2CH_2Al groups, whereas the polymerization reaction proceeds by a series of addition reactions of isoalkyl($R^1R^2CHCH_2$)–metal bonds, except in the very first step, leading to mixtures of polymers of different molecular weights. In the ZACA reaction, both absolute and relative configurations are critically important, whereas only diastereochemistry, i.e., tacticity, is critically important in the alkene polymerization. Finally, some seemingly related Zr-catalyzed asymmetric reactions of alkenes with alkylmagnesium reagents and chiral zirconocene catalysts are also known, but these alkylmagnesium reactions are discrete from the ZACA reaction in the overall transformation, synthetic scope, and, most certainly, mechanism. Most critically, methylmagnesiums cannot be used in the alkylmagnesium reactions, and n-propyl- and higher alkylmagnesium reagents generally lead to rather low product yields [13].

Results
ZACA–Pd-Catalyzed Cross-coupling Tandem Processes for the Synthesis of Deoxypolypropionates and Related Compounds

Despite the somewhat moderate *ee* figures associated with methylalumination, the development of a few synthetic protocols and the exploitation of the well-known principle of statistical enantiomeric amplification have led to some efficient, selective, and practical processes for the synthesis of both α-monoheterofunctional and α,ω-diheterofunctional deoxypolypropionates and related compounds containing two or more asymmetric carbon atoms [5–10]. The principle of statistical enantiomeric amplification predicts that, in cases where there are two asymmetric carbon atoms in the target compounds, *use or generation of two asymmetric carbon centers of a combined total enantiomeric excess of ≥ 160% will lead to the overall enantiomeric purity of ≥ 98% ee and maximally possible yields of ≥ 80%*. The presence of three or more asymmetric carbon centers in the target molecules will make it possible to readily synthesize them in enantiomerically very pure forms (≥ 99.9% *ee*).

Several other breakthroughs leading to the current level of development of the ZACA-based asymmetric synthetic method include (i) the realization that Me-branched chiral compounds can be synthesized by the ZACA reaction by at least three discrete routes [4, 5] (Scheme 2), as exemplified by the results shown in Scheme 3 [5]; (ii) the unexpected finding that the presence of a free hydroxy group readily permits diastereomeric separation at the C2 and C4 positions relative to OH by ordinary chromatography [5–10]; and (iii) the development of a "one-pot" ZACA–Pd-catalyzed cross-coupling tandem process [9, 10] (Scheme 4).

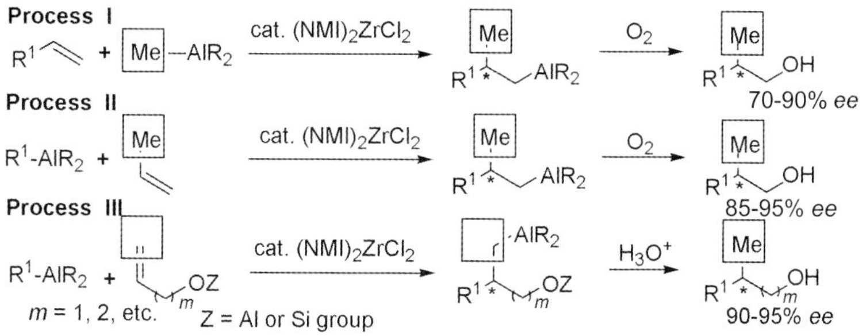

Scheme 2 Three processes for the synthesis of Me-branched 1-alkanols.

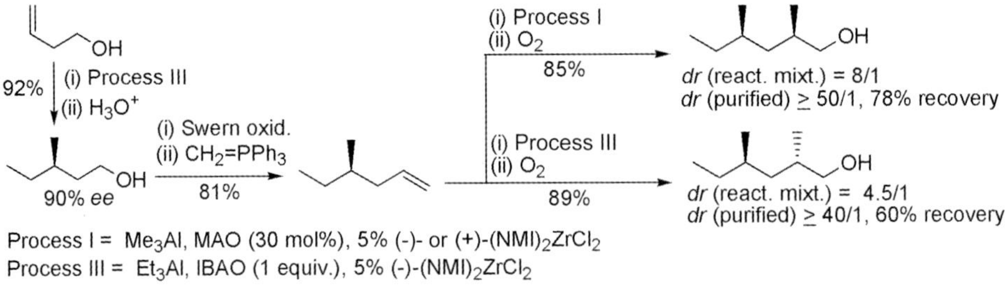

Scheme 3 Combined use of Processes I and III for the synthesis of 2,4-dimethyl-1-hexanols.

Process I = Me$_3$Al, MAO (30 mol%), 5% (-)- or (+)-(NMI)$_2$ZrCl$_2$
Process III = Et$_3$Al, IBAO (1 equiv.), 5% (-)-(NMI)$_2$ZrCl$_2$

For the synthesis of α,ω-diheterofunctional deoxypolypropionates, silyl-protected 2-methyl-4-penten-1-ols were initially prepared in 4 steps from the so-called Roche ester [6]. However, a couple of more efficient and potentially economical protocols based on the ZACA reaction of styrene [9] and allyl alcohol [10] have been developed, as shown in Schemes 5 and 6.

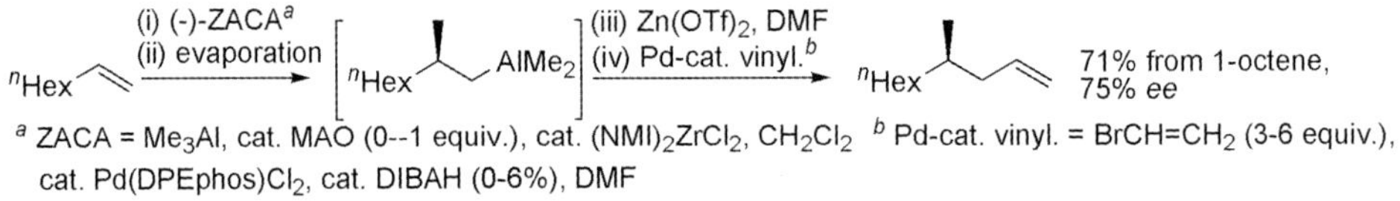

^{a} ZACA = Me$_3$Al, cat. MAO (0–1 equiv.), cat. (NMI)$_2$ZrCl$_2$, CH$_2$Cl$_2$ ^{b} Pd-cat. vinyl. = BrCH=CH$_2$ (3-6 equiv.), cat. Pd(DPEphos)Cl$_2$, cat. DIBAH (0-6%), DMF

Scheme 4 "One-pot" ZACA–Pd-catalyzed vinylation for iterative homologation of deoxypolypropionates.

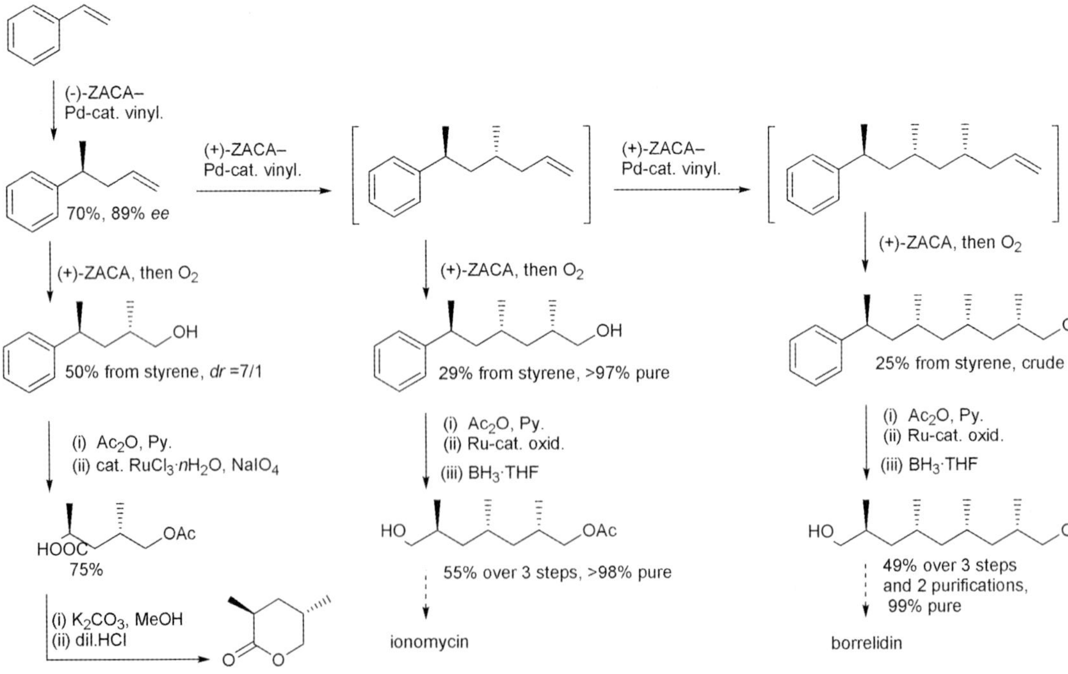

Scheme 5 Styrene-based protocol for the synthesis of α,ω-diheterofunctional deoxypolypropionates.

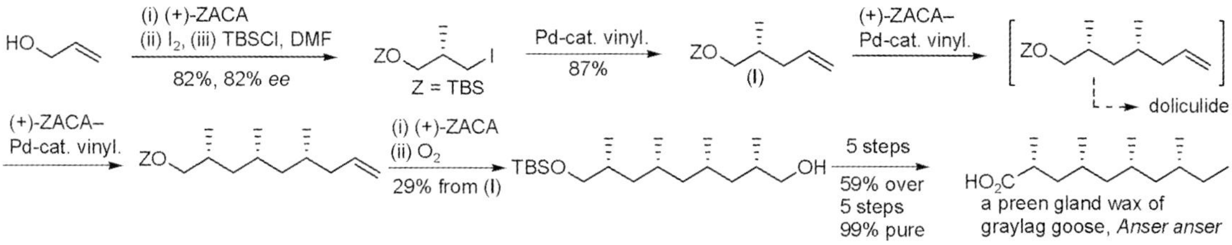

Scheme 6 Allyl alcohol-based route to α,ω-diheterofunctional deoxypolypropionates.

Synthesis of 2-Methyl- or 2-Ethyl-1-alkanols Not Readily Purifiable by Chromatography or Recrystallization

Until recently, one of the major pending issues was how to readily purify chiral compounds that contain just one asymmetric carbon center. Reduced isoprenoids, such as vitamins E and K [3, 4], in which two adjacent asymmetric centers are sufficiently remote from each other are also difficult to purify. Lipase-catalyzed selective acetylation has been shown to be generally efficient for purification of 2-methyl-1-alkanols or even 2-ethyl-1-alkanols. The combined use of the ZACA reaction of 75–95 % ee and the lipase-catalyzed selective acetylation of (S)-2-methyl-1-alkanols permits not only the production of (R)-2-methyl-1-alkanols of ≥98 % ee in high recovery (60–80 %) but also the preparation of the acetates of (S)-2-methyl-1-alkanols of similarly high ee figures in good yields (Figure 1), although the applicability of the latter is more limited.

[a] Isolated yields of ≥97-98% pure compounds based on the starting alkenes. [b] Obtained after deacetylation.

Figure 1 Preparation of pure 2-methyl-1-alkanols via ZACA–lipase-catalyzed acetylation.

CV of Ei-ichi Negishi

Ei-ichi Negishi grew up in Japan and received his bachelor's degree from the University of Tokyo (1958). He then joined a chemical company, Teijin. In 1960 he came to the University of Pennsylvania on a Fulbright Scholarship and obtained his Ph.D. degree in 1963. He returned to Teijin but joined Professor H. C. Brown's Laboratories at Purdue as a Postdoctoral Associate in 1966. He was appointed Assistant to Professor Brown in 1968. He went to Syracuse University as Assistant Professor in 1972 and was promoted to Associate Professor in 1976. In 1979 he was invited back to Purdue University as Full Professor. In 1999 he was appointed the inaugural H. C. Brown Distinguished Professor of Chemistry. Various awards he has received include a Guggenheim Fellowship (1987), the 1996 A. R. Day Award, a 1997 Chemical Society of Japan Award, the 1998 ACS Organometallic Chemistry Award, an Alexander von Humboldt Senior Researcher Award, Germany (1998–2001), and the 2000 RSC Sir E. Frankland Prize Lectureship, UK.

Selected Publications

1. D. Kondakov, E. Negishi, *J. Am. Chem. Soc.* **1995**, *117*, 10771–10772. Zirconium-catalyzed enantioselective methylalumination of monosubstituted alkenes.
2. D. Kondakov, E. Negishi, *J. Am. Chem. Soc.* **1996**, *118*, 1571–1572. Zirconium-catalyzed enantioselective alkylalumination of monosubstituted alkenes proceeding via noncyclic mechanism.
3. S. Huo, E. Negishi, *Org. Lett.* **2001**, *3*, 3253–3256. A convenient and asymmetric protocol for the synthesis of natural products containing chiral alkyl chains via Zr-catalyzed carboalumination of alkenes. syntheses of phytol and vitamins E and K.
4. S. Huo, J. Shi, E. Negishi, *Angew. Chem. Int. Ed.* **2002**, *41*, 2141–2143. A new protocol for the enantioselective synthesis of methyl-substituted alkanols and their derivatives through a hydroalumination/zirconium-catalyzed alkylalumination tandem process.
5. E. Negishi, Z. Tan, B. Liang, T. Novak, *Proc. Natl. Acad. Sci. USA* **2004**, *101*, 5782–5787. An efficient and general route to reduced polypropionates via Zr-catalyzed asymmetric CC bond formation.
6. Z. Tan, E. Negishi, *Angew. Chem. Int. Ed.* **2004**, *43*, 2911–2914. An efficient and general method for the synthesis of α, ω-difunctional reduced polypropionates by Zr-catalyzed asymmetric carboalumination: synthesis of the scyphostatin side chain.
7. M. Magnin-Lachaux, Z. Tan, B. Liang, E. Negishi, *Org. Lett.* **2004**, *6*, 1425–1427. Efficient and selective synthesis of siphonarienolone and related reduced polypropionates via Zr-catalyzed asymmetric carboalumination.
8. X. Zeng, F. Zeng, E. Negishi, *Org. Lett.* **2004**, *6*, 3245–3248. Efficient and selective synthesis of 6, 7-dehydrostipiamide via Zr-catalyzed asymmetric carboalumination and Pd-catalyzed cross-coupling of organozincs.
9. T. Novak, Z. Tan, B. Liang, E. Negishi, *J. Am. Chem. Soc.* **2005**, *127*, 2838–2839. All catalytic, efficient, and asymmetric synthesis of α, ω-difunctional reduced polypropionates via "one-pot" Zr-catalyzed asymmetric carboalumination Pd-catalyzed cross-coupling tandem process.
10. B. Liang, T. Novak, Z. Tan, E. Negishi, *J. Am. Chem. Soc.* **2006**, *128*, 2770–2771. Catalytic, efficient, and syn-selective construction of deoxypolypropionates and other chiral compounds via Zr-catalyzed asymmetric carboalumination of allyl alcohol.
11. G. Erker, M. Aulbach, M. Knickmeier, D. Wingbermühle, C. Krüger, M. Nolte, S. Werner, *J. Am. Chem. Soc.* **1993**, *115*, 4590–4601. The role of torsional isomers of planarly chiral nonbridged bis(indenyl)metal type complexes in stereoselective propene polymerization.
12. P. Wipf, S. Ribe, *Org. Lett.* **2000**, *2*, 1713–1716. Water /MAO acceleration of the zirconocene-catalyzed asymmetric methylalumination of α-olefin.
13. For pertinent references, see: (a) Ref. [5]. (b) E. Negishi, *Dalton Trans.* **2005**, *5*, 827–848. A quarter of a century of explorations in organozirconium chemistry.

Bisoxazolines – a Privileged Ligand Class for Asymmetric Catalysis

Andreas Pfaltz, Department of Chemistry, University of Basel, Switzerland

Background

In the mid 1980s, we developed a new class of chiral C_2-symmetric N,N-ligands, the semicorrins **1** (Figure 1) [1A, 2]. Both the structure and synthesis of these ligands were inspired by Eschenmoser's work on corrins. For various reasons we felt that the semicorrins would be ideally suited for the enantiocontrol of metal-catalyzed reactions.

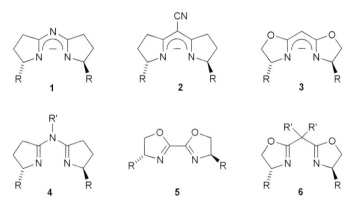

Figure 1 The semicorrin-bisoxazoline ligand family.

The two substituents at the stereogenic centers are held in close proximity to the coordination site by the rigid ligand scaffold and, therefore, should have a strong and direct influence on the stereochemical course of a metal-catalyzed process. In addition, because these substituents are derived from the carboxyl group of pyroglutamic acid, their structure can be readily altered. In this way, the ligand structure can be adjusted to the specific requirements of a particular reaction.

Asymmetric Synthesis – The Essentials.
Edited by Mathias Christmann and Stefan Bräse
Copyright © 2007 WILEY-VCH Verlag GmbH & Co. KGaA, Weinheim
ISBN: 978-3-527-31399-0

Results

Semicorrins

Two successful applications, enantioselective Cu-catalyzed cyclopropanation and Co-catalyzed conjugate reduction of α,β-unsaturated carboxylic esters and amides (Scheme 1), clearly demonstrated the potential of the semicorrins.

The promising results obtained with semicorrins prompted us and a number of other research groups to develop additional, structurally related ligands such as the aza-semicorrins **2** and **4** and the bisoxazolines **3**, **5**, and **6** (Figure 1) [1, 2]. All these ligands provide a chiral environment around a coordinated metal ion, which is similar to that of the semicorrins. Ligands **1–6** can be divided into two distinct classes, anionic and neutral ligands. The former (**1**, **2**, and **3**) possess an electron-rich conjugated π-system and, therefore, are expected to act as π- and σ-donors, which reduce the electrophilicity of a metal ion. Thus, for applications requiring an electrophilic metal ion, neutral analogs such as **4**, **5**, and **6** are better suited.

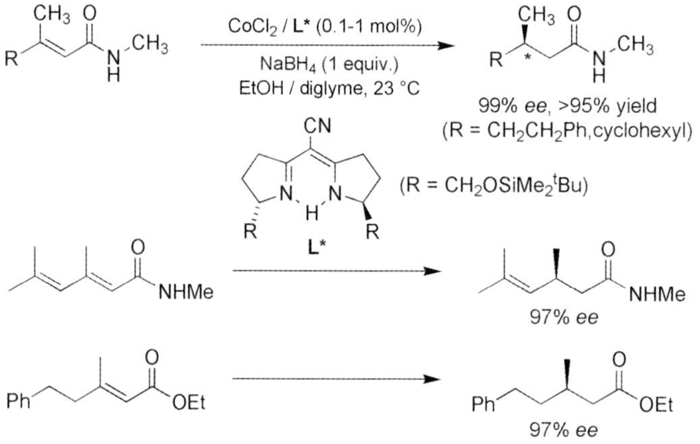

Scheme 1 Enantioselective conjugate reduction with cobalt-semicorrin catalysts.

C$_2$-Symmetric Bisoxazolines

Bisoxazolines such as **3**, **5**, and **6** are attractive ligands because chiral oxazoline units can be easily prepared in enantiomerically pure form starting from readily available amino alcohols. The potential of C$_2$-symmetric bisoxazolines was recognized around 1990 by many researchers, including the groups of Masamune, Evans, Corey, Lehn, Helmchen, Onishi, Dai, and our group [1]. As a consequence of this sudden interest in this class of ligands, the list of applications of bisoxazoline-metal complexes has rapidly grown [3R, 4R]. In particular the neutral bisoxazolines **6** (BOX ligands), now belong to the most widely used ligands in asymmetric catalysis. They have been successfully employed in an impressive range of different enantioselective metal-catalyzed processes such as cyclopropanation, Lewis acid-catalyzed cycloadditions and ene reactions, Mukaiyama aldol reactions, Michael additions, as well as oxidations and reductions.

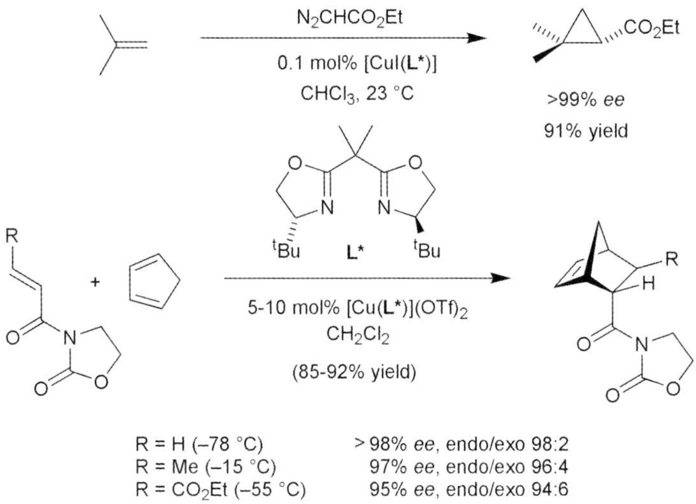

Scheme 2 Enantioselective cyclopropanation and Diels-Alder reaction with Cu-BOX catalysts (Evans et al. [1, 5]).

Evans and coworkers found that cationic copper(I)-BOX complexes are highly efficient catalysts for the asymmetric cyclopropanation of olefins with diazo compounds that outperform neutral complexes with anionic semicorrin-type ligands [1]. Subsequently they discovered that analogous Cu(II)-BOX complexes are remarkably active Lewis acids, which catalyze Diels-Alder reactions and many other Lewis acid-promoted transformations with high enantioselectivity [3R, 4R, 5, 6].

In addition to the ligands shown in Figure 1, many other types of C_2-symmetric bisoxazolines have been developed, such as aza-bisoxazolines with a bridging N-atom as in ligands 1 and 4 [7], as well as tri- and tetradentate variants and ligands containing two metal binding sites [2, 3R, 4R]. Among these, Nishiyama's tridentate bisoxazolines (PyBOX; Figure 2), which are derived from pyridine-2,6-biscarboxylate, should be mentioned as particularly useful ligands [8R]. What all these ligands have in common is that they are readily prepared and, because of their modular nature, the structures can be easily varied.

In addition, since the pioneering work of Brunner on chiral pyridine-oxazoline ligands [9] many classes of nonsymmetrical chiral oxazoline-based ligands have been reported including P,N-, S,N- and other mixed-donor ligands [4R]. During the last 10 years we have extensively studied phosphino-oxazoline (PHOX) ligands (Figure 2), which have emerged as new class of privileged ligands [2, 10, 11].

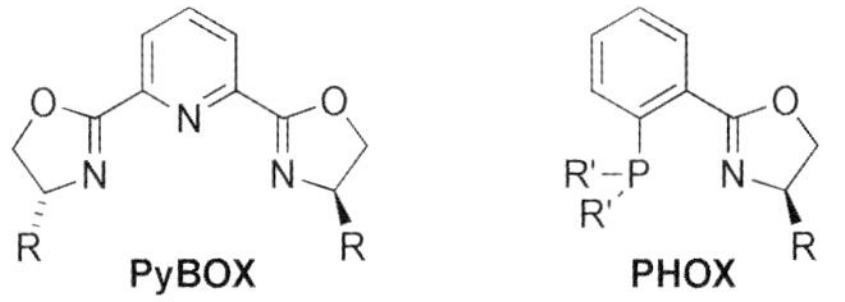

Figure 2 Structures of PyBOX and PHOX ligands.

Boron-bridged Bisoxazolines

Recently, we have added a new member to the bisoxazoline family, anionic bisoxazoline ligands **8** (BoraBOX) that contain a tetrasubstituted boron atom bridging the two oxazoline rings [12]. In contrast to the known anionic ligands such as **1**, **2**, and **3**, the negative charge is located in the backbone and the geometry of the bridging atom is tetrahedral, as in the neutral BOX ligand **6**. BoraBOX ligands are readily accessible by reaction of a metalated oxazoline with a dialkyl- or diaryl-haloborane (Scheme 3). The potential of this ligand class was demonstrated in the Cu-catalyzed monobenzoylation of *meso* 1,2-diols (Scheme 4) [12]. The best results were obtained with BoraBOX derivative **9a**, which outperformed analogous BOX ligands [13] in this reaction.

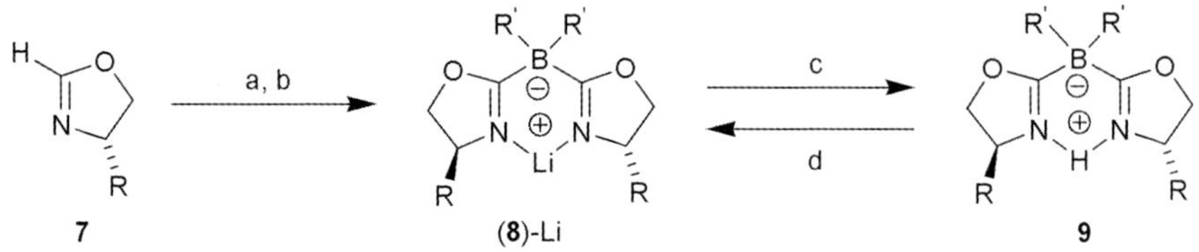

Scheme 3 Synthesis of BoraBOX ligands. (a) *tert*BuLi, THF, −78 °C; (b) R′$_2$BX (X=Cl, Br), toluene, −78 °C; (c) silica gel, EtOAc/hexanes/NEt$_3$; (d) BuLi, THF, 0 °C.

Conclusion

During the last 15 years, C_2-symmetric bisoxazolines have evolved into one of the most efficient and most widely used ligand classes in asymmetric catalysis. Because they are readily available from simple precursors and the substituents at the stereogenic centers can be easily modified, we expect that many further applications of these ligands will be found.

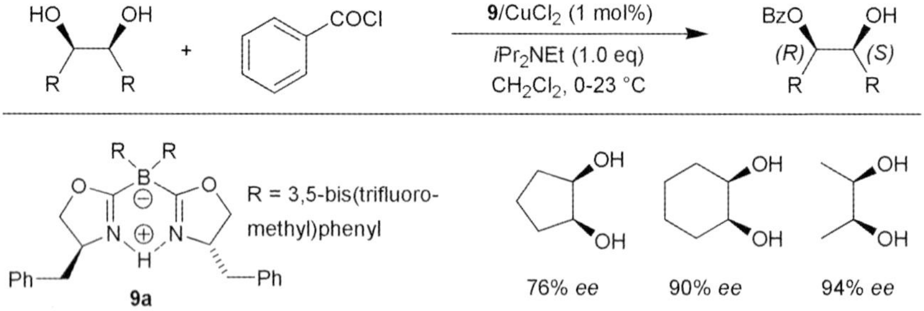

Scheme 4 Enantioselective monobenzoylation of *meso* 1,2-diols.

CV of Andreas Pfaltz

Andreas Pfaltz was born in Basel, Switzerland, in 1948. He studied at the ETH in Zürich where he received his Ph.D. in 1978, working under the direction of Albert Eschenmoser. After postdoctoral research with Gilbert Stork at Columbia University, he returned to the ETH where he was appointed 'Privatdozent' in 1987. From 1990 to 1995 he was Professor of Organic Chemistry at the University of Basel. After four years as a director at the Max-Planck-Institut für Kohlenforschung in Mülheim an der Ruhr, he returned to the University of Basel in 1999. He received the Horst Pracejus Prize of the German Chemical Society and the Prelog Medal of the ETH in 2003.

Selected Publications

1A. A. Pfaltz, *Acc. Chem. Res.* **1993**, *26*, 339–345. *Chiral Semicorrins and Related Nitrogen Heterocycles as Ligands in Asymmetric Catalysis.*

2. A. Pfaltz, *Synlett* **1999**, 835–842. *From Corrin Chemistry to Asymmetric Catalysis – A Personal Account.*

3R. A. K. Gosh, P. Mathivanan, J. Capiello, *Tetrahedron: Asymmetry* **1998**, *9*, 1–45. C_2-*Symmetric Chiral Bis(oxazoline)-Metal Complexes in Catalytic Asymmetric Synthesis.*

4R. H. A. McManus, P. J. Guiry, *Chem. Rev.* **2004**, *104*, 4151–4502. *Recent Developments in the Application of Oxazoline-Containing Ligands in Asymmetric Catalysis.*

5. J. S. Johnson, D. A. Evans, *Acc. Chem. Res.* **2000**, *33*, 325–335. *Chiral Bis(oxazoline) Copper(II) Complexes: Versatile Catalysts for Enantioselective Cycloaddition, Aldol, Michael, and Carbonyl Ene Reactions.*

6. K. A. Jørgensen, M. Johannsen, S. Yao, H. Audrain, J. Thorhauge, *Acc. Chem. Res.* **1999**, *32*, 605–613. *Catalytic Asymmetric Addition Reactions of Carbonyls. A Common Catalytic Approach.*

7. A. Gissibl, M. G. Finn, *Org. Lett.* **2005**, *7*, 2325–2328. *Cu(II)-Aza(bisoxazoline)-Catalyzed Asymmetric Benzoylations.*

8R. G. Desimoni, G. Faita, P. Quadrelli, *Chem. Rev.* **2003**, *103*, 3119–3154. *Pyridine-2,6-bis(oxazolines), Helpful Ligands for Asymmetric Catalysis.*

9. H. Brunner, U. Obermann, P. Wimmer, *J. Organomet. Chem.* **1986**, *316*, C1–C3. *Enantioselektive Phenylierung von cis-Cyclohexan-1,2-diol und meso-Butan-2,3-diol.*

10. G. Helmchen, A. Pfaltz, *Acc. Chem. Res.* **2000**, *33*, 336–345. *Phosphinooxazolines – A New Class of Versatile, Modular P,N-Ligands for Asymmetric Catalysts.*

11. J. M. J. Williams, *Synlett* **1996**, 705–710. *The Ups and Downs of Allylpalladium Complexes in Catalysis.*

12. C. Mazet, V. Köhler, A. Pfaltz, *Angew. Chem.* **2005**, *117*, 4966–4969; *Angew. Chem. Int. Ed. Engl.* **2005**, *44*, 4888–4891. *Chiral Boron-Bridged Bisoxazolines: Readily Available Anionic Ligands for Asymmetric Catalysis.*

13. Y. Matsumura, T. Maki, S. Murakami, O. Onumura, *J. Am. Chem. Soc.* **2003**, *125*, 2052–2053. *Copper Ion-Induced Activation and Asymmetric Benzoylation of 1,2-Diols: Kinetic Chiral Molecular Recognition.*

Enantioselective Cycloaddition Reactions Catalyzed by Hydrogen Bonding

Viresh H. Rawal and Avinash N. Thadani, University of Chicago, USA

Background

The term "hydrogen bond" was coined in 1931 by Linus Pauling in his seminal work on the nature of the chemical bond. One of the central forces in nature, this non-covalent interaction is observed universally. Individually, hydrogen bonds are relatively weak when compared to covalent bonds, but collectively they are of enormous importance. Hydrogen bonds constitute the "glue" that holds water molecules together and sustains the three-dimensional structures of large biomolecules. As such, they play a crucial role in the function of large biomolecules such as enzymes. Despite their profound impact on the structure and function of living systems, hydrogen bonds have been little utilized for the promotion of chemical reactions. Indeed, it is only within the past five years that chemists have successfully begun to exploit this ubiquitous force for the catalysis of various chemical reactions. We summarize below some of the key developments in our program in hydrogen bond catalyzed enantioselective reactions.

Our first foray in hydrogen bond catalysis was inspired by a lecture given at Ohio State University (1991) by the late Professor Margaret C. Etter from the University of Minnesota. Based on her work on the design of ordered crystalline solids, we examined the use of diaryl-ureas as hydrogen bond catalysts for a Diels-Alder (DA) reaction. Although the study validated the basic concept, the rate accelerations were modest (ca. 2-fold rate increase with 2% catalyst), and further studies in this area were put off for a few years. Our current work in hydrogen bond catalysis arose from our studies of aminosiloxy dienes **1**, particularly as applied to the total synthesis of alkaloids, such as tabersonine **5** (Scheme 1) [1–3].

In parallel with the DA work, we also investigated hetero Diels-Alder (HDA) reactions of amino-siloxy diene **1**. The diene was found to be exceptionally reactive to hetero-dienophiles, and underwent HDA reactions with various aldehydes at room temperature. Significantly, in contrast to nearly all reported carbonyl HDA reactions, those with diene **1** proceeded in the absence of any added catalyst (Scheme 2). Subsequent treatment of the intermediate cycloadducts **6** with acetyl chloride afforded the corresponding dihydro-4-pyrones **7** in good overall yields. Further studies of this reaction revealed a pronounced solvent effect. The reaction

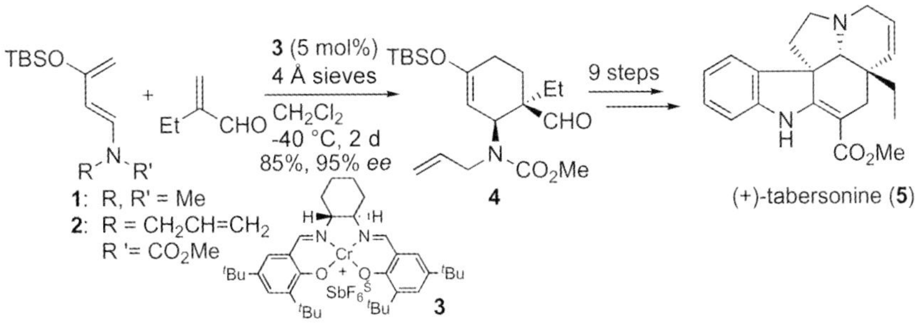

Scheme 1 Application of amino-siloxy dienes in natural product synthesis.

between **1** and anisaldehyde was significantly faster in chloroform (~10–30 ×) than in other common aprotic organic solvents, even polar solvents. The rate difference was greater still in alcohols. The higher rates in chloroform and alcohols were attributed to the presence of hydrogen bonds between the H-bond donating solvents and the carbonyl oxygen. This Lewis acid type of activation lowers the LUMO energy of the heterodienophile and, thereby, reduces the HOMO–LUMO gap for the cycloaddition. This H-bond promoted cycloaddition works even for simple ketones, which are generally considered to be unreactive in HDA reactions. The choice of the alcoholic solvent for the reaction proved critical. More acidic alcohols, such as methanol, gave a faster reaction, but the solvolytic lability of diene **1** precluded their use. The HDA reactions of a range of cyclic, acyclic, and heterecyclic ketones and **1** were successfully carried out in 2-butanol, with minimal diene decomposition (Scheme 3) [4].

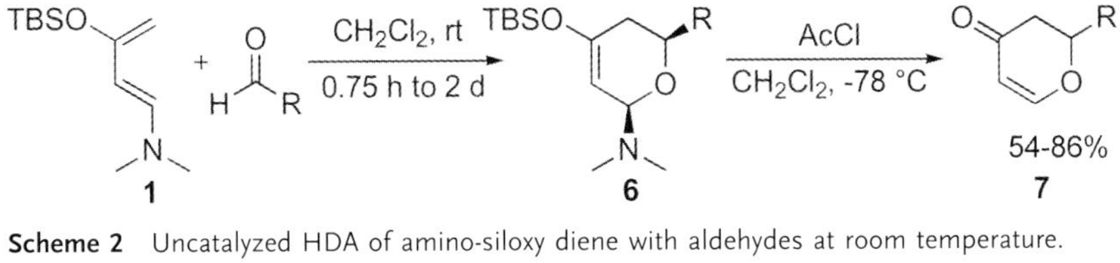

Scheme 2 Uncatalyzed HDA of amino-siloxy diene with aldehydes at room temperature.

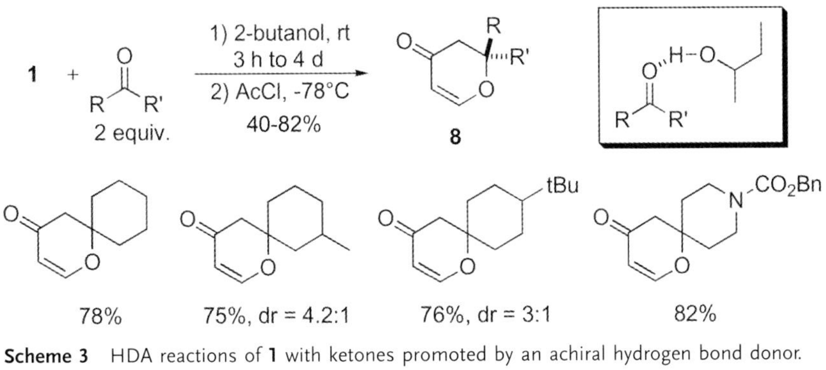

Scheme 3 HDA reactions of **1** with ketones promoted by an achiral hydrogen bond donor.

Strategy

The next logical step in our program was to examine the use of chiral hydrogen bond donors as catalysts for enantioselective reactions. After a brief examination of common alcohols, we focused on TADDOLs (α,α,α′,α′-tetraaryl-2,2-dimethyl-1,3-dioxolan-4,5-dimethanol), the tartrate derived class of diols popularized by Seebach and coworkers [5]. These C2-symmetric chiral diols form inclusion complexes with a wide variety of hydrogen bond acceptors. In the solid state, TADDOLs appear locked in well-defined conformations, with one of the hydroxyl groups internally-hydrogen bonded to the other, leaving the other hydroxyl group available for intermolecular hydrogen bonding (Figure 1). The opposing pairs of aryl groups exist in pseudo-axial and pseudo-equatorial orientations.

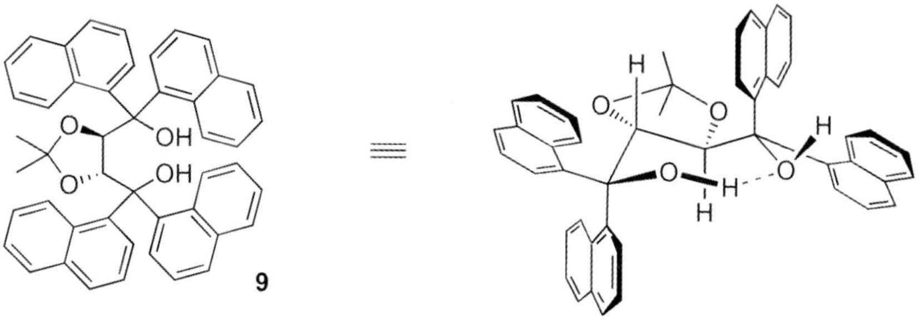

Figure 1 Internally hydrogen-bonded conformation of 1-naphthyl TADDOL **9**.

Results

We initially examined commercially available TADDOLs as hydrogen bond catalysts for the HDA reaction between **1** and benzaldehyde (Scheme 4). The best results were obtained with 1-naphthyl TADDOL **9**, which afforded the desired dihydropyranone product in 70 % yield and >98 % *ee*. Other aromatic aldehydes also afforded high enantioselectivities [6].

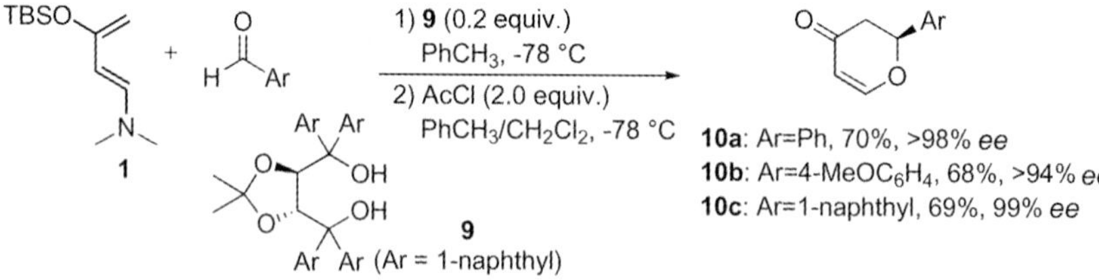
1) 9 (0.2 equiv.)
PhCH3, -78 °C
2) AcCl (2.0 equiv.)
PhCH3/CH2Cl2, -78 °C
10a: Ar=Ph, 70%, >98% ee
10b: Ar=4-MeOC6H4, 68%, >94% ee
10c: Ar=1-naphthyl, 69%, 99% ee
Ar Ar (Ar = 1-naphthyl)

Scheme 4 TADDOL-catalyzed HDA reactions between **1** and aromatic aldehydes.

The encouraging results of the HDA reactions were rationalized using the crystal structures of TADDOLs as a starting point. The internal hydrogen bond was expected to render the second hydroxyl group more acidic, hence a stronger hydrogen bond acceptor. The resulting hydrogen-bonded carbonyl group – an electron deficient π-bond – may be further stabilized by an electrostatic, donor–acceptor interaction with the electron-rich distal ring of the naphthalene unit. This stabilizing interaction serves not only to increase the population of the activated aldehyde, but also to block one face of the aldehyde.

Application to Other Reactions

Based on the successful results summarized above, we have explored the use of hydrogen bonding for the catalysis of several other enantioselective reactions. For example, we have found that the Diels-Alder reactions of diene **1** are also promoted by TADDOL **9** (Figure 6) [7]. We have also found that the vinylogous Mukaiyama aldol reaction of silyl-dienolate **12** and ethyl glyoxalate can be catalyzed by various hydrogen bond donors. The best results were again obtained using TADDOL **9** [8, 9].

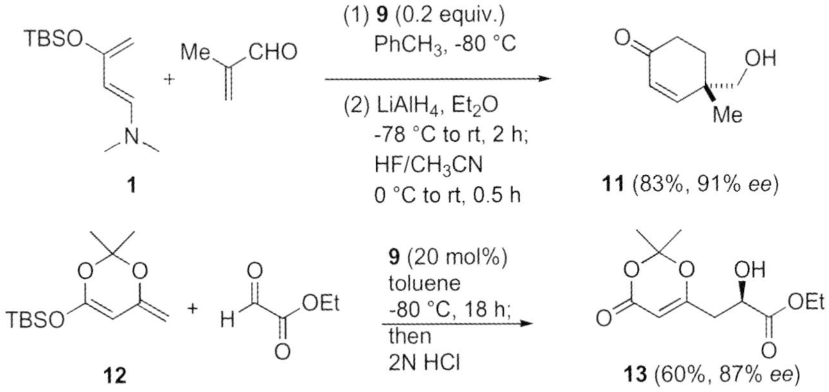

Figure 2 Hydrogen bonding catalyzed Diels-Alder and vinylogous Mukaiyama aldol reactions.

Conclusions and Future Perspectives

Enantioselective reactions have traditionally been catalyzed by metal-based Lewis acids. The results summarized above show that many enantioselective reactions can also be catalyzed by simple, chiral hydrogen bond donors, wherein the proton functions as the Lewis acid, in place of the metal. While it represents to be a new frontier in asymmetric catalysis, hydrogen bond catalysis involves well-established principles of catalysis and reactivity. For example, the strength of the catalyst reflects the pK_a of the acidic hydrogen. Also, given that the free ligands of many effective chiral Lewis acid catalysts possess an acidic hydrogen, and the fact that the hydrogen occupies approximately the same position as the metal, then it follows that the metal-free ligands will have a good chance of promoting asymmetric reactions. Based on this analysis, the prospects for this field are very good indeed.

CV of Viresh Rawal

Viresh Rawal was born in India in 1958 and emigrated with his family to the United States in 1968, settling in Connecticut. He received his B.S. degree from the University of Connecticut and his Ph.D. from the University of Pennsylvania (with M. P. Cava). From 1986 to 1988 he was a postdoctoral associate at

Columbia University in the laboratories of Professor Gilbert Stork. He began his independent research career at Ohio State University in 1988 and was promoted to Associate Professor in 1994. In the autumn of 1995 he moved to the University of Chicago, where he is Professor of Chemistry.

Selected Publications

1. S. A. Kozmin, V. H. Rawal, *J. Am. Chem. Soc.* **1999**, *121*, 9562–9573. *Chiral Amino Siloxy Dienes in the Diels-Alder Reaction: Applications to the Asymmetric Synthesis of 4-Substituted and 4,5-Disubstituted Cyclohexenones, and the Total Synthesis of (–)-a-Elemene.*
2. Y. Huang, T. Iwama, V. H. Rawal, *Org. Lett.* **2002**, *4*, 1163–1166. *Broadly Effective Enantioselective Diels-Alder Reactions of 1- Amino-Substituted-1,3-Butadienes.*
3. S. A. Kozmin, T. Iwama, Y. Huang, V. H. Rawal, *J. Am. Chem. Soc.* **2002**, *124*, 4628–4641. *An Efficient Approach to Aspidosperma Alkaloids via [4+2] Cycloadditions of Aminosiloxydienes: Stereocontrolled Total Synthesis of (+/-)-Tabersonine. Gram-Scale Catalytic Asymmetric Syntheses of (+)-Tabersonine and (+)-16-Methoxytabersonine. Asymmetric Syntheses of (+)-Aspidospermidine and (-)- Quebrachamine.*
4. Y. Huang, V. H. Rawal, *J. Am. Chem. Soc.* **2002**, *124*, 9662–9663. *Hydrogen-Bond-Promoted Hetero-Diels-Alder Reactions of Unactivated Ketones.*
5. A. K. Beck, D. Seebach, A. Heckel, *Angew. Chem. Int. Ed.* **2001**, *40*, 92–138. *TADDOLs, Their Derivatives, and TADDOL Analogs: Versatile Chiral Auxiliaries.*
6. Y. Huang, A. K. Unni, A. N. Thadani, V. H. Rawal, *Nature* **2003**, *424*, 146. *Hydrogen Bonding by a Chiral Alcohol to an Aldehyde Catalyzes Highly Enantioselective Reactions.*
7. A. N. Thadani, A. R. Stankovic, V. H. Rawal, *Proc. Natl. Acad. Sci.U.S.A.* **2004**, *101*, 5846–5850. *Enantioselective Diels-Alder Reactions Catalyzed by Hydrogen Bonding.*
8. V. B. Gondi, M. Gravel, V. H. Rawal, *Org. Lett.* **2005**, *7*, 5657–5660. *Hydrogen bond catalyzed Enantioselective Vinylogous Mukaiyama Aldol Reaction.*
9. A. K. Unni, N. Takenaka, H. Yamamoto, V. H. Rawal, *J. Am. Chem. Soc.*, **2005**, *127*, 1336–1337. *Axially Chiral Biaryl Diols Catalyze Highly Enantioselective Hetero-Diels-Alder Reactions Through Hydrogen Bonding.*

Direct Catalytic Asymmetric Aldol-Tishchenko Reaction

Masakatsu Shibasaki and Takashi Ohshima, The University of Tokyo, Japan

Background

The aldol reaction is generally regarded as one of the most efficient carbon–carbon bond-forming reactions. The development of a range of catalytic asymmetric aldol-type reactions has proven to be a valuable contribution to asymmetric synthesis. In all of these conventional asymmetric aldol-type reactions, however, preconversion of the ketone moiety to a more reactive species **2** such as an enol silyl ether, enol methyl ether, or ketene silyl acetal is an unavoidable necessity (Scheme 1). Thus, the development of a direct catalytic asymmetric aldol reaction is highly desirable in terms of atom economy. In 1997, we achieved success in performing the direct catalytic asymmetric aldol reactions of aldehydes with unmodified ketones using heterobimetallic asymmetric catalysis [1]. List et al. [2] and Trost et al. [3] also reported direct asymmetric aldol reactions using L-proline or a chiral semi-crown Zn complex as catalyst.

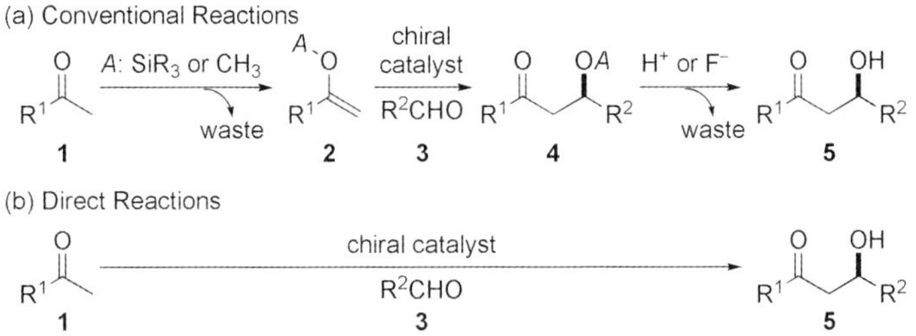

Scheme 1 Aldol-type addition of latent enolates and direct aldol reactions of unmodified ketones.

Most of the catalytic systems reported to date are unfortunately limited to rather simple donors such as methyl ketones, α-hydroxymethyl ketones, and easily enolizable aliphatic aldehydes. Direct aldol reactions of methylene ketones **6**, such as ethyl ketones (R^3 = Me) [4], are viewed as a formidable synthetic challenge because of poor participation of the resulting aldolates **7** in catalyst turnover and a strong tendency toward retro-aldol reactions (Scheme 2).

Asymmetric Synthesis – The Essentials.
Edited by Mathias Christmann and Stefan Bräse
Copyright © 2007 WILEY-VCH Verlag GmbH & Co. KGaA, Weinheim
ISBN: 978-3-527-31399-0

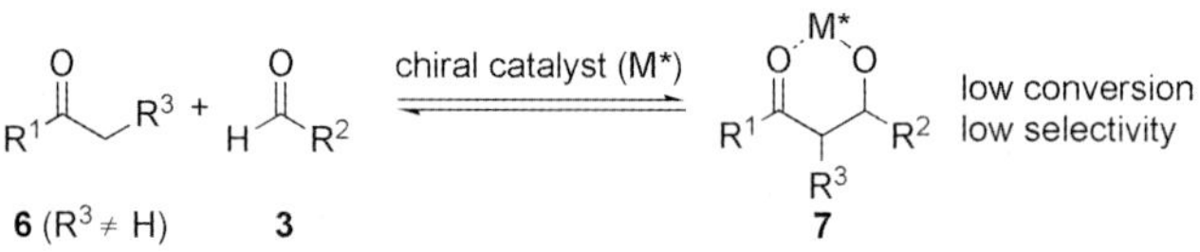

Scheme 2 Direct aldol reaction of methylene ketones **6** ($R^3 \neq H$).

Strategy

Considering the usefulness of the corresponding aldol product of ethyl ketones for the synthesis of 2-methyl-1,3-polyol arrays, we examined various reaction conditions using heterobimetallic asymmetric catalysts [4]; however, all attempts were unsatisfactory. In order to prevent rapid retro-aldol reaction of the resulting aldolates, we planned to couple an irreversible process with the reversible aldol reaction. Inspired by the earliest report of an SmI_2 catalyzed Tishchenko reaction of β-hydroxy ketone by Evans et al. [6], and an yttrium-salen complex catalyzed cross aldol-Tishchenko reaction by Morken et al. [7], we hypothesized that the metalated aldolates **7** derived from our lanthanoid-based heterobimetallic catalyst might be activated for the addition of another aldehyde molecule (Scheme 3). Accordingly, this activation–addition step would provide Evans intermediate **8**, which exhibits the appropriate orientation to rapidly undergo [3,3] bond reorganization to provide the Tishchenko adduct **9**. Therefore, we expected that this addition–rearrangement sequence would allow us to prevent the competitive retro-aldol reaction and provide an attractive platform for the development of highly enantio- and diastereoselective aldol-type adducts [8, 9].

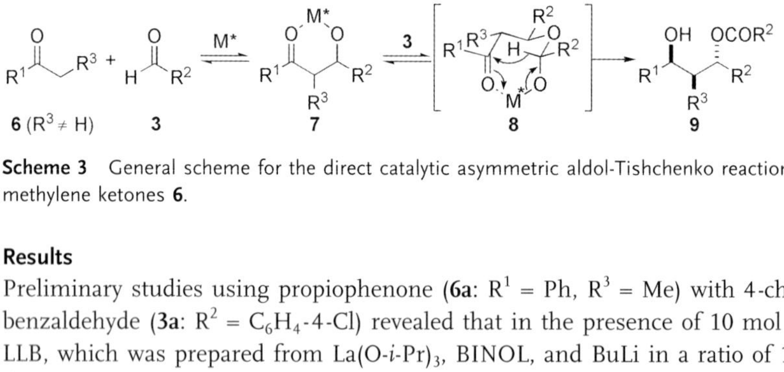

Scheme 3 General scheme for the direct catalytic asymmetric aldol-Tishchenko reaction of methylene ketones **6**.

Results

Preliminary studies using propiophenone (**6a**: $R^1 = Ph$, $R^3 = Me$) with 4-chlorobenzaldehyde (**3a**: $R^2 = C_6H_4\text{-4-Cl}$) revealed that in the presence of 10 mol % of LLB, which was prepared from $La(O\text{-}i\text{-}Pr)_3$, BINOL, and BuLi in a ratio of 1:3:3, the anticipated sequential addition proceeded to provide the desired Tishchenko product **9aa** with excellent diastereoselectivity and moderate enantiocontrol (>98/2 dr and 64% ee) although the catalytic efficiency was unsatisfactory (ca. 1:1 mixture of the aldol product and Tishchenko product, total 20% yields). To improve the reactivity, additive effects of metal salts, which might competitively coordinate and decrease the pKa of the ketone, were examined and 30 mol % LiOTf provided the optimal reaction efficiency and high selectivity. Alternatively, the same catalyst system as LLB·3LiOTf can be prepared by the addition of 6 equiv of BuLi to a 1:3 ratio mixture of $La(OTf)_3$ and BINOL. This new catalyst preparation method improved enantioselectivity. The commercial availability of La(OTf)$_3$ and high tolerance to air and water also enhance the worth of this catalyst system. Because even a slight excess of BuLi resulted in decreased enantioselectivity, the 1:3:5.6 La(OTf)$_3$/BINOL/BuLi catalyst system was set as the standard condition.

The scope and limitations of various substrates are summarized in Table 1. Under the optimized conditions, the direct catalytic asymmetric aldol-Tishchenko reaction of a variety of both aromatic aldehydes **3** and aromatic ketones **6** smoothly proceeded to give the Tishchenko product **9** and, after hydrolysis using NaOMe in MeOH, the corresponding diols **10** were obtained in up to 96% isolated yield and up to 95% *ee*. The use of either aliphatic aldehydes or ketones, unfortunately, resulted in unsatisfactory results. It is noteworthy that the reaction proceeded with the same efficiency even when propyl ketone **6h** (entry 16) and butyl ketone **6i** (entry 17) were used, achieving the asymmetric direct aldol-type reaction of propyl and butyl ketones for the first time [8, 9].

Table 1 Direct catalytic asymmetric aldol-Tishchenko reaction of various substrates.

Entry	Ketone 6 R^1	R^3	Aldehyde 3 R^2	Time (h)	ee of 10 (%)	Yield of 10 (%)
1	**6b**: C$_6$H$_4$-4-CF$_3$	Me	**3a**: C$_6$H$_4$-4-Cl	60	93	95
2	**6b**: C$_6$H$_4$-4-CF$_3$	Me	**3b**: C$_6$H$_4$-4-Br	48	95	96
3	**6b**: C$_6$H$_4$-4-CF$_3$	Me	**3c**: C$_6$H$_4$-4-F	72	92	85
4	**6b**: C$_6$H$_4$-4-CF$_3$	Me	**3d**: C$_6$H$_4$-4-Me	94	92	67
5	**6b**: C$_6$H$_4$-4-CF$_3$	Me	**3e**: Ph	84	91	95
6	**6b**: C$_6$H$_4$-4-CF$_3$	Me	**3f**: C$_6$H$_4$-3-Br	48	86	92
7	**6b**: C$_6$H$_4$-4-CF$_3$	Me	**3g**: C$_6$H$_4$-3-OMe	72	85	65
8	**6b**: C$_6$H$_4$-4-CF$_3$	Me	**3h**: 2-naphthyl	80	88	67
9	**6b**: C$_6$H$_4$-4-CF$_3$	Me	**3i**: 3-furyl	84	93	77
10	**6b**: C$_6$H$_4$-4-CF$_3$	Me	**3j**: 3-thienyl	84	94	82
11	**6c**: C$_6$H$_4$-4-Br	Me	**3b**: C$_6$H$_4$-4-Br	48	85	70
12	**6d**: C$_6$H$_4$-3-Cl	Me	**3a**: C$_6$H$_4$-4-Cl	48	84	60
13	**6e**: C$_6$H$_3$-3,4-Cl$_2$	Me	**3a**: C$_6$H$_4$-4-Cl	48	88	81
14	**6f**: C$_6$H$_3$-3,5-Cl$_2$	Me	**3a**: C$_6$H$_4$-4-Cl	48	85	73
15	**6g**: C$_6$H$_3$-3,5-F$_2$	Me	**3a**: C$_6$H$_4$-4-Cl	48	87	77
16	**6h**: C$_6$H$_4$-4-CF$_3$	Et	**3b**: C$_6$H$_4$-4-Br	90	88	90
17	**6i**: C$_6$H$_4$-4-CF$_3$	Pr	**3b**: C$_6$H$_4$-4-Br	90	87	88

Mechanistic Studies

Thorough inspection of the relation between the aldol product and the Tishchenko product, as well as their stereoselectivities, offered us a reasonable mechanistic explanation as depicted in Scheme 4. Metal enolate reacts reversibly with aldehyde, yielding all possible isomeric metal aldolates **7**. An *anti*-aldolate proceeds through a bicyclic transition state **8** to give **11**. A *syn*-aldolate can undergo a similar reaction with a slower rate through transition state **8'**, the alkyl group of which occupies an energetically unfavorable axial position. In such cases, a fast isomerization from *syn*-aldolate to *anti*-aldolate might surpass the rate of the Tishchenko reaction. Thus, a *syn*-aldolate might isomerize to *anti*-aldolate through a retro-aldol reaction and undergo the Tishchenko reaction via the more favorable transition state **8**, giving rise to *anti*-aldol *anti*-Tishchenko product **9** in high efficiency and diastereoselectivity [8].

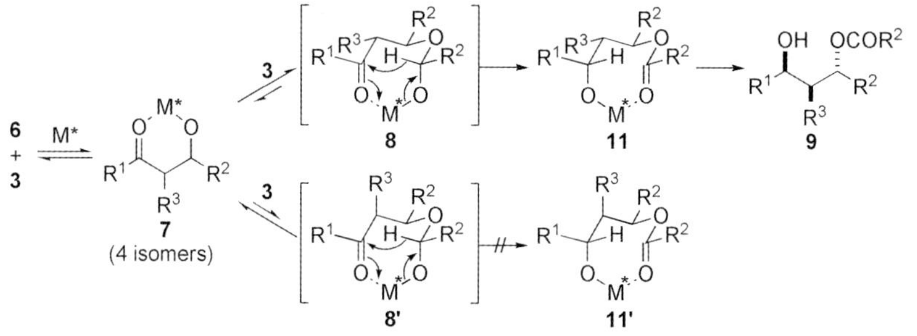

Scheme 4 Proposed mechanism for direct catalytic asymmetric aldol-Tishchenko reaction.

We also investigated the role of LiOTf in the direct catalytic asymmetric aldol-Tishchenko reaction and found that LiOTf promoted a dynamic structural change of LLB to generate a novel binuclear La_2Li_4(binaphthoxide)$_5$ complex (**12**), whose structure was determined by X-ray crystallography (Scheme 5) [9]. Moreover, we demonstrated that the self-assembled lanthanum complex **12** dynamically changed its structure to that of LLB (by the addition of Li_2(binol)) and an active species of the asymmetric catalysis (by the addition of Li_2(binol) and LiOTf).

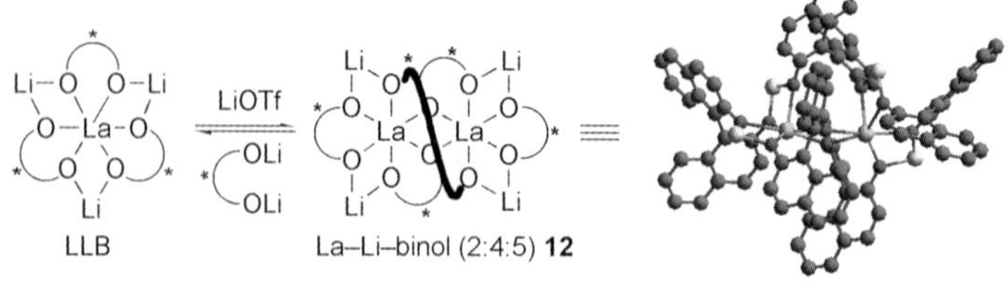

Scheme 5 Dynamic self-assembling of the lanthanum complexes and crystal structure of $La_2Li_4\{(R)\text{-binol}\}_5(thf)_8$ (**12**) (THF molecules are omitted for clarity).

Conclusions and Future Perspectives

As part of ongoing studies of direct aldol reactions, we studied a direct aldol reaction of ethyl ketones and established the Tishchenko reaction as one of the useful methods for overcoming the retro-aldol reaction problem for aromatic donors and acceptors. It is particularly noteworthy that this sequential process constructed three contiguous asymmetric centers in a highly enantio- and diastereoselective manner. Moreover, this is the first example of a direct aldol-Tishchenko reaction of propionate equivalent. Further studies on broadening the substrate scope to aliphatic donors and acceptors, useful precursors for the synthesis of polypropionate natural products, are currently ongoing.

CV of Masakatsu Shibasaki

Masakatsu Shibasaki received his Ph.D. from the University of Tokyo in 1974 under the direction of the late Professor Shun-ichi Yamada before doing postdoctoral studies with Professor E. J. Corey at Harvard University. In 1977 he returned to Japan and joined Teikyo University as an associate professor. In 1983 he moved to Sagami Chemical Research Center as a group leader, and in 1986 took up a Professorship at Hokkaido University, before returning to the University of Tokyo as a Professor in 1991. He has received the Pharmaceutical Society of Japan Award for Young Scientists (1981), Inoue Prize for Science (1994), Fluka Prize (Reagent of the Year, 1996), the Elsevier Award for Inventiveness in Organic Chemistry (1998), the Pharmaceutical Society of Japan Award (1999), the Molecular Chirality Award (1999), the Naito Foundation Research Prize for 2001 (2002), ACS Award (Arthur C. Cope Senior Scholar Award) (2002), the National Prize of Purple Ribbon (2003), the Toray Science Award (2004), and Japan Academy Prize (2005).

CV of Takashi Ohshima

Takashi Ohshima received his Ph.D. from The University of Tokyo in 1996 under the direction of Professor Masakatsu Shibasaki. In the following year, he joined Otsuka Pharmaceutical Co., Ltd. for one year. After two years as a postdoctoral fellow at The Scripps Research Institute with Professor K. C. Nicolaou (1997–1999), he returned to Japan and joined Professor Shibasaki's group at The University of Tokyo as an Assistant Professor. In 2005, he moved to Osaka University, where he is currently Associate Professor of Chemistry. He has received the Fujisawa Award in Synthetic Organic Chemistry (2001) and the Pharmaceutical Society of Japan Award for Young Scientists (2004).

Selected Publications

1. Y. M. A. Yamada, N. Yoshikawa, H. Sasai, M. Shibasaki, *Angew. Chem. Int. Ed. Engl.* **1997**, *36*, 1871–1873. *Direct Catalytic Asymmetric Aldol Reactions of Aldehydes and Unmodified Ketones.*
2. B. List, R. A. Lerner, C. F. Barbas III, *J. Am. Chem. Soc.* **2000**, *122*, 2395–2396. *Proline-Catalyzed Direct Asymmetric Aldol Reactions.*
3. B. M. Trost, H. Ito, *J. Am. Chem. Soc.* **2000**, *122*, 12003–12004. *Direct Catalytic Enantioselective Aldol Reaction via a Novel Catalyst Design.*
4. R. Mahrwald, *Org. Lett.* **2000**, *2*, 4011–4012. *Titanium(IV) Alkoxide Ligand Exchange with α-Hydroxy Acids: The Enantioselective Aldol Addition.*
5R. M. Shibasaki, N. Yoshikawa, *Chem. Rev.* **2002**, *102*, 2187–2209. *Lanthanide Complexes in Multifunctional Asymmetric Catalysis.*
6. D. A. Evans, A. H. Hoveyda, *J. Am. Chem. Soc.* **1990**, *112*, 6447–6449. *Samarium-catalyzed Intramolecular Tishchenko Reduction of β-Hydroxy Ketones. A Stereoselective Approach to the Synthesis of Differentiated Anti 1,3-Diol Monoesters.*
7. C. M. Mascarenhas, S. P. Miller, P. S. White, J. P. Morken, *Angew. Chem. Int. Ed. Engl.* **2001**, *40*, 601–603. *First Catalytic Asymmetric Aldol-Tishchenko Reaction - Insight into the Catalyst Structure and Reaction Mechanism.*
8. V. Gnanadesikan, Y. Horiuchi, T. Ohshima, M. Shibasaki, *J. Am. Chem. Soc.* **2004**, *126*, 7782–7783. *Direct Catalytic Asymmetric Aldol-Tishchenko Reaction.*
9. Y. Horiuchi, V. Gnanadesikan, T. Ohshima, H. Masu, K. Katagiri, Y. Sei, K. Yamaguchi, M. Shibasaki, *Chem. Eur. J.* **2005**, *11*, 5195–5204. *Dynamic Structural Change of the Self-Assembled Lanthanum Complex Induced by Lithium Triflate for Direct Catalytic Asymmetric Aldol-Tishchenko Reaction.*

Asymmetric Heck and other Palladium-catalyzed Reactions

Lutz F. Tietze and Florian Lotz, University of Göttingen, Germany

Background

Pd-catalyzed transformations are some of the most important synthetic methods with the great advantage of tolerating many functional groups, having low toxicity and being highly efficient as well as sensible to our environment [1R]. Using chiral ligands high enantioselectivities can be obtained [2R]. Moreover, Pd-catalyzed transformations can be used in domino processes allowing the formation of complex molecules starting from simple substrates [3R].

One of the most widely used Pd-catalyzed transformations is the Heck reaction, in which aryl- and alkenylhalides or -triflates react with alkenes. However, there is a major disadvantage of this transformation related to the low selectivity in the formation of the alkene moiety from the intermediate Pd-species by β-hydride elimination, if β- and β'-hydrogens are present. We have shown that for intramolecular reactions a high regioselectivity is obtained using allylsilanes as the terminal alkene component [4].

Enantioselective Heck Reactions with Allylsilanes as the Alkene Component

The use of allylsilanes as the alkene component in Heck reactions allowed the regioselective formation of tertiary stereogenic centers from acyclic alkenes for the first time. By careful choice of the reaction conditions and the catalyst, either of vinyl- or a trimethylsilylvinyl-substituted carbo- or heterocycles can be prepared. Using (S)-BINAP (*ent*-8) (Figure 1) as ligand, the vinyl-substituted tetraline (S)-2 was formed predominantly from 1 (2:3 = 83:17) with an *ee* = 90%; double bond isomers were not observed (Scheme 1) [4]. The procedure has been applied for the synthesis of the sesquiterpene 7-demethyl-2-methoxy-calamene 4 with 92% *ee* for the formation of the tetraline skeleton employing (R)-BINAP (8) [5].

In the reaction of substrates containing a nitrogen in the tether to give *N*-heterocyclic compounds, BINAP (8) was not the catalyst of choice leading to tetrahydroisoquinolines 6a/7a and benzazepanes 6b/7b from 5a/5c in only 72% and 64% *ee*, respectively (Scheme 2).

Asymmetric Synthesis – The Essentials.
Edited by Mathias Christmann and Stefan Bräse
Copyright © 2007 WILEY-VCH Verlag GmbH & Co. KGaA, Weinheim
ISBN: 978-3-527-31399-0

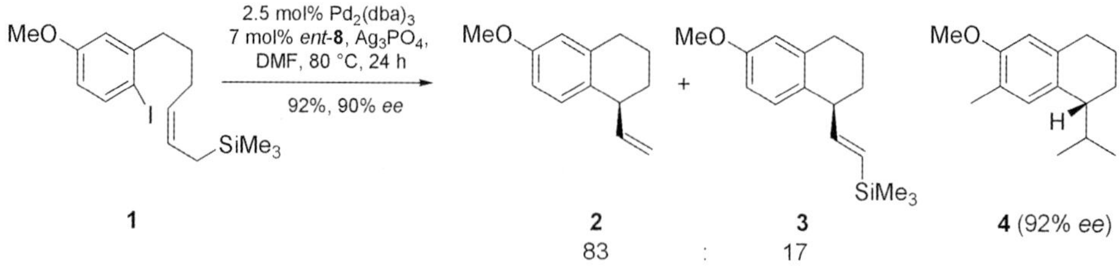

Scheme 1 Enantioselective synthesis of tetralines.

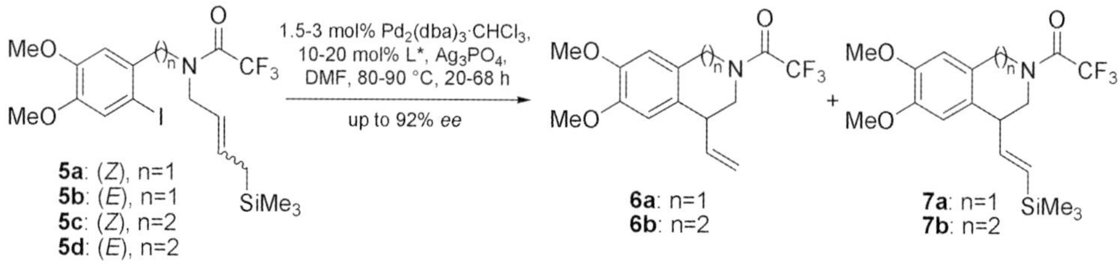

Scheme 2 Enantioselective synthesis of N-heterocycles.

However, the enantioselectivity could be significantly enhanced by employing the chiral ligands TMBTP (**9**) and BITIANP (**10**) [6]. Heck reaction of the (Z)-allylsilane **5a** in the presence of the chiral ligand **9** gave the vinyl-substituted tetrahydroisoquinoline (S)-**6a** exclusively in 73 % yield and 84 % ee [7]. Using the (E)-allylsilane **5b**, however, the selectivity was reduced, leading to a mixture of 56 % of (R)-**6a** with 56 % ee and 9 % of (R)-**7a** with 76 % ee.

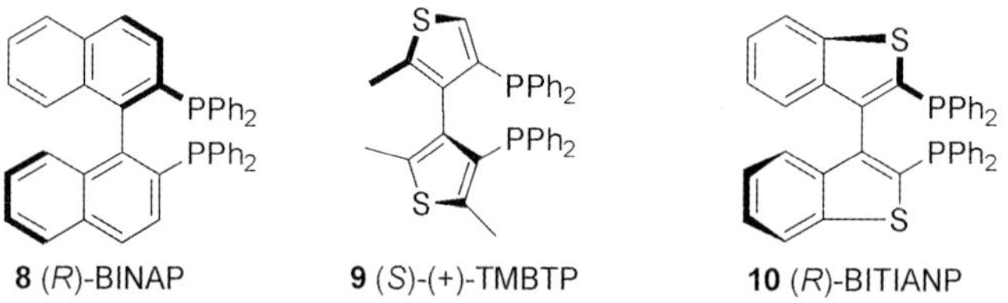

Figure 1 Ligands for the enantioselective Heck reaction.

The best results were obtained for the Pd-catalyzed transformation of **5c** to give the benzazepane (S)-**6b** exclusively with 71 % yield and 92 % ee using **9** as ligand. Under the same conditions but using **10** as ligand, **5d** led to a mixture of 66 % of (S)-**7b** and 21 % of (S)-**6b** with 91 % and 60 % ee, respectively.

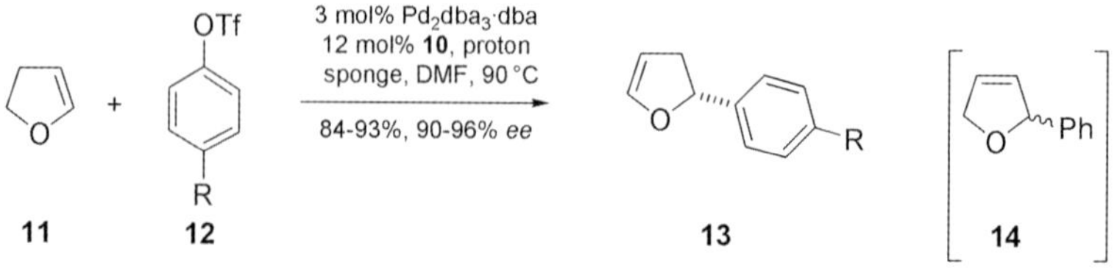

a: R=H, b: R=Cl, c: R=OMe, d: R=CN, e: R=COMe

Scheme 3 Pd-catalyzed enantioselective intermolecular arylation of 2,3-dihydrofuran (**11**).

The ligand **10** could also be employed successfully in the intermolecular Heck reaction of dihydrofuran (**11**) (Schemes 3 and 4) and N-substituted-2-pyrroline **17** (Schemes 5 and 6).

The Pd^0-catalyzed reaction of **11** with the triflate **12a** in the presence of **10** led to **13a** in 84% yield and 90% ee within 18 h (Scheme 3) [8]. The short reaction time is noteworthy, since other chiral ligands require much longer time and give a mixture of **13** and **14** [9]. Substituted phenyl triflates containing either electron-donating or electron-withdrawing groups can also be employed to yield the corresponding derivatives **13b–e** in 84–93% yield and 90–96% ee, with the 4-cyanophenyltriflate (**12d**) giving the best results. In addition, cyclohexenyltriflate (**15**) was used to give **16** with 76% yield and 86% ee (Scheme 4) [8].

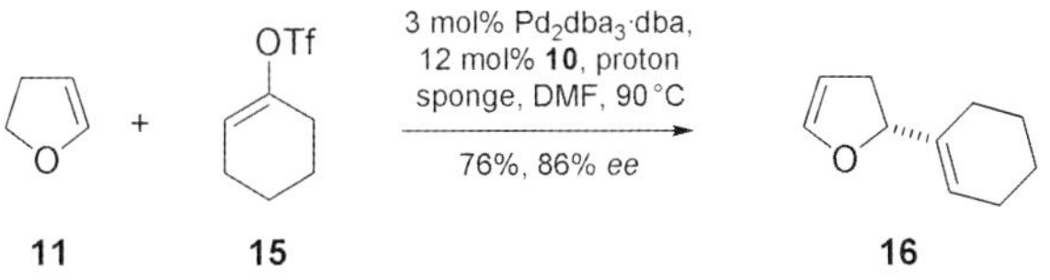

Scheme 4 Pd-catalyzed enantioselective intermolecular alkenylation of 2,3-dihydrofuran (**11**).

The reaction of the N-substituted 2-pyrroline **17** with aryltriflates **12** in the presence of the ligand (R)-BITIANP led to the 5-aryl-2-pyrrolines **18** in excellent regioselectivity (**18**:**19** = 31:1–18:1), high enantioselectivities (93–95% ee) and good yields (80–92%) (Scheme 5) [10]. The use of the chiral ligand BINAP (**8**) and (S)-TMBTP (**9**) was less successful. Thus, transformation of **17** and **12a** in the presence of **8** gave a mixture of **18a** and **19a** with 2% ee for **18** and 0% ee for **19**.

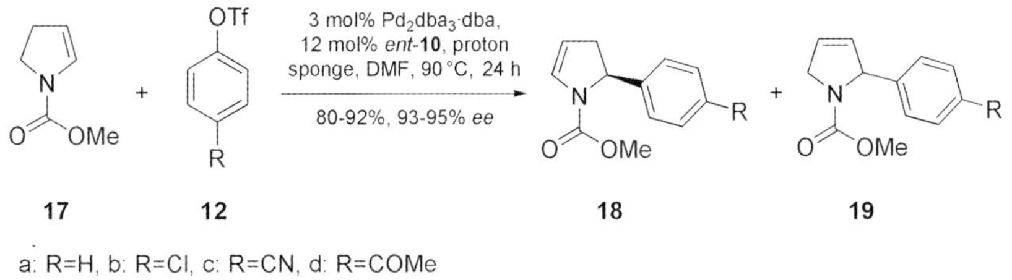

a: R=H, b: R=Cl, c: R=CN, d: R=COMe

Scheme 5 Enantioselective reaction of N-substituted 2-pyrroline **17** with aryltriflates.

Reaction of the cyclohexenyltriflate (**15**) with **17** in the presence of (S)-BITIANP (ent-**10**) led to **20** with 91% ee and 73% yield together with **21** (4:1) (Scheme 6) [10].

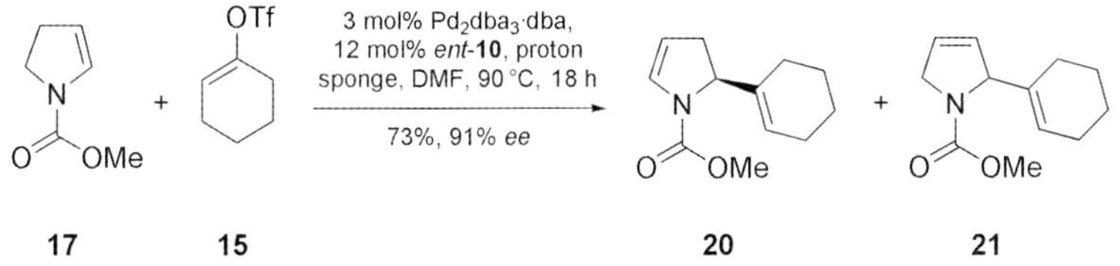

Scheme 6 Enantioselective reaction of N-substituted 2-pyrroline (**17**) with cyclohexenyl triflate (**15**).

Note that for the reaction of **17** with **12** and **17** with **15** a reaction temperature of 90 °C could be employed allowing a reaction time of 18–24 h. Noteworthy, the superiority of BITIANP (**10**) in comparison to BINAP (**8**) is presumably due to electronic and steric reasons. Due to the relatively electron-rich benzothiophene moiety an enhancement of the oxidative addition takes place and the phosphorus atoms are more strongly bound to the relatively electron poor palladium(II)-atom.

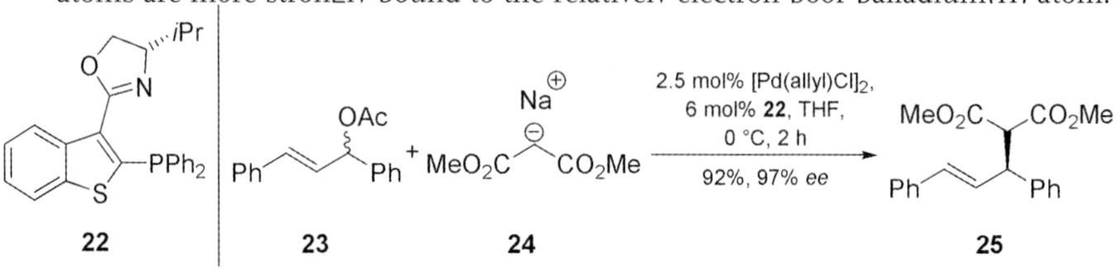

Scheme 7 Enantioselective allylation of **24** with **23** in the presence of **22**.

We have also prepared several other chiral ligands containing a thiophene moiety (Figure 2) and used them for the Pd-catalyzed nucleophilic substitution and the formation of aromatic ethers. As a standard transformation we investigated the reaction of racemic allylacetate (**23**) with the sodium salt of dimethyl malonate (**24**) [11]. The most successful ligand was **22** with an isopropyloxazoline moiety, which allowed the formation of **25** within 2 h in 92 % yield and 97 % ee (Scheme 7). The corresponding ligand containing a *tert*-butyl group instead of the isopropyl group gave lower enantioselectivities.

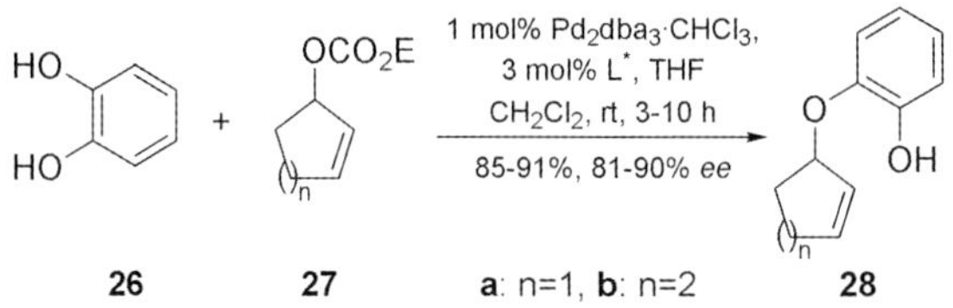

Scheme 8 Enantioselective allylation of catechol with **27a** and **27b** in the presence of the ligands **29** and **30**.

The thiophene-containing ligands **29** and **30** were used for the enantioselective allylation of catechol with cyclohexenyl- and cyclopentenylcarbonate. Reaction of **26** with **27a** in the presence of ligand **29** gave **28a** with 85 % yield and 81 % ee, whereas reaction of **26** and **27b** in the presence of ligand **30** led to **28b** with 91 % yield and 90 % ee (Scheme 8) [12].

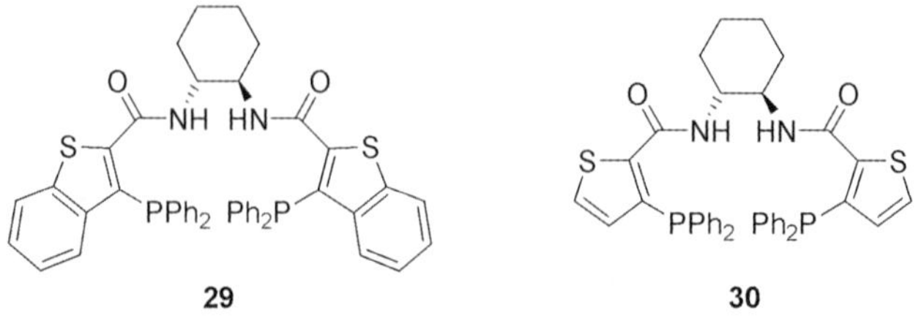

Figure 2 Thiophene-containing ligands for the allylation of catechol.

CV of L. F. Tietze

L. F. Tietze studied chemistry at the universities of Kiel and Freiburg with a Ph.D. in Kiel in 1968 under the guidance of B. Franck. He was a research associate at MIT, Cambridge, USA with G. Büchi and at the University of Cambridge, GB with A. Battersby. After his habilitation in Münster (1975), he was appointed as a Professor at the University of Dortmund (1978), and shortly after as Professor and Director at the University of Göttingen.

CV of F. Lotz

F. Lotz studied chemistry at he university of Göttingen with his diploma degree in July 2004. He is now performing his Ph.D. under the guidance of L. F. Tietze on Pd-catalyzed domino reactions.

Selected Publications

1R. *Handbook of Organopalladium Chemistry for Organic Synthesis*, E. Negishi, (Ed.) Wiley, Hoboken, NJ, **2002**.

2R. L. F. Tietze, H. Ila, H. Bell, *Chem. Rev.* **2004**, *104*, 3453–3516. *Enantioselective Palladium-Catalyzed Transformations.*

3R. (a) L. F. Tietze, *Chem. Rev.* **1996**, *96*, 115–136. *Domino Reactions in Organic Synthesis.* (b) L. F.Tietze, K. M. Sommer, J. Zinngrebe, F. Stecker, *Angew. Chem. Int. Ed. Engl.* **2005**, *44*, 257–259. *Palladium-Ccatalyzed Enantioselective Domino Reaction for the Efficient Synthesis of Vitamin E.* (c) L. F. Tietze, N. Rackelmann, *Pure Appl. Chem.* **2004**, 1967–1983. *Domino Reactions in the Synthesis of Heterocyclic Natural Products and Analogs.*

4. (a) L. F. Tietze, K. Kahle, T. Raschke, *Chem. Eur. J.* **2002**, *8*, 401–407. *Efficient Synthesis of Tetrasubstituted Alkenes by Allylsilane-Terminated Domino-Heck Double Cyclization.* (b) L. F. Tietze, K. Heitmann, T. Raschke, *Synlett* **1997**, 35–37. *Enantioselective Total Synthesis of a Natural Norsesquiterpene of the Calamenene Group by a Silane-Terminated Intramolecular Heck Reaction.* (c) L. F. Tietze, R. Schimpf, *Angew. Chem. Int. Ed. Engl.* **1994**, *33*, 1089–1091. *Regio- and Enantioselective Silicon-Terminated Intramolecular Heck Reactions.*

5. (a) L. F. Tietze, T. Raschke, *Liebigs Ann. Chem.* **1996**, 1981–1987. (b) L. F. Tietze, T. Raschke, *Synlett* **1995**, 597–598. *Enantioselective Total Synthesis of a Natural Norsesquiterpene of the Calamenene Group by a Silane-Terminated Intramolecular Heck Reaction.*

6. T. Benincori, E. Cesarotti, O. Piccolo, F. Sannicolò, *J. Org. Chem.* **2000**, *65*, 2043–2047. *2,2,5,5-Tetramethyl-4,4-bis(diphenylphoshino)-3,3-bithiophene: A New, Very Efficient, Easily Accessible, Chiral Biheteroaromatic Ligand for Homogeneous Stereoselective Catalysis.*

7. L. F. Tietze, K. Thede, R. Schimpf, F. Sannicolò, *Chem. Commun.* **2000**, 583–584. *Enantioselective Synthesis of Tetrahydroisoquinolines and Benzazepines by Silane Terminated Heck Reactions with the Chiral Ligands (+)-TMBTP and (R)-BITIANP.*

8. L. F. Tietze, K. Thede, F. Sannicolò, *Chem. Commun.* **1999**, 1811–1812. *Regio- and Enantioselective Heck Reactions of Aryl and Alkenyl Triflates with the New Chiral Ligand (R)-BITIANP.*

9. (a) G. Trabesinger, A. Albinati, N. Feiken, R. W. Kunz, P. S. Pregosin, M. Tschoerner, *J. Am. Chem. Soc.* **1997**, *119*, 6315–6323. *Enantioselective Homogeneous Catalysis and the "3,5-Dialkyl Meta-Effect". MeO-BIPHEP Complexes Related to Heck, Allylic Alkylation, and Hydrogenation Chemistry.* (b) O. Loiseleur, M. Hayashi, N. Schmees, A. Pfaltz, *Synthesis* **1997**, 1338–1345. *Enantioselective Heck Reactions Catalyzed by Chiral Phosphinooxazoline-Palladium Complexes.* (c) K. K. Hii, T. D. W. Claridge, J. M. Brown, *Angew. Chem. Int. Ed. Engl.* **1997**, *109*, 984–987. *Intermediates in the Intermolecular, Asym-*

metric Heck Arylation of Dihydrofurans.
(d) S. Hillers, S. Sartori, O. Reiser, *J. Am. Chem. Soc.* **1996**, *118*, 2087–2088. *Dramatic Increase of Turnover Numbers in Palladium-Catalyzed Coupling Reactions Using High-Pressure Conditions.*

10. L. F. Tietze, K. Thede, *Synlett* **2000**, 1470–1472. *Highly Regio- and Enantioselective Heck Reactions of N-Substituted 2-Pyrroline with the New Chiral Ligand BITIANP.*

11. L. F. Tietze, J. K. Lohmann, *Synlett* **2002**, 2083–2085. *Synthesis of Novel Chiral Thiophene-, Benzothiophene- and Benzofuran-Oxazoline Ligands and their Use in the Enantioselective Pd-catalyzed Allylation.*

12. L. F. Tietze, J. K. Lohmann, C. Stadler, *Synlett* **2004**, 1113–1116. *Synthesis of Novel Highly Active Thiophene and Benzothiophene Containing Diphosphine Ligands and their Use in the Asymmetric Allylation of Catechol.*

Asymmetric Catalysis with Chiral Acid

Hisashi Yamamoto, The University of Chicago, USA

Background

In 1988, the ASI workshop on "Selectivities in Lewis Acid Promoted Reactions" was held in Greece. During the symposium, I proposed the mechanism of our propargylation process using allenyl bronic ester [1]. In an enantioface differentiating reaction, the chiral nucleophile was added to the carbonyl group of aldehydes, thus allowing the preparation of the chiral alcohols [2]. Based on the anti-coplanar complex structure, we have postulated the rotation of the O–C bond prior to C–C bond formation.

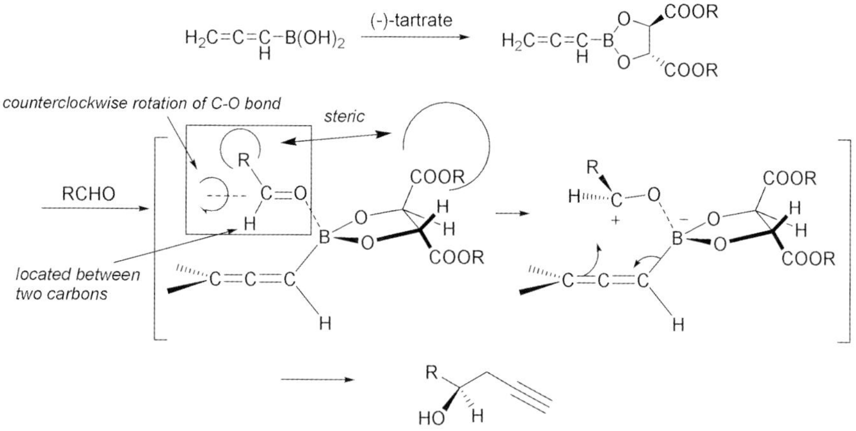

Figure 1 Rotation of C–O bond after coordination of Lewis acid reagent.

The reaction scheme as shown in Figure 1 demonstrates that the symmetry element on the metal center does have a significant effect on the direction of the C–O rotation and thus on the asymmetric induction of the reaction. After these considerations, we initiated our projects for development of the Lewis acid catalyst which has C_n symmetry elements.

Asymmetric Synthesis – The Essentials.
Edited by Mathias Christmann and Stefan Bräse
Copyright © 2007 WILEY-VCH Verlag GmbH & Co. KGaA, Weinheim
ISBN: 978-3-527-31399-0

Results

C_2 Symmetric Chiral Lewis Acid Catalysts

Based on these considerations, the chiral Lewis acid catalysts, which have the C_2 symmetry element was designed and tested for various asymmetric synthesis. Thus, in 1985 we reported the zinc reagent and in 1988 the bulky aluminum reagent [3, 4]. The zinc reagent was used for asymmetric cyclization of unsaturated aldehyde and the aluminum reagent was used for asymmetric hetero-Diels-Alder reaction with Danishefsky diene. Both reagents effectively discriminate the enantioface of aldehydes.

Since then, we and other groups have reported various kinds of chiral Lewis acid catalysts which have C_2 symmetry elements and all of them have proven quite effective for asymmetric carbon–carbon bond forming processes [5]. Not only main group metal catalysts but also transition metal catalysts having the C_2 symmetric structure can be used for asymmetric synthesis via selective activation of carbonyls (Scheme 1) [6].

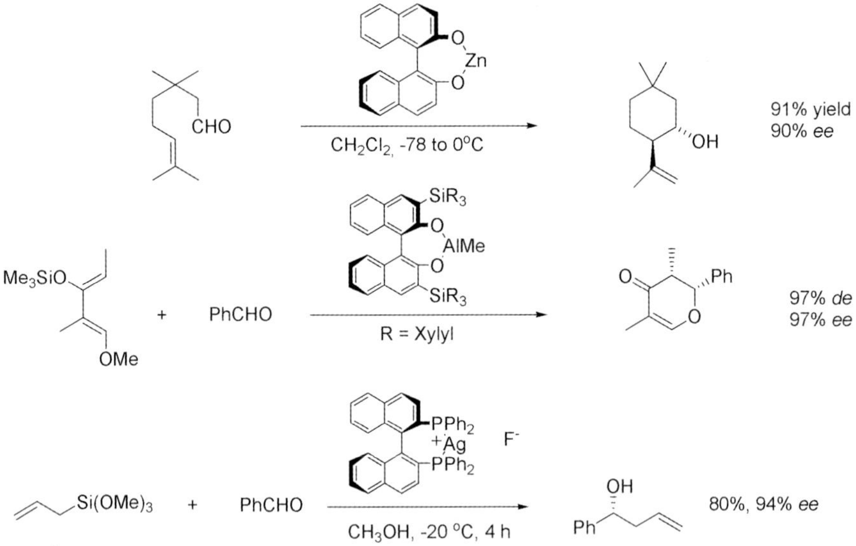

Scheme 1

Combined Acid Catalyst

Another effective way to design acid catalysis is the "combined acids system" [7]. The concept of combined acids, which can be classified into Brønsted acid assisted Lewis acid (BLA), Lewis acid assisted Lewis acid (LLA), Lewis acid assisted Brønsted acid (LBA), and Brønsted acid assisted Brønsted acid (BBA), can be a useful tool for designing asymmetric catalysis, because combining such acids will bring out their inherent reactivity by associative interaction, and also provide a more organized structure, which allows the securing of an effective asymmetric environment (Figure 2).

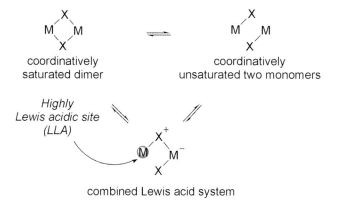

Figure 2 Association of Lewis acid (LM).

Recently Negishi showed that coordinatively unsaturated monomers are far more Lewis acidic than doubly bridged coordinatively saturated dimers [8].

The mono-coordinated complex, however, can be generated and is even more Lewis acidic than the monomer through the formation of a singly bridged dimer. This species is the combined acid catalyst.

BLA: In 1988 we first reported a chiral boron catalyst based on a tartaric acid ligand [9]. The high reactivity of the tartaric acid derived catalyst may originate from intramolecular hydrogen bonding of the terminal carboxylic acid to the alkoxy oxygen. Another boron based BLA was proposed to achieve high selectivity through the double effect of intramolecular hydrogen bonding interaction and attractive π–π donor–acceptor interaction in the transition state [10].

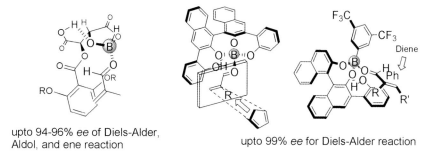

upto 94-96% ee of Diels-Alder, Aldol, and ene reaction

upto 99% ee for Diels-Alder reaction

Figure 3 Examples of BLA catalysts

LLA: Reactive Lewis acid-assisted Lewis acid (LLA) catalysts are relatively well known. Electron-deficient metal compounds can be further activated as electrophiles through hetero- and homodimeric associative interaction. However, its full recognition as a synthetically powerful tool does not yet appear to be widespread. It may be further extended to include asymmetric catalysis design. Shown in Scheme 2 is one example of LLA of a chiral boron reagent activated by various achiral Lewis acids including $SnCl_4$, $AlCl_3$, $FeCl_3$, and others [11].

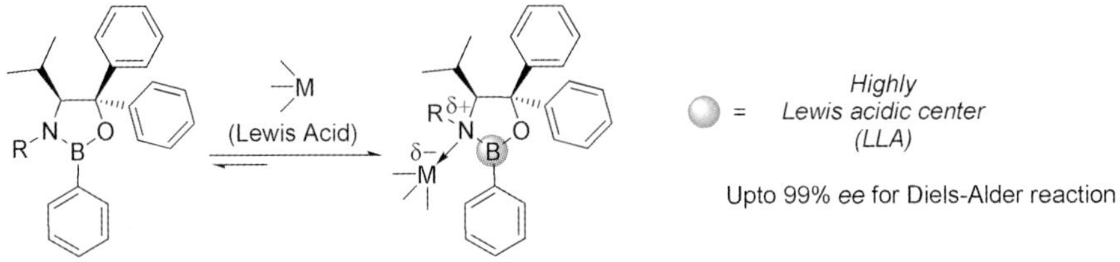

Asymmetric Catalysis with Chiral Acid

Scheme 2

Upto 99% ee for Diels-Alder reaction

○ = Highly Lewis acidic center (LLA)

LBA: Combining Lewis acids and Brønsted acids to give Lewis acid-assisted Brønsted acid (LBA) catalysts can provide an opportunity to design a unique chiral proton [4, 7], that is, the coordination of a Lewis acid to the hetero atom of the Brønsted acid could significantly increase its original acidity (Scheme 3) [12].

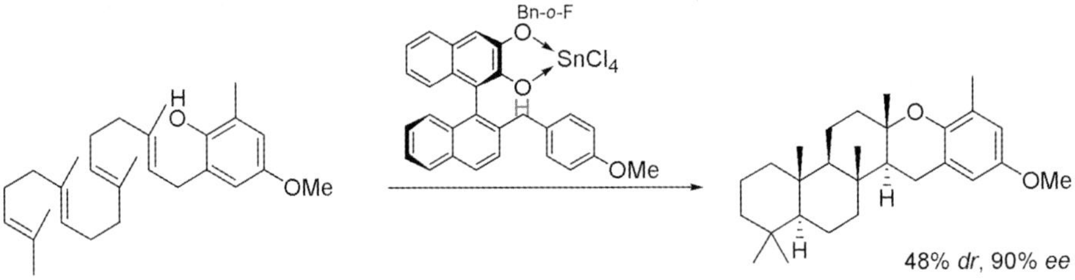

48% dr, 90% ee

Scheme 3

BBA: Hydrogen bonding can frequently be observed inside enzymes, and such a weak interaction has a crucial role in organizing their three-dimensional structure. Additionally, the hydrogen bonding is often involved in the reaction inside the active site of an enzyme. Such an elegant device could be applicable to asymmetric catalysis. Especially for Brønsted acid catalysis, the design of these catalysts would result not only in formation of a highly organized chiral cavity but also in an increase in the Brønsted acidity of the terminal proton in a much milder way compared to that of the LBA system. Shown in Scheme 4 are recent examples of such BBA for our nitroso Aldol reaction [13].

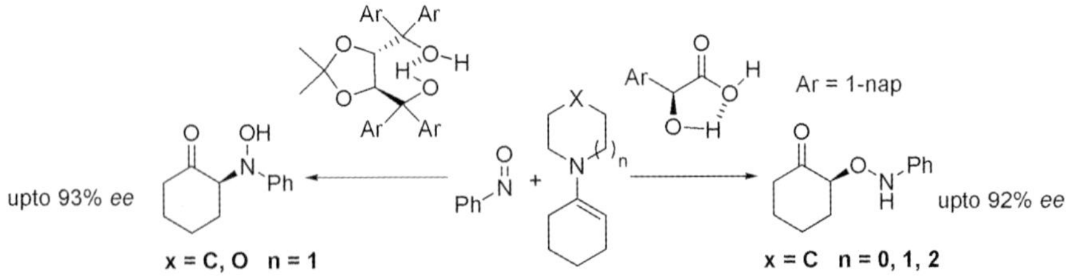

upto 93% ee   x = C, O   n = 1

Ar = 1-nap   upto 92% ee   x = C   n = 0, 1, 2

Scheme 4

CV of Hisashi Yamamoto

Hisashi Yamamoto was born on July 16, 1943 in Kobe, Japan, and received his B.S. from Kyoto University in 1967 where he got his start in research in Professor H. Nozaki's laboratory. He received his Ph. D. from Harvard under the mentorship of Professor E. J. Corey. He started his academic career as Instructor and then Lecturer at Kyoto University, and in 1977 was appointed Associate Professor of Chemistry at the University of Hawaii. In 1980 he moved to Nagoya University where he became Professor in 1983. In 2002, he more to the University of Chicago as a full professor. He is a recipient of numerous awards including the Chemical Society of Japan Award for Young Chemists (1977), IBM Award (1989), Houkou Award (1991), Chunichi Award (1992), Prelog Medal (1993), Chemical Society of Japan Award (1995), Toray Science Award (1997), Max-Tishler Prize (1998), Le Grand Prix de la Foundation Maison de la Chimie (2002), Tetrahedron Chair (2002), National Prize of Purple Medal (Japan) (2002), Molecular Chirality Award (2003), Fellow of American Association for the Advancement of Science (2003), and Yamada Prize (2004).

Selected Publications

1. H. Yamamoto, in *"Selectivities in Lewis Acid Promoted Reactions"*, D. Schinzer (Ed.), Ch. 15, pp. 281–294, NATO ASI Series, Kluwer Academic Publishers, Dordrecht, **1989**. *Chiral Lewis Acid Catalysts. Organoaluminum and Boron Reagents.*
2. R. Haruta, M. Ishiguro, N. Ikeda, H. Yamamoto, *J. Am. Chem. Soc.*, **1982**, *104*, 7667–7668. *Chiral allenylboronic esters: A practical reagent for enantioselective carbon-carbon bond formation.*
3. S. Sakane, K. Maruoka, H. Yamamoto, *Tetrahedron Lett.*, **1985**, *26*, 5535. *Asymmetric cyclization of unsaturated aldehydes catalyzed by a chiral Lewis acid.*
4. K. Maruoka, T. Itoh, T. Shirasaka, H. Yamamoto, *J. Am. Chem. Soc.* **1988**, *110*, 310. *Asymmetric hetero-Diels-Alder reaction catalyzed by chiral organoaluminum reagent.*
5. *Lewis Acids in Organic Synthesis*, H. Yamamoto (Ed.),Vols. 1 and 2, Wiley-VCH, Weinheim, **2000**
6. A. Yanagisawa, H. Kageyama, Y. Nakasuka, , K. Asakawa, Y. Matsumoto, H. Yamamoto, *Angew. Chem. Int. Ed. Enl.*, **1999**, *38*, 3701. *Enantioselective addition of allylic trimethoxysilanes to aldehydes catalyzed by p-tol-BINAP*AgF.*
7. H. Yamamoto, K. Futatsugi, *Angew. Chem. Int. Ed. Engl.* **2005**, *44*, 1924–2942. *Designer Acids, Combined Acid Catalysis for Asymmetric Synthesis*
8. E. Negishi, *Chem. Eur. J.* **1999**, *5*, 411–420. *Principle of Activation of Electrophiles by Electrophiles through Dimeric Association – Two Are Better than One*
9. K. Furuta, Y. Miwa, K. Iwanaga, H. Yamamoto, *J. Am. Chem. Soc.* **1988**, *110*, 6254. *Acyloxyborane: An activating device for carboxylic acids.*
10. K. Ishihara, H. Yamamoto, *J. Am. Chem. Soc.* **1994**, *116*, 1561. *Brønsted acid assisted chiral Lewis acid (BLA) catalyst for asymmetric Diels-Alder reaction.*
11. K. Futatsugi and H. Yamamoto, *Angew. Chem. Int. Ed. Engl.* **2004**, *44*, 1484–1487. *Oxazaborolidine-Derived Lewis Acid Assisted Lewis Acid as a Moisture-Tolerant Catalyst for Enantioselective Diels-Alder Reactions.*
12. H. Ishibashi, K. Ishihara, H. Yamamoto, *J. Am. Chem. Soc.* **2004**, *126*, 11122–11123. *A New Artificial Cyclase for Polyprenoids: Enantioselective Total Synthesis of (-)-Chromazonarol, (+)-8-epi-Puupehedione, and (-)-11-Deoxytaondiol Methyl Ether.*
13. N. Momiyama, H. Yamamoto, *J. Am. Chem. Soc.*, **2005**, *127*, 1080–1081. *Brønsted Acid Catalysis of Achiral Enamine for Regio- and Enantioselective Nitroso Aldol Synthesis.*

Part III

Biocatalysis and Organocatalysis: Asymmetric Synthesis Inspired by Nature

Benjamin List, Max-Planck-Institut für Kohlenforschung, Germany

Introduction

Asymmetric synthesis has likely existed since the beginning of life. Nature's asymmetric catalysts evolved over billions of years and are generally remarkably effective and selective. Chemists have long tried to use and optimize enzymes for their purposes. As a result, biocatalysis is becoming of increasing relevance in asymmetric synthesis and is currently often used, even for large scale industrial applications. There is little less to expect than an even brighter future for biocatalysis. In addition to using and manipulating enzymes, however, chemists also seek to understand their mechanisms and to use this information in the design of small molecule catalysts. This is apparent in the area of organocatalysis, where certain reactions can be catalyzed by both enzymes and small organic molecules. With some right however, it may also be argued that transition metal catalysis is inspired by enzymes since about half of all enzymes are metalloenzymes, catalysts that utilize a metal ion as the catalytic principle. In a sense therefore, asymmetric synthesis as a whole has been "inspired by nature".

The Aldol Reaction as an Example

As an illustration one can consider the aldol reaction. In nature there are two enzyme types, class I and class II aldolases. [1, 2] Both of these enzymes catalyze direct asymmetric aldol reactions of unmodified carbonyl compounds to give β-hydroxycarbonyl compounds. Class I aldolases on the one hand are metal-free catalysts that utilize Lewis-base catalysis to bring about aldolization. The reaction is initiated via iminium ion formation of the aldol donor with a primary amino group from a Lysine residue. This lowers the LUMO energy and facilitates α-deprotonation to give an enamine. Its reaction with the aldol acceptor and hydrolysis then gives the aldol product (Scheme 1).

Class II aldolases on the other hand are metalloenzymes containing a zinc(II)-ion in the active site. The function of this Lewis-acid cofactor is to coordinate to the aldol donor carbonyl, which again lowers the LUMO energy and facilitates α-deprotonation, resulting in a zinc enolate. Its reaction with the acceptor carbo-

Asymmetric Synthesis – The Essentials.
Edited by Mathias Christmann and Stefan Bräse
Copyright © 2007 WILEY-VCH Verlag GmbH & Co. KGaA, Weinheim
ISBN: 978-3-527-31399-0

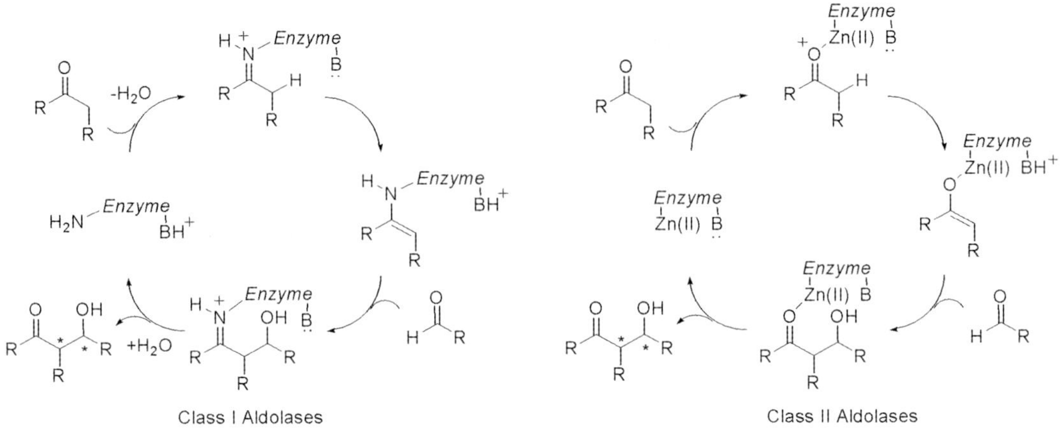

Scheme 1 Mechanism of the aldol reaction catalyzed by class I and class II aldolases.

nyl compound provides the aldol adduct after departure from the enzyme for further turnover (Scheme 1).

Asymmetric Chemical Aldolizations

Initially, most chemical enantioselective aldolizations involved the use of chiral auxiliaries. The first enantioselective intermolecular aldol reaction was discovered by Dieter Enders and his group in 1978 (Scheme 2) [3]. For example, treatment of a chiral acetone hydrazone derived from proline with a base and isobutyraldehyde gave the desired product in 40% ee.

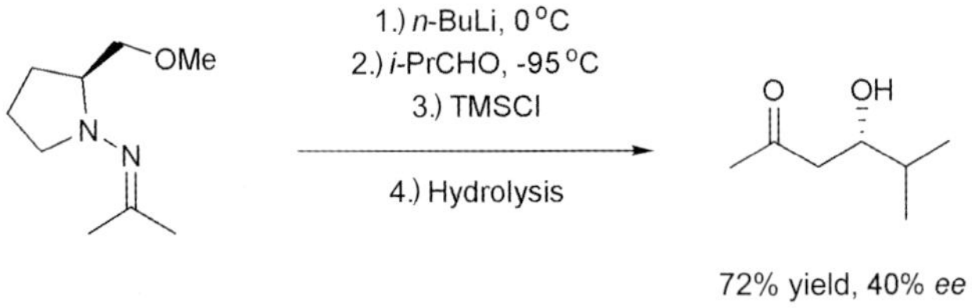

Scheme 2 Enders' asymmetric aldol reaction.

Many other chiral auxiliaries for asymmetric aldol reactions have been developed since then [4], most notably the often used Evans auxiliary [5] (Scheme 3). Commonly, amino alcohol derived propionyl oxazolidinones are enolized with nBu$_2$BOTf and then treated with aldehydes giving *syn*-aldol products in exceptionally high diastereoselectivities. The products can be easily converted into useful aldols.

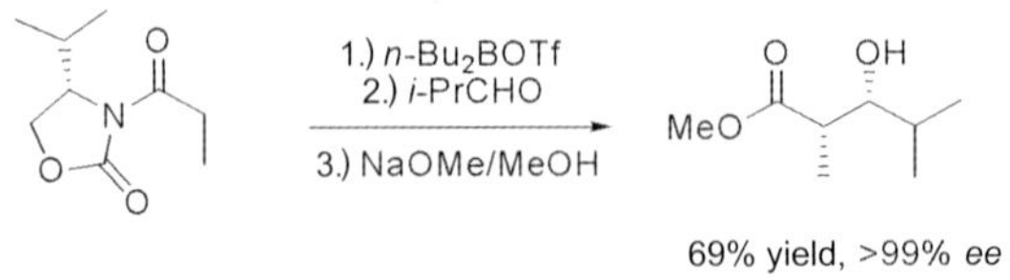

Scheme 3 Evans' asymmetric aldol reaction.

While auxiliary and substrate controlled reactions had been both pioneering and incredibly useful for the synthesis of diverse aldol products, it became apparent that asymmetric aldol reactions would be even more useful if they were catalytic in the required chiral reagent. With this and the advent of the Mukaiyama aldol reaction in mind, several elegant chiral Lewis acid catalyzed variants have been developed by various groups [6]. For example, Carreira et al. developed a titanium catalyzed highly enantioselective Mukaiyama aldol reaction [7]. Treatment of ketene acetals with aldehydes in the presence of chiral titanium catalyst gives the desired aldols in high *ee*s (Scheme 3).

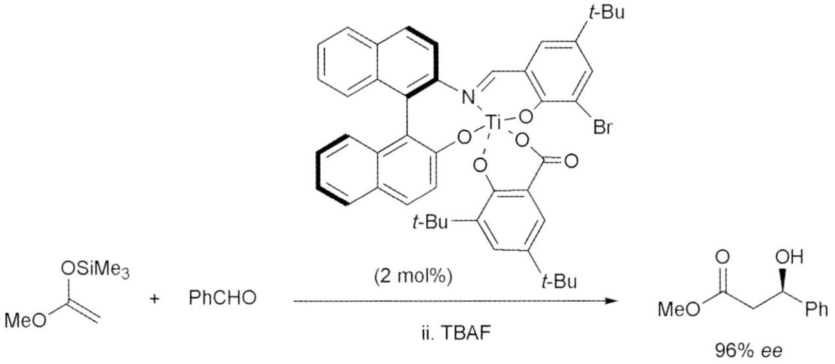

Scheme 4 Carreira's asymmetric Mukaiyama aldol reaction.

In the late 1990s a further improvement seemed desirable. In contrast to the enzyme-catalyzed aldol reactions, the corresponding chemical asymmetric aldolizations all required preformation of the enolate component. Prior to the aldolization, the donor carbonyl compound needed to be derivatized using stoichiometric amounts of reagents in an additional chemical operation. Not only for the sake of atom economy but also in the light of practicality issues, Nature's approach utilizing the carbonyl components in their unmodified form seemed even more attractive. Shibasaki et al. developed the first efficient enantioselective direct asymmetric aldol reaction of ketones with aldehydes [8]. Treating an excess of the ketone donor with the aldehyde in the presence of a chiral lanthanum catalyst gave the desired aldol products, often with high *ee*s (Scheme 5). This catalyst may be considered to be a class II aldolase mimic.

The realization that not only enzymes but also small molecules can catalyze the direct aldol reaction sparked a renewed interest in the use of amines and amino acids as asymmetric catalysts of the aldol reaction. This strategy was first realized in the Hajos–Parrish–Eder–Sauer–Wiechert reaction, a proline-catalyzed intramolecular aldolization [9]. The first examples of a direct asymmetric intermolecular aldol reaction were published in early 2000 (Scheme 6) [10]. In these reactions proline functions as a class I aldolase mimic [11].

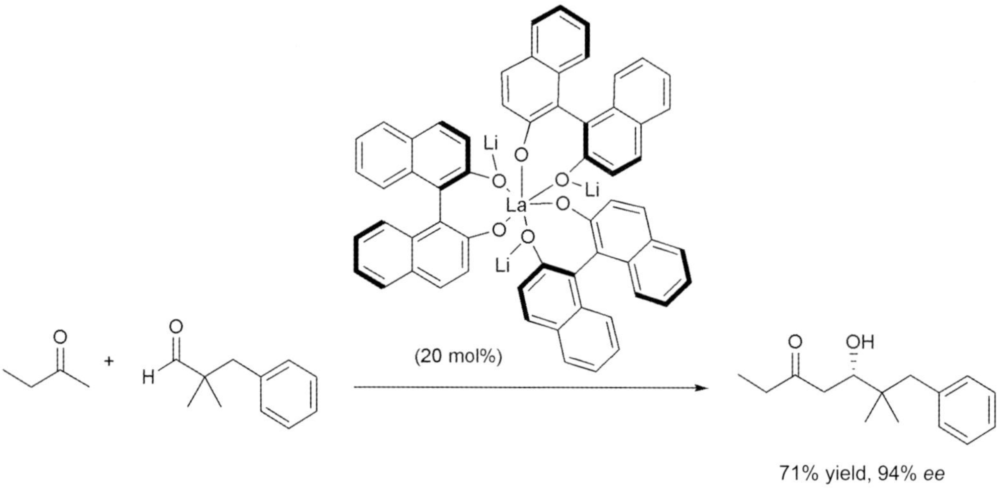

Scheme 5 Shibasaki's direct asymmetric aldol reaction.

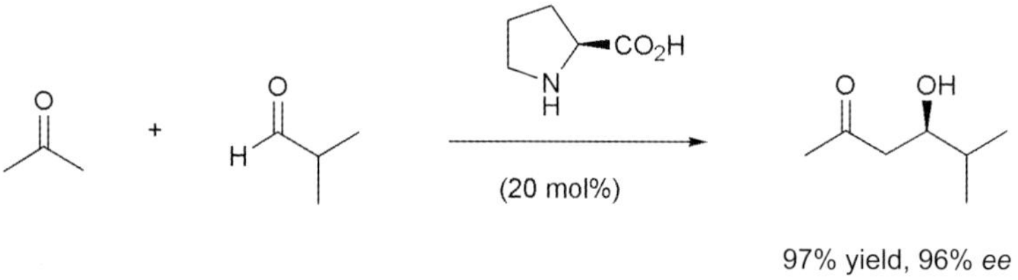

Scheme 6 Proline-catalyzed direct asymmetric aldol reaction.

Conclusions and Outlook

I have discussed the development of the asymmetric aldol reaction from its roots in nature, over the first examples of man-made reagents for the asymmetric synthesis of aldols, to its most recent chemical variants. Interestingly, these latest developments increasingly begin to resemble what nature has already accomplished long before the first human beings arose on this planet. So the question of whether or not the development of asymmetric synthesis has been inspired by nature can be answered with a resounding "yes".

Taking a leaf out of nature's book has been tremendously beneficial for the asymmetric synthesis of chiral compounds. However, in parallel to our efforts in mimicking nature, we should also continue going a step further and design completely new strategies for asymmetric synthesis. Ultimately, our efforts should enable us to create our own artificial enzymes, powerful asymmetric catalysts for new chemical reactions beyond nature.

CV of Benjamin List

Benjamin List was born in Frankfurt, Germany, in 1968. He studied Chemistry at the Freie-University of Berlin and obtained his Ph.D. working on the development of a new synthetic strategy towards Vitamin B_{12} with Johann Mulzer in 1997 at the University of Frankfurt. After postdoctoral studies at the Scripps Research Institute he became an Assistant Professor at the same institute in 1999. In 2003 he moved back to Germany to take a position as group leader at the Max-Planck-Institut für Kohlenforschung in Mülheim. In 2004 he also became an Honorary Professor at the University of Cologne. In 2005 he was promoted to become a Director at the Max-Planck-Institut für Kohlenforschung. His group is mostly interested in developing new strategies for organic synthesis.

References

1. B. L. Horecker, O. Tsolas, C.-Y. Lai, in *The Enzymes*, Vol. VII, P. D. Boyer (Ed.) Academic Press, New York, **1975**, p. 213.
2. T. D. Machajewski, C.-H. Wong, *Angew. Chem. Int. Ed. Engl.* **2000**, *39*, 1352–1374. *The catalytic asymmetric aldol reaction.*
3. H. Eichenauer, E. Friedrich, W. Lutz, D. Enders, *Angew. Chem.* **1978**, *90*, 219–220. *Regiospecific and enantioselective aldol reactions.*
4. J. Seyden-Penne, *Chiral auxiliaries and ligands in asymmetric synthesis*, Wiley, New York, **1995**.
5. D. A. Evans, J. Bartroli, T. L. Shih, *J. Am. Chem. Soc.* **1981**, *103*, 2127–2129. *Enantioselective aldol condensations. 2. Erythroselective chiral aldol condensations via boron enolates.*
6. H. Gröger, E. M. Vogl, M. Shibasaki, *Chem Eur J.* **1998**, *4*, 1137–1141. *New catalytic concepts for the asymmetric aldol reaction.*
7. E. M. Carreira, R. A. Singer, W. Lee, *J. Am. Chem. Soc.* **1994**, *116*, 8837–8838. *Catalytic, Enantioselective Aldol Additions with Methyl and Ethyl Acetate O-Silyl Enolates: A Chiral Tridentate Chelate as a Ligand for Titanium(IV).*
8. N. Yoshikawa, Y. M. A. Yamada, J. Das, H. Sasai, M. Shibasaki, *J. Am. Chem. Soc.* **1999**, *121*, 4168–4178. *Direct catalytic asymmetric aldol reaction.*
9. (a) Z. G. Hajos, D. R. J. Parrish, *J. Org. Chem.* **1974**, *39*, 1615–1621. *Asymmetric synthesis of bicyclic intermediates of natural product chemistry.* (b) U. Eder, G. Sauer, R. Wiechert, *Angew. Chem., Int. Ed. Engl.* **1971**, *10*, 496–497. *New Type of Asymmetric Cyclization to Optically Active Steroid CD Partial Structures.*
10. B. List, R. A. Lerner, C. F. Barbas III, *J. Am. Chem. Soc.* **2000**, *122*, 2395–2396. *Proline-catalyzed direct asymmetric aldol reactions.*
11. H. Gröger, J. Wilken, *Angew. Chem. Int. Ed. Engl.* **2001**, *40*, 529–532. *The application of L-proline as an enzyme mimic and further new asymmetric syntheses using small organic molecules as chiral catalysts.*

Enantioselective Photochemical Reactions

Thorsten Bach, Technische Universität München, Germany

Background

The generation of the first stereogenic center in a molecule is the key step for the preparation of enantiomerically pure compounds. The development of enantioselective reactions in conventional ('thermal') chemistry has been breathtaking in recent years and this book summarizes many of the major achievements.

Photochemistry has to some extent lagged behind this development. The control of facial diastereoselectivity in photochemical reactions is reasonably well understood: Auxiliary-induced diastereoselectivity has been employed for the photochemical preparation of enantiomerically pure compounds, substrate-induced diastereoselectivity for the preparation of single stereoisomers starting from chiral pool building blocks. We have for example used a diastereoselective [2+2]-photocycloaddition reaction as a key step in the synthesis of kelsoene (1) [1]. (Scheme 1.)

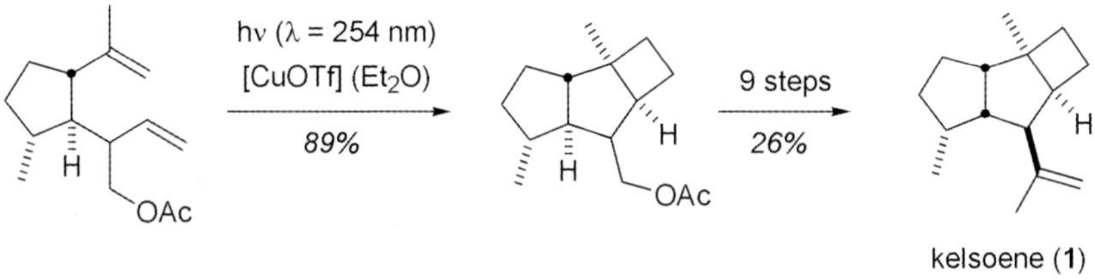

Scheme 1 Diastereoselective photochemical key step in the synthesis of kelsoene (1).

Attempts to achieve enantioselective reactions in solution photochemistry by the use of a chiral reagent (e.g. circularly polarized light), a chiral complexing agent, or a chiral catalyst have seen limited success. Either the enantiomeric excess (*ee*) achieved was high but yields remained low, or the enantioselectivity was low in high-yielding reactions.

Concept

In the quest for enantioselective photochemical reactions we considered chiral hydrogen-bonded complexing agents as powerful tools. While previous work had relied on nonspecific and less directed interactions between the complexing agent and the substrate [2], hydrogen bonds appealed to us as efficient noncovalent controlling devices. Compound (+)-**2** (Figure 1), which is readily available from 1,3,5-trimethylcyclohexane-1,3,5-tricarboxylic acid (Kemp's triacid) [3], and its enantiomer *ent*-(−)-**2** turned out to be excellent complexing agents to achieve enantioselective photochemical reactions of amides and lactams.

Figure 1 Structure of the chiral complexing agent **2** and its enantiomer *ent*-**2**.

The key issues relevant to the success of these reagents and some examples for their use in the synthesis of enantiomerically pure compounds are presented in this short account. More recently, we have started to develop compounds, in which the passive tetrahydronaphthalene shield is replaced by a bulky catalytically active unit. Results of this endeavor will be described in the last section.

Results

Compounds with a secondary lactam or amide linkage can bind to the complexing agents (+)-**2** or (−)-**2** via two hydrogen bonds. Alternatively, the compounds can dimerize, while for steric reasons the enantiomerically pure compound **2** does not dimerize. Remarkably, the association constant of many lactams to **2** appears to be significantly higher than their dimerization constant. Quinolone, as an example, binds to complexing agent **2** at 293 K in toluene with an association constant of 580 M^{-1}. These figures, which have been verified by NMR titration, microcalorimetry, and CD titration, are typical for a successful binding, which guarantees at low temperature a prevalence of bound vs. unbound substrate and a high *ee* in any subsequent reaction. This feature is illustrated by the [2+2]-photocycloaddition of 4-allyloxy-2-quinolone (**3**) leading to the cyclobutane product **4** in 93% *ee* [4, 5] (Scheme 2).

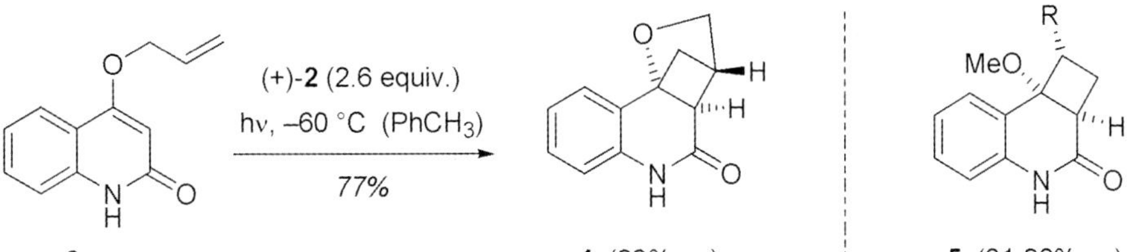

Scheme 2 Enantioselective intra- and intermolecular [2+2]-photocycloaddition to 4-alkoxy-2-quinolones.

The recovery yield of the complexing agent is high (> 90%). The enantioface differentiation is easy to understand based on the association of substrate **3** to complexing agent (+)-**2** as discussed above. Similar reactions can be conducted intermolecularly. 4-Methoxy-2-quinolone, for example, was converted into the cyclobutanes **5** (Scheme 2) by an intermolecular [2+2]-photocycloaddition to various terminal alkenes [4, 6], which proceeded equally well as the intramolecular reaction.

The efficiency of the complexing agent is enhanced if the product of the photochemical reaction dissociates due to steric reasons. Binding sites are liberated and the ratio of complexing agent to substrate increases in the corresponding reaction. The photoinitiated Diels-Alder reaction of *ortho*-quinodimethane **8** serves to illustrate this point (Scheme 3) [7, 8].

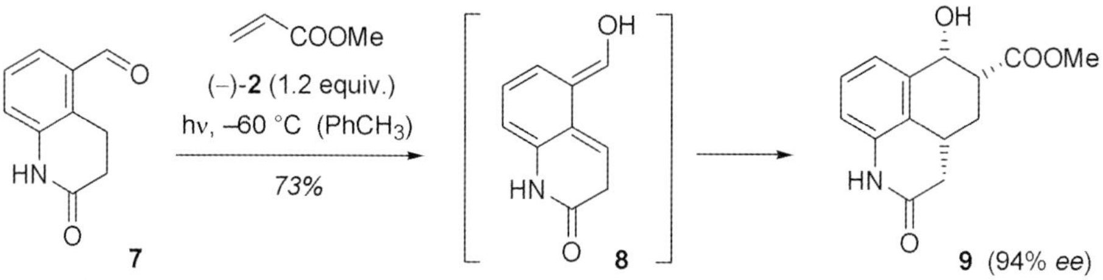

Scheme 3 Enantioselective Diels-Alder reaction of the photochemically generated *ortho*-quinodimethane **8**.

Even with 1.2 equiv of complexing agent (−)-**2** an enantioselectivity of 94% ee was achieved at −60 °C. Apparently, as proven for a related product [8], the tricyclic *endo*-product **9** does not bind effectively to the complexing agent. The binding of substrate **7** becomes more effective and the *ee* increases while the reaction proceeds.

The concept of lactam and even amide binding is generally applicable and not restricted to quinolones or dihydroquinolones. Other substrates have been successfully converted into enantiomerically enriched products in diverse photochemical reactions. Examples include a [6π]-photocyclization to **10**, a Norrish–Yang-cyclization to **11**, a [4+4]-photocycloaddition to **12** and a [4π]-photocyclization to **13** (Figure 2) [2]. The newly formed bonds are indicated by shaded lines.

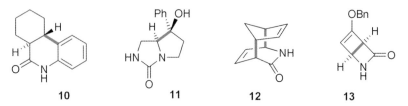

Figure 2 Products **10-13** of enantioselective photochemical reactions in solution.

Perspectives

Apart from synthetic applications of the enantioselective photochemical reactions, two lines of research appear particular attractive for further work. First, the concept of a chiral complexing agent can be applied to other than photochemical reactions. In this context we have recently shown that hydrogen atom transfer reactions can be conducted enantioselectively. As an example, the intramolecular radical reaction of iodide **14**, which proceeds via radical **15**, resulted in the enantioselective formation of piperidone **16** (Scheme 4) [9]. The enantioface differentiation can be explained by invoking a coordination of intermediate **15** to the chiral complexing agent (+)-**2**. The enantioselectivity is concentration dependent and showed a maximum at 100 mM with 88% ee (78% yield). At higher concentrations side reactions took over and yields decreased. Further work on other radical cyclizations is in progress.

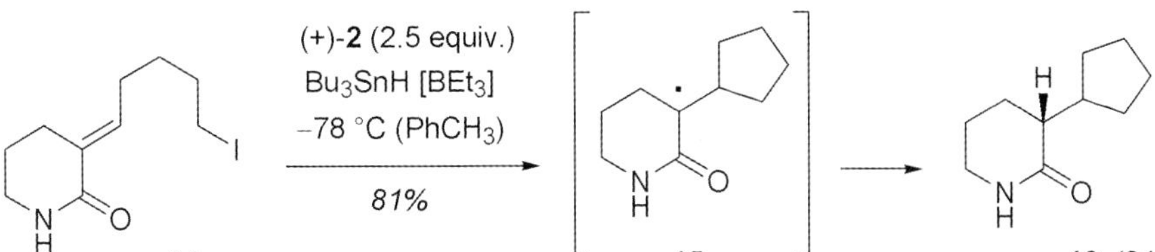

Scheme 4 Enantioselective BEt$_3$-initiated radical reaction of iodide **14** to piperidone **16**.

In a second area of interest, we aim at catalytic photochemical reactions in solution. Due to the distance dependence of sensitization processes a catalytic cycle for enantioselective photochemical reactions can be easily formulated. The premises for such a cycle are (i) an exclusive sensitization in the complex of substrate and catalyst, (ii) a high enantioface differentiation, (iii) a reasonable binding constant of the substrate to the catalyst, and ideally (iv) a low binding constant of the product to the catalyst. In the first catalyst generation the sensitizing unit, for instance a benzoylated biphenyl, was linked to the binding site via an ester or an amide bond [e.g. (+)-**17**, Figure 3]. It was shown that the association of lactams to these compounds was not sufficiently high. Despite considerable success, the concept was changed and organocatalysts were devised in which the sensitizing unit was linked to the binding site via an oxazole. The first active catalyst of this generation, (−)-**18**, is depicted in Figure 3. It allowed a catalytic photoinduced electron transfer (PET) cyclization of a 4-aminoethyl-substituted quinolone, which proceeded in enantioselectivites up to 70% ee and in 52–64% yield [10].

It was shown for the first time that chiral multiplication is possible in a photochemical process. Even with 5 mol% catalyst, 20% ee could be achieved (61% yield).

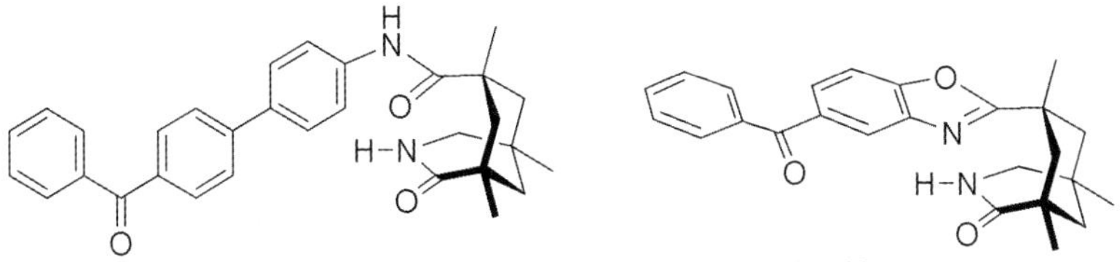

Figure 3 Structure of the enantiomerically pure chiral benzophenone catalysts (+)-**17** and (−)-**18**.

Of course, there is much room for improvement. Racemic background reactions need to be suppressed, the efficiency of sensitization has to be increased and other test systems need to be devised. Research in these directions is currently being pursued.

CV of Thorsten Bach

Thorsten Bach was born in Ludwigshafen/Rhein in 1965. He studied chemistry at the University of Heidelberg and at the University of Southern California (Diplomarbeit with George A. Olah). Subsequently, he joined the group of Manfred T. Reetz as a Kekulé fellow and received his Ph.D. in 1991 from the University of Marburg. After a postdoctoral stay at Harvard University with David A. Evans he started his independent research at the University of Münster in 1992. In 1997 he became Professor of Organic Chemistry at the University of Marburg and in April 2000 he was appointed Full Professor of Organic Chemistry at the Technische Universität München. His awards include the Dozentenstipendium des Fonds der Chemischen Industrie (1997), the AstraZeneca Research Award in Organic Chemistry (2001), the Novartis European Young Investigator Award (2003), and the Degussa Prize for Chirality in Chemistry (2006).

Selected Publications

1. T. Bach, A. Spiegel, Synlett **2002**, 1305–1307. Stereoselective Total Synthesis of the Tricyclic Sesquiterpene (±)-Kelsoene by an Intramolecular Cu(I)-Catalyzed [2+2]-Photocycloaddition Reaction.
2. B. Grosch, T. Bach, in Molecular and Supramolecular Photochemistry, Vol. 11: Chiral Photochemistry, Y. Inoue, V. Ramamurthy (Eds.), Dekker, New York, **2004**, pp. 315–340. Template Induced Enantioselective Photochemical Reactions in Solution.
3. T. Bach, H. Bergmann, B. Grosch, K. Harms, E. Herdtweck, Synthesis **2001**, 1395–1405. Synthesis of Enantiomerically Pure 1,5,7-Trimethyl-3-azabicyclo[3.3.1]nonan-2-ones as Chiral Host Compounds for Enantioselective Photochemical Reactions in Solution.
4. T. Bach, H. Bergmann, B. Grosch, K. Harms, J. Am. Chem. Soc. **2002**, 124, 7982–7990. Highly Enantioselective Intra- and Intermolecular [2+2]-Photocycloaddition Reactions of 2-Quinolones Mediated by a Chiral Lactam Host. Host-Guest Interactions, Product Configuration, and the Origin of the Stereoselectivity in Solution.
5. T. Bach, H. Bergmann, K. Harms, Angew. Chem. **2000**, 112, 2391–2393. Angew. Chem. Int. Ed. Engl. **2000**, 39, 2302–2304. Enantioselective Intramolecular [2+2]-Photocycloaddition Reactions in Solution.
6. T. Bach, H. Bergmann, J. Am. Chem. Soc. **2000**, 122, 11525–11526. Enantioselective Intermolecular [2+2]-Photocycloaddition Reactions of Alkenes and a 2-Quinolone in Solution.
7. B. Grosch, C. Orlebar, E. Herdtweck, W. Massa, T. Bach, Angew. Chem. **2003**, 115, 3822–3824. Angew. Chem. Int. Ed. **2003**, 42, 3693-3696. Highly Enantioselective Diels-Alder Reactions of a Photochemically Generated o-Quinodimethane with Olefins.
8. B. Grosch, C. N. Orlebar, E. Herdtweck, M. Kaneda, T. Wada, Y. Inoue, T. Bach, Chem. Eur. J. **2004**, 10, 2179–2189. Enantioselective [4+2]-Cycloaddition Reaction of a Photochemically Generated o-Quinodimethane. Mechanistic Details, Association Studies, and Pressure Effects.
9. T. Aechtner, M. Dressel, T. Bach, Angew. Chem. **2004**, 116, 5974–5976. Angew. Chem. Int. Ed. Engl. **2004**, 43, 5849–5851. Hydrogen Bond Mediated Enantioselectivity of Radical Reactions.
10. A. Bauer, F. Westkämper, S. Grimme, T. Bach, Nature **2005**, 436, 1139–1140. Catalytic enantioselective reactions driven by photoinduced electron transfer.

Asymmetric Catalysis via Dynamic Kinetic Resolution

Jan-E. Bäckvall, Stockholm University, Sweden

Background

Kinetic resolution (KR) and dynamic kinetic resolution (DKR) are important methods for the synthesis of enantiomerically pure compounds. The drawback with KR is that a maximum of 50% of the starting material can be used to give product. This problem can be circumvented if the substrate can be racemized during the resolution process and this leads to DKR [1R]. The principles of KR and DKR are shown in Scheme 1. An efficient DKR is obtained if $k_{rac} > k_F \gg k_S$.

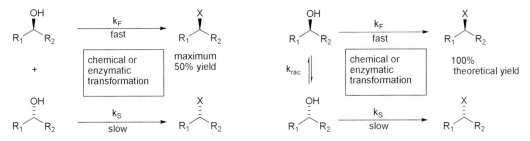

Scheme 1 Comparison of kinetic resolution (KR) and dynamic kinetic resolution (DKR) with alcohol as the substrate.

A successful DKR relies on the combination of a highly enantioselective kinetic resolution with an efficient racemization process. The kinetic resolution can be carried out with enzyme catalysis or chemocatalysis (e.g. transition metal catalysis). Racemization can be obtained in different ways including thermal racemization, base-catalyzed racemization and acid-catalyzed racemization. For compounds bearing an acidic proton (e.g. α to ketone and ester) base-catalyzed racemization is the method of choice. More recently transition metal-catalyzed racemization (of e.g. alcohols and amines) has become important.

An early example of non-enzymatic DKR was described in 1979 by Tai and coworkers who studied nickel-catalyzed hydrogenation of an oxobutyrate. This led to a moderate dynamic resolution and stimulated further studies in this

Asymmetric Synthesis – The Essentials.
Edited by Mathias Christmann and Stefan Bräse
Copyright © 2007 WILEY-VCH Verlag GmbH & Co. KGaA, Weinheim
ISBN: 978-3-527-31399-0

area. More recently, enzyme-based DKRs have been developed. Noyori has provided a mathematical treatment of DKR.

Combined Enzyme and Metal Catalysis for Efficient DKR

The use of biocatalysts for asymmetric transformations has become increasingly important in recent years. In KR most applications are done with enzyme catalysis and here lipases predominate. The fact that lipases can be used in organic solvents with improved chemo- regio- and enantioselectivity has dramatically increased their use. Combination of enzymatic resolution with base-catalyzed racemization of the starting material was described by the groups of Sih and Drueckhammer. Williams demonstrated the compatibility of enzymes and metal catalysts in one pot via palladium-catalyzed racemization of allylic acetates with concomitant enzymatic hydrolysis of the latter acetate leading to DKR. In 1997, we reported the first efficient DKR of alcohols using a ruthenium catalyst and a lipase [2].

Results

Reaction of alcohols with acyl donor **1** in the presence of *Candida antarctica* lipase B and ruthenium catalyst **2** at 70 °C resulted in a DKR to give the corresponding acetate of the alcohol in good to high yields and in excellent enantiomeric excess (Scheme 2) [2, 3].

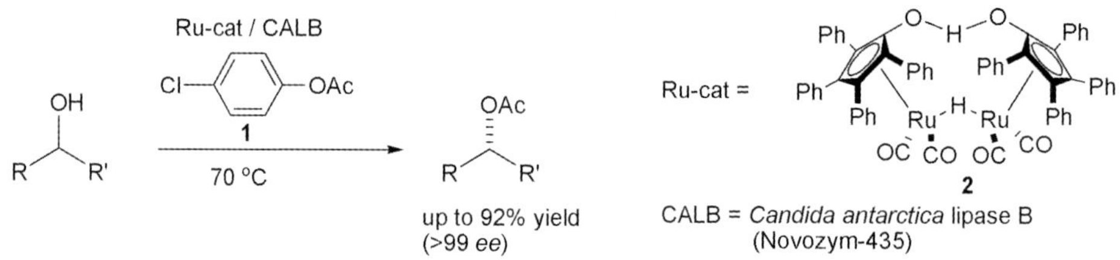

Scheme 2 Chemoenzymatic dynamic kinetic resolution.

The combination of enzyme- and transition metal-catalyzed reactions was subsequently successfully applied to the DKR of different functionalized alcohols that are useful building blocks for the synthesis of high-value compounds. Some examples are given in Scheme 2. Chloroalcohols, azidoalcoholes, cyanoalcohols and various hydroxy acid derivatives as well as hydroxyphosphonates were transformed into enantiomerically pure acetates via the DKR protocol developed. For example, enantioselective synthesis of β-hydroxy acid derivatives was achieved via a one-pot aldol reaction-dynamic kinetic resolution [4]. The β-chloroalcohol products are useful precursors for epoxides [5].

The first generation of chemoenzymatic DKR has been reviewed [6R].

With the first generation of a catalytic system using catalyst **2** and a lipase, an elevated temperature (e.g. 70 °C) and reaction times of 24–48 h were required. Recently, a highly efficient racemization catalyst was discovered in our group. Ruthenium catalyst **3** was found to racemize alcohols in less than 10 min at

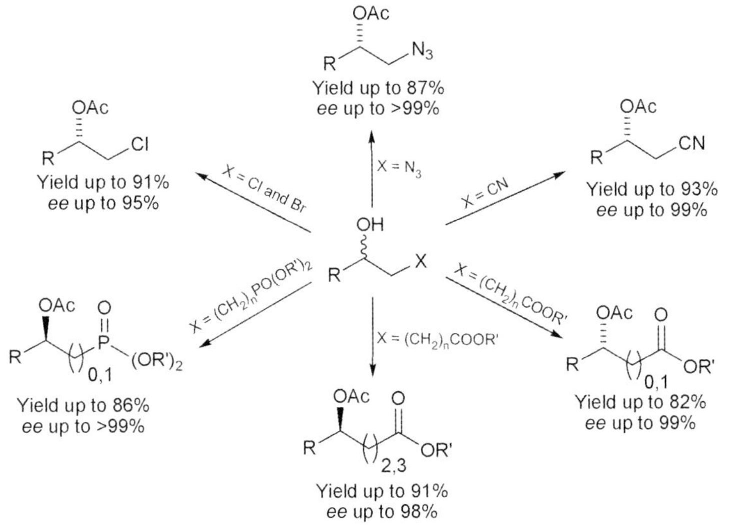

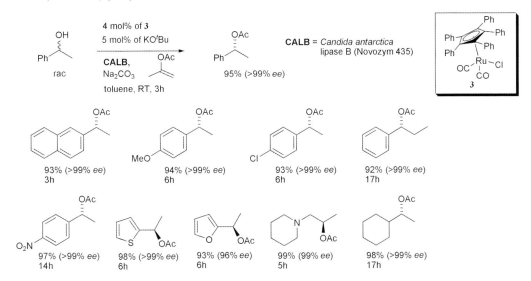

Scheme 3 Chemoenzymatic DKR of functionalized alcohols.

room temperature. The combination of this racemization catalyst with a lipase led to a highly compatible system for DKR of alcohols. With this new system, DKR of a wide variety of alcohols can be carried out at room temperature to give enantiomerically pure products (> 99% *ee*) in high yields in short reaction times (Scheme 4) [7, 8].

Scheme 4 Efficient chemoenzymatic DKR at room temperature.

In all the lipase-based systems developed for DKR of alcohols the (R)-configuration of the product is obtained. This has to do with the fact that essentially all lipases known follow the so-called Kazlauskas' rule where (R)-alcohols react faster than (S)-alcohols. Recently we combined the new catalyst 3 with an (S)-selective enzyme, a protease, which allows the preparation of alcohol products with the complementary configuration in a room temperature DKR with short reaction times [9].

Extension to DKR of Amines

Very recently, we have developed a DKR of primary amines by combining a ruthenium racemization catalyst and a lipase [10]. In this way, a racemic amine can be transformed to the corresponding amide in high yields and high ee. Reaction of racemic 1-phenylethylamine with isopropylacetate in the presence of catalyst 4 and the lipase (CALB) at 90 °C produced the N-acetylated amine in 90 % yield and > 98 % ee (Scheme 5). The examples with other amines show that the reaction is general.

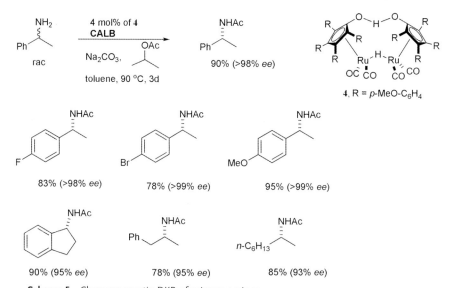

Scheme 5 Chemoenzymatic DKR of primary amines.

Conclusions and Future Perspectives

Over the past eight years, we have designed efficient chemoenzymatic methods for dynamic kinetic resolution (DKR) of alcohols. The recent efficient racemization catalyst 3 that works at room temperature led to a highly compatible DKR system that transforms a racemic alcohol into its ester in nearly quantitative yield and >99 % ee in short reaction times at room temperature. By combining this catalyst with a protease a highly efficient (S)-selective DKR process was obtained. Very recently, the DKR was extended to primary amines. Although the present approach via combination of an enzyme and a transition metal has so far been used only for alcohols and amines, extension to other substance classes should be possible. Future extensions will involve DKR of allenes with an axial chirality.

CV of Jan-E. Bäckvall

Jan-Erling Bäckvall was born in Malung, Sweden, in 1947. He received his Ph.D. from the Royal Institute of Technology, Stockholm, in 1975 under the guidance of Prof. B. Åkermark. After postdoctoral work (1975–1976) with Prof. K. B. Sharpless at Massachusetts Institute of Technology he joined the faculty at the Royal Institute of Technology. He was appointed Professor of Organic Chemistry at Uppsala University in 1986. In 1997, he moved to Stockholm University where he is currently Professor of Organic Chemistry. He is a member of the Royal Swedish Academy of Sciences and the Finnish Academy of Science and Letters. His current research interests include the use of transition metals in selective organic transformations, enzyme chemistry, and development of new and mild oxidation methodology. Recently efficient systems for dynamic kinetic resolution of alcohols based on combined ruthenium and enzyme catalysis have been developed in his laboratory.

Selected Publications

1R. F. F. Huerta, A. B. E. Minidis, J. E. Bäckvall, *Chem. Soc. Rev.* **2001**, *30*, 321–331. *Racemisation in Asymmetric Synthesis. Dynamic Kinetic Resolution and Related Processes in Enzyme and Metal Catalysis.*

2. A. L. E. Larsson, B. A. Persson, J. E. Bäckvall, *Angew. Chem.* **1997**, *109*, 1256–1258. *Angew. Chem. Int. Ed. Engl.* **1997**, *36*, 1211–1212. *Enzymatic Resolution of Alcohols Coupled with Ruthenium-Catalyzed Racemization of the Substrate Alcohol.*

3. B. A. Persson, A. L. E. Larsson, M. Le Ray, J. E. Bäckvall, *J. Am. Chem. Soc.* **1999**, *121*, 1645–1650. *Ruthenium- and Enzyme-Catalyzed Dynamic Kinetic Resolution of Secondary Alcohols.*

4. F. F. Huerta, J. E. Bäckvall, *Org. Lett.* **2001**, *3*, 1209–1212. *Enantioselective Synthesis of β-Hydroxy Acid Derivatives via a One-Pot Aldol Reaction-Dynamic Kinetic Resolution.*

5. O. Pàmies, J. E. Bäckvall, *J. Org. Chem.* **2002**, *67*, 9006–9010. *Chemoenzymatic Dynamic Kinetic Resolution of β-Halo Alcohols. An Efficient Route to Chiral Epoxides.*

6R. O. Pàmies, J. E. Bäckvall, *Chem. Rev.* **2003**, *103*, 3247–3262. *Combination of Enzymes and Metal Catalysts. A Powerful Approach in Asymmetric Catalysis.*

7. B. Martín-Matute, M. Edin, K. Bogár, J. E. Bäckvall, *Angew. Chem. Int. Ed. Engl.* **2004**, *43*, 6535–6539. *Highly Compatible Metal and Enzyme Catalysts for Efficient Dynamic Kinetic Resolution of Alcohols at Ambient Temperature.*

8. B. Martín-Matute, M. Edin, K. Bogár, F. B. Kaynak, J. E. Bäckvall, *J. Am. Chem. Soc.* **2005**, *127*, 8817–8825. *Combined Ruthenium(II) and Lipase Catalysis for Efficient Dynamic Kinetic Resolution of sec-Alcohols. Insight into a New Racemization Mechanism.*

9. L. Borén, B. Martín-Matute, Y. Xu, A. Córdova, J. E. Bäckvall, *Chem. Eur. J.* **2006**, *12*, 225–232. *(S)-Selective Kinetic Resolution and Chemoenzymatic Dynamic Kinetic Resolution of Secondary Alcohols.*

10. J. Paetzold, J. E. Bäckvall, *J. Am. Chem. Soc.* **2005**, *127*, 17620–17621. *Chemoenzymatic Dynamic Kinetic Resolution of Primary Amines.*

Catalytic Asymmetric Epoxidation of Enones and Related Compounds

Albrecht Berkessel, University of Cologne, Germany

General

The (asymmetric) synthesis of the epoxides **2**/*ent*-**2**, carrying electron withdrawing groups, is in most cases effected by oxygen transfer to the α,β-unsaturated and prochiral precursors **1** (Scheme 1). An alternative exists in the (formal) carbene addition to aldehydes which is discussed in the chapter by Aggarwal, page 52.

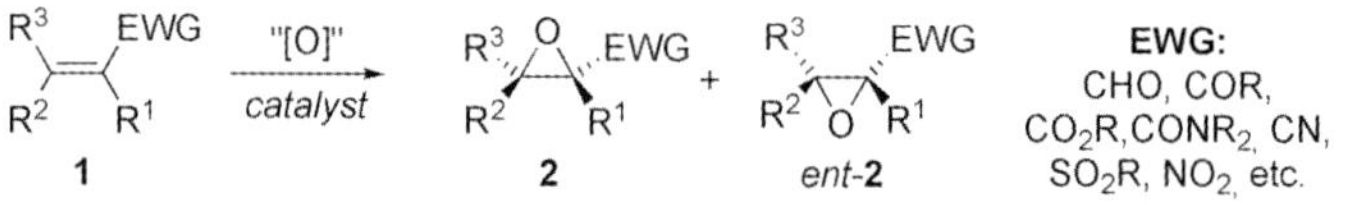

Scheme 1 Epoxidation of α,β-unsaturated carbonyl compounds, nitriles, sulfones, nitro compounds etc.

The oxidant may be either *electrophilic* or *nucleophilic* in nature (Scheme 2). Due to the relatively low electron density of the π-system of enones, enoates etc., only the most electrophilic oxidants (dioxiranes, oxo-complexes of high-valent transition metals) give satisfactory results. The epoxidation by nucleophilic oxidants (e.g. hydrogen peroxide or hydroperoxides in the presence of base) is well known as the *Weitz–Scheffer* reaction, and proceeds via a β-peroxyenolate intermediate (Scheme 2). Bifunctional catalysis may involve the activation of *both* the enone/enoate acceptor and the nucleophilic oxidant.

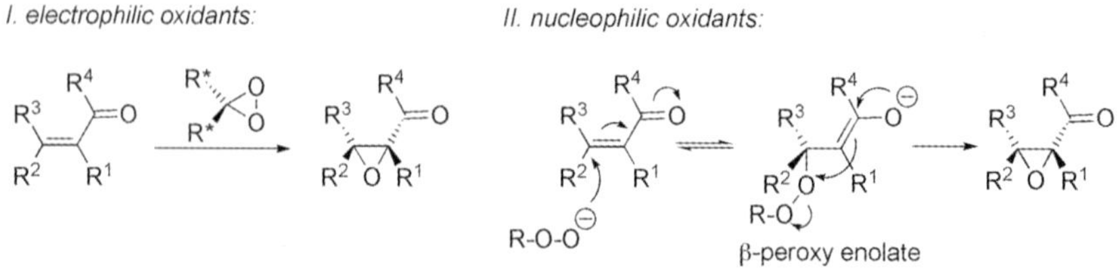

Scheme 2 Electrophilic and nucleophilic epoxidation of electron-poor C=C-double bonds.

Asymmetric Synthesis – The Essentials.
Edited by Mathias Christmann and Stefan Bräse
Copyright © 2007 WILEY-VCH Verlag GmbH & Co. KGaA, Weinheim
ISBN: 978-3-527-31399-0

Electrophilic Catalytic Epoxidation

Chiral ketone catalysts: Scheme 3 shows the structures of the chiral ketones 3–5 by Shi, Armstrong and Yang which have been used successfully in the asymmetric epoxidation of *E*-enones and enoates (up to 97% *ee*) [1R]. Typically, oxone is used for the *in situ* generation of the active dioxiranes from the chiral ketone precursors.

Metal-based catalysis: Manganese salen complexes such as Jacobsen's catalyst 6 [2R] can be used for the asymmetric epoxidation of *cis*-cinnamates with NaOCl (Scheme 3). Some *trans*-epoxides may be formed as side products.

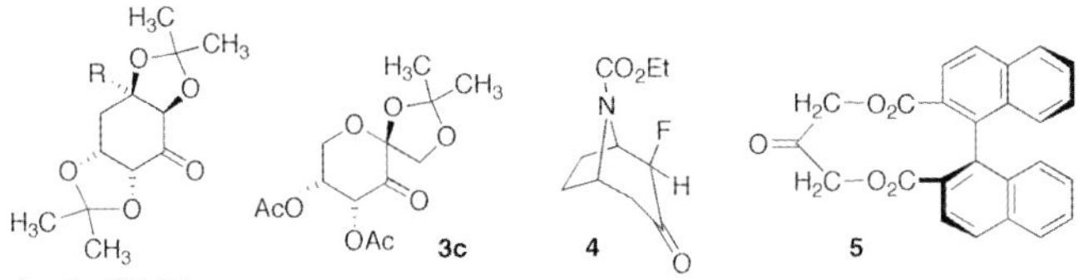

3a: R = CH$_2$OAc
3b: R = C(CH$_3$)$_2$OH

Figure 1 Selected ketone catalysts for the asymmetric epoxidation of enones and enoates.

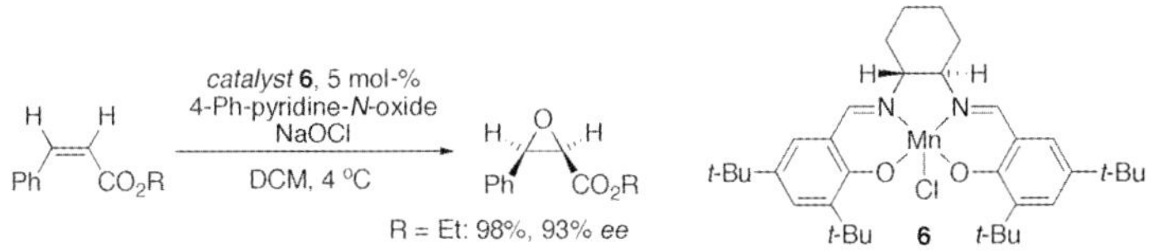

Scheme 3 Manganese catalysts for the asymmetric epoxidation of Z-cinnamates.

Nucleophilic Catalytic Epoxidation

Metal-based catalysis: Shibasaki et al. introduced the La-BINOL-Ph$_3$AsO catalyst (generated from La(O-*i*-Pr)$_3$, BINOL and Ph$_3$AsO in a 1:1:1-ratio) which allows the epoxidation of a variety of *E*-enones 7 in high yields and *ee* (Scheme 5) [3]. A related Yb-complex catalyzes the epoxidation of Z-enones with up to 96% *ee*, and with negligable formation of *trans*-epoxide [4]. For the epoxidation of α,β-unsaturated amides 8, a Sm-BINOL-Ph$_3$AsO (1:1:1) complex proved best (Scheme 4) [5]. Shibasaki et al. furthermore introduced the combination of ytterbium with the ligand 10 and Ph$_3$AsO for the highly enantioselective epoxidation of a variety of enoates 9 [6].

Jackson et al. developed an alternative catalyst system, based on Mg-tartrate complexes that catalyze the epoxidation of *E*-enones in up to 95 % *ee* [7, 8]. Enders et al. discovered that in the presence of chiral aminoalcohols such as *N*-methylpseudoephedrine, *E*-enones can be converted to the corresponding *trans*-epoxides (up to 99% *ee*) by treatment with diethylzinc in the presence of oxygen [9, 10].

Peptide catalysts: Chalcones and related enones can be epoxidized with alkaline hydrogen peroxide in the presence of poly-amino acids (Juliá-Colonna epoxidation, Scheme 5) [1]. Based on mechanistic studies by Berkessel et al. [11, 12], catalysis is effected by hydrogen bonding of the substrates to the N-terminus of the helical peptide catalyst: Two NH-bonds activate the enone towards nucleophilic

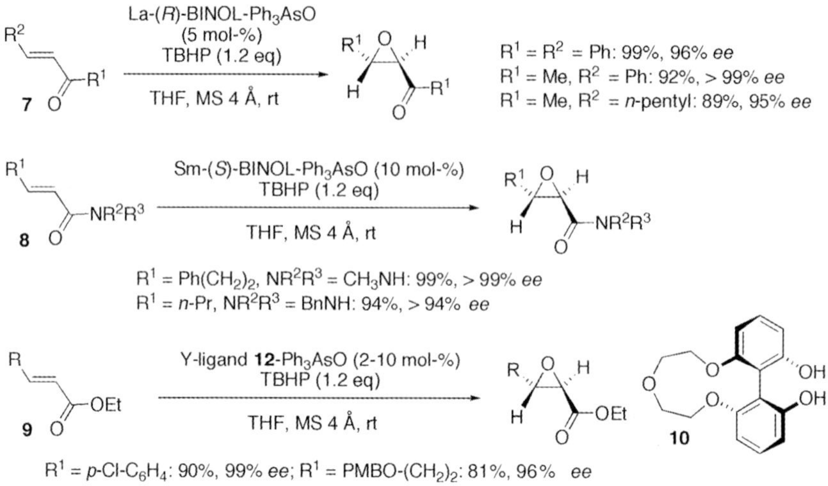

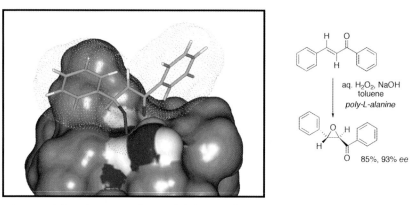

Scheme 4 Epoxidation of enones, amides and enoates by lanthanide catalysts.

Scheme 5 Juliá-Colonna epoxidation of chalcone. Left: Hydrogen bonding of the intermediate β-peroxyenolate to the peptide catalyst (Berkessel et al., [11, 12]). Right: The original Juliá-Colonna reaction.

attack, whereas a third H-bond orients the hydroperoxide nucleophile for selective addition to one of the enantiotopic faces of the enone. Berkessel et al. furthermore showed that for TentaGel-bound catalysts, enantioselectivities > 95 % *ee* can be achieved with as little as five *L*-Leu residues.

Phase-transfer catalysis: Figure 2 shows the phase-transfer catalysts by Corey, Lygo (**11, 12**) and Arai (**13**) which allow the epoxidation of a variety of acyclic enones with alkaline hydrogen peroxide or hypohalites with enantioselectivities of up to 95 % [1]. Even higher selectivities (up to 99 % *ee*), at low catalyst loading, were achieved by Maruoka et al., using the chiral ammonium salt **14** [13].

Quinones are notoriously „difficult" substrates, with menaquinone (vitamin K$_3$, **15**) serving as the touchstone for the development of novel epoxidation methods.

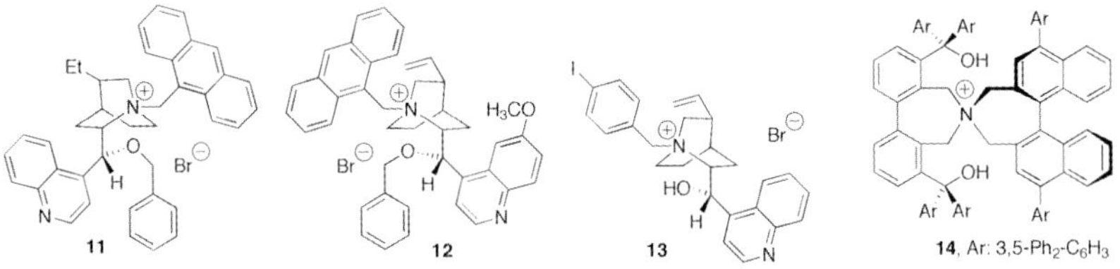

Figure 2 Phase-transfer catalysts for the asymmetric epoxidation of enones.

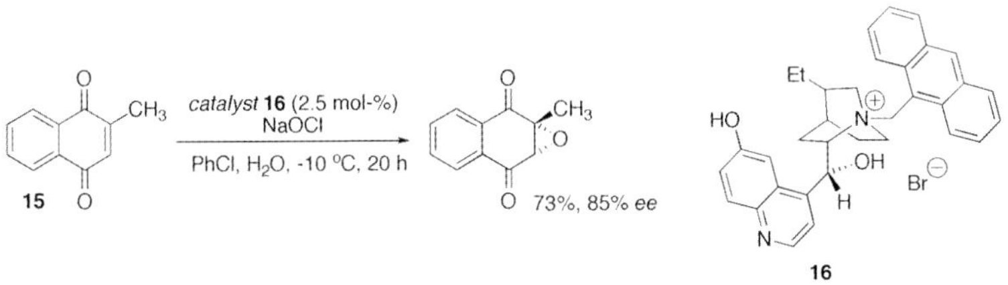

Scheme 6 Phase-transfer catalyzed asymmetric epoxidation of menaquinone (vitamin K_3, **15**).

Berkessel et al. achieved the highest *ee* reported so far for the epoxidation of **15** by using the phase-transfer catalyst **16**, in combination with NaOCl as oxidant (Scheme 7) [14]. The latter catalyst is readily available in three steps from quinine.

Conclusions and Future Perspectives
Highly enantioselective catalytic epoxidation methods have been developed for acyclic enones, enoates, and some enoate derivatives. Cyclic enones and quinones remain "difficult" substrates. A broadly applicable method for the catalytic asymmetric epoxidation of the latter continues to be a highly desirable goal for future research.

CV of Albrecht Berkessel
Albrecht Berkessel was born in 1955 and obtained his Diplom in 1982 at the University of Saarbrücken. For his Ph.D. studies, he moved to the laboratory of Professor Waldemar Adam at the University of Würzburg. In 1985, he obtained his for mechanistic studies on the photochemistry of divinyl ethers. In 1985, he joined the research group of Professor Ronald Breslow at Columbia University, New York, to work on functionalized cyclodextrins as enzyme models and on the mechanism of biotin action. In 1986, he returned to Germany to start independent research on the mechanisms of nickel enzymes from methanogenic archaea. His habilitation at the University of Frankfurt/Main (associated to Professor Gerhard Quinkert) was completed in 1990. In 1992, he became Associate Professor at the University of Heidelberg. Since 1997, he has been a Full Professor of Organic Chemistry at the University of Cologne. His research interests center around various aspects of catalysis, such as mechanism and method development in metal-based catalysis and in particular organocatalysis, biomimetic chemistry, as well as combinatorial methods for catalyst development.

Selected Publications

1R. A. Berkessel, H. Gröger, *Asymmetric Organocatalysis*, Wiley-VCH, Weinheim, **2005**, Ch.10.2.

2R. E. N. Jacobsen, A. Pfaltz, H. Yamamoto (Eds.), *Comprehensive Asymmetric Catalysis*, Springer, Berlin, **1999**.

3. T. Nemoto, T. Ohshima, K. Yamaguchi, M. Shibasaki, *J. Am. Chem. Soc.* **2001**, *123*, 2725-2732. *Catalytic Asymmetric Epoxidation of Enones Using La-BINOL-Triphenylarsine Oxide Complex: Structural Determination of the Asymmetric Catalyst.*

4. S. Watanabe, T. Arai, H. Sasai, M., Bougauchi, M. Shibasaki, *J. Org. Chem.* **1998**, *63*, 8090–8091. *The First Catalytic Enantioselective Synthesis of cis-Epoxyketones from cis-Enones.*

5. T. Nemoto, H. Kakei, V. Gnandesikan, S. Tosaki, T. Ohshima, M. Shibasaki, *J. Am. Chem. Soc.* **2002**, *124*, 14544–14545. *Catalytic Asymmetric Epoxidation of α,β-Unsaturated Amides: Efficient Synthesis of β-Aryl α-Hydroxy Amides using a One-Pot Tandem Catalytic Asymmetric Epoxidation-Pd-Catalyzed Epoxide Opening Process.*

6. H. Kakei, R. Tsuji, T. Ohshima, M. Shibasaki, *J. Am. Chem. Soc.* **2005**, *127*, 8962–8963. *Catalytic Asymmetric Epoxidation of α,β-Unsaturated Esters Using an Yttrium-Biphenyldiol Complex.*

7. C. L. Elston, R. F. W. Jackson, S. J. F. MacDonald, P. J. Murray, *Angew. Chem.* **1997**, *109*, 379–381; *Angew. Chem. Int. Ed.* **1997**, *36*, 410–412. *Asymmetric Epoxidation of Chalcones with Chirally Modified Lithium and Magnesium tert-Butyl Peroxides.*

8. O. Jacques, S. J. Richards, R. F. W. Jackson, *Chem. Commun.* **2001**, 2712–2713. *Catalytic Asymmetric Epoxidation of Aliphatic Enones Using Tartrate-Derived Magnesium Alkoxides.*

9. D. Enders, J. Zhu, G. Raabe, *Angew. Chem.* **1996**, *108*, 1827–1829; *Angew. Chem. Int. Ed.* **1996**, *35*, 1725–1728. *Asymmetric Epoxidation of Enones With Oxygen in the Presence of Diethylzinc and (R,R)-N-Methylpseudoephedrine.*

10. D. Enders, J. Zhu, L. Kramps, *Liebigs Ann./Recueil* **1997**, 1101–1113. *Zinc-Mediated Asymmetric Epoxidation of α-Enones.*

11. A. Berkessel, N. Gasch, K. Glaubitz, C. Koch, *Org. Lett.* **2001**, *3*, 3839–3842. *Highly Enantioselective Enone Epoxidation Catalyzed by Short Solid Phase-Bound Peptides: Dominant Role of Peptide Helicity.*

12. A. Berkessel, B. Koch, C. Toniolo, M. Rainaldi, Q. B. Broxterman, B. Kaptein, *Biopolymers: Pept. Sci.* **2006**, *84*, 90–96. *Asymmetric Enone Epoxidation by Short Solid-Phase Bound Peptides: Further Evidence for Catalyst Helicity and Catalytic Activity of Individual Peptide Strands.*

13. T. Ooi, D. Ohara, M. Tamura, K. Maruoka, *J. Am. Chem. Soc.* **2004**, *126*, 6844–6845. *Design of New Phase-Transfer Catalysts with Dual Function for Highly Enantioselective Epoxidation of α,β-Unsaturated Ketones.*

14. A. Berkessel, M. Guixa, F. Schmidt, submitted for publication.

Kinetic Investigations of the Soai Autocatalytic Reaction

Donna G. Blackmond, Imperial College, United Kingdom

Background

The origin of biological homochirality has intrigued scientists ever since the importance of L-amino acids and D-sugars was first recognized. In a theoretical paper, Frank [1] showed that if one hand of a primitive asymmetric catalyst could act to replicate itself and, at the same time, act to suppress replication of its opposite enantiomer, this would provide a "simple and sufficient life model" to explain how homochirality could have developed from an initial small imbalance of enantiomers. Experimental confirmation of this theory came with Soai's discovery that the alcohol product of the alkylation of pyrimidyl aldehydes (Eq. (1)) catalyzes its own production at a much greater rate than it does the production of its enantiomer [2].

The groups of Blackmond and Brown have carried out extensive studies of mechanistic aspects of this reaction, both jointly [3] and separately [4]. This body of work has led to a coherent mechanistic model of the Soai reaction that includes the following important features: (i) the alkanol product of the reaction is significantly driven towards dimerization; (ii) heterochiral and homochiral dimers are formed stochastically with approximately equal thermodynamic stability; and (iii) homochiral dimers are implicated as the active species in autocatalytic reactions in which product enantiomeric excess is amplified.

This review highlights the importance of combining accurate *in situ* kinetic studies with detailed spectroscopic characterization of species in developing a fundamental mechanistic picture of this complex and intriguing reaction system.

Kinetic Studies using Reaction Calorimetry

Autocatalytic processes exhibit complex temporal reaction progress profiles, featuring time-dependent product selectivities and catalyst concentrations. One key to understanding such systems lies in finding accurate methods for analyzing the progress of these reactions. Mechanistic studies by our group focus on detailed reaction progress kinetic analysis using *in situ* tools that provide virtually continuous temporal rate profiles [5]. Our experimental technique of choice in many cases is reaction calorimetry. This technique relies on the accurate measurement of the heat evolved or consumed when chemical transformations occur. Consider a catalytic reaction proceeding in the absence of side reactions or other thermal effects. The energy characteristic of the transformation – the heat of reaction, ΔH_{rxn} – is manifested each time a substrate molecule is converted to a product molecule. This thermodynamic quantity serves as the proportionality constant between the evolved heat and the reaction rate. The fractional heat evolution is identical to the fraction conversion of the limiting substrate. Eq. (2) demonstrates these relationships.

$$q = \Delta H_{rxn} \cdot [\text{reaction volume}] \cdot \text{rate}; \quad \text{fraction conversion} = \frac{\int_0^t q(t)dt}{\int_0^{t(\text{final})} q(t)dt} \quad (2)$$

The primary data obtained in a reaction monitored by reaction calorimetry is a kinetic profile of heat flow as a function of time. This is shown in Figure 1a for the Soai reaction carried out using 10 mol% alkanol catalyst present as enantiopure, racemic, and 43% *ee* [3]. These data may be converted to fraction conversion vs. time using Eq. (1), as shown in Figure 1b. A third kinetic plot may be obtained by combining the first two plots. By removing time as an explicit variable, we obtain a plot of normalized rate (rate(*t*)/max rate) vs. fraction conversion in the plot in Figure 1c. Such a plot represents the quantities found to the left and right in a reaction rate equation, and therefore is called a *"graphical rate equation"* [5].

The wealth of information contained in these reaction progress profiles enabled us to propose a mechanistic model for the Soai reaction. Figure 1a shows that the maximum rate of the reaction carried out using enantiopure catalyst was double that with the racemic catalyst. Figure 1c shows that the maximum rate was attained at identical fraction conversion for these two reactions, while for the 43% *ee* catalyst the maximum rate was shifted by ca. 5% toward higher conversion. Note that neither of these features of the data may be easily extracted from the data plotted as fraction conversion vs. time in Figure 1b, the most common type of kinetic profile.

Models for rationalizing nonlinear effects in catalytic reactions invoke the formation of dimeric species, with the dimers themselves acting as catalysts or serving as an inactive reservoir (Scheme 1) [6]. Homochiral and heterochiral dimerization equilibrium constants can be combined to give K_{dimer}, a parameter that reveals the relative proportions of heterochiral to homochiral dimers. These models may be extended to autocatalytic reactions [3].

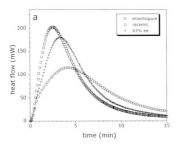 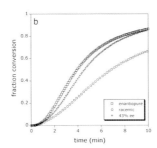 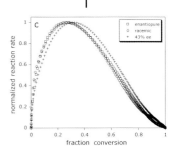

Figure 1 Reaction calorimetric monitoring of the Soai reaction using 10 mol% enantiopure, 43% *ee* and racemic alkanol **1** as catalyst. a) reaction heat flow vs. time; b) fraction conversion vs. time; c) normalized rate vs. fraction conversion.

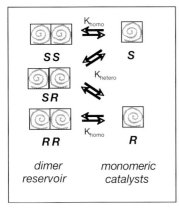 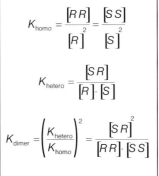 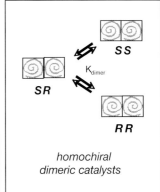

Scheme 1 Models for rationalizing asymmetric amplification in autocatalytic reactions.

Models that invoke monomer as catalyst cannot simultaneously satisfy both of the experimental observations noted above. For a system exhibiting a low value of K_{dimer}, the maximum rate for racemic and enantiopure monomeric catalysts occurs at a similar conversion, but the racemic rate approaches that of enantiopure; at higher values of K_{dimer}, the enantiopure:racemic rate ratio approaches 2:1, but the rate maximum does not occur at the same conversion in the two cases. Only in the limiting case where $K_{hetero} \gg K_{homo}$ and both are very large – meaning that K_{dimer} will be very large – can a monomer-active model approximately account for the experimental observations. In that case the monomer model predicts that for a racemic system, most of the product will be present as heterochiral dimers, leaving only a small fraction of product available to act further as a monomeric autocatalyst.

The simplest dimeric model that accounts for our experimental observations as described above dictates that the system must consist predominantly of dimers (K_{homo} and K_{hetero} are both large) that have approximately equal stability ($K_{hetero} = 2 \cdot K_{homo}$, or $K_{dimer} = 4$), and that the heterochiral dimer is inactive or significantly less active than the homochiral dimers. Kinetic modelling of the reaction rate data shown in Figure 1 corroborated this model, as is shown in Figure 2 [3c]. In addi-

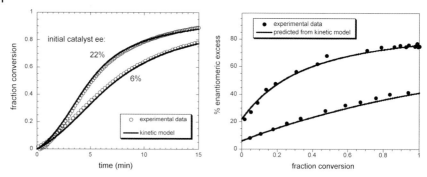

Figure 2 Comparison of experimental kinetic data and dimer model calculations for reactions at 22% and 6% initial catalyst enantiomeric excess.

tion, the model accurately predicts the evolution of enantiomeric excess for reactions initiated with alkanol catalyst at different ee, based solely on the rate data.

The stochastic distribution of dimer species predicted by the kinetic model was verified experimentally by ^{1}H NMR spectroscopic characterization of solutions of enantiopure and racemic alkanols [3, 4e]. Solutions of enantiopure alkanol revealed only homochiral dimer species. Racemic mixtures reveal heterochiral and homochiral dimers present in a 52:48 ratio, within experimental error of the 50:50 ratio predicted by the kinetic model with K_{dimer} = 4. The importance of this observation must be stressed. The veracity of a model developed from kinetic investigations alone requires corroboration from other methods, since a unique fit to a single mechanism is unlikely. Spectroscopic and structural identification of species predicted by a kinetic model can add significant weight to a mechanistic proposal and may help to eliminate mechanistic possibilities. The observation of a stochastic distribution of dimers in the present case rules out the possibility of a monomer-active model for asymmetric amplification in the Soai reaction under these conditions.

This mechanistic model highlights another important point contrasting this autocatalytic Soai reaction with similar catalytic dialkylzinc alkylations of aldehydes that exhibit asymmetric amplification. The equal stability of homochiral and heterochiral dimers (K_{dimer} = 4) observed in the Soai reaction can lead to significant amplification in a dimer-active *autocatalytic* reaction, where the same value of K_{dimer} in a monomer-active *catalytic* reaction would lead to *no* amplification of product enantiomeric excess. Indeed, asymmetric amplification of the levels observed in diethylzinc alkylations of benzaldehyde [6b] require substantial bias toward the heterochiral dimer, giving values of K_{dimer} = 5000.

In summary, a strong body of evidence suggests that the alkanol reaction product dimerizes to form the active catalyst in the Soai reaction. Homochiral and heterochiral dimers are formed without preference, but only the homochiral dimers are active as catalysts. The combination of *in situ* kinetic tools and spectroscopic characterization of the catalyst resting state presents a powerful approach to the mechanistic study of complex reaction networks.

CV of Donna G. Blackmond

Donna G. Blackmond is Professor of Chemistry, Professor of Chemical Engineering, and Chair in Catalysis at Imperial College London. Her research interests include kinetic aspects of complex catalytic reactions, especially asymmetric catalysis.

References

1. F. C. Frank, *Biochim. Biophys. Acta* **1953**, *11*, 459–463. Spontaneous Asymmetric Synthesis.
2. (a) K. Soai, T. Shibata, H. Morioka, K. Choji, *Nature* **1995**, *378*, 767–768. Asymmetric Autocatalysis and Amplification of Enantiomeric Excess of a Chiral Molecule. (b) T. Shibata, K. Choji, T. Hayase, Y. Aizu, K. Soai, *Chem. Commun.* **1996**, 1235–1236. Asymmetric Autocatalytic Reaction of 3-Quinolylalkanol with Amplification of Enantiomeric Excess. (c) T. Shibata, H. Morioka, T. Hayase, K. Choji, K. Soai, *J. Am. Chem. Soc.* **1996**, *118*, 471–472. Enantioselective Catalytic Asymmetric Automultiplication of Chiral Pyrimidyl Alcohol.
3. D. G. Blackmond, C. R. McMillan, S. Ramdeehul, A. Schorm, J. M. Brown, *J. Am. Chem. Soc.* **2001**, *123*, 10103–10104. Origins of Asymmetric Amplification in Autocatalytic Alkylzinc Additions.
4. (a) D. G. Blackmond, *Adv. Synth. Catal.* **2002**, *344*, 156–158. Description of the Condition for Asymmetric Amplification in Autocatalytic Reactions. (b) I. D. Gridnev, J. M. Serafimov, H. Quiney, J. M. Brown, *Org. Biomol. Chem.* **2003**, *1*, 3811–3819. Reflections on Spontaneous Asymmetric Synthesis by Amplifying Autocatalysis. (c) F. G. Buono, D. G. Blackmond, *J. Am. Chem. Soc.* **2003**, *125*, 8978–8979. Kinetic Evidence for a Tetrameric Transition State in the Asymmetric Autocatalytic Alkylation of Pyrimidyl Aldehydes. d) F. G. Buono, H. Iwamura, D. G. Blackmond, *Angew. Chem. Int. Ed.* **2004**, *43*, 2099–2103. Physical and Chemical Rationalization for Asymmetric Amplification in Autocatalytic Reactions. (e) I. D. Gridnev, J. M. Serafimov, J. M. Brown, *Angew. Chem. Int. Ed.* **2004**, *43*, 4884–4887. Solution structure and Reagent Binding of the Zinc Alkoxide Catalyst in the Soai Asymmetric Autocatalytic Reaction. (f) I. D. Gridnev, J. M. Brown, *Proc. Nat. Acad. Science USA* **2004**, *101*, 5727–5731. Asymmetric Autocatalysis: Novel Structures, Novel Mechanism? (g) D. G. Blackmond, *Proc. Nat. Acad. Science USA* **2004**, *101*, 5732–5736. Asymmetric Autocatalysis and its Implications for the Origin of Homochirality.
5. D. G. Blackmond, *Angew. Chemie Int. Ed.* **2005**, *44*, 4302–4320. Reaction Progress Kinetic Analysis: A Powerful Methodology for Mechanistic Studies of Complex Catalytic Reactions.
6. (a) C. Girard, H. B. Kagan, *Angew. Chem. Int. Ed. Engl.* **1998**, *37*, 2923–2959. Nonlinear Effects in Asymmetric Synthesis and Stereoselective Reactions: Ten Years of Investigation. (b) M. Kitamura, S. Suga, H. Oka, R. Noyori, *J. Am. Chem. Soc.* **1998**, *120*, 9800–9809. Quantitative Analysis of the Chiral Amplification in the Amino Alcohol-Promoted Asymmetric Alkylation of Aldehydes with Dialkylzincs.

Planar-chiral Heterocycles as Enantioselective Organocatalysts

Gregory Fu, Massachusetts Institute of Technology, USA

Background and Design

Because of the tremendous versatility of 4-(dimethylamino)pyridine (DMAP) as a nucleophilic catalyst [1–4], in 1995 we initiated a program directed at developing an effective chiral variant. Surprisingly, at that time there were no examples of applications of chiral DMAP derivatives in asymmetric catalysis.

Molecules that possess an improper axis of symmetry (e.g., a mirror plane) are achiral. Recognizing that DMAP has two mirror planes, we decided to generate a chiral variant by eliminating these symmetry elements (Figure 1) [5].

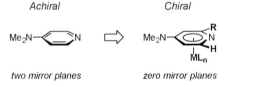

Figure 1 Achiral DMAP vs. a planar-chiral derivative of DMAP.

At the time that we began our studies, there had been no reports of applications of planar-chiral [6] heterocycles in asymmetric catalysis. Indeed, there was only one description of even an effort to obtain such a compound in an enantiopure form [7].

Applications

Upon synthesizing several planar-chiral derivatives of DMAP (e.g., **1–4**), we have investigated their utility in a variety of transformations. For our initial study, we chose to examine the catalytic process for which DMAP is best known: the acylation of alcohols by anhydrides. Specifically, we explored the kinetic resolution of secondary alcohols. We were pleased to discover that planar-chiral DMAP derivatives effectively resolve a range of alcohols with useful selectivity factors (Figure 2) [8].

Asymmetric Synthesis – The Essentials.
Edited by Mathias Christmann and Stefan Bräse
Copyright © 2007 WILEY-VCH Verlag GmbH & Co. KGaA, Weinheim
ISBN: 978-3-527-31399-0

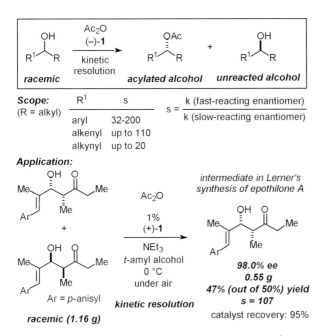

Figure 2 The kinetic resolution of alcohols catalyzed by a planar-chiral derivative of DMAP, including an application to the generation of an intermediate employed in the synthesis of enantiopure epothilone A.

Planar-chiral DMAP catalysts can also achieve the kinetic resolution of amines via an acylation process (Figure 3) [9]. Prior to this investigation, such resolutions had only been accomplished with enzymatic catalysts.

We have also examined the utility of planar-chiral DMAP derivatives as catalysts for enantioselective reactions of ketenes. For example, we have achieved asymmetric Staudinger syntheses of highly substituted β-lactams from ketenes and imines. Thus, either *cis* or *trans* β-lactams can be generated, depending on the protecting group on nitrogen (Figure 4) [10, 11]. We speculate that reactions of *N*-tosyl and *N*-triflyl imines lead to opposite stereochemical outcomes due to different reaction mechanisms.

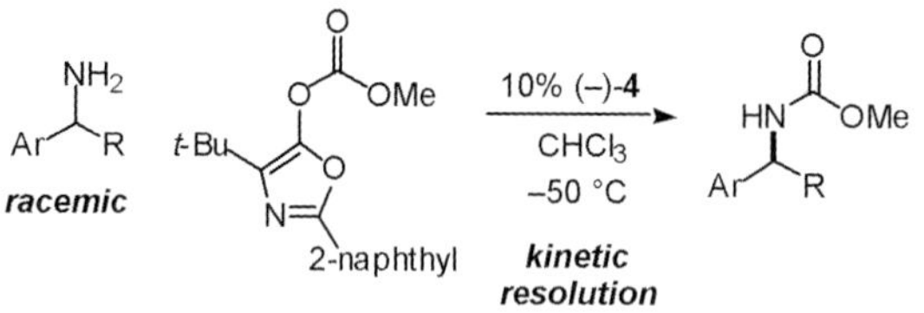

Figure 3 The kinetic resolution of amines.

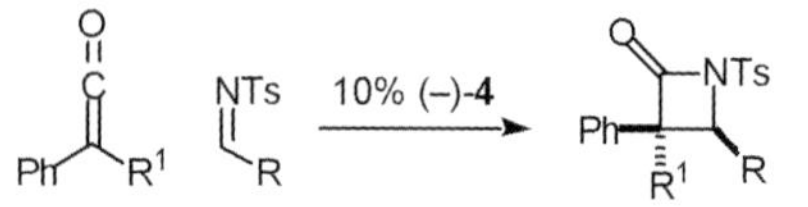

Ar	R	s
Ph	Me	12
o-tol	Me	16
1-naphthyl	Me	27
Ph	Et	16
4-(OMe)C$_6$H$_4$	Me	11
4-(CF$_3$)C$_6$H$_4$	Me	13

R^1	R	dr	ee (%)	yield (%)
i-Bu	cyclopropyl	15:1	89	88
i-Bu	Ph	8:1	98	88
i-Bu	2-phenylethenyl	10:1	98	95
Et	cyclopropyl	10:1	98	98
Et	2-furyl	9:1	95	97

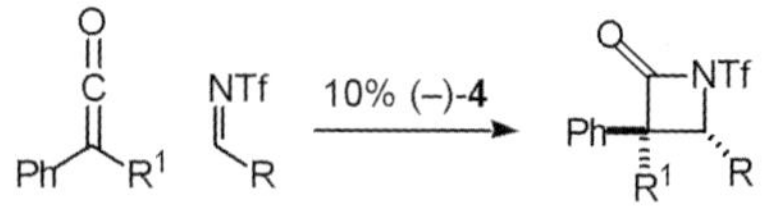

R^1	R	dr	ee (%)	yield (%)
Et	Ph	6:1	63	60
Me	Ph	50:1	81	83
i-Bu	Ph	30:1	63	72
Me	4-(CF$_3$)C$_6$H$_4$	30:1	69	80
Me	o-tolyl	4:1	99	89

Figure 4 Staudinger synthesis of *cis* and *trans* β-lactams.

Furthermore, we have explored the catalytic asymmetric addition of amines to ketenes. At the time of our investigation, there had been no success in addressing this challenge. We were therefore pleased to determine that planar-chiral catalyst 4 can accomplish this transformation with good enantioselectivity (Figure 5) [12].

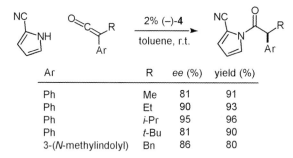

Ar	R	ee (%)	yield (%)
Ph	Me	81	91
Ph	Et	90	93
Ph	i-Pr	95	96
Ph	t-Bu	81	90
3-(N-methylindolyl)	Bn	86	80

Figure 5 Addition of amines to ketenes.

On the basis of mechanistic studies, we concluded that this reaction likely proceeds through chiral Brønsted acid, rather than nucleophilic, catalysis (Figure 6). This mode of reactivity was unanticipated, and it has added a new dimension to the use of planar-chiral heterocycles in asymmetric catalysis.

We hoped to establish that protonated planar-chiral heterocycles could serve as enantioselective Brønsted-acid catalysts for an array of other processes. We therefore investigated a coupling that was designed to favor this mode of reactivity: the

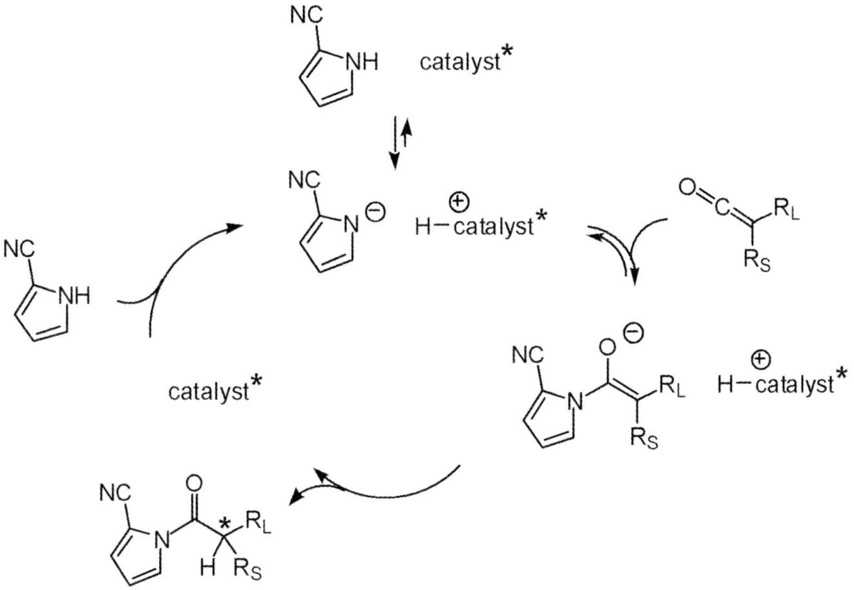

Figure 6 Proposed mechanism for the addition of amines to ketenes catalyzed by a planar-chiral DMAP derivative.

reaction of phenols with ketenes to generate aryl esters. We were gratified that the desired transformation proceeds with high selectivity for the predicted enantiomer, presumably via the illustrated ion pair (Figure 7) [13].

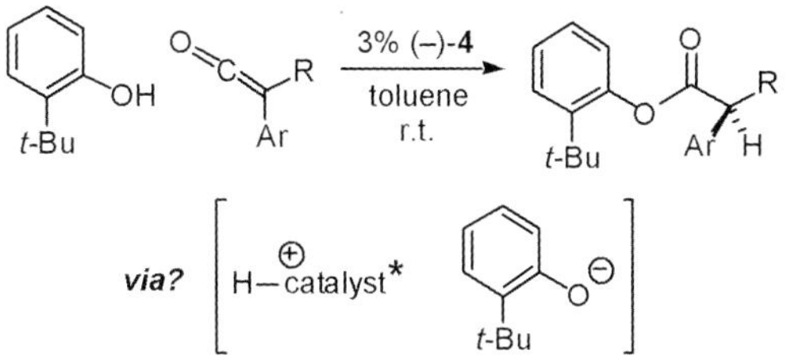

Ar	R	ee (%)	yield (%)
Ph	Me	79	87
Ph	Et	91	89
Ph	i-Pr	91	66
o-tol	Et	92	84
3-thienyl	i-Pr	79	94

Figure 7 Addition of phenols to ketenes.

Conclusions and Future Perspectives

Planar-chiral heterocycles serve as highly enantioselective organocatalysts (nucleophilic and Brønsted acid) for a variety of useful processes. We believe that the investigations outlined in this account represent only the "tip of the iceberg" with regard to their utility, and we look forward to pursuing a wide range of studies that will substantiate this hypothesis.

CV of Gregory C. Fu

Gregory C. Fu was born in Galion, Ohio in 1963. He received a B.S. degree from MIT in 1985, where he worked in the laboratory of Professor Barry Sharpless. After earning a Ph.D. from Harvard in 1991 under the guidance of Professor David Evans, he spent two years as a postdoctoral fellow with Professor Robert Grubbs at Caltech. In 1993, he returned to MIT, where he was promoted to Professor of Chemistry in 1999. He serves as an associate editor of the *Journal of the American Chemical Society*, and he is the recipient of the 2006 Mukaiyama Award of the Society of Synthetic Organic Chemistry of Japan, the 2004 Corey Award of the American Chemical Society, and the 2001 Springer Award in Organometallic Chemistry.

Selected Publications

1. W. Steglich, G. Höfle, *Angew. Chem., Int. Ed. Engl.* **1969**, *8*, 981. N,N-Dimethyl-4-pyridinamine, A Very Effective Acylation Catalyst.
2. L. M. Litvinenko, A. I. Kirichenko, *Dokl. Akad. Nauk SSSR, Ser. Khim.* **1967**, *176*, 97–100. Basicity and Stereospecificity in Nucleophilic Catalysis by Tertiary Amines.
3R. A. C. Spivey, S. Arseniyadis, *Angew. Chem., Int. Ed.* **2004**, *43*, 5436–5441. Nucleophilic Catalysis by 4-(Dialkylamino)pyridines Revisited – The Search for Optimal Reactivity and Selectivity.
4R. R. Murugan, E. F. V. Scriven, *Aldrichimica Acta* **2003**, *36*, 21–27. Applications of Dialkylaminopyridine (DMAP) Catalysts in Organic Synthesis.
5. G. C. Fu, *Acc. Chem. Res.* **2000**, *33*, 412–420. Enantioselective Nucleophilic Catalysis with "Planar-Chiral" Heterocycles.
6. R. S. Cahn, C. Ingold, V. Prelog, *Angew. Chem., Int. Ed. Engl.* **1966**, *5*, 385–415. Specification of Molecular Chirality.
7. K. Bauer, H. Falk, K. Schlögl, *Angew. Chem., Int. Ed. Engl.* **1969**, *8*, 135. Optically Active 2-Methylazaferrocene.
8. S. Bellemin-Laponnaz, J. Tweddell, J. C. Ruble, F. M. Breitling, G. C. Fu, *Chem. Commun.* **2000**, 1009–1010. The Kinetic Resolution of Allylic Alcohols by a Non-Enzymatic Acylation Catalyst; Application to Natural Product Synthesis.
9. S. Arai, S. Bellemin-Laponnaz, G. C. Fu, *Angew. Chem. Int. Ed.* **2001**, *40*, 234–236. Kinetic Resolution of Amines by a Non-Enzymatic Acylation Catalyst.
10. B. L. Hodous, G. C. Fu, *J. Am. Chem. Soc.* **2002**, *124*, 1578–1579. Enantioselective Staudinger Synthesis of β-Lactams Catalyzed by a Planar-Chiral Nucleophile.
11. E. C. Lee, B. L. Hodous, E. Bergin, C. Shih, G. C. Fu, *J. Am. Chem. Soc.* **2005**, *127*, 11586–11587. Catalytic Asymmetric Staudinger Reactions to Form β-Lactams: An Unanticipated Dependence of Diastereoselectivity on the Choice of the Nitrogen Substituent.
12. B. L. Hodous, G. C. Fu, *J. Am. Chem. Soc.* **2002**, *124*, 10006–10007. Enantioselective Addition of Amines to Ketenes Catalyzed by a Planar-Chiral Derivative of PPY: Possible Intervention of Chiral Brønsted-Acid Catalysis.
13. S. L. Wiskur, G. C. Fu, *J. Am. Chem. Soc.* **2005**, *127*, 6176–6177. Catalytic Asymmetric Synthesis of Esters from Ketenes.

An Organocatalytic Approach to Optically Active Six-membered Rings

Karl Anker Jørgensen, Aarhus University, Denmark

Background

Organocatalysis has emerged as a new and important field in chemistry for the construction of chiral compounds [1–3]. A variety of different new asymmetric reactions has been developed using chiral amines as the catalyst [1R–3R].

For the enantioselective α-functionalization of aldehydes and ketones a large number of different secondary amine-based catalysts have been developed. The mechanism for the α-functionalization of aldehydes and ketones is based on the formation of a chiral enamine intermediate (Scheme 1). The chiral substituent of the secondary amine catalyst determines the stereoselective approach of the electrophile to the nucleophilic carbon atom in the enamine intermediate.

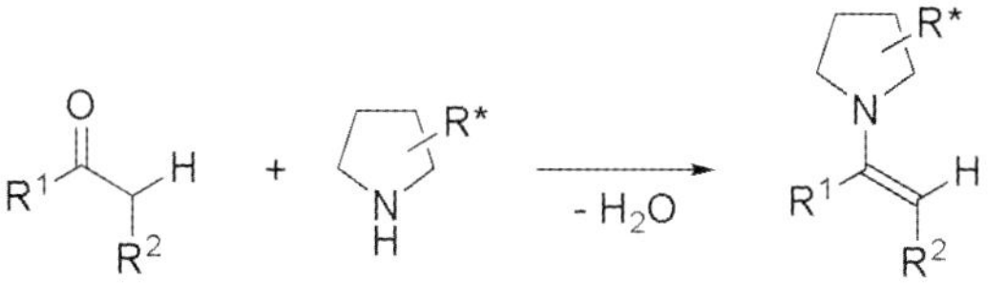

Scheme 1 Formation of enamine for the catalytic enantioselective α-functionalization of aldehydes and ketones.

Secondary amine-based catalysts can also activate α,β-unsaturated compounds via an iminium-ion intermediate leading to enantioselective Michael reaction, as the chiral substituent in the catalyst shields one of the faces of the carbon–carbon double bond (Scheme 2).

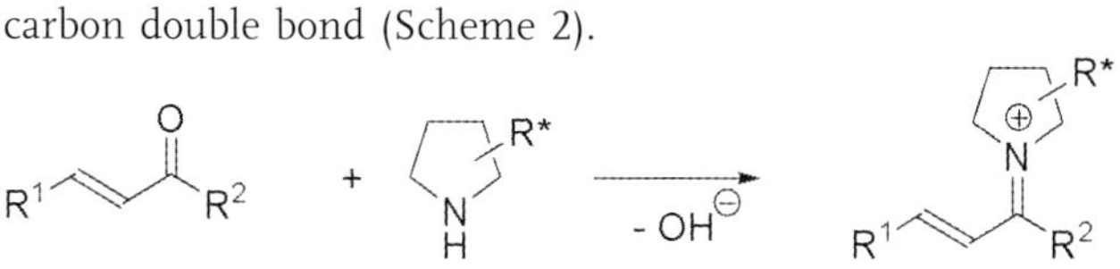

Scheme 2 Formation of iminium ion for the catalytic enantioselective Michael addition.

We envisioned that one could use the two types of intermediates in Schemes 1 and 2 to develop reactions other than "the expected" enantioselective α-functionalization of aldehydes and ketones, and additions to the β-carbon atom in α,β-un-

Asymmetric Synthesis – The Essentials.
Edited by Mathias Christmann and Stefan Bräse
Copyright © 2007 WILEY-VCH Verlag GmbH & Co. KGaA, Weinheim
ISBN: 978-3-527-31399-0

saturated compounds. In this chapter, the catalytic enantioselective inverse-electron-demand hetero-Diels-Alder (HDA) reaction, based on the chiral enamine intermediate [4], and the diastereo- and enantioselective asymmetric domino Michael-aldol reaction of β-ketoesters to α,β-unsaturated ketones [5] will be outlined.

Strategy

The strategy for the organocatalytic enantioselective inverse-electron-demand HDA reaction is presented in Scheme 3: The chiral enamine generated from the aldehyde and a chiral pyrrolidine derivative acts as an electron-rich alkene and undergoes an enantioselective HDA reaction with enones via the catalytic cycle outlined in Scheme 1 [4]: *In situ* generation of a chiral enamine **2** from the chiral pyrrolidine **1** and an aldehyde **3** followed by a stereoselective HDA reaction with enone **4** to give the aminal **5**. Hydrolysis of **5** produces the optically active hemiacetal **6** and releases the chiral catalyst **1** to complete the catalytic cycle.

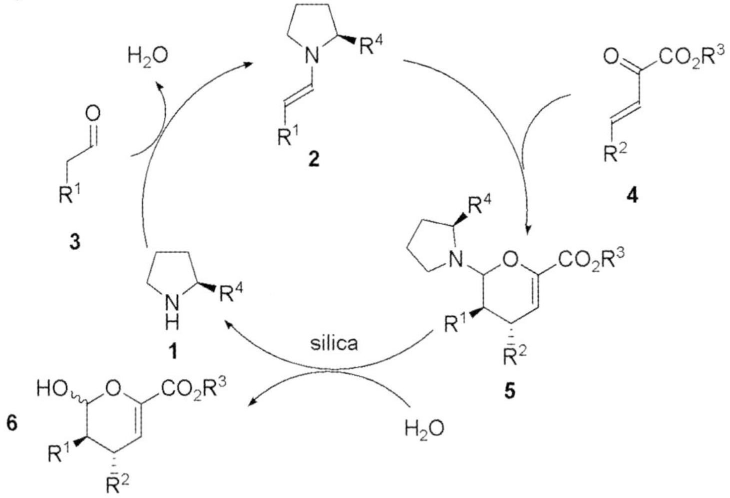

Scheme 3 Catalytic cycle for the organocatalytic hetero-Diels-Alder reaction.

The second approach to optically active six-membered rings is the organocatalytic asymmetric domino Michael-aldol reaction of β-ketoesters **8** with unsaturated ketones affording optically active cyclohexanones **10**, having up to four chiral centers, with excellent enantio- and diastereoselectivity [5]. The mechanistic approach is outlined in Scheme 4 and is a combination of the two mechanisms in Schemes 1 and 2. The first step is the intermolecular Michael addition of the β-ketoester **8** to the iminium-ion intermediate **7**, followed by an aldol reaction of, probably the intermediate **9**, to give **10**.

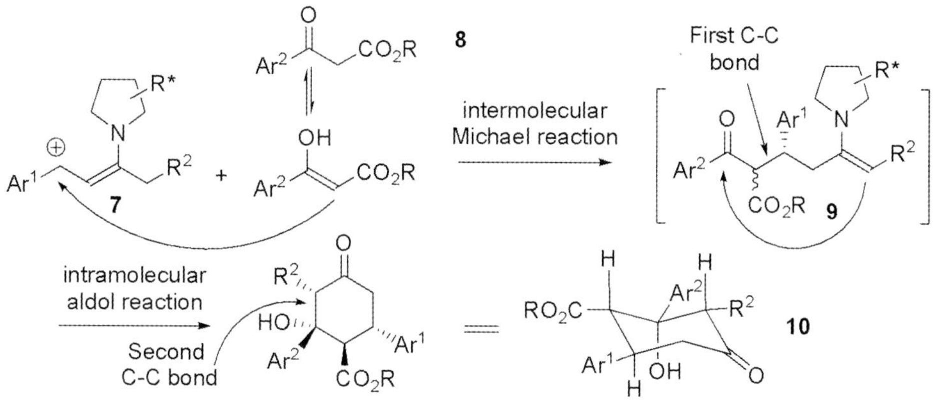

Scheme 4 Domino-Michael-aldol reaction of activate β-ketoesters with α,β-unsatutrated enones.

Results

According to the catalytic cycle in Scheme 3 for the inverse-electron-demand HDA reaction, a mixture of the two anomers of the hemiacetal **6** is formed. In was found than oxidation of the anomers with pyridine chlorochromate (PCC) yielded the corresponding lactone **11** as a single diastereomer. A screening of different secondary amine-based catalysts showed that (S)-2-[bis-(3,5-dimethyl-phenyl)-methyl]-pyrrolidine gave the best results in terms of yield and stereoselectivity (Scheme 5).

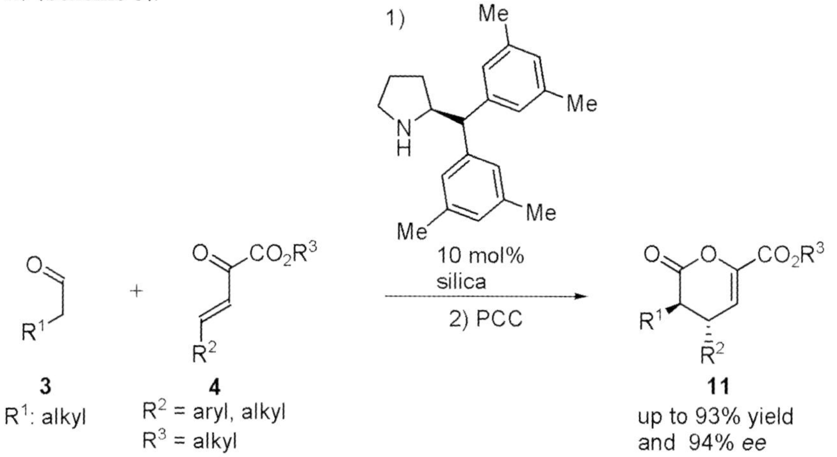

Scheme 5 Organocatalytic synthesis of optically active lactones by inverse-electron-demand HDA reactions.

The scope of the enantioselective organocatalytic inverse-electron-demand HDA reaction was demonstrated by the reaction of various aldehydes with β,γ-unsaturated-α-keto esters, having aromatic and alkyl substituents in the γ-position. The reaction proceeded smoothly with both very high diastereo- and enantioselectivity.

The aliphatic aldehydes were also varied and different substitution patterns were allowed. In general, good yields were obtained (62–93 %) and the enantioselectivities were as high as 94 % ee.

The observed stereochemistry is consistent with the transition state model outlined in Scheme 6. The electronic properties of the enamine govern the regioselectivity, while the 2-diarylmethyl substituent on the pyrrolidine ring shields the Si-face of the enamine. Thus, the 2-diarylmethyl substituent of the controls the addition of the enone to the Re-face of the alkene in an endo-selective fashion.

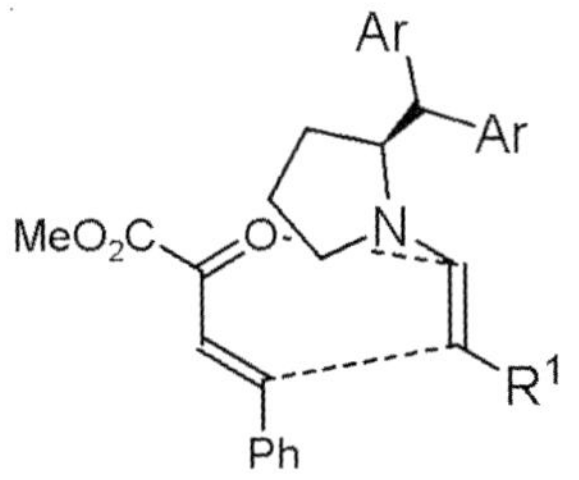

Scheme 6 Proposed transition state for the inverse-electron-demand HDA reaction via an enamine intermediate.

The phenylalanine derived imidazolidine catalyst **12** has proven to be a powerful catalyst for the catalytic asymmetric Michael reaction of unsaturated ketones with various Michael donors and here we present its application to the catalytic asymmetric domino Michael-aldol reaction of α,β-unsaturated ketones **13** with β-ketoesters **8** to produce optically active cyclohexanones **14** with up to four stereocenters (Scheme 7).

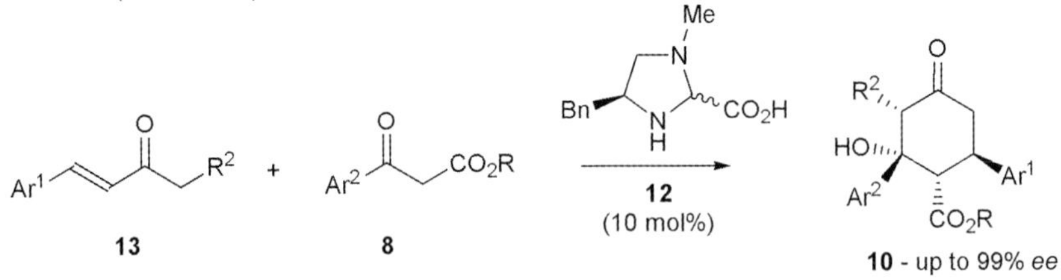

Scheme 7 Organocatalytic synthesis of optically active cyclohexanones.

Various α,β-unsaturated ketones **13** having aromatic and heteroaromatic β-substituents (Ar^1) all reacted well with the β-ketoesters, affording cyclohexanones **10** in moderate to good yields and with excellent diastereo- and enantioselectivities. For the β-ketoesters only one diastereomer could be observed and the enantiomeric excess was in the range from 83-99 % ee.

Synthesis of γ- and ε-lactones

The optically active cyclohexanones **10** formed by the asymmetric domino Michael-aldol reaction leads to a simple procedure for the formation of optically active γ- and ε-lactones. Scheme 8 shows the Baeyer-Villiger oxidation of **10** af-

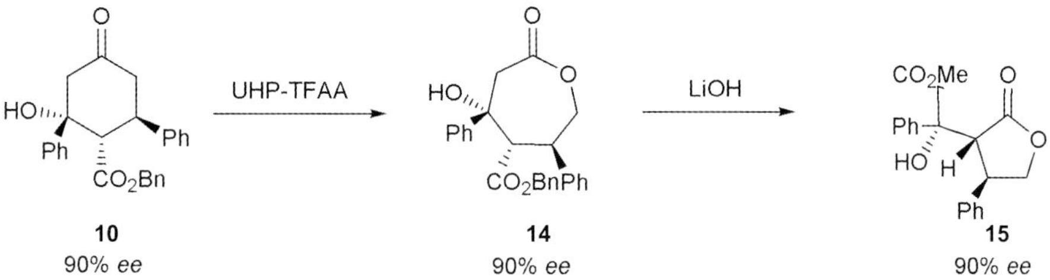

Scheme 8 Formation of optically active -lactone and γ-lactone from the optically active cyclohexanone formed by the asymmetric domino Michael-aldol reaction.

fording first the ε-lactone **14**. Treatment of the ε-lactone **14** with lithium hydroxide produces then the γ-lactone **15** maintaining the enantiomeric excess.

Conclusions and Future Perspectives

Two new approaches for the formation of optically active six-membered rings based on organocatalysis have been outlined. The first is an inverse-electron-demand hetero-Diels-Alder reaction of the enamine intermediate with β,γ-unsaturated-α-keto esters. These reactions proceed in high yields and with excellent diastereo- and enantioselectivity. The other approach is based on a combination of iminium ion and enamine intermediates and is the asymmetric domino Michael-aldol reaction of α,β-unsaturated ketones with β-ketoesters to produce optically active cyclohexanones with up to four stereocenters in excellent diastereo- and enantioselectivities. Furthermore, the use of the optically active cyclohexanones formed was demonstrated for the synthesis of optically active γ- and ε-lactones.

CV of Karl Anker Jørgensen

Karl Anker Jørgensen was born in Århus, Denmark in 1955. He obtained his Ph.D. in 1984 and Doctor of Science 1989 from Aarhus University. In 1985 he was a post-doc with Professor Roald Hoffmann, Cornell University where he was introduced to catalysis. Karl Anker Jørgensen was appointed as Assistant Professor at Aarhus University 1985, as Associate Professor 1988 and as Professor 1992. Since 1997 he is also the Director of the Center for Catalysis at Aarhus University.

Selected Publications

1R. P. L. Dalko, L. Moisan, *Angew. Chem., Int. Ed.* **2004**, *43*, 5138. In the Golden Age of Organocatalysis.

2R. A. Berkessel, H. Gröger, *Asymmetric Organocatalysis*; Wiley-VCH, Weinheim, Germany, 2004.

3R. J. Seayed, B. List, *Org. Biomol. Chem.* **2005**, *3*, 719. Asymmetric Organocatalysis.

4. K. Juhl, K. A. Jørgensen, *Angew. Chem., Int. Ed.* **2003**, *42*, 1498. The First Organocatalytic Enantioselective Inverse-Electron-Demand Hetero-Diels-Alder reaction.

5. N. Halland, P. S. Aburel, K. A. Jørgensen, *Angew. Chem., Int. Ed.* **2004**, *43*, 1272. Highly Enantio- and Diastereoselective Organocatalytic Asymmetric Domino Michael-Aldol Reactions of β-Ketoesters and α,β-Unsaturated Ketones.

Non-linear Effects in Asymmetric Catalysis

Henri B. Kagan, Paris-Sud University, Orsay, France

Background
Enantiomerically Impure Chiral Auxiliaries

Asymmetric synthesis is realized by various strategies, which always involve the help of a chiral auxiliary. Let us consider reactions using either a chiral reagent or a chiral catalyst. In both cases the chiral source originates from a natural compound (terpenes, amino acids, etc) and results from the resolution of a racemic mixture. If the chiral auxiliary is difficult to get with 100% *ee*, it can, nevertheless, be used in a preliminary screening. One can calculate the efficiency (ee_{max}) of the enantiopure reagent or catalyst by correcting the enantiomeric excess of the product ($ee_{product}$) by the enantiomeric excess of the chiral auxiliary (ee_{aux}). The calculation is easily done by using Eq. (1) which shows that the enantiomeric excess of the product is proportional to ee_{max} and to ee_{aux}. This equation is simple to establish (the *ee*s are taken between 0 and 1), using the assumption that there is additivity of the effects of the enantiomeric catalysts in their mixture. It is this assumption which is not general, as discussed in the next section.

$$ee_{product}(\%) = ee_{max} \cdot ee_{product} \cdot 100 \qquad (1)$$

Non-linear Effects

We have questioned the generality of Eq. (1) in 1986 in a joint work with Agami [1]. If the classical Eq. (1) is satisfied, then a plot $ee_{product} = f(ee_{max})$ must generate a straight line (**I**, in Scheme 1). When there are some interactions between the enantiomeric catalysts or, more generally, if several chiral auxiliaries are involved in some molecular species, then the situation is not necessarily the same when the catalyst is enantiopure or not. For example new diastereomeric species can be produced for non-enantiopure auxiliaries. This is easily understood for catalysts involving two chiral ligands, symbolized as ML_2 (M : metal, L : ligand). The enantiopure ligand L_R or L_S will produce the two homochiral catalysts ML_RL_R or ML_SL_S. An enantioimpure ligand is expected to form *in situ* an additional heterochiral (meso-type) catalyst ML_RL_S. This diastereomeric catalyst will produce a racemic product with a rate which may be different to that of the two homochiral com-

Asymmetric Synthesis – The Essentials.
Edited by Mathias Christmann and Stefan Bräse
Copyright © 2007 WILEY-VCH Verlag GmbH & Co. KGaA, Weinheim
ISBN: 978-3-527-31399-0

plexes. Consequently the plot $ee_{product} = f(ee_{max})$ will deviate from linearity, giving curves **II** or **III** (Scheme 1). Curve **II**, located above **I**, is characteristic of a positive non-linear effect (abbreviated as (+)-NLE), while **III** is a case of a negative non-linear effect ((−)-NLE). If $ee_{product}$ is higher than expected by the linear correlation, this beneficial situation has been named "asymmetric amplification" by Oguni et al. It can be useful for synthetic applications.

We published the first examples of non-linear effects (Scheme 2), in the Sharpless epoxidation of geraniol (curve **A**) and the enantioselective oxidation of *p*-tolyl methyl sulfide into the corresponding sulfoxide (curve **B**)[1]. A third example, a weak (−)NLE (a proline-catalyzed intramolecular aldol reaction) was later shown by List et al. to be indeed linear. Many cases of NLE have been displayed during the last two decades, as summarized below. Several reviews on NLE have been published by us [2R–7R] or by others.

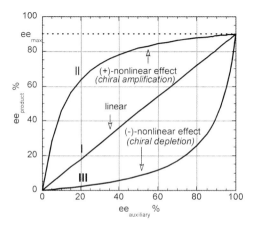

Scheme 1 Linear and non-linear relationships.

Results
Examples of Asymmetric Amplifications

Various types of transformations catalyzed by *chiral organometallics* have been associated with NLE, such as ene reactions, Diels-Alder or hetero Diels-Alder reactions, 1,4-additions on conjugated esters or ketones, oxidations, allylic substitutions, among others. One of the most spectacular (+)-NLE was reported in 1989 by Noyori et al. in the addition of Et_2Zn to benzaldehyde catalyzed by DAIB, an aminoalcohol prepared from camphor. In Ref. [7R] are listed many examples of reactions displaying asymmetric amplification. Non-linear effects with *chiral organocatalysts*, are very rare, an example of asymmetric amplification in a proline-catalyzed reaction was discussed in 2004 by D. Blackmond, see her chapter in this book, page 181.

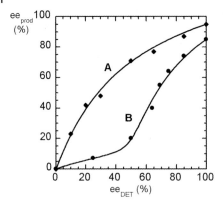

Scheme 2 Sharpless epoxidation of geraniol (**A**) and p-tolyl methyl sulfide oxidation (**B**) from Ref. [1]. In both cases the chiral auxiliary is (R,R)-diethyl tartrate (DET).

Kinetic Models

We used simple kinetic models to analyze some features of non-linear effects [1, 8]. In Scheme 3 is indicated the simplest model that we first discussed [1]. A catalyst of the type ML_2 can be built from a metal complex M and chiral ligands L_R and L_S. One assumes an equilibrium between the three possible complexes (equilibrium constant K), with the reactivities of the complexes being expressed by the pseudo-first order rate constants k_{RR}, k_{SS} and k_{RS}. The evaluation of the enantiomeric excess of the product ($ee_{product}$) coming from the three competitive routes is easy to do by taking as parameters the relative rate constants $g = k_{RS}/k_{RR}$ and the relative amounts of heterochiral- and meso-complexes $\beta = z/(x+y)$. Equation (2) is obtained:

$$ee_{product}(\%) = ee_{max} \cdot ee_{aux} \cdot (1+\beta)/(1+g\beta) \cdot 100 \qquad (2)$$

If $\beta = 0$ (absence of meso catalyst) or if $g = 1$, Eq. (2) is transformed into Eq. (1): the plot $ee_{product} = f(ee_{aux})$ is linear. In the general case there will be non-linearity. Asymmetric amplification will occur for $g < 1$, with its maximum for $g = 0$. (–)-NLE will occur if $g > 1$. The parameter β can be expressed as a function of the equilibrium constant K, allowing one to draw curves for various values of g.

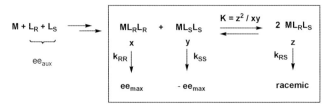

Scheme 3 The ML_2 model [1, 8].

The ML_2 model has been generalized to ML_n cases ($n = 3$, 4 or higher), giving interesting predictions [8]. A second simple model has been also proposed, the "reservoir" model [8]. It is based on the *in situ* trapping of some chiral auxiliary of racemic composition into catalytically inactive dimeric or oligomeric forms.

Applications of Non-linear Effects

The absence or the presence of NLE is often used in *mechanistic studies* in asymmetric catalysis. A NLE is information which can be correlated to aggregation phenomena of organometallic species or to other phenomena. Kinetic measurements may shed light on the origin of NLE as demonstrated by Blackmond in several systems (see her chapter in this book, page 181). The absence of NLE does not necessarily mean absence of aggregation or of ML_2 catalyst [7R].

Asymmetric amplification ((+)-NLE)) is especially useful when a chiral ligand is difficult to obtain enantiopure. Asymmetric amplification is also important in autocatalytic reactions, where the reaction product is a catalyst of its own formation (see the chapter by K. Soai). The dark facet of asymmetric amplification is that it is usually associated with a slow down of the reaction (with respect to the rate with enantiopure catalyst). This arises from the formation of species (poor catalysts or not active) which store racemic ligands.

Conclusion and Perspectives

The concepts which are at the foundation of non-linear effects can be found in areas which are not directly related to enantioselective catalysis. *Chiral reagents* may, in principle, give non-linear phenomena if they are used in excess or if the reaction is analyzed at low conversion. We studied the reduction of acetophenone by Ipc_2BCl, prepared from α-pinene of various ee [9, 10]. A strong asymmetric amplification was observed. The mixture of *pseudo-enantiomeric catalysts* giving products of opposite configurations should behave similarly to mixtures of enantiomeric catalysts. We tested this concept in the Sharpless dihydroxylation of a dibromo-stilbene by OsO_4 in the presence of binary mixtures of various alkaloids [11]. The use of enantioimpure catalysts in *kinetic resolution* has also been studied but cannot be considered here because of lack of space [12].

The concepts of non-linear effects in catalytic or stoichiometric enantioselective synthesis are now widely used and are of interest both in mechanistic studies and in synthesis.

CV of Henri Kagan

Henri B. Kagan was born in Boulogne-Billancourt, France in 1930. He studied in Paris (Sorbonne and Ecole Nationale Supérieure de Chimie de Paris). He obtained a Ph.D. at Collège de France under the supervision of Jean Jacques. He was associated in the Collège de France with Professor A. Horeau before moving to Université Paris-Sud where he is presently emeritus Professor. He developed research in various areas of organic chemistry, namely organic stereochemistry, asymmetric synthesis, asymmetric catalysis, chiral bidentate phosphines, organic reactions using SmI_2. He was the recipient of many awards including the Prelog

Medal, the August-Wilhelm-von-Hofmann medal, the Tetrahedron Prize, the Grand Prix de la Fondation de la Maison de la Chimie (shared with H. Yamamoto), the Wolf Prize (shared with R. Noyori and K. B. Sharpless), the Yamada Prize, the Nagoya medal for Organic Chemistry, the Ryoji Noyori Prize and the Bower award of the Franklin Institute. He is a member of the French Academy of Science.

Selected Publications

1. C. Puchot, O. Samuel, E. Dunach, S. Zhao, C. Agami, H. B. Kagan, *J. Am. Chem. Soc.* **1986**, *108*, 2353–2357. *Nonlinear Efects in Asymmetric Synthesis. Examples in Asymmetric Oxidation and Aldolization Reactions.*

2R. H. B. Kagan, C. Girard, D. Guillaneux, D. Rainford, O. Samuel, S. H. Zhao, S. Y. Zhang, *Acta Chem. Scand.*, **1996**, *50*, 345–352. *Nonlinear Effects in Asymmetric Catalysis : Some Recent Aspects,*

3R. C. Girard, H. B. Kagan, *Angew. Chem. Int. Ed. Engl.* **1998**, *37*, 2922–2959. *Nonlinear Effects in Asymmetric Synthesis and Stereoselective Reactions : Ten Years of Investigation.*

4R. H. B. Kagan, T. O. Luukas in *Comprehensive Asymmetric Catalysis*, E. N. Jacobsen, A. Pfaltz, H. Yamamoto (Eds.), Springer-Verlag, Berlin, **1999**, Vol. 1, pp.101–118. *Nonlinear Effects and Autocatalysis,*

5R. H. B. Kagan, D. Fenwick, *Topics Stereochem.*, **1999**, *22*, 257-296. *Asymmetric Amplification*

6. H. B. Kagan, *Synlett* **2001**, 888–900. *Nonlinear Effects in Asymmetric Catalysis: A Personal Account.*

7R. H. B. Kagan, *Adv. Synth. Catal.* **2001**, *343*, 227–233. *Practical Consequences of Non-Linear Effects in Asymmetric Synthesis,*

8. D. Guillaneux, S. H. Zhao, O. Samuel, D. Rainford, H. B. Kagan, *J. Am. Chem. Soc.* **1994**, *116*, 9430–9439. *Nonlinear Effects in Asymmetric Catalysis,*

9. C. Girard, H. B. Kagan, *Tetrahedron: Asymmetry* **1995**, *6*, 1881–1884. *Nonlinear Effects in the Reduction of Acetophenone by Diisopinocampheylchloroborane: Influence of the Reagent Preparation.*

10. C. Girard, H. B. Kagan, *Tetrahedron: Asymmetry* **1997**, *8*, 3851–3854. *Nonlinear Effects in the Reduction of Acetophenone by Diisopinocampheyl Chloroborane: Influence of Stoichiometry of the Reagent.*

11. S. Zhang, C. Girard, H. B. Kagan, *Tetrahedron: Asymmetry* **1995**, *6*, 2637–2640. *Nonlinear Effects Involving Two Competing Pseudo-Enantiomeric Catalysts: Example in Asymmetric Dihydroxylation of Olefins*

12. T. O. Luukas, C. Girard, D. Fenwick, H. B. Kagan, *J. Am. Chem. Soc .* **1999**, *121*, 9299–9306. *Kinetic Resolution when the Chiral Auxiliary is not Enantiomerically Pure: Normal or Abnormal Behavior.*

Asymmetric Organocatalysis

Steven V. Ley, University of Cambridge, England

Background

Organocatalysts can have many advantages over their traditional metal-mediated counterparts [1R]. While the proline-based organocatalytic system is a well-studied example, this highly enantioselective system has limitations, especially its reliance on polar solvents [2R]. We reasoned that as tetrazoles have been used as a bioisostere for carboxylic acids, with pK_a values of around 5, they might provide a more soluble alternative. We therefore prepared the tetrazole 5-pyrrolidin-2-yltetrazole **1** (and later its enantiomer **2**) from proline and used it initially in a series of Mannich-type additions of carbonyl compounds to *N*-PMP-protected α-amino ethyl glyoxalates: benchmarking it against proline itself [3]. We found that **1** compared very favorably to proline, with the added value that the reactions were faster and could be performed in nonpolar solvents (Scheme 1).

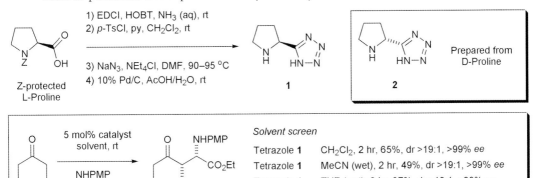

Scheme 1 Synthesis of tetrazole catalyst and its benchmark testing against proline.

The importance of this particular organocatalyst became evident as further studies were undertaken.

Results

The first such study of the tetrazole catalyst was of Michael-type reactions of ketones with nitro-olefins such as that shown in Scheme 2 [4]. Both the proline and tetrazole systems were compared specifically in the reaction of cyclohexanone and β-nitrostyrene; while each performed efficiently in DMSO, in both cases enhanced enantioselectivity could be gained by performing reactions in methanol [5]. In addition, by adding increasing proportions of isopropyl alcohol (IPA) to methanol we were able to optimise both the yield and *ee* of products and tetrazole **1** outperformed proline in all cases. Optimum conditions required equal portions of ethanol and IPA with 15 mol % of catalyst at room temperature for 24 h. Tetrazole catalyst **2** gave compounds of opposite stereoselectivity, relative to **1**, with comparable yields as expected. Tetrazole as a catalyst shows clear advantages over proline in both alcoholic solvent systems and dichloromethane.

Tetrazole catalyst **1** has also demonstrated itself as an improvement over proline, in the asymmetric addition of nitroalkanes to cyclic and acyclic enones: a reaction that proved scalable, providing enantiomeric excesses of up to 98 % using only 2 equivalents of nitroalkane [6]. We conducted these reactions in the presence of the achiral *meso* base 2,5-dimethylpiperazine, a method previously developed for proline-catalysed reactions [7]. This methodology has recently been extended to the asymmetric addition of malonates to cyclic and acyclic enones [8].

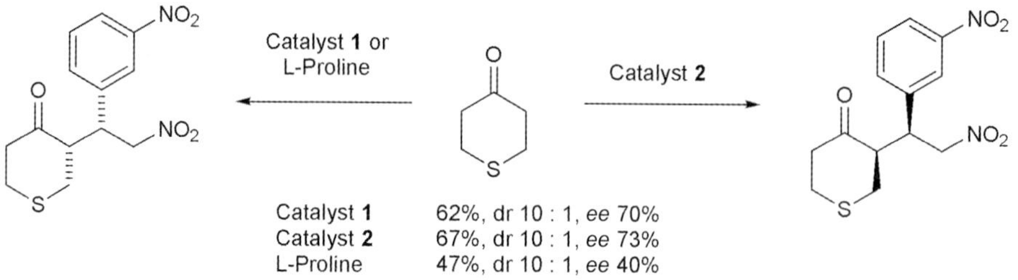

Catalyst 1   62%, dr 10 : 1, ee 70%
Catalyst 2   67%, dr 10 : 1, ee 73%
L-Proline   47%, dr 10 : 1, ee 40%

Reagents and conditions: Ketone (1.5 eq.), 15 mol% catalyst, β-3-dinitrostyrene, EtOH/IPA (1:1), rt, 24 h

Scheme 2 Organocatalytic Michael addition of enamines derived from ketones to nitro-olefins.

Following our initial publication of the tetrazole catalyst **1**, its application in asymmetric aldol reactions has been studied thoroughly by others as well as by ourselves and is discussed in a full paper [9].

Extension of our organocatalyst programme included preparation of similarly acidic acyl sulfonamide catalysts **3** and **4**, also derived from proline; these again confer both solubility and catalytic activity [9]. A number of significant observations were made when the earlier Mannich-type reaction was revisited with the sulfonamide system (Scheme 3). Firstly, in comparison to tetrazole **1** which produced excellent results in terms of yield and *ee* with as little as 1 mol %, the sulfonamide catalysts performed best at 20 mol %; while no reaction was observed with proline. Secondly, whether the sulfonamide catalyst bore a methyl or a phenyl group, the difference appeared to have little effect on the outcome of the reactions; judging by the overall pattern of yields and *ee* of the products. Interest-

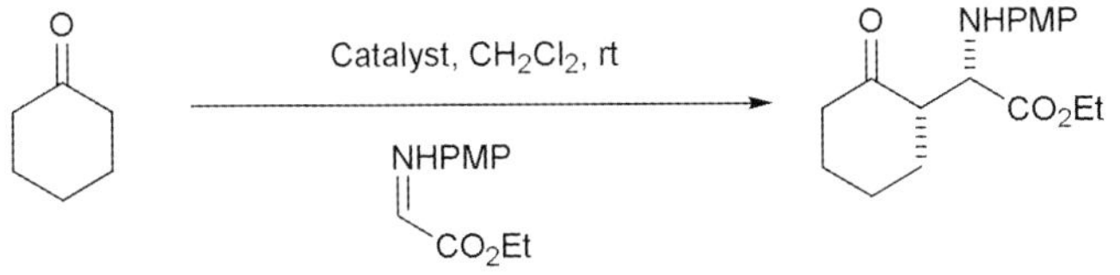

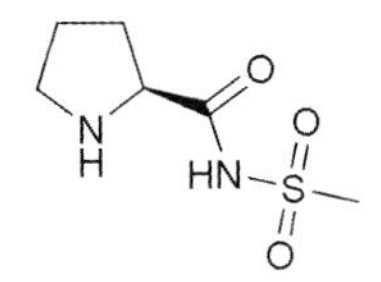

	mol %	Time/h	Yield (%)	dr (syn:anti)	ee (%)
Catalyst 1	5	2	65	>19 : 1	>99
L-Proline	5	2	0	>19 : 1	>99
Catalyst 1	1	16	70	>19 : 1	>99
Catalyst 3	20	24	82	>19 : 1	96
Catalyst 4	20	24	75	>19 : 1	>99

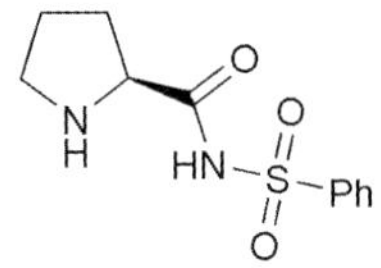

Scheme 3 Improved asymmetric Mannich-type reaction using new acyl sulfonamide and tetrazole organocatalysts.

ingly, **3** and **4** failed to catalyse the analogous nitro-Michael additions such as that shown in Scheme 2.

We also prepared a version of **1** in which the separation between the two nitrogen-containing rings was extended by one methylene unit and created the homoproline tetrazole catalyst **5** [10]. This modified catalyst led to increased enantioselectivity, as shown in Scheme 4. For example, using 15 mol% of catalyst in an equal part ethanol and IPA solvent system for the reaction of cyclohexanone with β-dinitrostyrene over 24 hours at room temperature, the product was afforded in 52% yield (51% ee) with proline, 80% (62% ee) with **1** and 88% (91% ee) with **5** (Scheme 4).

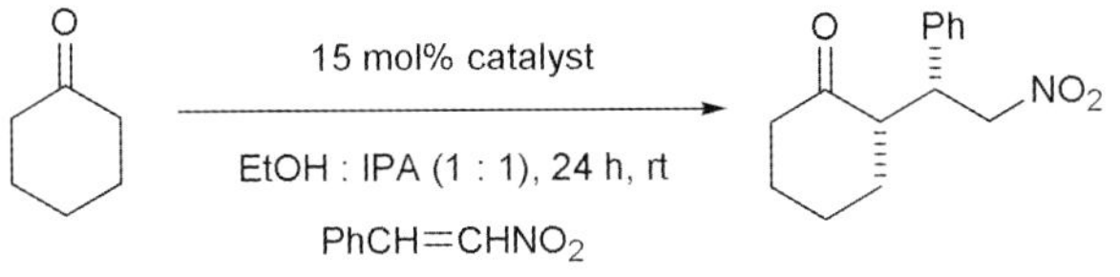

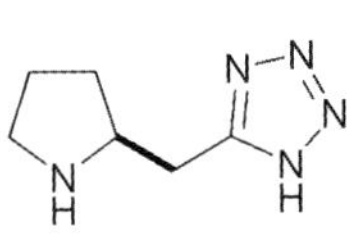

Catalyst	Yield (%)	dr (syn:anti)	ee (%)
L-Proline	52	>19 : 1	51
Tetrazole **1**	80	>19 : 1	62
Homo-Tetrazole **5**	88	>19 : 1	91

Scheme 4 New catalyst **5** for use in the asymmetric Michael addition of carbonyl compounds to nitro-olefins.

As part of our ongoing research into organocatalysis, we have also been investigating cyclopropanation reactions. Catalytic methods that are based on the reaction of ylides with electron-deficient alkenes have been under explored.

Ylides are known reagents for cyclopropanation reactions, where the ylide is usually prepared in a separate step. With an abundance of available tertiary amines, we were interested in developing a one-pot approach to cyclopropanation using ammonium ylide-catalysis. Our initial stoichiometric investigation identified a one-pot process in which the ylide was generated from the ammonium salt *in situ* (Scheme 5) [11]. The absence of reaction by-products was noteworthy. Indeed, after a simple aqueous work-up, the cyclopropanes are essentially pure by ^{1}H NMR spectroscopy and LCMS. We developed this into a catalytic process based on the postulated catalytic cycle (Scheme 5b). We were able to achieve good results using catalytic base and, for example, found the analogous reaction to proceed in 69% yield (Scheme 5a, ii).

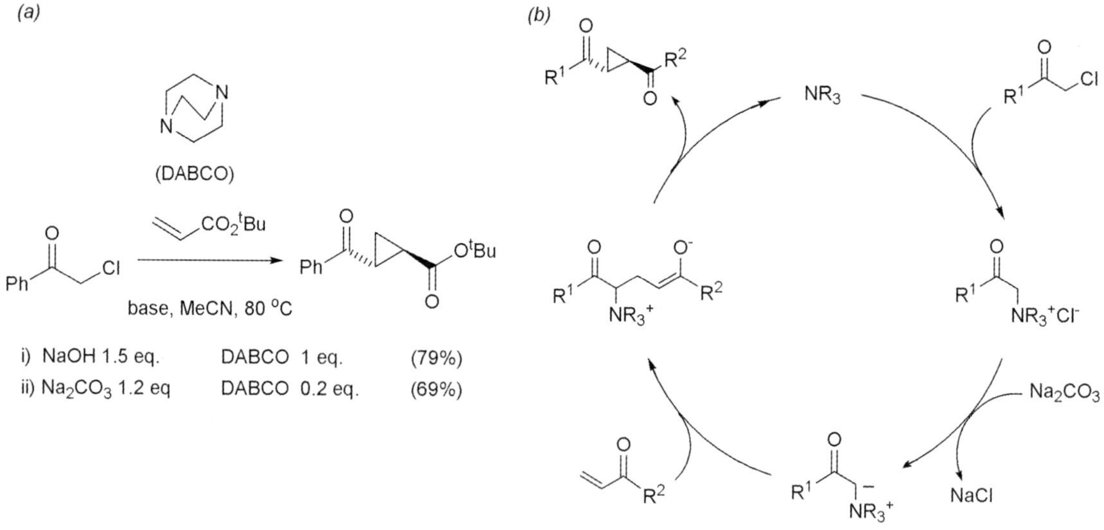

Scheme 5 (a) Stoichiometric 'one-pot' cyclopropanation.
(b) Proposed catalytic cycle for cyclopropanation.

These studies evolved, first by way of an intramolecular process [12], then into a general intermolecular enantioselective organocatalytic cyclopropanation reaction via ammonium ylides. We developed a process that produced a range of functionalized molecules with excellent diastereo- and enantioselectivity, and as either enantiomer [13]. We found cinchona alkaloids also catalysed the reactions; and by using their pseudoenantiomeric quinine or quinidine derivatives **6** and **7**, both enantiomers of the cyclopropanes could be readily accessed (Scheme 6).

Conclusions and Future Perspectives

In summary, we believe that asymmetric organocatalytic systems provide very attractive alternatives to metal-based catalysts, and they can be assured of an impact in organic synthesis programmes into the future.

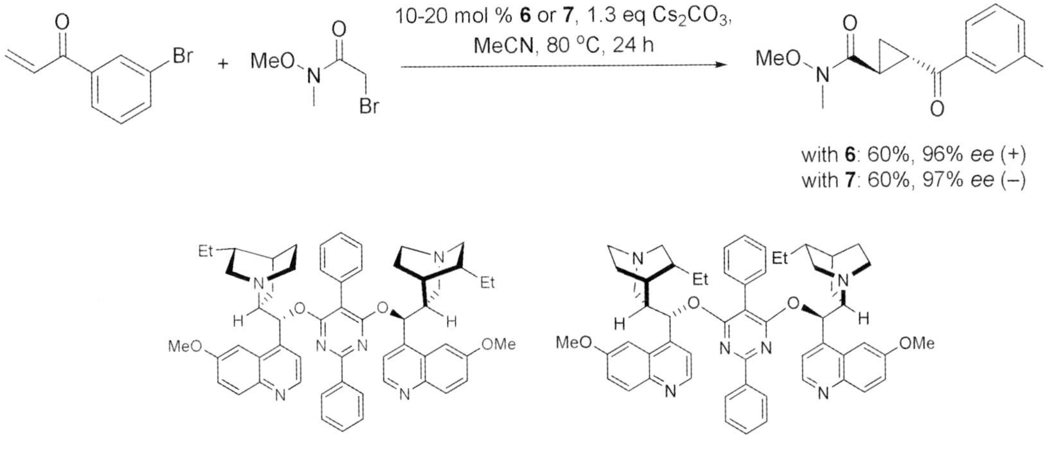

Scheme 6 One example of the enantioselective organocatalytic cyclopropanation using quinine or quinidine derivatives of the cinchona alkaloids.

CV of Steven V Ley

Steven Ley is the BP (1702) Professor of Organic Chemistry at the University of Cambridge and a Fellow of Trinity College. He was President of the Royal Society of Chemistry (2000–2002) and was made a Commander of the British Empire in January 2002. He is also a Fellow of both the Royal Society (London) and the Academy of Medical Sciences. His research has been recognised by major prizes and awards, notably, the Adolf Windaus Medal from the German Chemical Society, the Dr. Paul Janssen Prize for Creativity in Organic Synthesis, the Davy Medal from the Royal Society, the Ernest Guenther Award in the Chemistry of Natural Products (ACS), the August-Wilhelm-von Hofmann Medal, the Alexander-von-Humboldt Award, the Yamada-Koga Prize, the Messel Medal (SCI), the Robert Robinson Award and Medal (RSC) and the Nagoya Gold Metal. See http://leygroup.ch.cam.ac.uk.

Selected Publications

1R. P. I. Dalko, L. Moisan, *Angew. Chem. Int. Ed.* **2004**, *43*, 5138–5175. *In the Golden Age of Organocatalysis.*
2R. B. List, *Tetrahedron* **2002**, *58*, 5573–5590. *Proline-catalyzed asymmetric reactions.*
3. A. J. A. Cobb, D. M. Shaw, S. V. Ley, *Synlett* **2004**, 558–560 and references cited therein. *5-Pyrrolidin-2-yltetrazole: A New, Catalytic, More Soluble Alternative to Proline in an Organocatalytic Asymmetric Mannich-type Reaction.*
4. A. J. A. Cobb, D. A. Longbottom, D. M. Shaw, S. V. Ley, *Chem. Commun.* **2004**, 1808–1809 and the references cited therein. *5-Pyrrolidin-2-yltetrazole as an asymmetric organocatalyst for the addition of ketones to nitro-olefins.*
5. D. Enders, A. Seki, *Synlett* **2002**, 26–28. *Proline-Catalyzed Enantioselective Michael Additions of Ketones to Nitrostyrene.*
6. C. E. T. Mitchell, S.E. Brenner, S.V. Ley, *Chem. Commun.* **2005**, 5346–5348 and the references cited therein. *A versatile organocatalyst for the asymmetric conjugate addition of nitroalkanes to enones.*

7. S. Hanessian, V. Pharm, *Org Lett.* **2000**, *2*, 2975–2978. *Catalytic Asymmetric Conjugate Additions of Nitroalkanes to Cycloalkenones.*
8. K. R. Knudsen, C. E. T. Mitchell, S. V. Ley. *Chem. Commun.* **2006**, 66–68. *Organocatalytic asymmetric conjugate addition of malonates to enones using a proline tetrazole catalyst.*
9. A. J. A. Cobb, D. M. Shaw, D. A. Longbottom, J. B. Gold, S. V. Ley. *Org. Biomol. Chem.* **2005**, *3*, 84–96 and references cited therein. *Organocatalysis with proline derivatives: improved catalysts for the asymmetric Mannich, nitro-Michael and aldol reactions.*
10. C. E. T. Mitchell, A. J. A. Cobb, S. V. Ley, *Synlett* **2005**, 611–614. *A Homo-Proline Tetrazole as an Improved Organocatalyst for the Asymmetric Michael Addition of Carbonyl Compounds to Nitro-Olefins.*
11. C. D. Papageorgiou, S. V. Ley, M. J. Gaunt, *Angew. Chem. Int. Ed.* **2003**, *42*, 828–831. *Organic-Catalyst-Mediated Cyclopropanation Reaction.*
12. N. Bremeyer, S. C. Smith, S. V. Ley, M. J. Gaunt, *Angew. Chem. Int. Ed.* **2004**, *43*, 2681–2684. *An Intramolecular Organocatalytic Cyclopropanation Reaction.*
13. C. D. Papageorgiou, M. A. Cubillo de Dios, S. V. Ley, M. J. Gaunt, *Angew. Chem. Int. Ed.* **2004**, *43*, 4641–4644. *Enantioselective Organocatalytic Cyclopropanation via Ammonium Ylides.*

Directed Evolution of Enzymes for Asymmetric Syntheses
Manfred T. Reetz, Max-Planck-Institut für Kohlenforschung, Mülheim/Ruhr, Germany

Background

Some time ago we proposed and implemented experimentally a fundamentally new approach to asymmetric enzyme catalysis [1]: Directed evolution of enantioselective enzymes as chiral catalysts in synthetic organic chemistry. Rather than relying on decisions based on structure, mechanism and molecular modeling, evolution is simulated in the test tube. The combination of appropriate molecular biological methods for random gene mutagenesis and expression coupled with efficient high-throughput screening systems [2R] for evaluating the enantioselectivity of thousands of mutant enzymes forms the basis of this new way of creating enantioselective catalysts (Figure 1). In the first cycle the most enantioselective mutant is identified by the *ee*-screening system which, as such, is not yet an evolutionary process. However, upon going through a second cycle of mutagenesis/screening, the evolutionary character is introduced. Such an "evolutionary pressure" can be maintained and increased, if necessary, by going through further cycles. The challenges in putting such an evolution-machine into practice revolve around the decision regarding the optimal choice of the currently available mutagenesis methods and the necessity of developing high-throughput *ee*-screening systems.

Strategy and Early Results

When we began our (ad)venture into this new area of research in the mid 1990s, a number of gene mutagenesis methods were known and were being applied by several groups in the quest to evolve mutant enzymes having enhanced stability and activity [3R]. However, optimal strategies on how to apply the molecular biological methods had not been implemented to a satisfactory level. Indeed, the development of efficient strategies for scanning through the virtually endless protein sequence space is crucial to any form of directed evolution, and certainly when attempting to evolve enantioselectivity. In our early work regarding the hydrolytic kinetic resolution of the racemic ester **1** catalyzed by the lipase from *Pseudomonas aeruginosa* (Figure 2), we first applied several cycles of error-prone polymerase chain reaction (epPCR), which led to sequential enhancement of en-

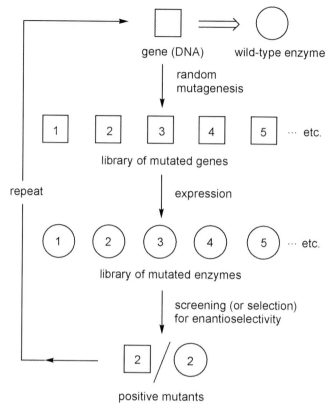

Figure 1 Strategy for directed evolution of an enantioselective enzyme.

antioselectivity [1], which is proof-of-principle. The positions of amino acid exchange were then identified by sequencing the gene, and these hot spots were then saturated, leading to the identification of highly improved mutants. DNA shuffling of the mutant genes did not lead to enhanced enantioselectivity. The application of higher mutation rate turned out to be beneficial, and DNA shuffling of the corresponding genes then furnished the most enantioselective mutant leading to a selectivity factor of $E = 51$ compared to $E = 1.1$ of the unengineered enzyme [4]. Moreover, cassette mutagenesis in a four amino acid region near the binding site in the form of a focused library was also successful [4].

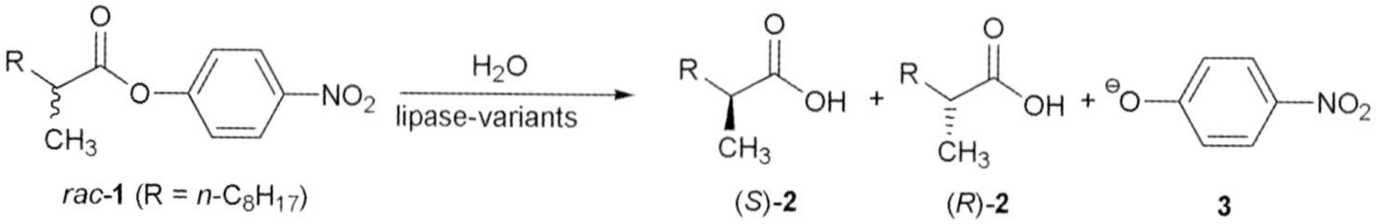

Figure 2 Lipase-catalyzed hydrolytic kinetic resolution of ester **1**.

A detailed theoretical analysis based on molecular modeling and density functional calculations of the best mutant demonstrated that directed evolution is also an excellent tool for learning about how enzymes function [5]. Moreover, high-throughput *ee*-screens were developed for a variety of reaction types [2R].

Generalizing the Concept

Using the methods and strategies outlined above, we turned to other enzymes including epoxide hydrolases [6] and monooxygenases such as cyclohexanone monooxygenase (CHMO) [7, 8]. The results of typical case studies are shown in Figure 3.

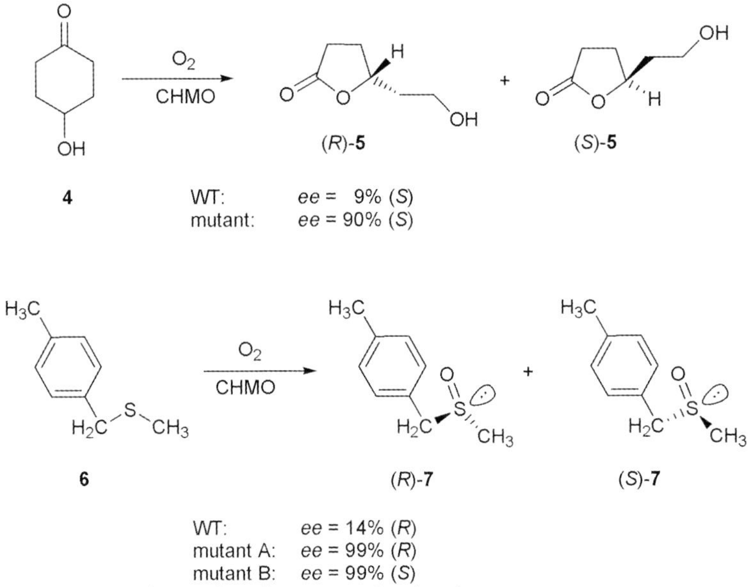

	WT:	ee = 9% (S)
	mutant:	ee = 90% (S)

	WT:	ee = 14% (R)
	mutant A:	ee = 99% (R)
	mutant B:	ee = 99% (S)

Figure 3 Typical examples of CHMO-catalyzed oxidations.

New Strategies

The strategies and methods that we developed during the first eight years of research in this area function well [4]. Nevertheless, we became convinced that it should be possible to develop even more efficient strategies for searching in protein sequence space, and proposed the Complete Active-Site Saturation Test (CAST) [9]. It entails two simple steps: (i) Identification of sets of two or three amino acids whose side-chains are located near the binding sites; and (ii) simultaneous randomization at each site comprising two or three amino acid positions with the formation of relatively small focused libraries. We first applied CASTing in order to solve a classical problem when using enzymes in organic chemistry, namely expanding the range of substrate acceptance of enzymes. We then introduced "reCASTing", which simply means iterative cycles of CASTing [10]. It is a crucial extension of the concept because evolutionary character is introduced. The mutant gene of the best enzyme in an initial CAST library is used as a template for another cycle of CASTing. We applied iterative CASTing and were able to enhance the enantioselectivity of an epoxide hydrolase by a factor of 25 [10].

Conclusions and Perspectives

Directed evolution of enantioselective enzymes has emerged as a fundamentally new and successful approach to asymmetric catalysis. Research in the development of strategies for probing protein sequence space has culminated in the implementation of iterative CASTing as a particularly effective method.

CV of Manfred T. Reetz

Manfred T. Reetz was born in Hirschberg, Germany in 1943. He obtained his PhD from the University of Göttingen, Germany, in 1969 under U. Schöllkopf. Following a postdoctoral stay at the University of Marburg (R. W. Hoffmann), he obtained his Habilitation there in 1974 before becoming Associate Professor of Organic Chemistry at the University of Bonn in 1978. From 1980 to 1991 he was Full Professor at the University of Marburg before moving to Mülheim/Ruhr, Germany, as director at the Max-Planck-Institut für Kohlenforschung. He has received a number of awards and honors, including the Otto-Bayer-Prize (1986), the Leibniz-Prize (1989), Fluka Reagent of the Year Award (1997), the Nagoya Gold Medal (2000), the Karl-Ziegler-Prize (2005) and membership of the prestigious Deutsche Akademie der Naturforscher Leopoldina.

Selected Publications

1. M. T. Reetz, A. Zonta, K. Schimossek, K. Liebeton, K.-E. Jaeger, *Angew. Chem.* **1997**, *109*, 2961–2963; *Angew. Chem., Int. Ed. Engl.* **1997**, *36*, 2830–2832. Creation of Enantioselective Biocatalysts for Organic Chemistry by in vitro Evolution.
2. M. T. Reetz, *Angew. Chem.* **2001**, *113*, 292–320; *Angew. Chem. Int. Ed.* **2001**, *40*, 284–310. Combinatorial and Evolution-Based Methods in the Creation of Enantioselective Catalysts.
3. F. H. Arnold, G. Georgiou (Eds.), *Directed Enzyme Evolution: Screening and Selection Methods*, Vol. 230, Humana Press, Totowa, N J, 2003.
4. M. T. Reetz, *Proc. Natl. Acad. Sci. U.S.A.* **2004**, *101*, 5716–5722. Controlling the Enantioselectivity of Enzymes by Directed Evolution: Practical and Theoretical Ramifications.
5. M. Bocola, N. Otte, K.-E. Jaeger, M. T. Reetz, W. Thiel, *ChemBioChem* **2004**, *5*, 214–223. Learning from Directed Evolution: Theoretical Investigations into Cooperative Mutations in Lipase Enantioselectivity.
6. M. T. Reetz, C. Torre, A. Eipper, R. Lohmer, M. Hermes, B. Brunner, A. Maichele, M. Bocola, M. Arand, A. Cronin, Y. Genzel, A. Archelas, R. Furstoss, *Org. Lett.* **2004**, *6*, 177–180. Enhancing the Enantioselectivity of an Epoxide Hydrolase by Directed Evolution.
7. M. T. Reetz, B. Brunner, T. Schneider, F. Schulz, C. M. Clouthier, M. M. Kayser, *Angew. Chem.* **2004**, *116*, 4167–4170; *Angew. Chem. Int. Ed.* **2004**, *43*, 4075–4078. Directed Evolution as a Method to Create Enantioselective Cyclohexanone Monooxygenases for Catalysis in Baeyer-Villiger Reactions.
8. M. T. Reetz, F. Daligault, B. Brunner, H. Hinrichs, A. Deege, *Angew. Chem.* **2004**, *116*, 4170–4173; *Angew. Chem. Int. Ed.* **2004**, *43*, 4078–4081. Directed Evolution of Cyclohexanone Monooxygenases: Enantioselective Biocatalysts for the Oxidation of Prochiral Thioethers.

9. M. T. Reetz, M. Bocola, J. D. Carballeira, D. Zha, A. Vogel, *Angew. Chem.* **2005**, *117*, 4264–4268; *Angew. Chem. Int. Ed.* **2005**, *44*, 4192–4196. *Expanding the Range of Substrate Acceptance of Enzymes: Combinatorial Active-Site Saturation Test.*

10. M. T. Reetz, L.-W. Wang, in part M. Bocola, *Angew. Chem.*, **2006**, *118*, 1258–1263; erratum: 2556; *Angew. Chem. Int. Ed.* **2006**, *45*, 1236–1241; erratum: 2494. *Directed Evolution of Enantioselective Enzymes: Iterative Cycles of CASTing as an Efficient Strategy for Probing Protein Sequence Space.*

Asymmetric Autocatalysis and Its Implications in the Origin of Chiral Homogeneity of Biomolecules

Kenso Soai and Tsuneomi Kawasaki, Tokyo University of Science, Tokyo, Japan

Background

Asymmetric autocatalysis is a reaction where the chiral product acts as a chiral catalyst for its own production (Scheme 1). Asymmetric autocatalysis is an efficient asymmetric synthesis because the process is an automultiplication of a chiral compound. Because the amount of catalyst increases during the reaction, decrease in the amount of catalyst and deterioration of the catalytic activity can be avoided in the ideal asymmetric autocatalysis. In addition, no process is required for the separation of the catalyst from the product because their structures are identical.

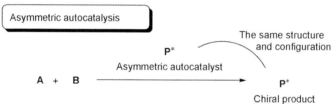

Scheme 1 Asymmetric autocatalysis.

Chiral homogeneity of biomolecules such as L-amino acids and D-sugars has been a long-standing puzzle. Several mechanisms have been proposed as the origins of the chirality of organic compounds. However, enantiomeric excesses (*ee*) of organic compounds induced by circularly polarized light (CPL) and quartz, for example, have usually been very low (< 2% *ee*). Therefore, an amplification process of the very low enantiomeric excess of organic compounds to a very high enantiomeric excess is necessary to correlate with the homochirality of organic compounds.

We describe here highly enantioselective asymmetric autocatalysis with amplification of enantiomeric excess and asymmetric autocatalysis triggered by CPL and by chiral substances such as quartz [1A, 2]. Spontaneous absolute asymmetric synthesis in conjunction with asymmetric autocatalysis is also described.

Asymmetric Synthesis – The Essentials.
Edited by Mathias Christmann and Stefan Bräse
Copyright © 2007 WILEY-VCH Verlag GmbH & Co. KGaA, Weinheim
ISBN: 978-3-527-31399-0

Results

We found that chiral 5-pyrimidyl alkanols are excellent asymmetric autocatalysts for the addition of iso-Pr$_2$Zn to pyrimidine-5-carbaldehyde [3]. In the reactions, zinc monoalkoxide of pyrimidyl alkanol works as an asymmetric autocatalyst. Among pyrimidyl alkanols, 2-alkynylpyrimidyl alkanol exhibits almost complete enantioselectivity and very high catalytic activity. When (S)-2-(2-*tert*-butylethynyl)-5-pyrimidyl alkanol with > 99.5 % *ee* was used as an asymmetric autocatalyst, (S)-alkanol with > 99.5 % *ee* was formed in a yield of > 99 % (Scheme 2).

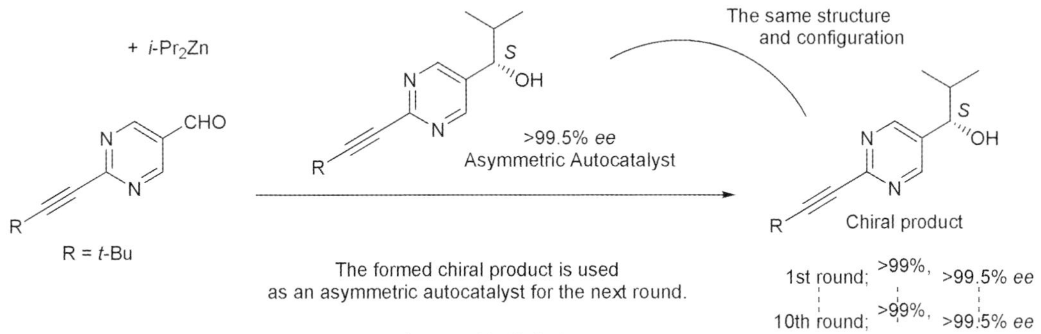

Scheme 2 Asymmetric autocatalysis of pyrimidyl alkohol.

Chiral Amplification by Asymmetric Autocatalysis

We found that asymmetric autocatalysis of a pyrimidyl alkanol exhibits amplification of enantiomeric excess [2]. When a pyrimidyl alkanol with low enantiomeric excess was used as an asymmetric autocatalyst, the enantiomeric excess of the product (including the original autocatalyst) was higher than that of the original catalyst. To take advantage of the asymmetric autocatalytic amplification of enantiomeric excess over non-autocatalytic amplification [4], the product of one run was used as an asymmetric autocatalyst for the next run. Thus, extremely low (ca. 0.00005 %) enantiomeric excess of pyrimidyl alkanol was amplified to near enantiopure (> 99.5 % *ee*) by consecutive asymmetric autocatalysis [5]. During the three consecutive asymmetric autocatalyses, the initially major (S)-enantiomer has automultiplied by a factor of 630 000, whereas the initially minor (R)-enantiomer has only automultiplied by a factor of less than 1000 (Scheme 3).

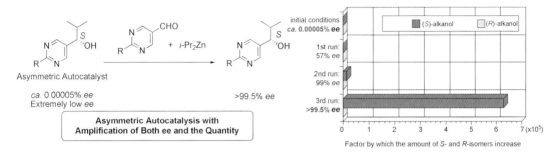

Scheme 3 Asymmetric autocatalysis with amplification of both *ee* and the quantity.

Asymmetric Autocatalysis as a Link Between the Origin of Chirality and Highly Enantioenriched Molecules

It is known that leucine and hexahelicene with low (< 2%) enantiomeric excess have been obtained from asymmetric photodegradation and photosynthesis, respectively, by using CPL. In the presence of leucine and hexahelicene with only < 2% ee as chiral triggers, the reaction of pyrimidine-5-carbaldehyde with iso-Pr$_2$Zn afforded enantioenriched pyrimidyl alkanol with an enhanced enantiomeric excess [6, 7]. Thus, the low enantiomeric excesses induced by CPL [8] have been linked to the high enantiomeric excess of pyrimidyl alkanol, for the first time, by asymmetric autocatalysis with amplification of chirality.

The direct asymmetric synthesis of chiral compounds with a high enantiomeric excess using CPL remains a challenge. Near enantiopure (> 99.5% ee) (S) or (R)-pyrimidyl alkanol is formed by asymmetric photodegradation of racemic pyrimidyl alkanol with the irradiation of left (l)- or right (r)-CPL, respectively, followed by asymmetric autocatalysis (Scheme 4) [9].

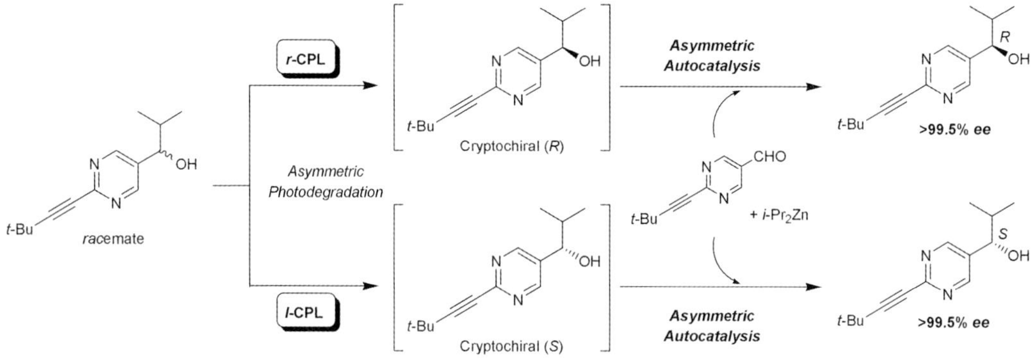

Scheme 4 Asymmetric autocatalysis triggert by circularly polarized light.

A chiral organic compound with high enantiomeric excess has been formed, using chiral inorganic [10, 11] and organic crystals [12] as the chiral auxiliary in conjunction with asymmetric autocatalysis. Quartz and sodium chlorate are chiral inorganic crystals. They exhibit either dextrorotatory (d) or levorotatory (l) enantiomorphs and act as chiral triggers of asymmetric autocatalysis. When pyrimidine carbaldehyde was reacted with iso-Pr$_2$Zn in the presence of d and l-quartz, (S) and (R)-pyrimidyl alkanols with very high enantiomeric excess were obtained (Scheme 5) [10]. The initial small enantiomeric imbalance of the formed (zinc alk-

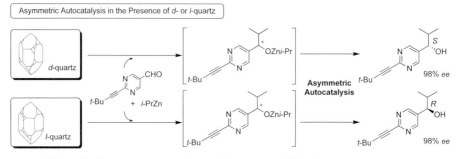

Scheme 5 Asymmetric autocatalysis in the presence of d- or l-quartz.

oxide of) pyrimidyl alkanol induced by chiral quartz has been amplified significantly by the subsequent asymmetric autocatalysis.

Spontaneous Absolute Asymmetric Synthesis

It is considered that small fluctuations in the ratio of the two enantiomers are present statistically if chiral molecules are produced from achiral starting materials in conditions under which the probability of formation of the enantiomers is equal. If the small fluctuation of chirality is amplified by the asymmetric autocatalysis, the chiral compound with above the detection level would be obtained.

Reaction of 2-alkynylpyrimidine-5-carbaldehyde with iso-Pr$_2$Zn in a mixed solvent of ether and toluene and the following one-pot asymmetric autocatalysis with amplification of enantiomeric excess afforded enantioenriched pyrimidyl alkanols (Scheme 6). The absolute configurations of the pyrimidyl alkanol formed exhibit an approximately stochastic distribution of S and R enantiomers (19 times formation of S and 18 times R) [13]. The approximate stochastic behavior in the formation of pyrimidyl alkanols forms one of the conditions necessary for absolute (spontaneous) asymmetric synthesis.

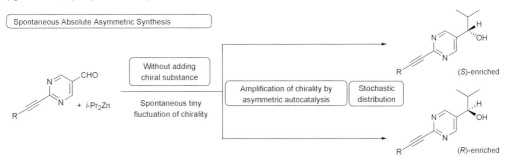

Scheme 6 Spontaneous absolute asymmetric synthesis.

Conclusions and Future Perspectives

Asymmetric autocatalysis is an efficient method of enantioselective synthesis. Asymmetric autocatalysis of pyrimidyl alkanols with amplification of chirality enables one to link the tiny enantiomeric imbalances induced by the proposed mechanisms of the origins of chirality to very high enantioenrichments of chiral compounds, and should play an essential role in clarifying the origin of chirality.

CV of Kenso Soai

Kenso Soai was born in Hiroshima, Japan in 1950. He studied at the University of Tokyo. He received his Ph.D. in 1979 after working with Teruaki Mukaiyama. After post-doctoral appointments at the University of North Carolina at Chapel Hill (Ernest L. Eliel), he joined the faculty of Tokyo University of Science in 1981 as a lecturer and began his independent research career. He was promoted to Associate Professor in 1986 and Professor in 1991. He has been the recipient of the Progress Award in Synthetic Organic Chemistry, the Chisso Award in Synthetic Organic Chemistry, the Inoue Prize for Science, the Organic Chemistry Award of the Chemical Society of Japan, the Merit Award of Tokyo Metropolitan for Science and Technology, the Molecular Chirality Award, the Medal of The Academy of Sciences, Literatures and Arts, Modena, the Synthetic Organic Chemistry Award, and the Chirality Medal.

CV of Tsuneomi Kawasaki

Tsuneomi Kawasaki was born in Yamanashi, Japan in 1976. He received his B.S. in 1999 and his Ph.D. in 2004 from the University of Tokyo, under the supervsion of Professor Takeshi Kitahara. He joined Soai's group at Tokyo Univesity of Science as a research associate in 2004. He received a Nissan Chemical Industry Award in Synthetic Organic Chemistry Japan in 2005. His research interests include synthetic organic chemistry.

Selected Publications

1A. K. Soai, T. Shibata, I. Sato, *Acc. Chem. Res.* **2000**, *33*, 382–390. *Enantioselective Automultiplication of Chiral Molecules by Asymmetric Autocatalysis.*
2. K. Soai, T. Shibata, H. Morioka, K. Choji, *Nature (London)* **1995**, *378*, 767–768. *Asymmetric Autocatalysis and Amplification of Enantiomeric Excess of a Chiral Molecule.*
3. T. Shibata, S. Yonekubo, K. Soai, *Angew. Chem. Int. Ed.* **1999**, *38*, 659–661. *Practically Perfect Asymmetric Autocatalysis using 2-Alkynyl-5-pyrimidylalkanol.*
4. For non-autocatalytic amplification, see the contribution by Kagan in this book.
5. I. Sato, H. Urabe, S. Ishiguro, T. Shibata, K. Soai, *Angew. Chem. Int. Ed.* **2003**, *42*, 315–317. *Amplification of Chirality from Extremely Low to Greater than 99.5% ee by Asymmetric Autocatalysis.*
6. T. Shibata, J. Yamamoto, N. Matsumoto, S. Yonekubo, S. Osanai, K. Soai, *J. Am. Chem. Soc.* **1998**, *120*, 12157–12158. *Amplification of a Slight Enantiomeric Imbalance in Molecules based on Asymmetric Autocatalysis. –The First Correlation between High Enantiomeric Enrichment in a Chiral Molecule and Circularly Polarized Light.*
7. I. Sato, R. Yamashima, K. Kadowaki, J. Yamamoto, T. Shibata, K. Soai, *Angew. Chem. Int. Ed.* **2001**, *40*, 1096–1098. *Asymmetric Induction by Helical Hydrocarbons: [6]- and [5]Helicenes.*
8. I. Sato, R. Sugie, Y. Matsueda, Y. Furumura, K. Soai, *Angew. Chem. Int. Ed.* **2004**, *43*, 4490–4492. *Asymmetric Synthesis of a Highly Enantioenriched Compound by Circularly Polarized Light Mediated with Photoequilibrium of Chiral Olefins in Conjunction with Asymmetric Autocatalysis.*
9. T. Kawasaki, M. Sato, S. Ishiguro, T. Saito, Y. Morishita, I. Sato, H. Nishino, Y. Inoue, K. Soai, *J. Am. Chem. Soc.* **2005**, *127*, 3274–3275. *Enantioselective Synthesis of Near Enantiopure Compound by Asymmetric Autocatalysis Triggered by Asymmetric Photolysis with Circularly Polarized Light.*
10. K. Soai, S. Osanai, K. Kadowaki, S. Yonekubo, T. Shibata, I. Sato, *J. Am. Chem. Soc.* **1999**, *121*, 11235–11236. *d- and l-Quartz-Promoted Highly Enantioselective Synthesis of a Chiral Compound.*
11. I. Sato, K. Kadowaki, K. Soai, *Angew. Chem. Int. Ed.* **2000**, *39*, 1510–1512. *Asymmetric Synthesis of an Organic Compound with High Enantiomeric Excess Induced by Inorganic Ionic Sodium Chlorate.*
12. T. Kawasaki, K. Jo, H. Igarashi, I. Sato, M. Nagano, H. Koshima, K. Soai, *Angew. Chem. Int. Ed.* **2005**, *44*, 2774–2777. *Asymmetric Amplification Using Chiral Co-crystal Formed from Achiral Organic Molecules by Asymmetric Autocatalysis.*
13. K. Soai, I. Sato, T. Shibata, S. Komiya, M. Hayashi, Y. Matsueda, H. Imamura, T. Hayase, H. Morioka, H. Tabira, J. Yamamoto, Y. Kowata, *Tetrahedron Asymmetry*, **2003**, *14*, 185–188. *Asymmetric Synthesis of Pyrimidyl Alkanol without Adding Chiral Substances by The Addition of Diisopropylzinc to Pyrimidine-5-carbaldehyde in Conjunction with Asymmetric Autocatalysis.*

Asymmetric Synthesis using Deoxyribose-5-phosphate Aldolase

Chi-Huey Wong and William A. Greenberg, The Scripps Research Institute, La Jolla, CA, USA

Background

Aldolases are a wide-ranging class of enzymes that catalyze asymmetric carbon–carbon bond-forming reactions with a variety of substrates. Several aldolase types have been used as catalysts for organic synthesis, and are the subject of earlier reviews [1R, 2R]. In this chapter we will focus on a specific enzyme, deoxyribose-5-phosphate aldolase (DERA), and discuss its catalytic mechanism, synthetic utility, and practical applications to the synthesis of epothilones and statin cholesterol-lowering drugs.

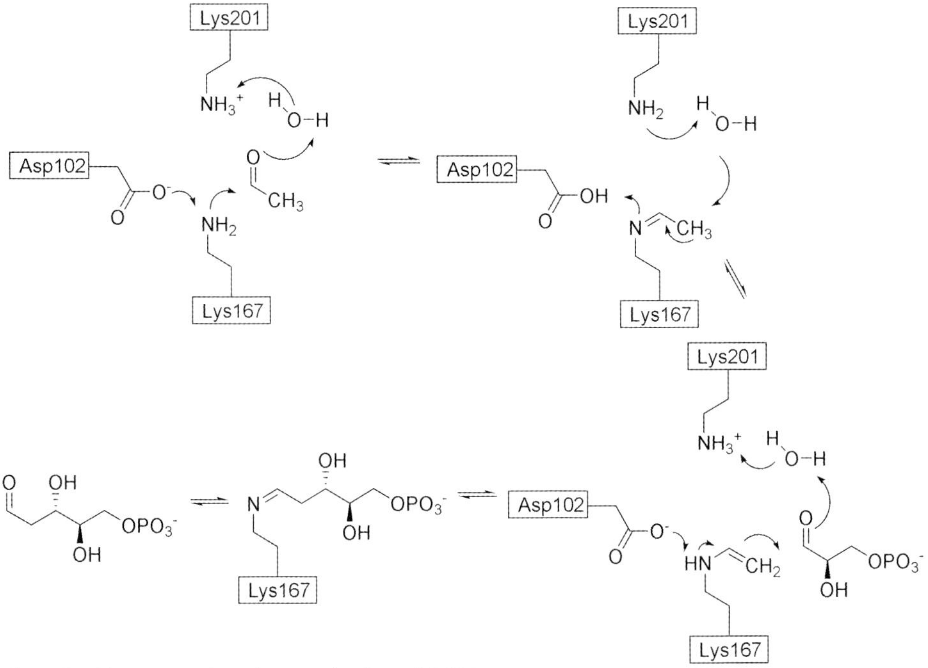

Figure 1 The proposed catalytic mechanism of DERA.

Asymmetric Synthesis – The Essentials.
Edited by Mathias Christmann and Stefan Bräse
Copyright © 2007 WILEY-VCH Verlag GmbH & Co. KGaA, Weinheim
ISBN: 978-3-527-31399-0

Results

Catalytic Mechanism

DERA is a class I aldolase that uses a covalent Schiff base intermediate to catalyze the aldol reaction between the donor, acetaldehyde, and the acceptor, D-glyceraldehyde-3-phosphate, to form 2-deoxyribose-5-phosphate (Figure 1). High resolution X-ray crystal structures of the enzyme bound to reaction intermediates have allowed a detailed mechanism to be proposed, including identification of an active-site water molecule that effects the key proton abstraction to form the nucleophilic enamine [3].

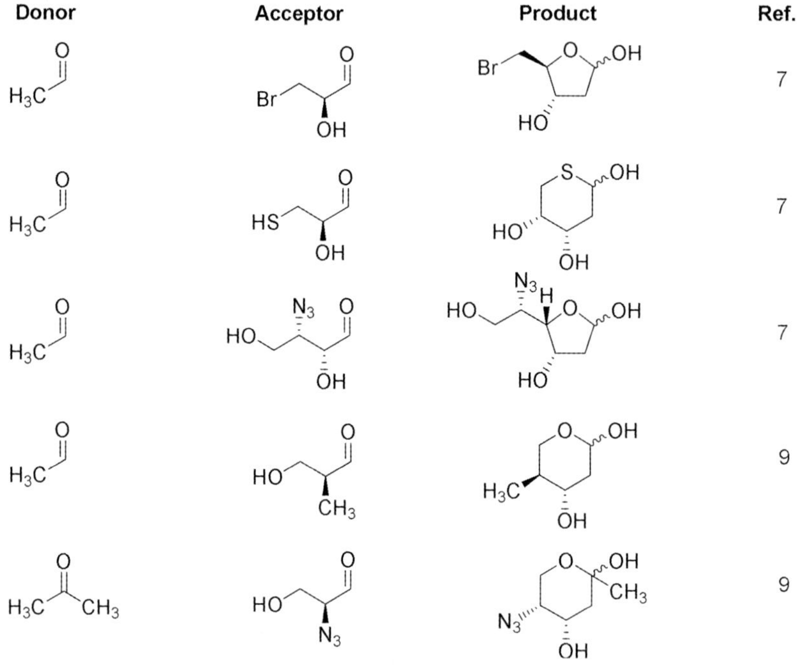

Figure 2 Selected examples of reactions performed with DERA.

Synthetic Applications

DERA from *E. coli* was made available as a synthetic tool when it was first overexpressed in our laboratory in 1989 [4]. Since then, it has been used in many applications [5–7]. The utility of DERA arises from its ability to tolerate a wide range of substrates. In common with other aldolase classes, the donor substrate requirement for acetaldehyde is quite strict, but the acceptor aldehyde can be dramatically modified while still retaining catalytic activity. Most notably, phosphorylation is not required. Some examples of DERA-catalyzed reactions are shown in Figure 2.

DERA has been found to efficiently catalyze sequential addition of multiple acetaldehyde equivalents to a substrate in a single pot, forming 1,3-polyol synthons [6, 7]. The product of the first aldol reaction serves as the substrate

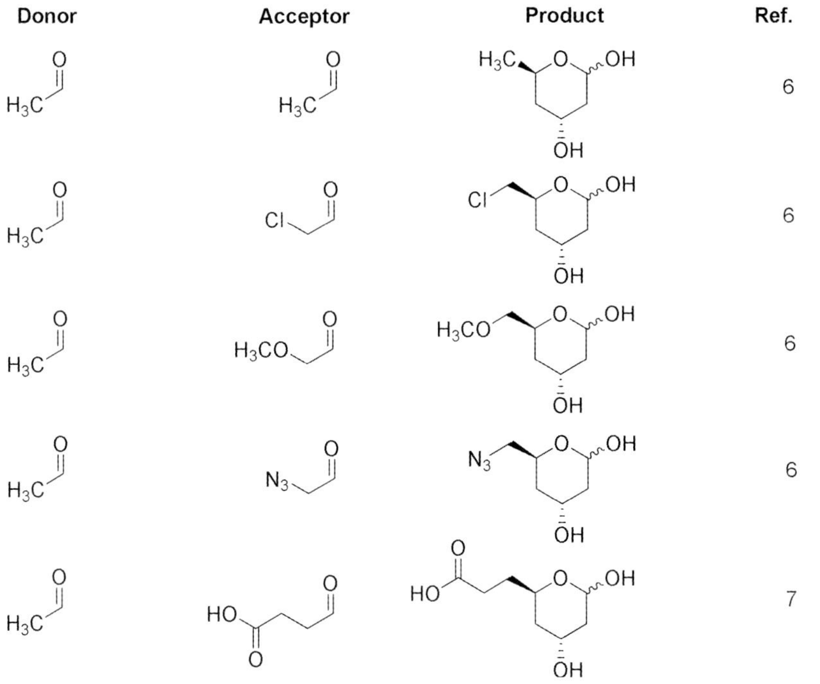

Donor	Acceptor	Product	Ref.
acetaldehyde	acetaldehyde	methyl-substituted dihydroxy tetrahydropyran hemiacetal	6
acetaldehyde	chloroacetaldehyde	chloromethyl-substituted dihydroxy tetrahydropyran hemiacetal	6
acetaldehyde	methoxyacetaldehyde	methoxymethyl-substituted dihydroxy tetrahydropyran hemiacetal	6
acetaldehyde	azidoacetaldehyde	azidomethyl-substituted dihydroxy tetrahydropyran hemiacetal	6
acetaldehyde	3-hydroxy-4-oxobutanoate	acetoxymethyl-substituted dihydroxy tetrahydropyran hemiacetal	7

Figure 3 Sequential one-pot aldol reactions catalyzed by DERA.

for the second addition. Cyclization to a stable hemiacetal stops the reaction after two additions. Figure 3 illustrates how complex molecules can be rapidly built from simple starting materials using this strategy.

The substrate specificity of DERA has been further broadened by rational mutagenesis of the enzyme, based on X-ray crystal structural information [8]. Directed evolution represents another method of changing the selectivity, catalytic efficiency, and enantioselectivity of the enzyme.

Application to the Synthesis of Epothilones and Statins

Recently, the power of DERA as a synthetic tool has been demonstrated by its application to the synthesis of the anticancer epothilones [9], and to the synthesis of the 1,3-diol side chain of statin cholesterol-lowering drugs such as atorvastatin [8, 10]. In the epothilone synthesis (Figure 4), DERA-catalyzed reactions were used to make building blocks for *both* fragments in a Suzuki coupling strategy to assemble the natural products.

The previously mentioned sequential one-pot aldol reaction affords products that are directly applicable to synthesis of the key 1,3-diol side-chain of statin drugs such as atorvastatin (Figure 5). It has therefore attracted attention from industrial process chemistry groups, who have focused on optimizing catalyst load, reaction rate, and volumetric productivity. This chemistry has now been scaled up to multi-kg levels.

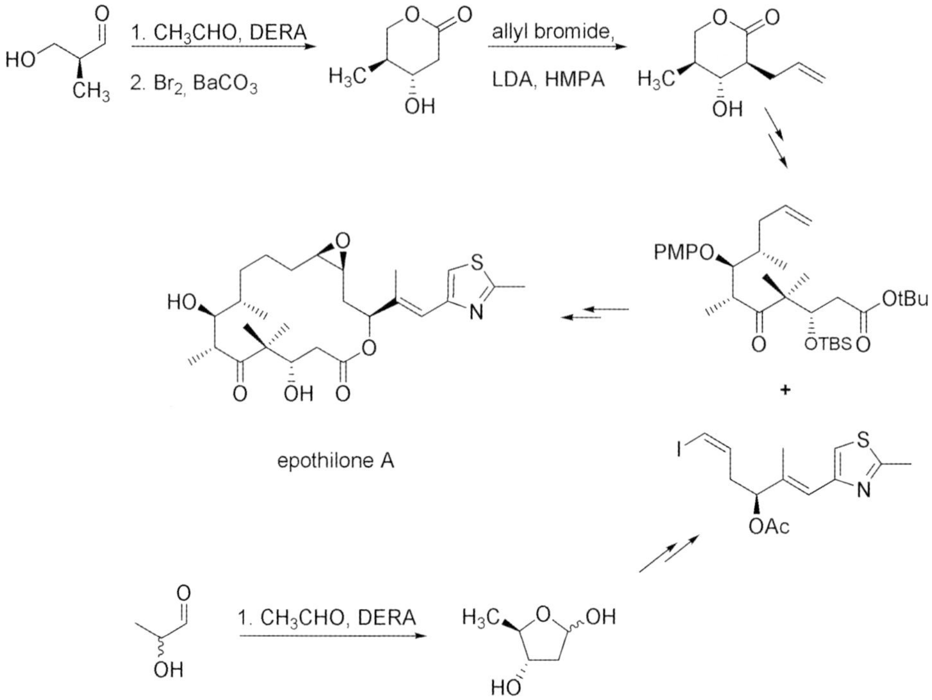

Figure 4 Use of DERA to assemble building blocks for epothilones.

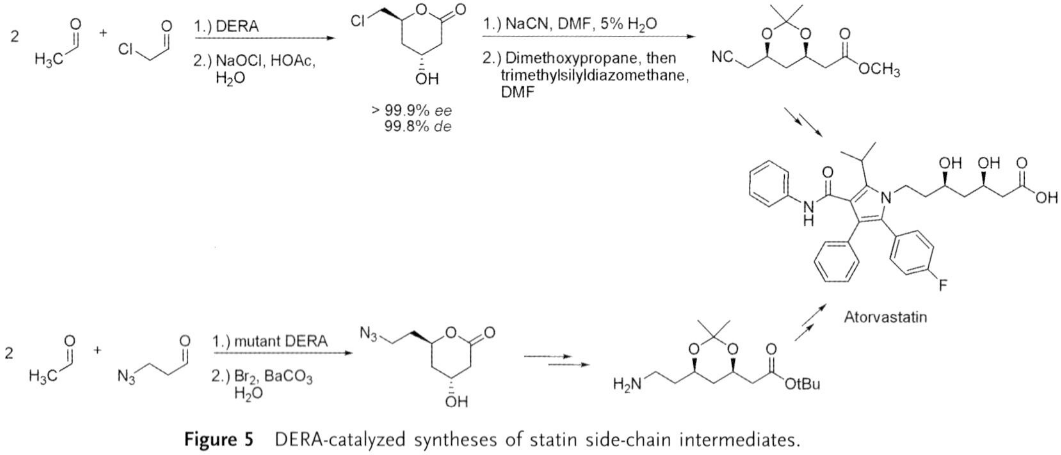

Figure 5 DERA-catalyzed syntheses of statin side-chain intermediates.

Conclusion

This short review has illustrated the versatility of a single aldolase type, deoxyribose-5-phosphate aldolase (DERA), as a catalyst for asymmetric synthesis of a variety of highly functionalized structures from simple starting materials. Other classes of aldolases, discussed in the cited reviews, are equally useful. The practical and large-scale applications of DERA demonstrate the enormous potential of aldolases as tools for the synthetic chemist.

CV of Chi-Huey Wong

Chi-Huey Wong is Professor and Ernest W. Hahn Chair in Chemistry at The Scripps Research Institute and also a member of the Skaggs Institute for Chemical Biology. His research interests are in the areas of bioorganic and synthetic chemistry and biocatalysis, including development of new synthetic chemistry based on enzymatic and chemo-enzymatic reactions, study of carbohydrate-mediated biological recognition, drug discovery, and development of carbohydrate microarrays for high-throughput screening and study of reaction mechanism. He received his B.S. and M.S. degrees from National Taiwan University, and Ph.D. in Chemistry from Massachusetts Institute of Technology. He is a member of Academia Sinica, Taipei, the American Academy of Arts and Sciences, and the National Academy of Sciences.

CV of William A. Greenberg

William A. Greenberg is an Assistant Professor in the Department of Chemistry at The Scripps Research Institute. He received his Ph.D. in Chemistry at the California Institute of Technology with Peter B. Dervan, and performed postdoctoral research with Chi-Huey Wong at Scripps. After several years at Diversa Corporation working in the fields of biocatalysis, process chemistry, and medicinal chemistry, he returned to Scripps in 2005.

Selected Publications

1R. T.D. Machajewski, C.-H. Wong, *Angew. Chem. Int. Ed.* **2000**, *39*, 1352–1374. *The catalytic asymmetric aldol reaction.*

2R. M.G. Silvestri, G. DeSantis, M. Mitchell, C.-H. Wong, *Top. Stereochem.* **2003**, *23*, 267–342. *Asymmetric aldol reactions using aldolases.*

3. A. Heine, G. DeSantis, J.G. Luz, M. Mitchell, C.-H. Wong, I.A. Wilson, *Science* **2001**, *294*, 369–374. *Observation of covalent intermediates in an enzyme mechanism at atomic resolution.*

4. C.F. Barbas, Y.-F. Wang, C.-H. Wong, *J. Am. Chem. Soc.* **1990**, *112*, 2013–2014. *Deoxyribose-5-phosphate aldolase as a synthetic catalyst.*

5. L. Chen, D. P. Dumas, C.-H. Wong, *J. Am. Chem. Soc.* **1992**, *114*, 741–748. *Deoxyribose-5-phosphate aldolase as a catalyst in asymmetric aldol condensation.*

6. H.J.M. Gijsen, C.-H. Wong, *J. Am. Chem. Soc.* **1994**, *116*, 8422–8423. *Unprecedented asymmetric aldol reactions with three aldehyde substrates catalyzed by 2-deoxyribose-5-phosphate aldolase.*

7. C.-H. Wong, E. Garcia-Junceda, L. Chen, O. Blanco, H.J.M Gijsen, D. H. Steensma, *J. Am. Chem. Soc.* **1995**, *117*, 3333–3339. *Recombinant 2-deoxyribose-5-phosphate aldolase in organic synthesis: use of sequential two-substrate and three-substrate aldol reactions.*

8. J. Liu, C.-C. Hsu, C.-H. Wong, *Tetrahedron. Lett.* **2004**, *45*, 2439–2441. *Sequential aldol condensation catalyzed by DERA mutant Ser238Asp and a formal total synthesis of atorvastatin.*

9. J. Liu, C.-H. Wong, *Angew. Chem. Int. Ed.* **2002**, *41*, 1404-1407. *Aldolase-catalyzed asymmetric synthesis of novel pyranose synthons as a new entry to heterocycles and epothilones.*

10. W.A. Greenberg, A. Varvak, S.R. Hanson, K. Wong, H. Huang, P. Chen, M.J. Burk, *Proc. Natl. Acad. Sci.* **2004**, *101*, 5788–5793. *Development of an efficient, scalable, aldolase-catalyzed process for enantioselective synthesis of statin intermediates.*

Part IV

Asymmetric Reactions in Total Synthesis

K. C. Nicolaou and Paul G. Bulger, The Scripps Research Institute, La Jolla, California, USA

Background

Through most of the history of organic chemistry, synthetic chemists had virtually no practical methods for the preparation of chiral molecules in enantiomerically (as opposed to diastereomerically) pure form, without resorting to resolution procedures or the judicious choice of optically pure starting materials from the 'chiral pool' [1R, 2R]. However, the development of effective, reliable asymmetric reactions over the last few decades now allows many of the fundamental organic transformations to be performed in an enantioselective fashion, and has consequently revolutionized organic synthesis.

Total synthesis has proven to be a challenging testing ground for asymmetric synthetic methodology, whilst at the same time providing the spur for the development of many new reactions and technologies [3R, 4R]. In this chapter we provide a 'top ten' of what we feel are the most synthetically useful and widely used asymmetric reactions in total synthesis, highlighting in each case one or more applications of the reaction to the synthesis of complex natural products. This list is entirely subjective (and in no particular order), yet will hopefully serve to underscore the impact and utility of asymmetric reactions in total synthesis.

Reactions
Aldol

The aldol reaction has become one of the most powerful methods available to synthetic chemists for the formation of carbon–carbon bonds in a regio-, stereo- and enantioselective manner [5R]. The reaction provides an excellent method for the control of acyclic stereochemistry, generating β-hydroxy carbonyl compounds which are themselves ripe for a variety of further stereoselective transformations. The most commonly used asymmetric aldol reactions employ chiral auxiliary-based systems and/or substrate control to dictate the stereochemical outcome. As an example of the former in action, Evans and coworkers utilized a 1,2-*syn*-selective aldol reaction between (1*S*,2*R*)-norephedrine-derived *N*-acyl oxazolidinone

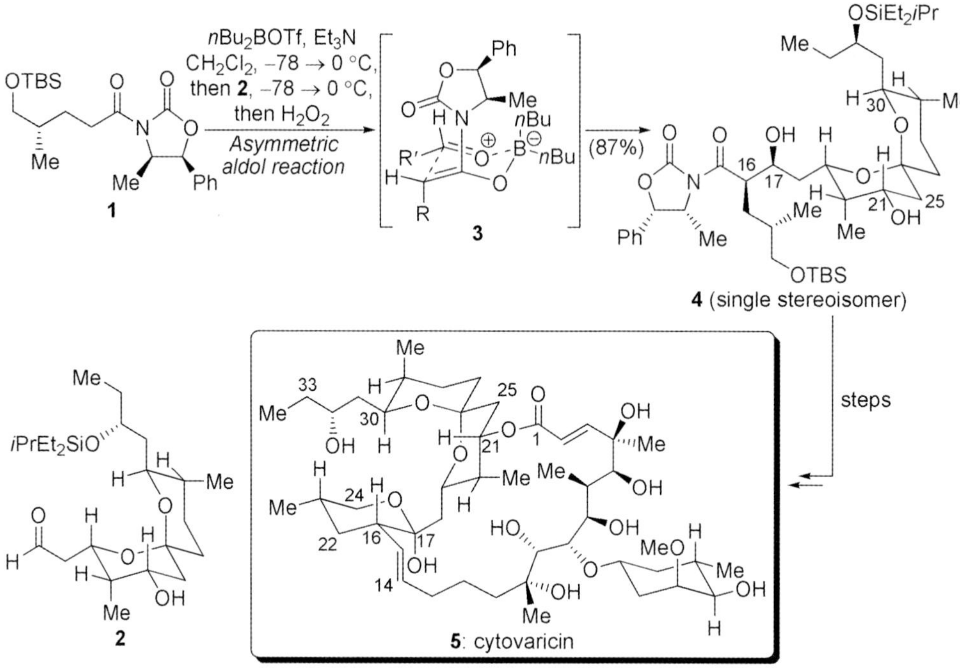

Scheme 1 Use of a chiral auxiliary-mediated asymmetric aldol fragment coupling reaction in the total synthesis of cytovaricin (**5**) [6].

1 and aldehyde **2** to generate the C16–C17 bond, and concomitantly to install the required stereocenters, in their total synthesis of cytovaricin (**5**, Scheme 1) [6].

Paterson and coworkers made use of the potential for substrate control in aldol addition reactions in their enantioselective synthesis of (+)-spongistatin 1 (**10**, Scheme 2) [7]. Thus, the reaction of aldehyde **6** with the E-boron enolate derived from ketone **7** led to the formation of 1,2-*anti* aldol adduct **9**, corresponding to that predicted by the Felkin–Ahn model, in excellent yield and with good diastereoselectivity.

Alkylation

Besides asymmetric aldol reactions, asymmetric alkylation reactions have arguably had the greatest impact on enantioselective carbon–carbon bond formation to date. The use of chiral auxiliary-derivatized carbonyl compounds dominates this field as well, and has reached a relatively advanced stage. The SAMP/RAMP hydrazone methodology pioneered by the Enders' group represents a particularly robust and general method for the indirect α-alkylation of aldehydes and ketones [8]. This methodology has found widespread application in enantioselective synthesis; for example, utilization by Nicolaou and coworkers en route to their total synthesis of swinholide A (**14**, Scheme 3) [9].

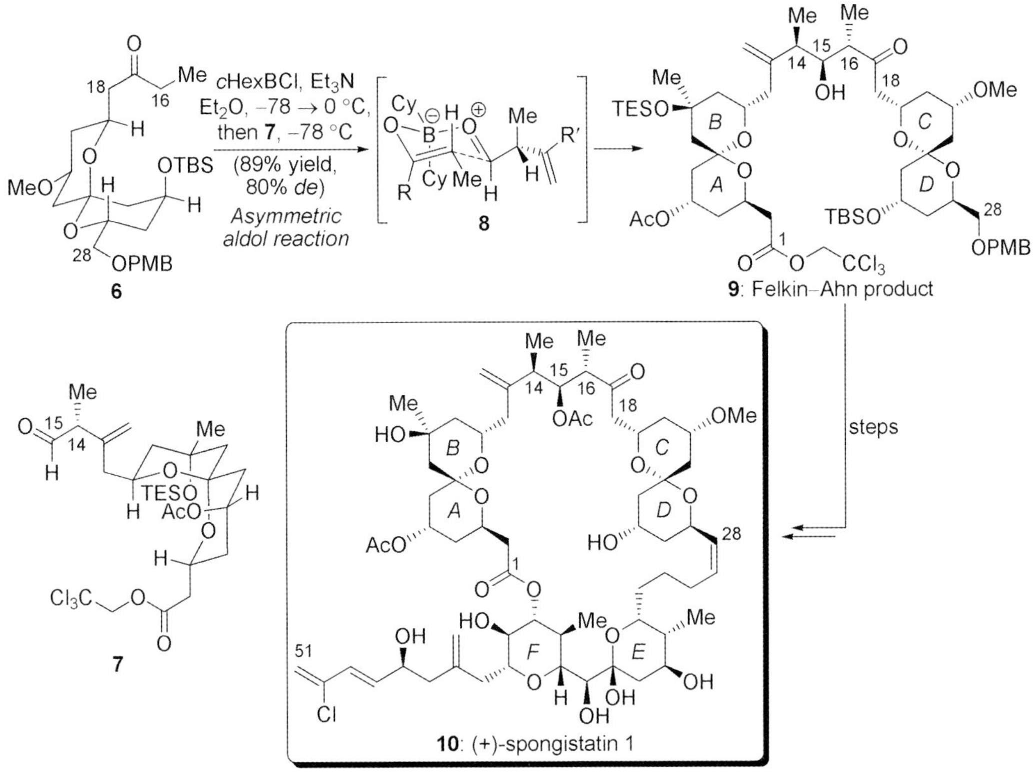

Scheme 2 Use of substrate control in a diastereoselective boron-mediated fragment coupling aldol reaction in the enantioselective synthesis of (+)-spongistatin 1 (**10**) [7].

As with the corresponding aldol processes, Evans-type N-acyl oxazolidinones have traditionally been the most commonly used ester/carboxylic acid surrogates in asymmetric α-alkylation reactions. However, a more recent method, developed by the Myers group, that is gaining popularity involves the alkylation of pseudoephedrine amides [10], an example of which is the efficient alkylation of amide **15** with iodide **16** (see **18**) in the enantioselective synthesis of (−)-dictyostatin (**19**, Scheme 4) by Curran and coworkers [11, 12].

Diels–Alder

Few reactions can lay claim to having enabled and shaped the art and science of total synthesis over the last few decades to the extent that the Diels–Alder reaction, in all its various guises, has done [13]. Many classic total syntheses have made use of the ability of this most venerable of organic reactions to generate not only a plethora of carbocyclic and heterocyclic ring systems, but also to fashion up to four contiguous stereocenters in a single step. Despite this, the issue of controlling the absolute (as opposed to relative) stereochemistry of the Diels–Alder reaction, particularly beyond substrate control, has only begun to be ad-

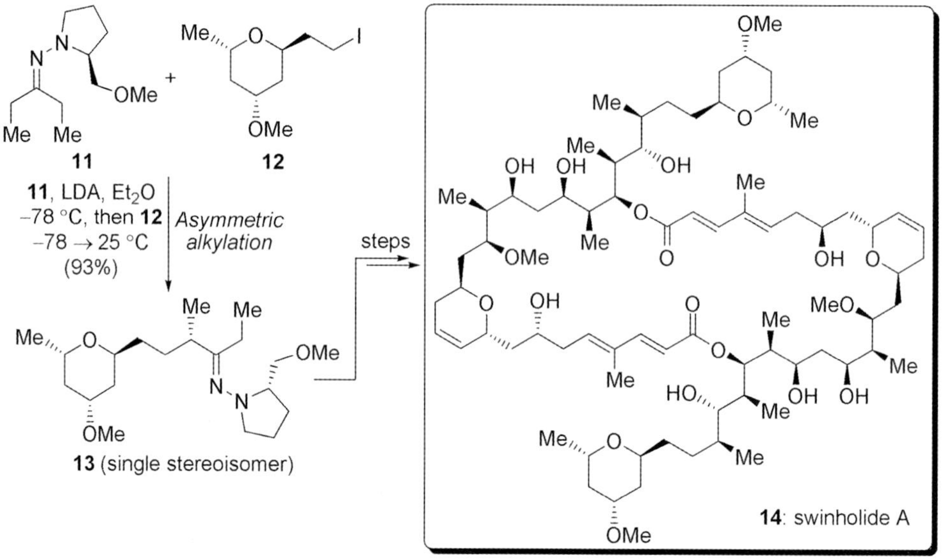

Scheme 3 Asymmetric alkylation using the SAMP/RAMP methodology as a key step in the total synthesis of swinholide A (**14**) [9].

dressed successfully in more recent years. Chiral Lewis acids provide a particularly appealing solution to this problem, allowing for true catalytic asymmetric reactions. Nicolaou and coworkers employed the BINOL-derived catalyst **22** [14] (Scheme 5) to promote cycloaddition between quinone **20** and silyloxydiene **21** in a regio- and enantioselective manner [15]. The methyl group-bearing stereocenter installed in cycloadduct **24** ultimately determined the stereochemical course of the remaining steps in the total synthesis of the target compound, colombiasin A (**26**).

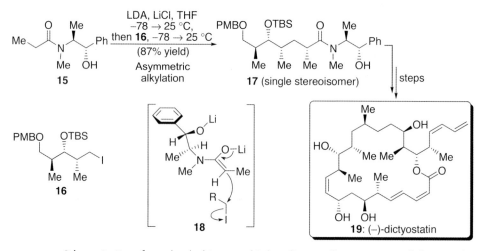

Scheme 4 Use of pseudoephedrine as a chiral auxiliary to effect asymmetric alkylation in the total synthesis of (−)-dictyostatin (**19**) [11].

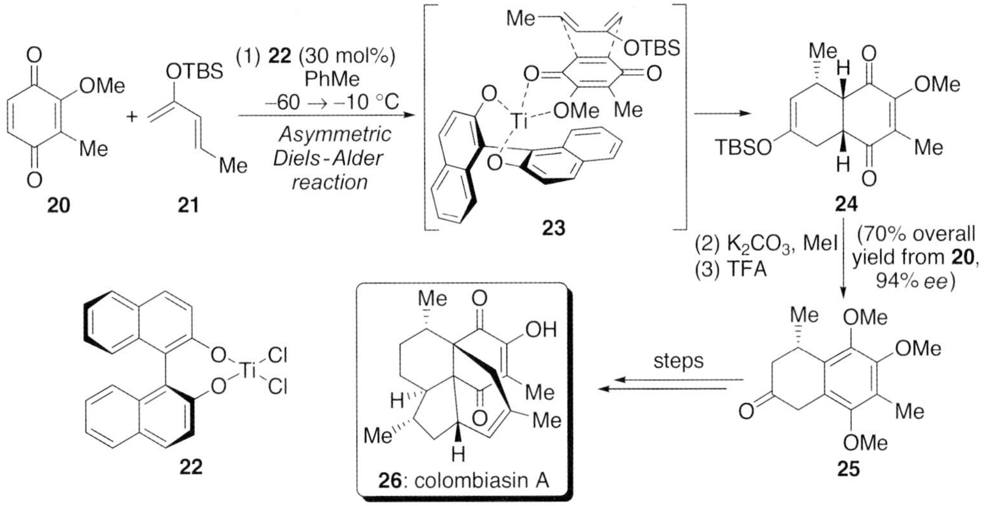

Scheme 5 Use of an asymmetric Lewis acid-catalyzed Diels–Alder reaction in the total synthesis of colombiasin A (**26**) [15].

The Jacobsen group made productive use of catalytic asymmetric hetero-Diels–Alder reactions in their total synthesis of (+)-ambruticin (**32**, Scheme 6), forming both pyran core ring systems using this emerging methodology [16]. Scheme 6 illustrates the construction of the 'left-hand' pyran ring by means of cycloaddition between silyloxydiene **27** and aldehyde **28**, catalyzed by the novel salen-type complex **59**.

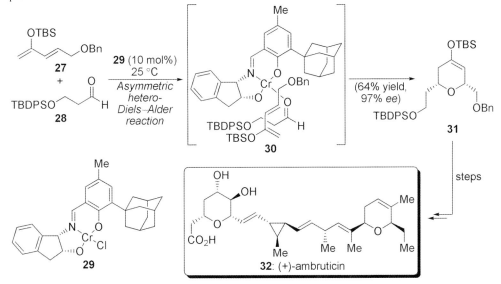

Scheme 6 Use of a chromium-catalyzed asymmetric hetero-Diels–Alder reaction in the total synthesis of (+)-ambruticin (**32**) [16].

Allylation

Enantioselective allylation reactions (and related processes such as crotylation reactions), especially the addition of chiral allylmetal reagents to aldehydes, have been a mainstay of asymmetric synthesis for over two decades. Notably, allylation reactions serve as useful substitutes for notoriously 'difficult' asymmetric aldol processes, such as acetate aldol reactions and 1,2-*anti* additions. Many allylation reagents and systems, both stoichiometric chiral auxiliary-based and catalytic methods, have been developed, offering a broad spectrum of reactivity and stereoselectivity [17]. Nevertheless, currently the most popular are the family of diisopinocampheylborane reagents first developed by the Brown group in the mid-1980s. Armstrong and Ogawa used these reagents no fewer than four times during the construction of the C1–C25 fragment in their total synthesis of calyculin C (**37**, Scheme 7) [18]. As an illustrative example, treatment of aldehyde **33** with enantiopure crotylborane **34** (prepared *in situ*) generated the 1,2-*anti* product **35** as a single stereoisomer. The Brown asymmetric allylation reaction has also been employed three times in a re-iterative and biomimetic sequence in the total synthesis of (–)-apicularen A [19].

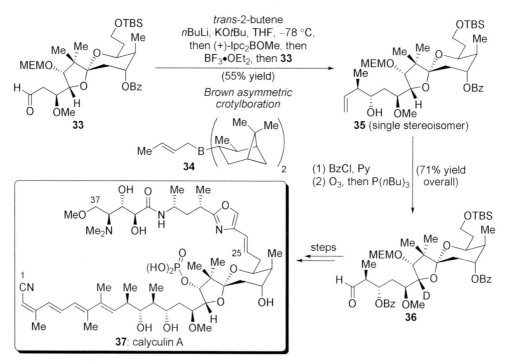

Scheme 7 Use of a Brown asymmetric crotylboration reaction in the total synthesis of calyculin C (**37**) [18].

Palladium-catalyzed Cross-coupling

In no field of organic chemistry has the impact of the development of palladium-catalyzed cross-coupling reactions been more keenly felt than in total synthesis [20]. Numerous variants of these processes exist, each assuming their own name and identity based on the nature of the reacting partners, and over the last 25 years they have collectively revolutionized the way chemists can approach synthetic problems. The two such processes which have seen the most use in asymmetric synthesis are the asymmetric allylic alkylation and intramolecular Heck reactions. A leading exponent of the latter has been the Overman group, one of whose crowning achievements has been the total synthesis of the alkaloid (−)-quadrigemine C (**41**, Scheme 8) [21]. One of the key steps in the route involved inducing *meso* intermediate **38** to participate in a set of tandem, reagent controlled asymmetric Heck cyclizations, leading to the desymmetrized intermediate **40** bearing two new oxindole rings. The Shibasaki group also pioneered and admirably applied the asymmetric Heck reaction in total synthesis [22].

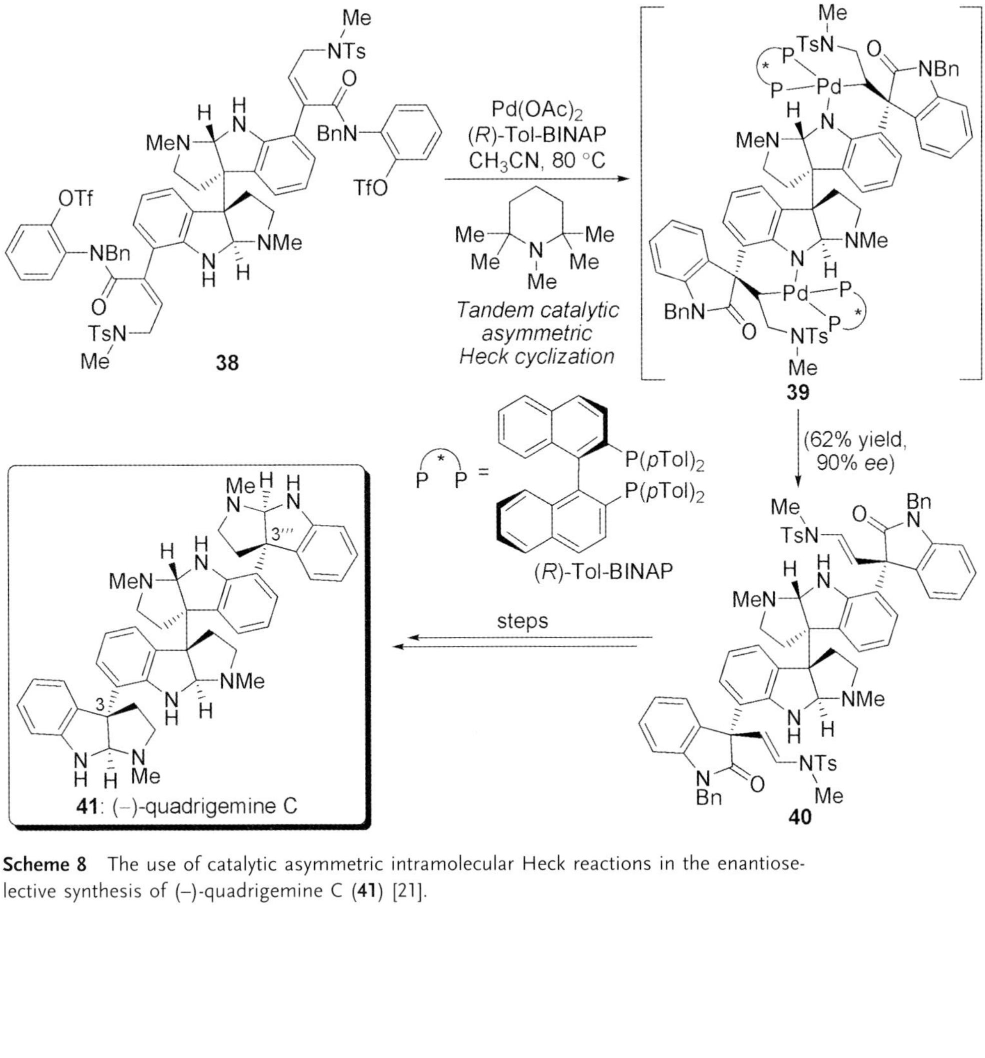

Scheme 8 The use of catalytic asymmetric intramolecular Heck reactions in the enantioselective synthesis of (−)-quadrigemine C (**41**) [21].

In contrast, it has been the Trost group which has largely pioneered the asymmetric allylic alkylation reaction. In their enantioselective synthesis of (−)-morphine (**48**, Scheme 9), the Trost group employed palladium-catalyzed cross-coupling reactions to create the entire carbon framework and form four of the five rings of the target compound [23]. The first of these reactions was the asymmetric allylic alkylation of phenol **42** with allylic trichloroacetate **43**, to provide allylic ether **47** in good yield and enantioselectivity. As an aside, it is certainly instructive to compare this synthetic route with the first total synthesis of (racemic) morphine reported by the Gates group 50 years earlier [24], in order to further appreciate the degree to which chemical synthesis has advanced as both a science and an art in the intervening period.

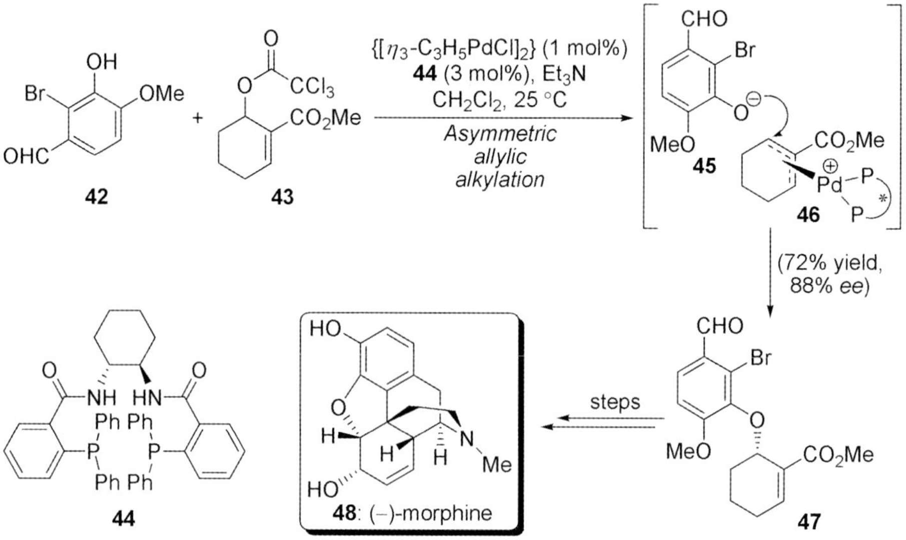

Scheme 9 Use of a palladium-catalyzed asymmetric allylic alkylation in the total synthesis of (−)-morphine (**48**) [23].

Epoxidation

The Sharpless asymmetric epoxidation (SAE) of allylic alcohols was first reported in 1980 [25], and refined into a truly general catalytic process a few years later [26]. To this day the SAE remains one of the most popular and widely used asymmetric reactions in total synthesis, in no small part due to the general, efficient and predictable nature of this protocol. In addition, numerous methods for the regio- and stereoselective opening of the oxirane ring in the 2,3-epoxy alcohol products using a wide variety of nucleophiles have been developed, further enhancing the utility of this process. The Nicolaou group made extensive use of the SAE/epoxide opening protocol in their studies on the synthesis of marine polyether neurotoxins. For example, in the landmark total synthesis of brevetoxin B (**53**, Scheme 10) [27, 28], this two-stage procedure was used to forge the F-, G- and (as is shown in Scheme 10) B-rings of the polycyclic target structure.

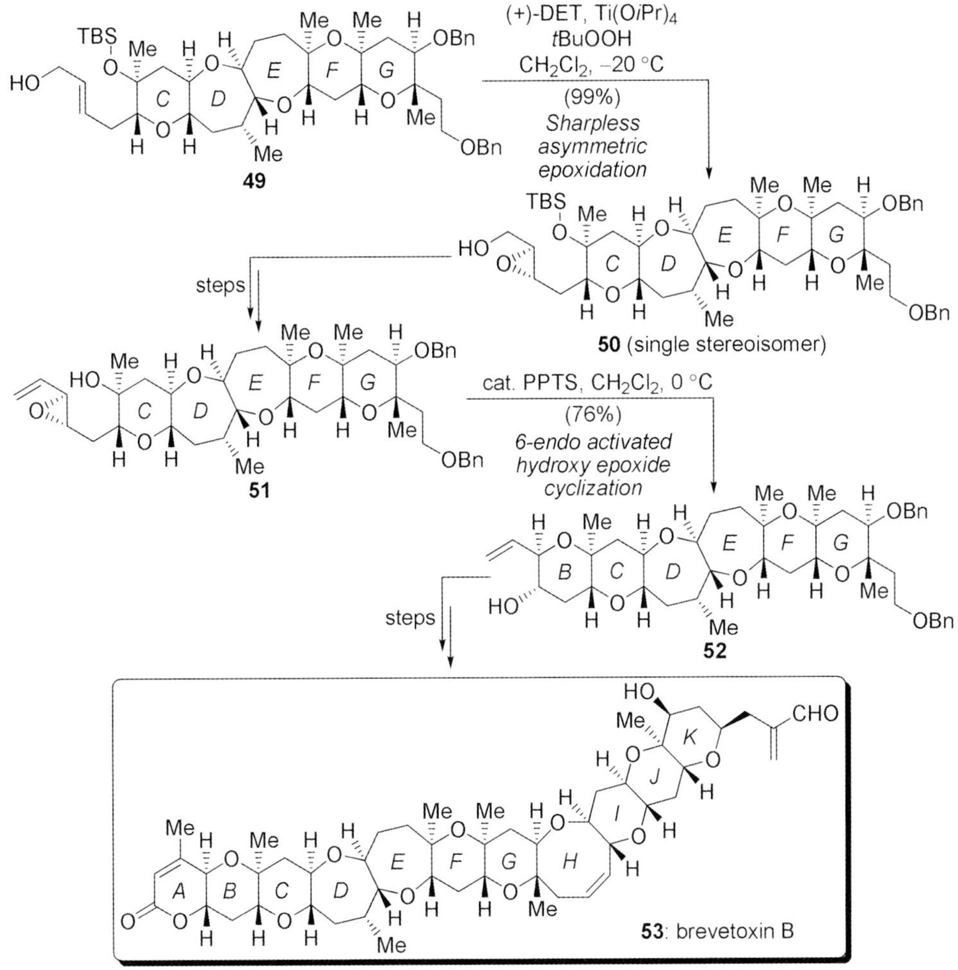

Scheme 10 Sharpless asymmetric epoxidation and stereoselective epoxide opening reactions in the total synthesis of brevetoxin B (**53**) [27, 28].

More recent developments in asymmetric epoxidation methodology have focused on the epoxidation of unfunctionalized olefins. The two most effective catalyst systems to have emerged in this regard are the Jacobsen–Katsuki epoxidation using Mn-salen complexes [29], and the Shi epoxidation employing a fructose-derived carbohydrate scaffold [30]. Nicolaou and coworkers effected the diastereoselective epoxidation of alkene **54** using the modified Mn-salen catalyst **55** (developed by the Katsuki group [31]) and NaOCl as the oxidant in their recent total synthesis of the thiopeptide antibiotic thiostrepton (**57**, Scheme 11) [32, 33].

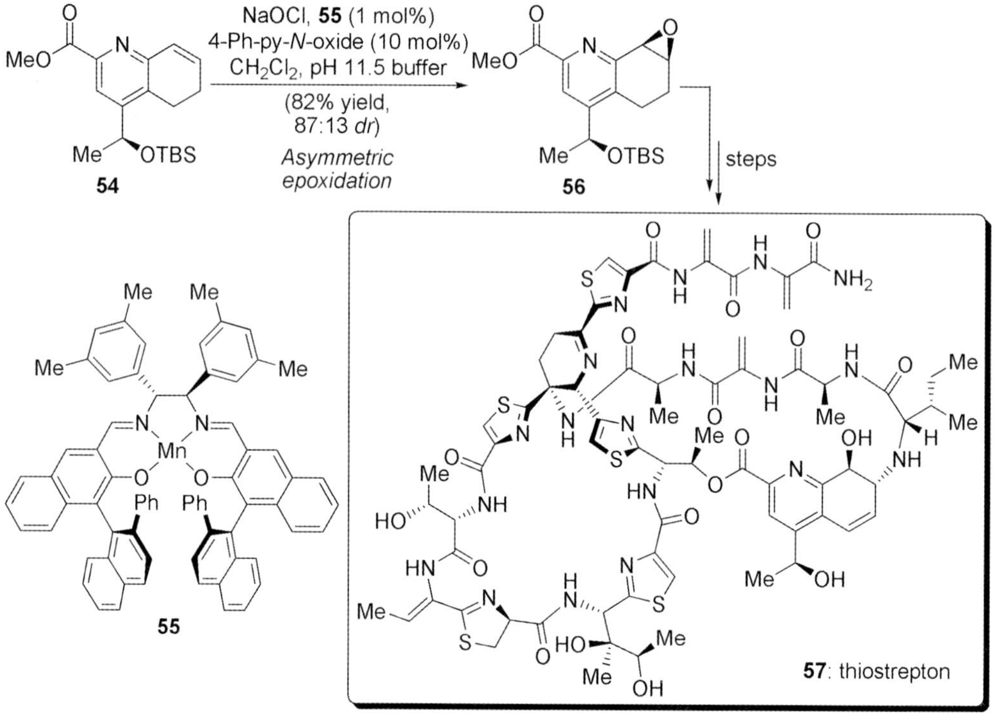

Scheme 11 The catalytic asymmetric epoxidation of a quinaldic acid moiety in the total synthesis of thiostrepton (**57**) [32, 33].

Dihydroxylation/aminohydroxylation

The asymmetric dihydroxylation (AD) [34] and aminohydroxylation (AA) [35] reactions, introduced and developed by the Sharpless group, represent two of the most valuable transformations for introducing heteroatom functionality and stereochemical complexity into organic molecules. 1,2-Diol or aminoalcohol motifs are embedded in a host of Nature's most remarkable natural products, and the ability to install these units in a single step from simple, unfunctionalized alkene precursors makes the AA and (in particular) the AD reactions treasured additions to the synthetic chemist's armory. The daunting molecular architecture of vancomycin (**64**, Scheme 12), the current antibiotic of last resort against MRSA bacteria, is replete with subunits that could potentially be derived from AD and/or AA reactions. Indeed, in their total synthesis of vancomycin (**64**), the Nicolaou group made extensive use of AD and AA chemistry in the elaboration of A-, C- and D-ring building blocks (Scheme 12) [36].

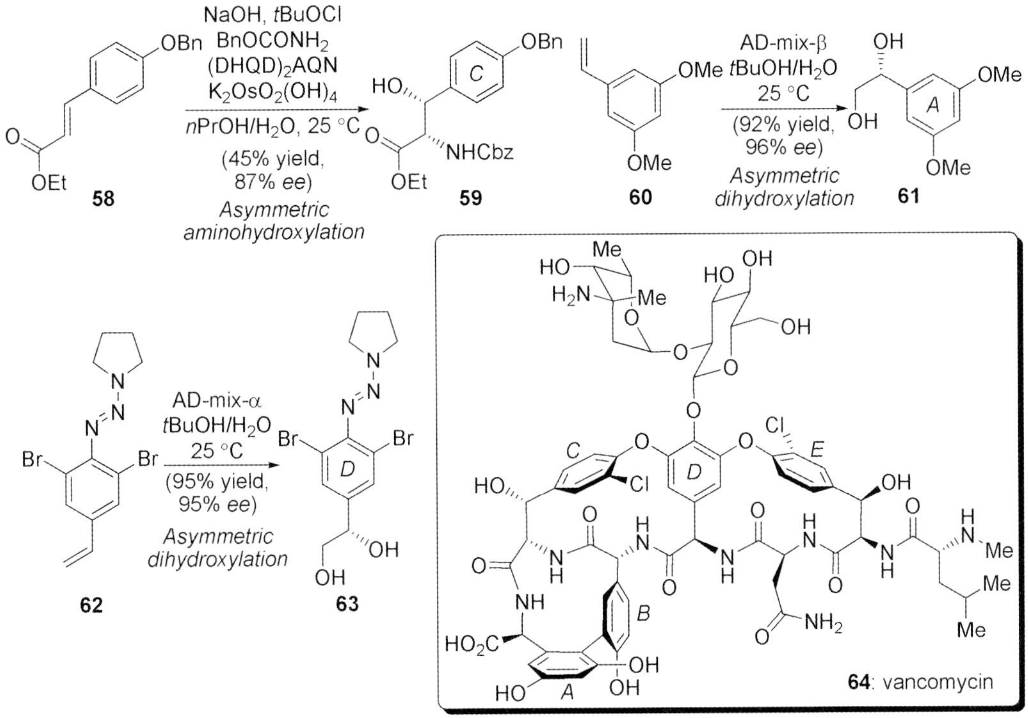

Scheme 12 Asymmetric dihydroxylation and asymmetric aminohydroxylation reactions in the total synthesis of vancomycin (**64**) [36].

Organocatalytic Reductions

The reduction of ketone carbonyl groups to the corresponding secondary alcohols represents one of the most fundamental processes in synthetic organic chemistry. When performed in an enantioselective manner, this process constitutes a valuable method for introducing asymmetry into molecules. Of the many methods which have been developed for the enantioselective reduction of ketones, the combination of a borane reducing agent and chiral oxazaborolidine catalyst (dubbed the 'CBS catalyst' in honor of its original protagonists, Corey, Bakshi and Shibata [37]) is perhaps the most routinely used today [38]. Thus, as shown in Scheme 13, reduction of ketone **65** using $BH_3 \cdot THF$ and (*R*)-CBS catalyst **66** proceeded diastereoselectively to afford allylic alcohol **68**, an important intermediate in the synthesis of various members of the prostaglandin family, including $PGF_{2\alpha}$ (**69**) [39].

Catalyst **66** (or its enantiomer) also finds use in the kinetic resolution of racemic ketones. For example, ketone **70** (Scheme 14) was resolved to 90% *ee* in 31% yield (maximum theoretical yield is 50%), by reduction using catecholborane and catalyst **66**, and subsequently elaborated to give (+)-CP-263,114 (**71**) by Shair and coworkers in their total synthesis of this stunning natural product [40].

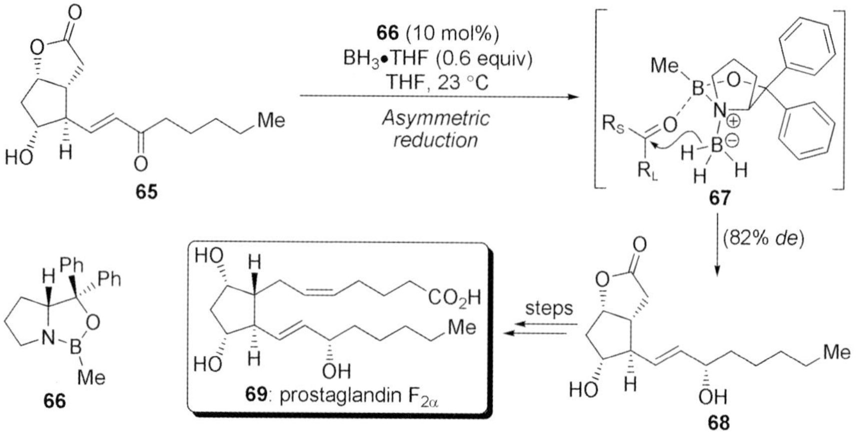

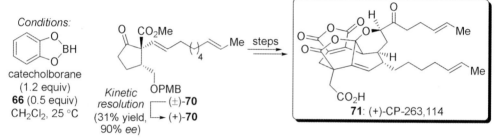

Scheme 13 A diastereoselective hydride reduction employing the CBS catalyst in the synthesis of a key prostaglandin intermediate (**68**) [39].

Scheme 14 Kinetic resolution of ketone **70** using the CBS catalyst (**66**) in the enantioselective synthesis of (+)-CP-263,114 (**71**) [40].

Metal-catalyzed Hydrogenations

The homogenous hydrogenation of ketone and alkene functional groups, catalyzed by chiral transition metal complexes, occupies an important niche in the repertoire of asymmetric synthesis. For example, asymmetric hydrogenation constitutes one of the most efficient means to access novel α-amino acids, being used industrially for the enantioselective production of L-DOPA (for the treatment of Parkinson's disease) and aspartame (an artificial sweetener). In the context of total synthesis, a recent application of asymmetric hydrogenation can be found in the total synthesis of the promising anticancer agent ecteinascidin 743 (**75**, Scheme 15) by the Corey group [41]. Thus, hydrogenation of cinnamate derivative **72** using the famous Monsanto catalyst **73** (developed by Knowles and coworkers [42]) proceeded under a moderate pressure (45 psi) of hydrogen to provide **74** in nearly quantitative yield and in an impressive 96% ee. The Noyori asymmetric reductions have also been widely employed in total synthesis, as demonstrated in the cases of gloeosporone [43] and the prostaglandins [44].

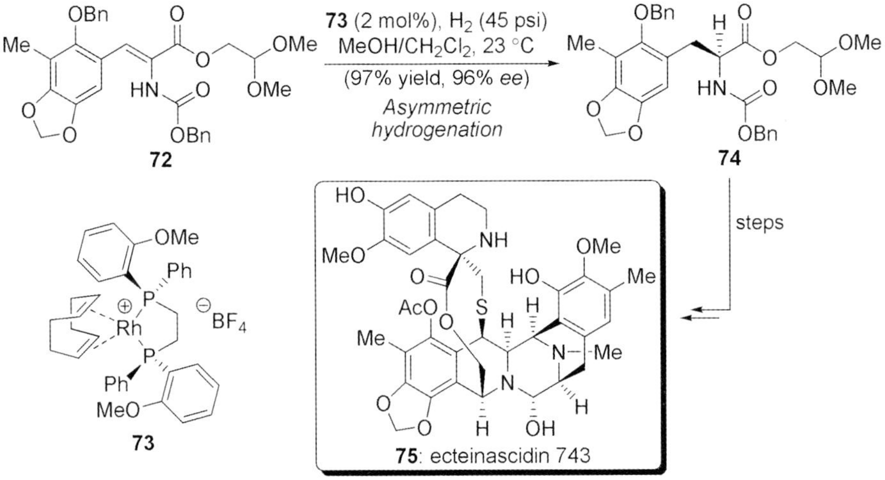

Scheme 15 Use of a rhodium-catalyzed asymmetric hydrogenation reaction in the enantiose-lective synthesis of ecteinascidin (**75**) [41].

Enzymatic

A large number of organic transformations have been shown to be amenable to enzymatic catalysis, including esterification/hydrolysis reactions, oxidations, reductions and aldol reactions [45]. In total synthesis, enzymatic reactions are most commonly used for the preparation of enantiomerically enriched building blocks, typically through resolution or desymmetrization processes, that are structurally small but stereochemically rich and suitable for further elaboration. For example, in their enantioselective synthesis of (−)-borrelidin (**78**, Scheme 16) [46], Omura and coworkers obtained acetate **77**, corresponding to the C5-C9 portion of the target molecule (**78**), in 72% yield and 97% ee by means of enzymatic desymmetrization of readily available *meso*-diol **76**.

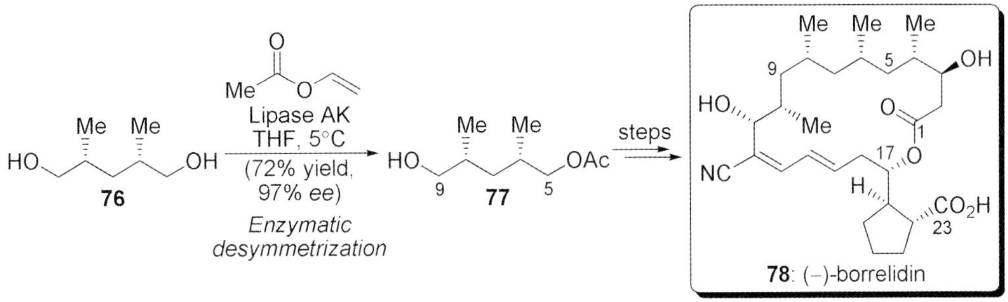

Scheme 16 Enzymatic desymmetrization of a *meso*-diol in the enantioselective synthesis of (−)-borrelidin (**78**) [46].

Conclusions

As can hopefully be gleaned from the above exemplars, the development of modern asymmetric reaction methodology allows today's chemists to approach the rational synthesis of complex natural products with an elegance and efficiency that was unthinkable even 25 years ago. Whilst great strides have been made during this period, the future undoubtedly holds the promise of even more remarkable advances in the field. In particular, the development of novel asymmetric catalysts, rather than stiochiometric chiral auxiliaries, that operate at high efficiency at low loading levels and with a broad substrate scope, will undoubtedly feature prominently in years to come. What can be certain, however, is that total synthesis will continue to play a central role in challenging and advancing the science of asymmetric synthesis.

CV of K. C. Nicolaou

K. C. Nicolaou was born in Cyprus and educated in the UK and USA. He is Chairman of the Department of Chemistry at The Scripps Research Institute where he holds the Darlene Shiley Chair in Chemistry and the Aline W. and L. S. Skaggs Professorship in Chemical Biology. He is also Distinguished Professor of Chemistry at the University of California, San Diego and Director of the Biopolis Chemical Synthesis Laboratory in Singapore. His impact on chemistry, biology and medicine flows from his works in chemical synthesis and chemical biology described in hundreds of publications and 55 patents. For his contributions to research and education, he was elected a Member of the Academy of Sciences, USA, a Fellow of the American Academy of Arts and Sciences, and a Foreign Member of the Academy of Athens, Greece, and has received numerous awards and prizes.

CV of Paul G. Bulger

Paul G. Bulger was born in London, England in 1978. He received his M. Chem in 2000 from the University of Oxford, completing his Part II project under the supervision of Dr. Mark G. Moloney. He remained at the University of Oxford for his graduate studies, obtaining his D.Phil in 2003 for research conducted, under the supervision of Professor Sir Jack Baldwin, towards the synthesis and biological investigation of complex natural products and their analogs. In the fall of 2003 he joined Professor K. C. Nicolaou's group at the Scripps Research Institute, where he is currently engaged in the total synthesis of novel marine macrolide natural products. His research interests encompass reaction mechanism and design, and their application to complex natural product synthesis and chemical biology.

Selected Publications

1R. J. D. Morrison, H. S. Mosher, *Asymmetric Organic Reactions*, Prentice-Hall, Englewood Cliffs, **1971**, p. 465. Reviewing the literature up to 1968, this book quoted fewer than 10 examples which produced compounds in >90% ee.

2R. S. Hanessian, *Total Synthesis of Natural Products: The "Chiron" Approach*, Pergamon Press, Oxford, **1983**, p. 291.

3R. K. C. Nicolaou, E. J. Sorensen, *Classics in Total Synthesis*, Wiley-VCH, Weinheim, **1996**, p. 821.

4R. K. C. Nicolaou, S. A. Snyder, *Classics in Total Synthesis II*, Wiley-VCH, Weinheim, **2003**, p. 658.

5R. *Modern Aldol Reactions, Vols 1&2*, R. Mahrwald (Ed.), Wiley-VCH, Weinheim, **2004**, p. 699.

6. D. A. Evans, S. W. Kaldor, T. K. Jones, J. Clardy, T. J. Stout, *J. Am. Chem. Soc.* **1990**, *112*, 7001–7031. Total Synthesis of the Macrolide Antibiotic Cytovaricin.

7. I. Paterson, D. Y.-K. Chen, M. J. Coster, J. L. Aceña, J. Bach, K. R. Gibson, L. E. Keown, R. M. Oballa, T. Trieselmann, D. J. Wallace, A. P. Hodgson, R. D. Norcross, *Angew. Chem. Int. Ed.* **2001**, *40*, 4055–4060. Stereocontrolled Total Synthesis of (+)-Altohyrtin A/Spongistatin 1.

8R. A. Job, C. F. Janeck, W. Bettray, R. Peters, D. Enders, *Tetrahedron* **2002**, *58*, 2253–2329. The SAMP-/RAMP-Hydrazone Methodology in Asymmetric Synthesis.

9. K. C. Nicolaou, K. Ajito, A. P. Patron, H. Khatuya, P. K. Richter, P. Bertinato, *J. Am. Chem. Soc.* **1996**, *118*, 3059–3060. Total Synthesis of Swinholide A.

10. A. G. Myers, B. H. Yang, H. Chen, L. McKinstry, D. J. Kopecky, J. L. Gleason, *J. Am. Chem. Soc.* **1997**, *119*, 6496–6511. Pseudoephedrine as a Practical Chiral Auxiliary for the Synthesis of Highly Enantiomerically Enriched Carboxylic Acids, Alcohols, Aldehydes, and Ketones.

11. Y. Shin, J.-H. Fournier, Y. Fukui, A. M. Brückner, D. P. Curran, *Angew. Chem. Int. Ed.* **2004**, *43*, 4634–4637. Total Synthesis of (−)-Dictyostatin: Confirmation of Relative and Absolute Configurations.

12. A similar alkylation was utilized by the Paterson group in their total synthesis of the same target. See I. Paterson, R. Britton, O. Delgado, A. Meyer, K. G. Poullennec, *Angew. Chem. Int. Ed.* **2004**, *43*, 4629–4633. Total Synthesis and Configurational Assignment of (−)-Dictyostatin, a Microtubule-Stabilizing Macrolide of Marine Sponge Origin.

13R. K. C. Nicolaou, S. A. Snyder, T. Montagnon, G. Vassilikogiannakis, *Angew. Chem. Int. Ed.* **2002**, *41*, 1668–1698. The Diels–Alder Reaction in Total Synthesis.

14. K. Mikami, Y. Motoyama, M. Terada, *J. Am. Chem. Soc.* **1994**, *116*, 2812–2820. Asymmetric Catalysis of Diels–Alder Cycloadditions by an MS-Free Binaphthol–Titanium Complex: Dramatic Effect of MS, Linear vs Positive Nonlinear Relationshop, and Synthetic Applications.

15. K. C. Nicolaou, G. Vassilikogiannakis, W. Mägerlein, R. Kranich, *Angew. Chem. Int. Ed.* **2001**, *40*, 2482–2486. Total Synthesis of Colombiasin A.

16. P. Liu, E. N. Jacobsen, *J. Am. Chem. Soc.* **2001**, *123*, 10772–10773. Total Synthesis of (+)-Ambruticin.

17R. S. E. Denmark, N. G. Almstead in *Modern Carbonyl Chemistry* (Ed. J. Otera), Wiley-VCH, Weinheim, **2000**, pp. 299–401. Allylation of Cabonyls: Methodology and Stereochemistry.

18. A. K. Ogawa, R. W. Armstrong, *J. Am. Chem. Soc.* **1998**, *120*, 12435–12442. Total Synthesis of Calyculin C.

19. K. C. Nicolaou, D. W. Kim, R. Baati, A. O'Brate, P. Giannakakou, *Chem. Eur. J.* **2003**, *9*, 6177–6191. Total Synthesis and Biological Evaluation of (−)-Apicularen A and Analogues Thereof.

20R. K. C. Nicolaou, P. G. Bulger, D. Sarlah, *Angew. Chem. Int. Ed.* **2005**, *44*, 4442–4489. Palladium-Catalyzed Cross-Coupling Reactions in Total Synthesis.

21. A. D. Lebsack, J. T. Link, L. E. Overman, B. A. Stearns, *J. Am. Chem. Soc.* **2002**, *124*, 9008–9009. Enantioselective Total Synthesis of Quadrigemine C and Psycholeine.

22. K. Kagechika, M. Shibasaki, *J. Org. Chem.* **1991**, *56*, 4093–4094. Asymmetric Heck Reaction: A Catalytic Asymmetric Synthesis of the Key Intermediate for $\Delta^{9(12)}$-Capnellene-$3\beta,8\beta,10\alpha$-triol and $\Delta^{9(12)}$-Capnellene-$3\beta,8\beta,10\alpha,14$-tetrol.

23. B. M. Trost, W. Tang, *J. Am. Chem. Soc.* **2002**, *124*, 14542–14543. *Enantioselective Synthesis of (–)-Codeine and (–)-Morphine.*
24. M. Gates, G. Tschudi, *J. Am. Chem. Soc.* **1952**, *74*, 1109–1110. *The Synthesis of Morphine.*
25. T. Katsuki, K. B. Sharpless, *J. Am. Chem. Soc.* **1980**, *102*, 5976–5978. *The First Practical Method for Asymmetric Epoxidation.*
26. R. M. Hanson, K. B. Sharpless, *J. Org. Chem.* **1986**, *51*, 1922–1925. *Procedure for the Catalytic Asymmetric Epoxidation of Allylic Alcohols in the Presence of Molecular Sieves.*
27. K. C. Nicolaou, E. A. Theodorakis, F. P. J. T. Rutjes, J. Tiebes, M. Sato, E. Untersteller, X.-Y. Xiao, *J. Am. Chem. Soc.* **1995**, *117*, 1171–1172. *Total Synthesis of Brevetoxin B. 1. CDEFG Framework.*
28. K. C. Nicolaou, F. P. J. T. Rutjes, E. A. Theodorakis, J. Tiebes, M. Sato, E. Untersteller, *J. Am. Chem. Soc.* **1995**, *117*, 1173–1174. *Total Synthesis of Brevetoxin B. 2. Completion.*
29R. T. Linker, *Angew. Chem. Int. Ed. Engl.* **1997**, *36*, 2060–2062. *The Jacobsen–Katsuki Epoxidation and Its Controversial Mechanism.*
30. Z. Tu, Z.-X. Wang, Y. Shi, *J. Am. Chem. Soc.* **1996**, *118*, 9806–9807. *An Efficient Asymmetric Epoxidation Method for trans-Olefins Mediated by a Fructose-Derived Ketone.*
31. H. Sasaki, R. Irie, T. Hamada, K. Suzuki, T. Katsuki, *Tetrahedron* **1994**, *50*, 11827–11838. *Rational Design of Mn-Salen Catalyst (2): Highly Enantioselective Epoxidation of Conjugated cis-Olefins.*
32. K. C. Nicolaou, B. S. Safina, M. Zak, A. A. Estrada, S. H. Lee, *Angew. Chem. Int. Ed.* **2004**, *44*, 5087–5092. *Total Synthesis of Thiostrepton, Part 1: Construction of the Dehydropiperidine/Thiazoline-Containing Macrocycle.*
33. K. C. Nicolaou, M. Zak, B. S. Safina, S. H. Lee, A. A. Estrada, *Angew. Chem. Int. Ed.* **2004**, *44*, 5092–5097. *Total Synthesis of Thiostrepton, Part 2: Construction of the Quinaldic Acid Macrocycle and Final Stages of the Synthesis.*
34R. H. C. Kolb, M. S. VanNieuwenhze, K. B. Sharpless, *Chem. Rev.* **1994**, *94*, 2483–2547. *Catalytic Asymmetric Dihydroxylation.*
35R. J. A. Bodkin, M. D. McLeod, *J. Chem. Soc., Perkin Trans. 1* **2002**, 2733–2746. *The Sharpless Asymmetric Aminohydroxylation.*
36. K. C. Nicolaou, H. J. Mitchell, N. F. Jain, N. Winssinger, R. Hughes, T. Bando, *Angew. Chem. Int. Ed.* **1999**, *38*, 240–244. *Total Synthesis of Vancomycin.*
37. E. J. Corey, R. K. Bakshi, S. Shibata, *J. Am. Chem. Soc.* **1987**, *109*, 5551–5553. *Highly Enantioselective Borane Reduction of Ketones Catalyzed by Chiral Oxazaborolidines. Mechanism and Synthetic Implications.*
38R. E. J. Corey, C. J. Helal, *Angew. Chem. Int. Ed.* **1998**, *37*, 1986–2012. *Reduction of Carbonyl Compounds with Chiral Oxazaborolidine Catalysts: A New Paradigm for Enantioselective Catalysis and a Powerful New Synthetic Method.*
39. E. J. Corey, R. K. Bakshi, S. Shibata, C.-P. Chen, V. K. Singh, *J. Am. Chem. Soc.* **1987**, *109*, 7925–7926. *A Stable and Easily Prepared Catalyst for the Enantioselective Reduction of Ketones. Applications to Multistep Syntheses.*
40. C. Chen, M. E. Layton, S. M. Sheehan, M. D. Shair, *J. Am. Chem. Soc.* **2000**, *122*, 7424–7425. *Synthesis of (+)-CP-263,114.*
41. E. J. Corey, D. Y. Gin, R. S. Kania, *J. Am. Chem. Soc.* **1996**, *118*, 9202–9203. *Enantioselective Total Synthesis of Ecteinascidin 743.*
42. W. S. Knowles, *Acc. Chem. Res.* **1983**, *16*, 106–112. *Asymmetric Hydrogenation.*
43. S. L. Schreiber, S. E. Kelly, J. A. Porco Jr., T. Sammakia, E. M. Suh, *J. Am. Chem. Soc.* **1988**, *110*, 6210–6218. *Structural and Synthetic Studies of the Spore Germination Autoinhibitor Gloeosporone.*
44. M. Suzuki, A. Yanagisawa, R. Noyori, *J. Am. Chem. Soc.* **1985**, *107*, 3348–3349. *An Extremely Short Way to Prostaglandins.*
45. *Enzyme Catalysis in Organic Synthesis: A Comprehensive Handbook*, K. Drauz, H. Waldmann (Eds.), Wiley-VCH, Weinheim, **1995**, p. 1050.
46. T. Nagamitsu, D. Takano, T. Fukuda, K. Otoguro, I. Kuwajima, Y. Harigaya, S. Omura, *Org. Lett.* **2004**, *6*, 1865–1867. *Total Synthesis of (–)-Borrelidin.*

Ring Rearrangement Metathesis (RRM) in Alkaloid Synthesis

Nicole Holub and Siegfried Blechert, Technische Universität Berlin, Germany

Background and Strategy

Ring rearrangement metathesis (RRM), combines different metathesis transformations in a single process and has proven to be a powerful method for the rapid construction of complex structures [1R, 2R]. These domino reactions can be either bond forming or bond breaking, but the subsequent reaction always occurs at the functionality formed in the previous step [3R]. This is challenging, as in classical organic synthesis the individual bonds in the target molecule are normally formed step by step, with increasing structural complexity.

RRM has been implemented in the synthesis of fused carbocycles [4], as well as heterocycles [5]. As RRM proceeds with transfer of stereocenters from the corresponding carbocycle into the metathesis product, this method offers a very effective way to synthesize chiral heterocycles from easily accessible precursors in one pot (Figure 1). Addition of further metathesis transformations to this ROM-RCM process allows the synthesis of higher functionalized heterocycles. This can be achieved by conducting a cross metathesis (CM) with the new double bond formed in the rearrangement product or by modifying R^2 to contain a double or triple bond so that a subsequent RCM reaction occurs.

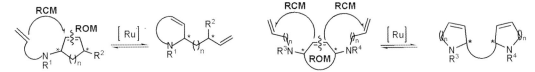

Figure 1 Application of RRM transformation for the synthesis of *N*-heterocycles.

The high flexibility and efficiency of this concept will be illustrated with a few selected piperidines and pyrrolizidines alkaloids, which have been synthesized in our group.

Asymmetric Synthesis – The Essentials.
Edited by Mathias Christmann and Stefan Bräse

ISBN: 978-3-527-31399-0

Results

Tetraponerines

Tetraponerines **T1–T8** (Figure 2) each contain three stereocenters and differ from each other in the side chain and stereochemistry at C-9, as well as the size of ring A. Due to these facts a general synthesis is challenging [6].

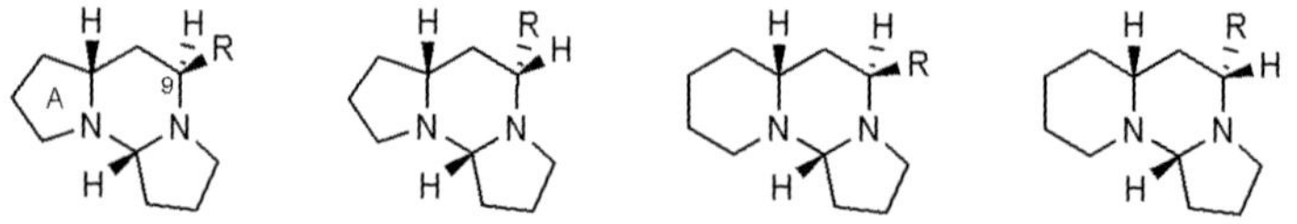

T1 R = C_3H_7
T5 R = C_5H_{11}

T2 R = C_3H_7
T6 R = C_5H_{11}

T3 R = C_3H_7
T7 R = C_5H_{11}

T4 R = C_3H_7
T8 R = C_5H_{11}

Figure 2 Structures of tetraponerines **T1–T8**.

The application of a ROM-RCM sequence for the synthesis of the monosubstituted piperidines **4**, **5** (Scheme 1) allows the flexible construction of the stereocenter at C-2, depending on the configuration of the utilized precursor **2**, **3**. Starting from *meso*- dicarbonate **1**, both *cis*- and *trans*-configurated precursors were synthesized via domino allylic alkylation or single allylic alkylation and subsequent Mitsunobu reaction. Introduction of the different side chains by CM failed, but was achieved by functionalization of the terminal double bond. The synthesis was completed by stereoselective acidic cyclization to give the tetraponerines **T4, T6–T8**.

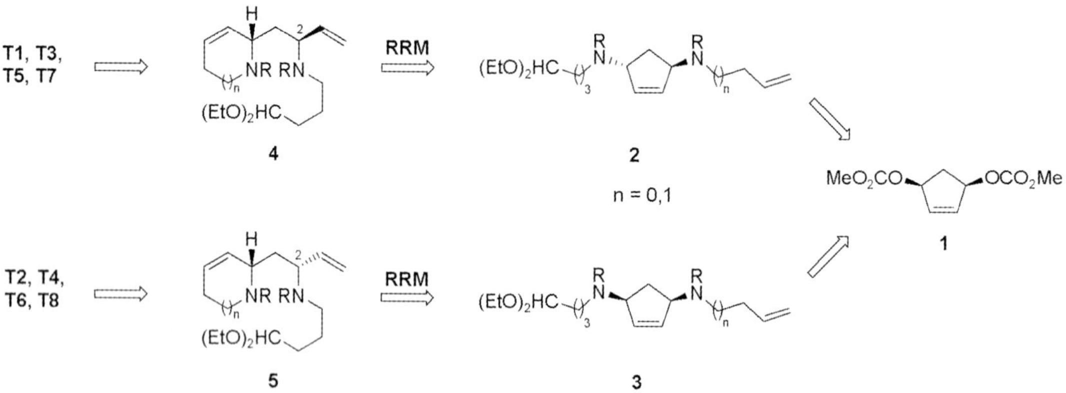

Scheme 1 Retrosynthetic analysis of the tetraponerines **T1–T8**.

Trans-195A

In the synthesis of (+)-*trans*-195A (Scheme 2) the ROM-RCM sequence was extended to the construction of disubstituted piperidines [7]. The stereoinformation in precursor **10** was therefore not only transferred from the carbocycle but also from the side chain into the 2,5-disubstituted tetrahydropyridine **11**.

The required amine **7** was synthesized by double copper-catalyzed epoxide ring-opening of **6** with Grignard reagents and subsequent substitution, whereas **9** was

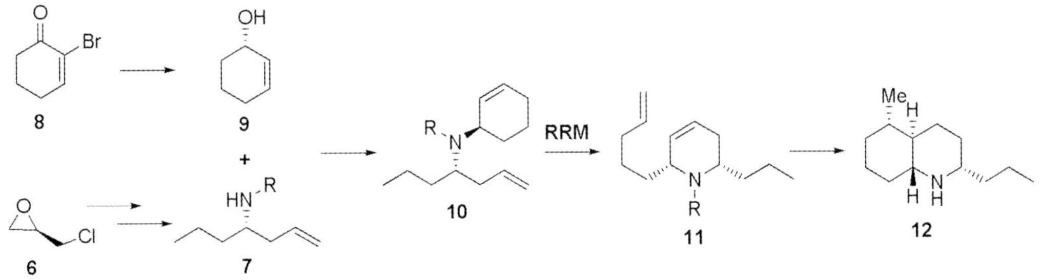

Scheme 2 Synthesis of *trans*-195A (**12**).

obtained with 99% *ee* via CBS reduction of prochiral ketone **8**. Following Mitsunobu reaction, RRM and zirconium mediated coupling of the two double bonds gave rise to the decahydrochinoline **12**. By comparison of the synthesized (+)-enantiomer with the isolated substance, the tentative absolute configuration of naturally occurring *trans*-195A could be determined to be (−)-2R,4aS,5R,8aS.

Lasubine II

The successful application of a domino reaction connecting a CM-process to the ROM-RCM sequence was achieved in the synthesis of quinolizidine alkaloid lasubine II (**17**) [8]. Starting from precursor **14**, this tandem process allows for the construction of the 2-substituted piperidine **16** with extended side chain in one step.

Chiral monoacetate **13** was chosen as starting material for the preparation of **14**, which could be transformed into the desired product **16** through a RRM process, followed by CM of the intermediate with the coupling partner **15**. Subsequent intramolecular Michael reaction and reduction led to (−)-lasubine II (**17**) (Scheme 3). Due to the free choice of the CM partner, this strategy offers a powerful method for the synthesis of other quinolizidine, indolizidine and pyrrolizidine alkaloids.

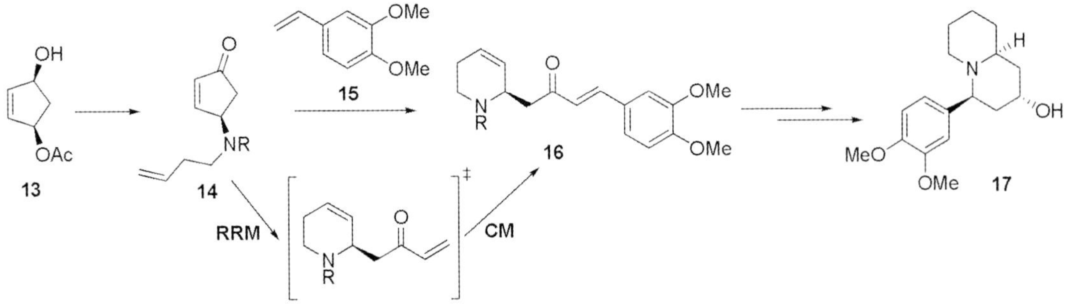

Scheme 3 RRM-CM domino reaction for the synthesis of **17**.

Astrophylline and Dendrochrysine

For the construction of two piperidine-rings in one step, the RRM process has been expanded by another RCM reaction and was successfully applied in the synthesis of astrophylline (**18**) [9]. The synthetic challenge of this alkaloid lies not only in the unsymmetrically bridged heterocycles, but also in the *cis*-cinnamoyl substituent which requires an orthogonal protecting group strategy. In contrast to other precursors, **19** contains one amine side chain which is not directly positioned at the carbocyle. This distance is necessary to achieve the non-C_2-symmetric bridged *bis*-piperidine structure after RRM and proves the flexibility of this method.

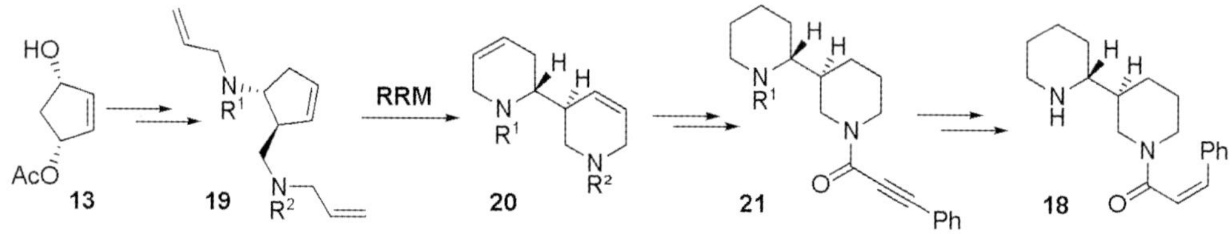

Scheme 4 Synthesis of astrophylline **18** by RCM-ROM-RCM transformation.

Chiral monoacetate **13** served again as the starting compound for the precursor **19**, which was transformed into *bis*-piperidine **20** via successful RRM. The selective deprotection, acylation and Z-selective Lindlar reduction gave alkaloid **18** (Scheme 4).

This RCM-ROM-RCM process was also successfully applied in the syntheses of cuscohygrine and dihydrocuscohygrine [10], where a seven-membered ring was rearranged to achieve the symmetrical, propyl-chain linked *bis*-pyrrolizidine structure, proving that the rearrangement is not limited to a particular ring size and that the distance of the linkage is flexible. The combination of these earlier results with the protecting group strategy in the synthesis of **18**, is now under investigation.

Dendrochrysine (**22**) occurs as the *cis*- or *trans*-cinnamoyl protected alkaloid, thus the selective hydrogenation of the triple bond in **26** should lead to both derivates. Chiral triole **23** was chosen as the starting material for the preparation of the precursor **24**, which was rearranged into the desired product **25** and will be further functionalized to give dendrochrysine (Scheme 5).

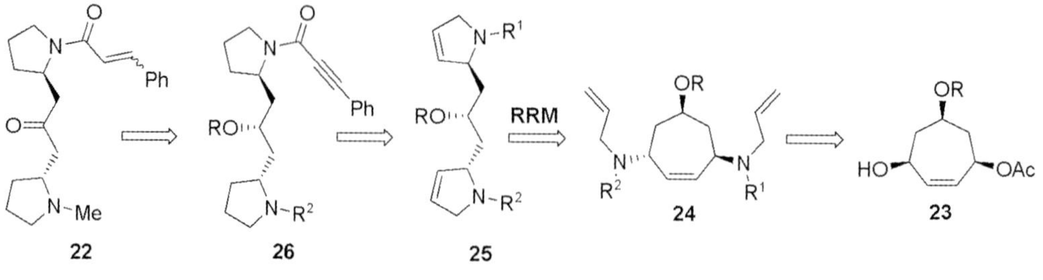

Scheme 5 Retrosynthetic analysis of **22**.

Conclusions and Future Perspectives

We have synthesized a number of alkaloids by application of RRM processes. Currently we are interested in the extension of this RRM principle to diastereoselective RRM processes. In these transformations the stereoinformation is not only transferred from a carbocycle or a side chain into the metathesis product, but a new stereocenter is also generated (Figure 3).

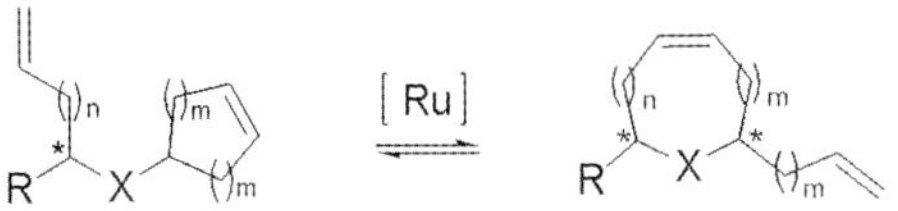

Figure 3 Principle of diastereoselective RRM.

This is the result of a chiral functional group R, positioned at the carbocycle or in the side chain of the precursor, which influences the stereoselectivity of the intermediate formed species and therefore the *cis/trans*-ratio of the obtained metathesis product. The application of this method in natural product synthesis is currently under investigation.

CV of Siegfried Blechert

Siegfried Blechert was born in Aalborg, Denmark, in 1946. He studied chemistry at the University of Hannover, Germany, and completed his Ph.D. under the supervision of Prof. E. Winterfeldt in 1974. After a research stay with Professor P Potier at Gif-sur-Yvette, France, he completed his habilitation at the University of Hannover in 1986 before taking up a professorship at Bonn University in 1986. In 1990 he accepted the Chair of the Organic Chemistry Department at the Technical University of Berlin, where he has been Head of the Chemistry Department since 2004.

Selected Publications

1R. S. J. Connon, S. Blechert, *Angew. Chem. Int. Ed. Engl.* **2003**, 1900–1923. Recent developments in olefin metathesis.

2R. M. Schuster, S. Blechert, *Angew. Chem. Int. Ed. Engl.* **1997**, 2037–2056. Olefin metathesis in organic chemistry.

3R. S. Basra, S. Blechert, *Strategies and Tactics in Org. Synth.* **2004**, 315–346. Ring Rearrengement Metathesis (RRM) – a new concept in piperidine and pyrrolidine synthesis.

4. R. Stragies, S. Blechert, *Synlett* **1998**, 169–170. Domino metathesis. A combined ring opening-, ring-closing-, and olefin metathesis

5. U. Voigtmann, S. Blechert, *Synthesis* **2000**, 893–898. Enantioselective synthesis of α,α'-disubstituted piperidines via ruthenium-catalyzed ring rearrangement.

6. R. Stragies, S. Blechert, *J. Am. Chem. Soc.* **2000**, 9584–9591. Enantioselective Synthesis of Tetraponerines by Pd- and Ru-Catalyzed Domino Reactions.

7. N. Holub, S. Blechert, *Org. Lett.* **2005**, 1227–1229. Total Synthesis of (+)-trans-195A.

8. M. Zaja, S. Blechert, *Tetrahedron* **2004**, 9629–9624. Concise enantioselective synthesis of (-) lasubine II.

9. M. Schaudt, S. Blechert, *J. Org. Chem.* **2003**, 2913–2920. Total Synthesis of (+)-Astrophylline.

10. C. Stapper, S. Blechert, *J. Org. Chem.* **2002**, 6456–6460. Total Synthesis of (+)-Dihydrocuscohygrin and Cuscohygrin.

Asymmetric Synthesis of Biaryls by the 'Lactone Method'

Gerhard Bringmann, Tanja Gulder, and Tobias A. M. Gulder,
University of Würzburg, Germany

Introduction

The stereoselective synthesis of compounds with one or more stereogenic centers is an important task. Axial chirality occurring in rotationally hindered biaryls of natural or synthetic origin was discovered in the early 20th century [1R]. With the recognition of the importance of axial chirality for the pharmacological properties of bioactive compounds [2R] and for their use in asymmetric synthesis [3], the need for diastereo- and enantioselective aryl–aryl coupling methods has grown steadily. In this chapter we present the 'lactone method' [4R], an efficient pathway to rotationally stable biaryls.

The Basic Concept of the 'Lactone Method'

For the regio- and stereoselective formation of, even sterically hindered, axially chiral biaryl systems we have developed a fundamentally new approach, the 'lactone method' [4R] (Scheme 1). It differs from most other coupling reactions by separating the two formal partial goals of atropisomer-selective synthesis: the C–C bond formation and the asymmetric induction.

The nonstereoselective Pd^{II} catalyzed aryl–aryl bond formation is achieved intramolecularly after pre-fixation of a bromonaphthoic acid **1** and a phenol **2** via an ester bridge as in **3**. This bridge fulfills several tasks: it pre-links the two building blocks, thus guaranteeing the formation of a 6-membered lactone of type **4** in excellent yields, even against highest steric hindrance (e.g. > 80 % yield for **4** with R = *tert*Bu), but the most significant (and innovative) function of the bridge is the dramatic lowering of the isomerization barrier at the newly created biaryl axis. Lactones **4** thus usually occur as racemic mixtures of rapidly interconverting atropo-enantiomers, (*M*)-**4** and (*P*)-**4**.

This opens the possibility for the ring cleavage of the lactones into configurationally stable biaryls to proceed atropisomer-selectively, in a separate step and with dynamic kinetic resolution. Thus, **4** can be opened highly atropo-diastereoselectively with chiral O- or N-nucleophiles (like, e.g., (*R*)-sodium 8-phenylmentholate (*R*)-**8**, giving > 98 % *de*), to deliver esters or amides **6**, or atropo-en-

Asymmetric Synthesis – The Essentials.
Edited by Mathias Christmann and Stefan Bräse
Copyright © 2007 WILEY-VCH Verlag GmbH & Co. KGaA, Weinheim
ISBN: 978-3-527-31399-0

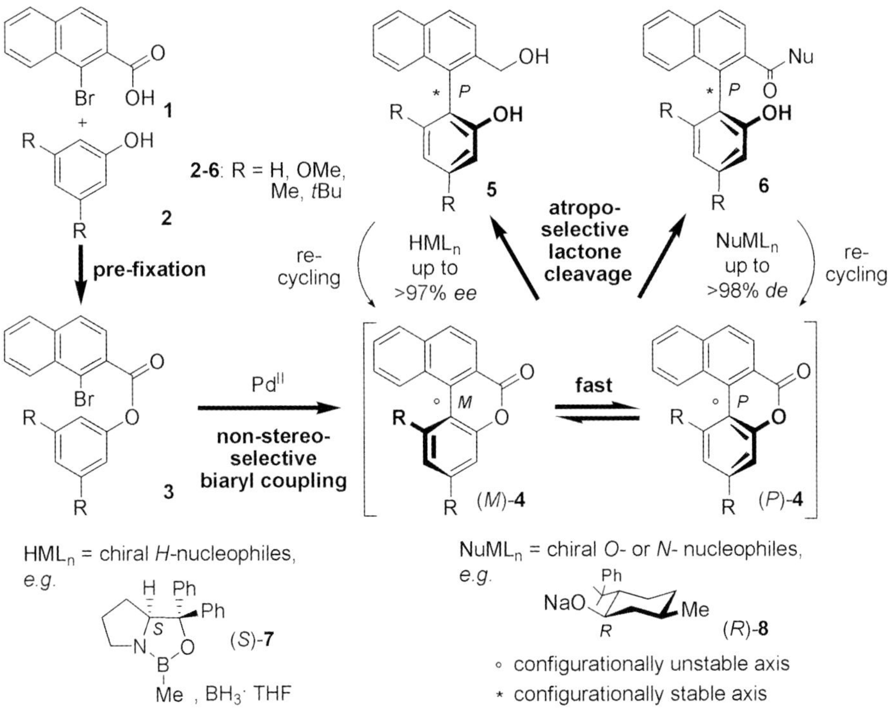

Scheme 1 The basic principle of the 'lactone concept': the dynamic kinetic resolution of configurationally unstable biaryl lactones **4**. Configurational assignments arbitrarily for R = H or alkyl.

antioselectively with chiral H-nucleophiles (like, e.g., using borane activated by oxazaborolidine (S)-**7**), providing alcohols **5** with > 97% ee. Another advantage is the optional preparation of either of the two atropisomers from the same, 'late', precursor **4**, by just using the enantiomeric reagent for the lactone ring cleavage. Moreover, even the minor atropisomeric by-products can easily be re-used, by cyclization back to the configurationally unstable lactone, which can then be ring-opened once again [5R]. The efficiency and practicability of the method has already been demonstrated in the total synthesis of ca. 30 natural products [5R, 6].

Application of the 'Lactone Concept' to Natural Product Synthesis

A rewarding synthetic target molecule is the antimalarial naphthylisoquinoline alkaloid dioncopeltine A (**14**) from the tropical liana *Triphyophyllum peltatum* [2R].

Its stereoselective total synthesis [7] is shown in Scheme 2. The bromoester **11** was 7,1'-coupled and O-deisopropylated to provide lactone **12** in 92% yield. For the stereochemical key step, the, now atropo-diastereoselective, ring opening, the chiral oxazaborolidine [(S)-**7**] borane system gave the best asymmetric induction, leading to the rotationally stable biaryl (M)-**13** in good optical yields (> 90%

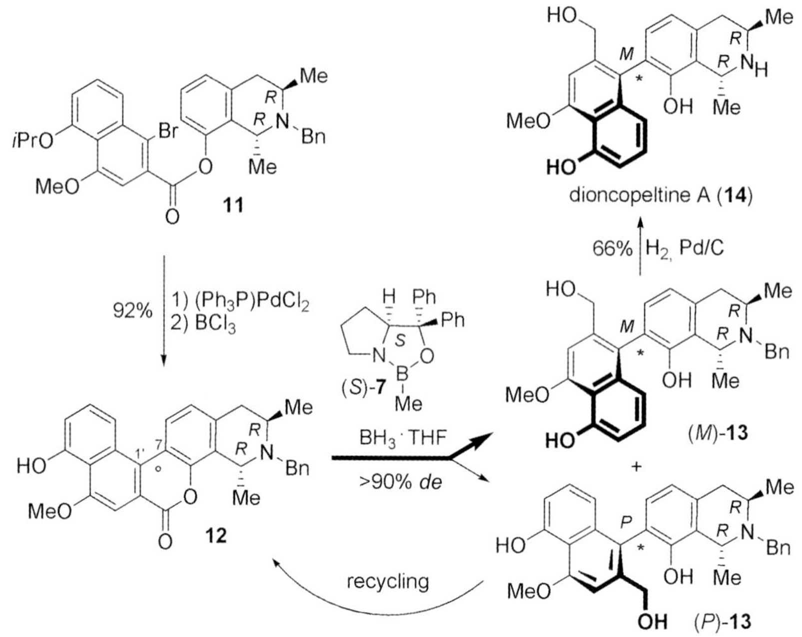

Scheme 2 Application of the 'lactone method' for the atroposelective synthesis of dioncopeltine A (**14**).

de, matched case). Its conversion into natural dioncopeltine A (**14**) was completed by N-debenzylation. The small amount of the undesired diastereomer (*P*)-**13**, obtained during the dynamic kinetic resolution of the lactone **12**, can be reused by re-cyclization to **12** (here with oxidation). If required, the *directed* synthesis of (*P*)-**13** succeeds by cleavage of **12** by using the enantiomeric *H*-nucleophile (*R*)-**7**, still with a good asymmetric induction (> 70% *de*, mismatched case).

The 'lactone method' for the construction of axially chiral natural product synthesis has also been used by other groups, *inter alia* Abe, Harayama et al. [8] for a formal total synthesis of (–)-steganone (**17**) (Scheme 3). The introduction of the chiral information at the biaryl axis in **16** was achieved with excellent chemical (97%) and optical (83% *ee*) yields using borane activated by (*S*)-**7**.

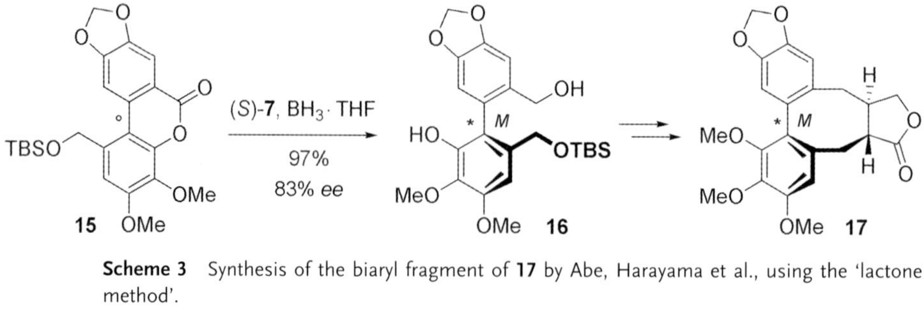

Scheme 3 Synthesis of the biaryl fragment of **17** by Abe, Harayama et al., using the 'lactone method'.

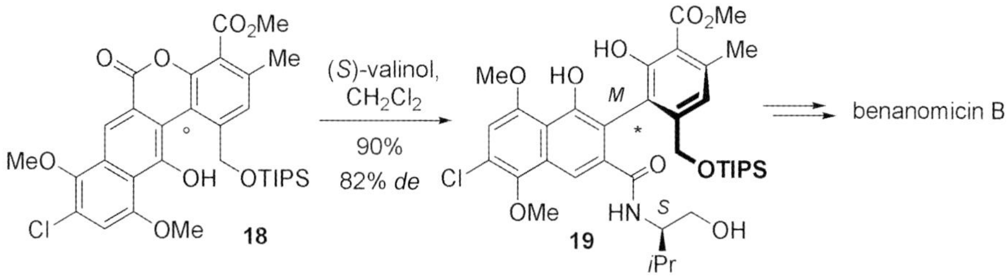

Scheme 4 Key step in the synthesis of benanomicin B by Suzuki et al., applying the 'lactone method'.

Recently, Suzuki et al. [9] have reported on the stereoselective ring opening of lactone **18** with (S)-valinol as an N-nucleophile, giving the M-configured biaryl **19** with 90% yield and 82% de (Scheme 4). Further transformation of the axial into central chirality completed the total synthesis of benanomicin B.

Conclusions and Further Perspectives

Biaryl lactones are versatile intermediates in the atroposelective synthesis of configurationally stable natural products and (not presented here) of C_1-, C_2-, and C_3-symmetric axially chiral reagents and ligands [5R, 10]. The method works under mild conditions, accepts most different functional groups, and gives access to a broad diversity of structurally complex biaryls with either axial configuration. The stereochemical information is introduced by the atroposelective ring cleavage with dynamic kinetic resolution, thus permitting complete conversion of the racemic lactone mixture into a stereochemically pure product. The method has also been extended to the enantioselective cleavage of 7-membered (and thus configurationally stable) biaryl lactones [11]. This variant is especially suited for the synthesis of constitutionally symmetric biaryls, for example mastigophorene B [12], a nerve growth stimulating sesquiterpene, since it does not require the availability of two different aryl precursors (bromoacid and phenolic moiety) as needed for the construction of the 6-membered lactones [6, 11].

CV of Gerhard Bringmann

Gerhard Bringmann, born in 1951, studied chemistry and biology in Gießen and Münster and received his Ph.D. with B. Franck in 1978. After postdoctoral studies with Sir D. H. R. Barton in Gif-sur-Yvette (France) from 1978 to 1979, and his habilitation in 1984 at the University of Münster, he was offered Full Professorships of Organic Chemistry at the Universities of Vienna and Würzburg, of which he accepted the latter in 1987. His research interests lie in the field of analytical, synthetic, and computational natural products chemistry, with particular emphasis on axially chiral biaryls. He received, *inter alia*, the Otto-Klung prize (1988) and the Prize for Good Teaching (1999) and was an Eli Lilly Lecturer (1999) and a Guest Professor at the École Supérieure de Chimie Physique Électronique in Lyon (CPE Lyon), France (2005).

Selected Publications

1R. R. Adams, H. C. Yuan, *Chem. Rev.* **1933**, *12*, 261–338. *The Stereochemistry of Diphenyls and Analogous Compounds.*

2R. G. Bringmann, C. Günther, M. Ochse, O. Schupp, S. Tasler, *Progress in the Chemistry of Organic Natural Products, Vol. 82*, W. Herz, H. Falk, G. W. Kirby, R. E. Moore (Eds.), Springer, Vienna, **2001**, pp. 1–249. *Biaryls in Nature.*

3. M. McCarthy, P. J. Guiry, *Tetrahedron* **2001**, *57*, 3809–3844. *Axially chiral bidentate ligands in asymmetric catalysis.*

4R. G. Bringmann, M. Breuning, S. Tasler, *Synthesis* **1999**, 525–558. *The Lactone Concept: An Efficient Pathway to Axially Chiral Natural Products and Useful Reagents.*

5R. G. Bringmann, A. J. Price Mortimer, P. A. Keller, M. J. Gresser, J. Garner, M. Breuning, *Angew. Chem.* **2005**, *117*, 5518–5563; *Angew. Chem. Int. Ed.* **2005**, *44*, 5384–5427. *Modern Concepts for the Atroposelective Synthesis of Axially Chiral Biaryls.*

6. G. Bringmann, D. Menche, *Acc. Chem. Res.* **2001**, *34*, 615–624. *Stereoselective Total Synthesis of Axially Chiral Natural Products via Biaryl Lactones.*

7. G. Bringmann, W. Saeb, M. Rübenacker, *Tetrahedron* **1998**, *55*, 423–432. *Directed Joint Total Synthesis of the three Naphthylisoquinoline Alkaloids Dioncolactone A, Dioncopeltine A, and 5'-O-Demethyldioncophylline A.*

8. H. Abe, S. Takeda, T. Fujita, K. Nishioka, Y. Takeuchi, T. Harayama, *Tetrahedron Lett.* **2004**, *45*, 2327–2329. *Enantioselective construction of biaryl part in the synthesis of stegane related compounds.*

9. K. Ohimori, M. Tamiya, M. Kitamura, H. Kato, M. Oorui, K. Suzuki, *Angew. Chem.* **2005**, *117*, 3939–3942; *Angew. Chem. Int. Ed.* **2005**, *44*, 3871–3874. *Regio- and Stereocontrolled Total Synthesis of Benanomicin B.*

10. G. Bringmann, R.-M. Pfeifer, C. Rummey, K. Hartner, M. Breuning, *J. Org. Chem.* **2003**, *68*, 6859–6863. *Synthesis of Enantiopure Axially Chiral C_3-Symmetric Tripodal Ligands and Their Application as Catalysts in Asymmetric Addition of Dialkylzinc to Aldehydes.*

11. G. Bringmann, J. Hinrichs, T. Papst, P. Henschel, K. Peters, E.-M. Peters, *Synthesis* **2001**, 155–167. *From Dynamic to Non-Dynamic Kinetic Resolution of Lactone-Bridged Biaryls: Synthesis of Mastigophorene B.*

12. Y. Fukuyama, Y. Asakawa, *J. Chem. Soc., Perkin Trans. 1* **1991**, 2737–2741. *Novel Neurotrophic Isocuparane-type Sesquiterpene Dimers, Mastigophorenes A, B, C and D, Isolated from Liverwort Mastigophora diclados.*

Asymmetric Synthesis of Merrilactone A

Samuel J. Danishefsky, Sloan-Kettering Institute for Cancer Research and Columbia University, USA

Background

Merrilactone A (**1**), isolated from the pericarps of the *Illicium merrillianum* plant [1, 2], is a member of a growing family of naturally occurring small molecules that have been identified as nonpeptidyl neurotrophic factors. These molecules, which mediate the growth and survival of neurons, may play an important therapeutic role in impeding the progress of a range of neurodegenerative disorders, including Alzheimer's, Parkinson's and Huntington's Diseases [3]. Our ongoing interest in the preparation and biological evaluation of molecules with potential application to neurodegenerative diseases has led us to assemble a small library of fully synthetic compounds with reported neurotrophic activity. This growing collection now includes tricycloillicinone [4], jiadifenin [5], NGA0187 [6], scabronine G methyl ester [7], and garsubellin A [8]. The synthesis of merrilactone A was accomplished first in the racemic series [9] and subsequently in asymmetric fashion [10]. We describe herein the asymmetric synthesis of merrilactone A.

Synthetic Strategy

The synthesis of racemic merrilactone A was completed in 20 steps in approximately 11% overall yield [9]. Although the first-generation synthetic route will not be discussed in detail in this account, we had determined that it would not, as such, be directly amenable to enantiocontrol. We did hope to retain some of the particularly elegant features of the later stages of the first-generation synthesis while working around the early awkwardnesses. With these considerations in mind, we selected iodolactone intermediate **7** from the first-generation synthesis as the target compound of our asymmetric route. With optically active compound **7** in hand, the synthesis of enantioenriched merrilactone A (**1**) would be completed as previously described.

In considering an asymmetric synthesis of **7**, we came to favor a somewhat unconventional degradative approach, commencing with *meso* compound **2**. Clearly, the development of a plan for the conversion of **2** to **7** would require careful consideration of issues of regio-, diastereo-, and enantiocontrol. Thus, in an overall sense, the equivalent carbons of the *meso* epoxide (a and b) would ultimately have to be differentially converted to an alkyl iodide (a) and a secondary alcohol (b) (see iodolactone **7**). Similarly, the two primary alcohols of compound **2** (c and d) would eventually be converted to a lactone, wherein one of the carbons would remain in the alcohol oxidation state (c) while the other would be oxidized (d). Furthermore, the differentiation of these equivalent functionalities would have to be accomplished in a regioselective fashion, such that the lactone carbonyl (d) would be proximally located to the secondary alcohol (b) in **7**.

We hoped that issues of enantiocontrol would be addressed through desymmetrization of epoxide **2** via asymmetric ring opening. The resultant diol, **3**, would be converted to intermediate **4**, which, upon exposure to regiospecific Baeyer-Villiger reaction, would suffer interpolation of an oxygen, as shown, followed by ring opening of the resultant lactone to afford an intermediate of the type **5**. The latter would be advanced to **6** and, thence, following iodolactonization, enantiomerically enriched intermediate **7** would be in hand. Ultimate conversion of **7** to merrilactone would first require elaboration of the alkyl iodide to a vinyl bromide of the type **8** followed by free radical-induced cyclization to afford **9**. Isomerization of the exocyclic olefin followed by an epoxidation and, thence, homo-Payne rearrangement would provide **1**. (Scheme 1.)

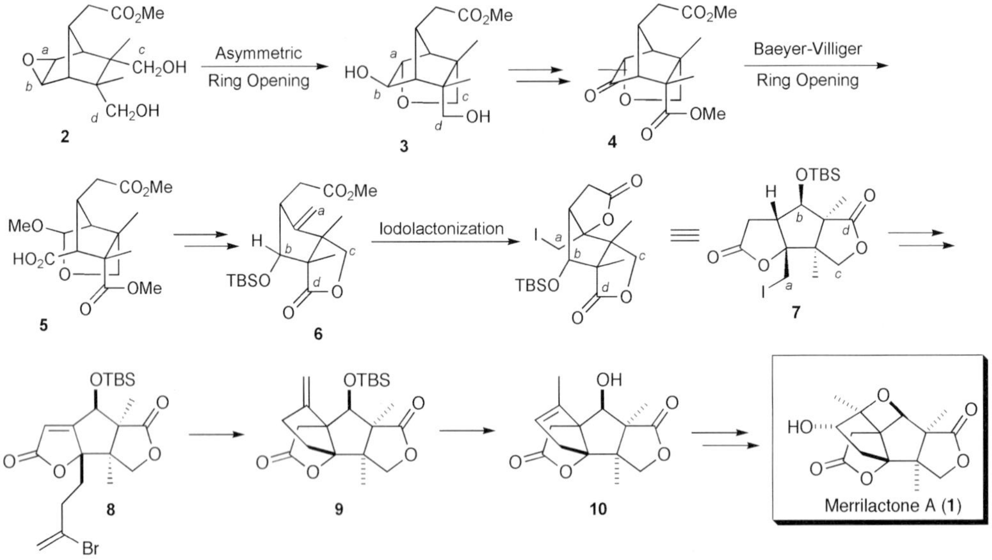

Scheme 1 Synthetic strategy toward optically active merrilactone A (**1**).

Synthesis of Merrilactone A

The first task would be the preparation of the *meso* epoxide **2**. Thus, thermal Diels-Alder reaction of **11** and **12** provided the *endo* adduct **13** in good yield. It should be noted that, although direct cycloaddition of **11** with 2,3-dimethylmaleic anhydride was unsuccessful, we were able to achieve the functional equivalent of this type of transformation through a diastereoselective C-alkylation of ester **13**. The resultant dimethylated adduct **14** was then converted to ester **16** in a straightforward manner, as shown. DMDO-mediated olefin epoxidation provided the key *meso* intermediate **2**. We were pleased to find that application of the Jacobsen asymmetric ring opening methodology to this substrate was very successful, providing the cyclic ether **3** in good yield and 86 % ee. (Scheme 2.) It should be noted that, in addition to allowing entry to the optically active series, this key ring opening protocol also served to provide a means for the regioselective chemodifferentiation between the homologous diol and epoxide carbons.

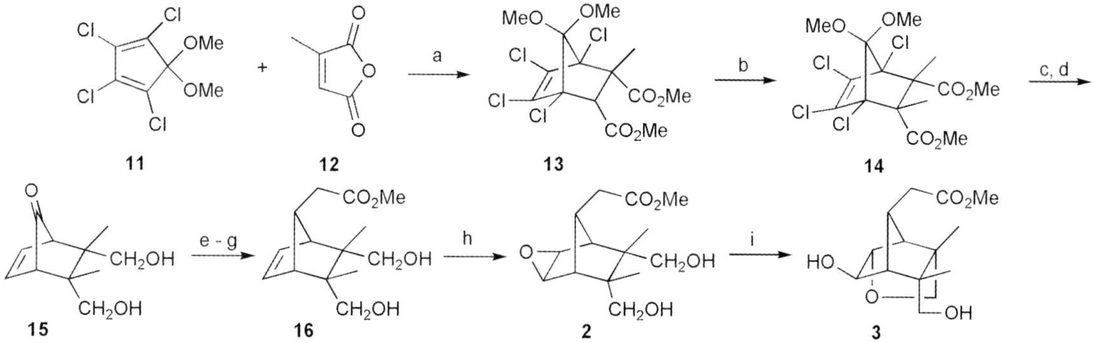

Scheme 2 (a) 180 °C, neat; then MeOH, reflux, PhH/MeOH, TMSCHN$_2$, 92 % for one-pot reaction; (b) LDA, HMPA, MeI, THF, −78 °C → RT, 95 %; (c) LAH, THF, reflux; (d) Na, NH$_3$, THF/EtOH, 72 % over two steps; (e) 2,2-dimethoxypropane, acetone, pTsOH; (f) NaH, (EtO)$_2$POCH$_2$CO$_2$Et, THF, 86 % over two steps; (g) Mg, MeOH, acidic workup, 77 %; (h) DMDO, CH$_2$Cl$_2$, 0.5–1 h; (i) (S,S)-[Co(III)(salen)]-OAc, −78 °C, two days; then − 25 °C, two days, THF, 86 % over two steps.

Diol **3** was readily converted to ketoester **4**. As expected, exposure of **4** to Baeyer-Villiger oxidation conditions provided the ring-opened adduct **5**. Execution of a "carboxy inversion" protocol allowed conversion of the carboxylic acid to a secondary alcohol (**17**), of course with retention of stereochemistry. The methoxy-tetrahydrofuran was then opened by dithiane trapping of the masked aldehyde with subsequent lactonization of the released alcohol, as shown (cf. **17** to **18**). Reduction of the dithiane was followed by elimination of the resulting primary alcohol under the Grieco protocol to afford the exocyclic olefin of intermediate **6**. The latter underwent diastereoselective iodolactonization to provide the key compound **7** in enantioenriched fashion. (Scheme 3.)

The next task was the elaboration of iodolactone **7** to cyclization precursor **8**. Accordingly, **7** was advanced to **20**, as shown. Conversion of **20** to the requisite vinyl bromide **8** was accomplished through an efficient three-step protocol, as shown. Thus, selenenylation of the α-position of the lactone was achieved through an intermediate silyl ketene acetal. Subsequent bromoselenenylation of the terminal olefin provided intermediate **21** which, upon exposure to O$_3$, underwent concurrent oxidative deselenation to afford the key cyclization precursor **8**.

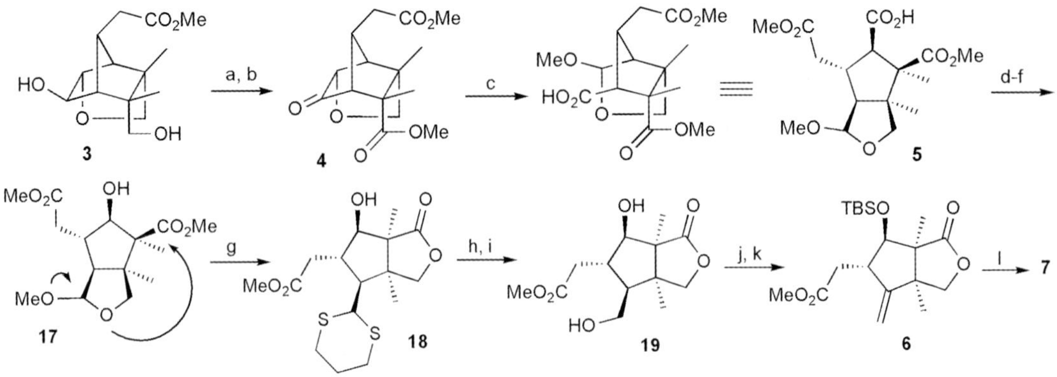

Scheme 3 (a) PDC, DMF; (b) K₂CO₃, MeI, acetone, reflux, 70% over two steps; (c) MMPP, MeOH, 0°C→RT, 88%; (d) DCC, mCPBA, 0°C->RT, 83%; (e) PhH, reflux; (f) K₂CO₃, MeOH, 70%; (g) BF₃·OEt₂, HS(CH₂)₃SH, CH₂Cl₂, 50%;
(h) PhI(OCF₃CO₂)₂, CH₃CN/H₂O, 50%; (i) NaBH₄, MeOH, 0°C; (j) o-NO₂C₆H₄SeCN, Bu₃P, THF, then H₂O₂ (30%), 86%; (k) TBSOTf, Et₃N, CH₂Cl₂, 76%; (l) LiOH, MeOH/H₂O; then I₂, saturated NaHCO₃/THF, 75%.

Happily, despite the considerable steric congestion of the substrate, intermediate **8** readily underwent radical-induced cyclization to afford the tetracyclic adduct **9** in excellent yield. Isomerization of the exocyclic olefin provided adduct **10**, which, upon exposure to mCPBA, underwent diastereoselective epoxidation to afford **22**. Finally, acid-induced homo-Payne rearrangement of **22** provided merrilactone A (**1**). (Scheme 4.)

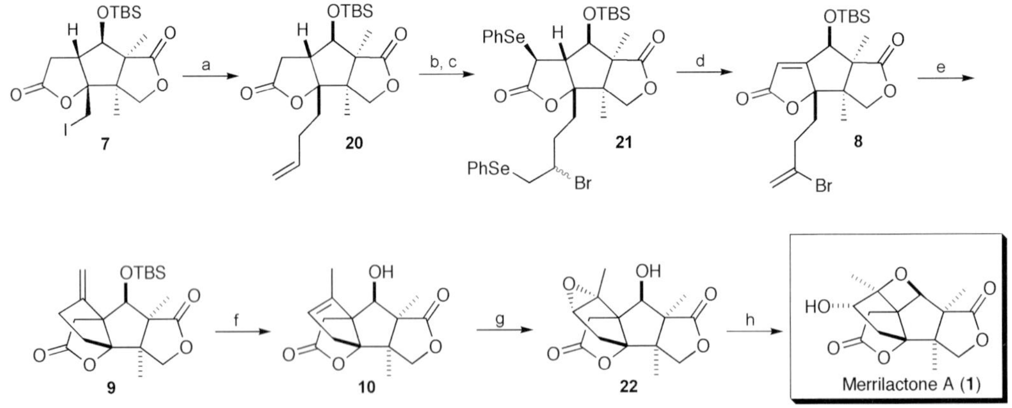

Scheme 4 Synthesis of merrilactone A. (a) allylSnBu₃, AIBN, PhH, 75%; (b) LiHMDS, TMSCl, PhSeCl; (c) PhSeBr, MeCN; (d) O₃, CH₂Cl₂, 1-hexene, PhH, NEt₃, reflux, 77% over three steps, (e) Bu₃SnH, AIBN, PhH, 90%; (f) TsOH·H₂O, PhH, reflux, 98%; (g) mCPBA, CH₂Cl₂, (α:β epoxide = 3.5:1); (h) TsOH·H₂O, CH₂Cl₂, rt, 71% over two steps.

In conclusion, the total synthesis of enantiomerically enriched merrilactone A has been described. Key features of this route include an asymmetric ring opening reaction, a regiospecific degradation pathway, and a free radical-induced cyclization.

CV of Samuel J. Danishefsky

Samuel Danishefsky completed his BS at Yeshiva University in 1956 and his Ph.D. at Harvard University with Peter Yates. After postdoctoral studies at Columbia University with Gilbert Stork, he began his independent academic career in 1963 at the University of Pittsburgh, where he became Professor in 1971. In 1980, he moved to Yale University, but returned to New York in 1993 as Professor of Chemistry at Columbia University and Kettering Chair at the Memorial Sloan-Kettering Cancer Center.

Selected Publications

1. J. Huang, R. Yokoyama, C. Yang, Y. Fukuyama, *Tetrahedron Lett.* **2000**, *41*, 6111–6114. *Merrilactone A, a novel neurotrophic sesquiterpene dilactone from Illicium merrillianum.*
2. J. Huang, C. Yang, M. Tanaka, Y. Fukuyama, *Tetrahedron* **2001**, *22*, 4691–4698. *Structures of merrilactones B and C, novel anislactone-type sesquiterpenes from Illicium merrillianum, and chemical conversion of anislactone B to merrilactone A.*
3. D. Dawbarn, S. J. Allen, *Neuropathol. Appl. Neurobiol.* **2003**, *29*, 211–230. *Neurotrophins and neurodegeneration.*
4. T. R. R. Pettus, M. Inoue, X.-T. Chen, S. J. Danishefsky, *J. Am. Chem. Soc.* **2000**, *122*, 6160–6168. *A fully synthetic route to the neurotrophic illicinones: Syntheses of tricycloillicinone and bicycloillicinone aldehyde.*
5. Y. S. Cho, D. A. Carcache, Y. Tian, Y.-M. Li, S. J. Danishefsky, *J. Am. Chem. Soc.* **2004**, *126*, 14358–14359. *Total synthesis of (±)-jiadifenin, a non-peptidyl neurotrophic modulator.*
6. Z. Hua, D. A. Carcache, Y. Tian, Y.-M. Li, S. J. Danishefsky, *J. Org. Chem.* **2005**, *70*, 9849–9856. *The synthesis and preliminary biological evaluation of a novel steroid with neurotrophic activity: NGA0187.*
7. S. P. Waters, Y. Tian, Y. Li, S. J. Danishefsky, *J. Am. Chem. Soc.* **2005**, *127*, 13514–13515. *Total synthesis of (−)-scabronine G, an inducer of neurotrophic factor production.*
8. D. R. Siegel, S. J. Danishefsky, *J. Am. Chem. Soc.* **2006**, *128*, 1048–1049. *Total synthesis of garsubellin A.*
9. V. B. Birman, S. J. Danishefsky, *J. Am. Chem. Soc.* **2002**, *124*, 2080–2081. *The total synthesis of (±)-merrilactone A.*
10. Z. Meng, S. J. Danishefsky, *Angew. Chem. Int. Ed.* **2005**, *44*, 1511–1513. *A synthetic pathway to either enantiomer of merrilactone A.*

Asymmetric Synthesis of Cyclic Ketal and Spiroaminal-Containing Natural Products

Craig J. Forsyth, University of Minnesota

Background

An array of biologically important natural products contains stereo-defined cyclic ketal and aminal moieties. While the genotype of such biogenetic cyclic ketals reflects enzymatic induction of secondary alcohol-derived configurations, their phenotype generally embodies a thermodynamically driven combination of equilibratable stereogenic centers and augmenting conformations. Examples of such natural products that have inspired new methods for asymmetric synthesis in our laboratories are illustrated in Figure 1. Reliable laboratory methods for stereoselective secondary carbinol installation have been used as a prelude to executing several innovative cyclic ketalization / aminalization strategies to access the key cyclic ketal and aminal moieties of these targets. Examples of the latter are highlighted here.

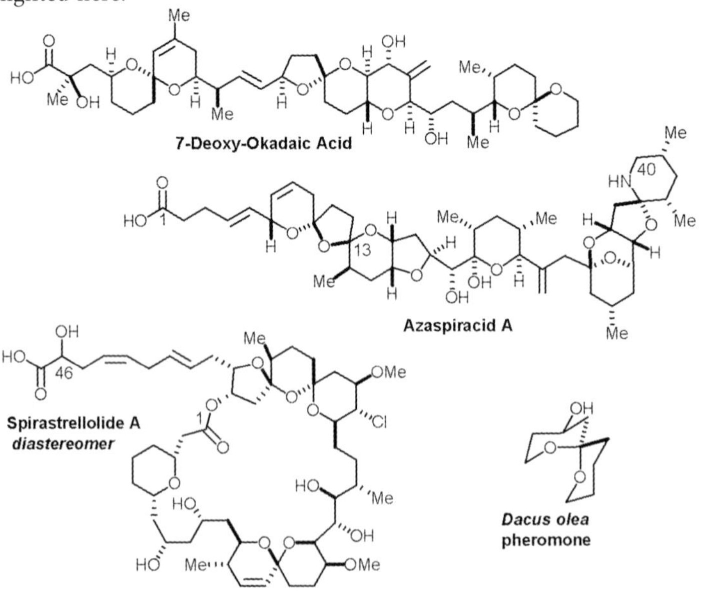

Figure 1 Representative ketal and spiroamnal-containing natural products.

Asymmetric Synthesis – The Essentials.
Edited by Mathias Christmann and Stefan Bräse
Copyright © 2007 WILEY-VCH Verlag GmbH & Co. KGaA, Weinheim
ISBN: 978-3-527-31399-0

Results

Okadaic Acid – Three Spiroketalization Methods

The natural products okadaic acid and 7-deoxy-okadaic acid (Figure 1) are the most potent inhibitors of serine-threonine phosphatases 1 and 2A known [1]. They contain three distinct spiroketal moieties centered at C8, C19, and C34. The C8 spiroketal was originally assembled in our total syntheses via a classic acid-induced dehydration of a α-α'-dihydroxy-ketone (**1**, Scheme 1), as precedented in this context by Isobe [2]. In this approach, the pre-installed configurations of the α-α'-hydroxy bearing carbons (C4 and C12) were expected to drive the doubly anomeric stabilized configuration of the resultant C8 spiroketal center. This was accomplished by prolonged treatment of a α-α'-bis-silyloxy-enone precursor to **1** with *p*-TsOH in benzene, which gave only the anticipated C8 spiroketal [3, 4]. An alternative approach based upon a double intramolecular hetero-Michael addition (DIHMA) has since been developed (Scheme 1). Here, an initial ynone (**3**) is susceptible to successive nucleophilic additions by C3 and C12 hydroxy groups, masked as labile TES ethers. Prolonged treatment of **3** with TsOH in benzene generated the C8 spiroketal **2** (ca. 80%, > 19:1 *ds*) bearing a ketone at C10, presumably via successive exo and endo cyclizations.

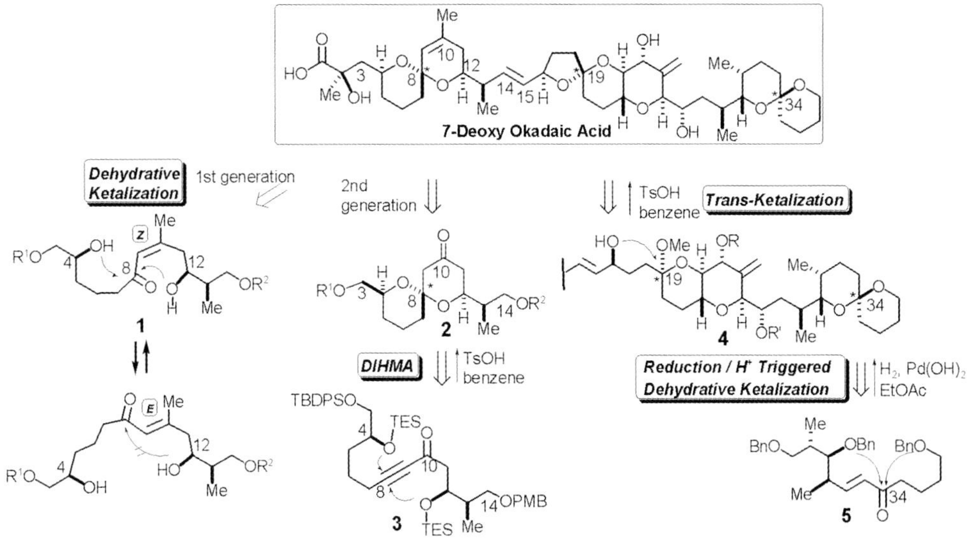

Scheme 1 Spiroketalization methods in the synthesis of 7-deoxy-okadaic acid.

The central C19 spiroketal was installed last via a diastereoselective reduction of a C16 ketone using Corey's CBS reagent followed by acid-induced transketalization of the C19 mixed methyl ketal **4** to generate an. 8:1 ratio of diastereomeric products [2]. It was established that the spiroketalization was highly stereocontrolled (> 19:1 *ds* at C19), reflecting a combination of conformational and anomeric effect stabilization, whereas the C16 stereocenter was generated in a ratio of ca. 8:1 from the CBS reduction [3].

A variation of the classic acid-induced dihydroxy ketone spiroketalization was used to assemble okadaic acid's C34 spiroketal. Hydrogenation of bis-benzyloxy-enone **5** (Scheme 1) with Pearlman's catalyst in ethyl acetate not only cleaved the benzyl ethers and reduced the alkene, but also led to spontaneous cyclizations to yield the natural product's thermodynamically favored spiroketal as the exclusive (> 19:1 ds) spiroketal product in high yield [2]. Here, the sterically demanding tertiary C29 substituent and pre-installed C30 stereogenic centers combine to define the resultant C34 configuration under thermodynamic control. Hence, we developed and applied three aptly suited approaches to form the C8, C19, and C34 stereogenic centers of okadaic acid and its analogs: DIHMA, transketalization, and a dehydrative spiroketalization, respectively.

Additional Ketals via Double Intramolecular Hetero Michael Additions

We have explored three variations of isomerization of dihyroxy-ynones to diastereoselectively generate intramolecular ketals. This process can be classified by the respective positioning of functionalization about the central ynone acceptor. First, placement of two simple heteroatom nucleophiles on the same α-branch of the ynone leads to a bridged ketal (α-α' monoketalization), as exemplified in closure of the 2,9-dioxabicyclo[3.3.1]nonane system of the azaspiracids (**8**, Scheme 2) [5]. The configuration of the newly formed ketal is geometrically defined by the pre-installed configurations of the nucleophilic hydroxy-bearing carbons.

Second, installation of simple heteroatom nucleophiles on each of the α and α' branches of an ynone may generate β-keto spiroketals (α-α' monoketalization), as illustrated in the formation of the C8 spiroketal of 7-deoxy okadaic acid and in the synthesis of the *Dacus olea* olive fly pheromone **11** [6]. Third, the strategic introduction of a ketone between the α branch heteroatom nucleophile and the ynone has allowed the rapid generation of trioxadispiroketals (α-α' bis-spiroketalization). Two examples are found in the syntheses of the azaspiracid A-D ring system (**14**) [7] and the trioxadispiroketal domain of spirastrellolide A (**17**) [8]. In the former case, acid-induced desilylation of the bis-TES ether **12** presumably allowed an initial exo-conjugate addition of the C13 ketone-derived oxygen to form the first spiroketal (**13**). The residual C6 hydroxy group could then add to the enone to close the second spiroketal. This thermodynamically-driven one-pot process provided the bis-spiroketal **14** with both ketals enjoying the benefit of the anomeric effect. A stepwise DIHMA process was required for spirastrellolide A. Here, a 5,6-spiroketal (**16**) could be formed from ynone **15** under basic reaction conditions, but final closure to **17** required subsequent treatment with acid (ca. 40% overall yield). Attempts to effect bis-spiroketalization of **15** under acidic conditions largely failed.

Synthesis of the Azaspiracid Spiroaminal

The emergence of the azaspiracid natural products prompted us to expand the repertoire of ketalization methods to spiroaminals. We initially employed a Yb(OTf)$_3$-induced transketalization, analogous to the C19 transketalization in okadaic acid [2], for the formation of the azaspiracid C36 spiroaminal center [5].

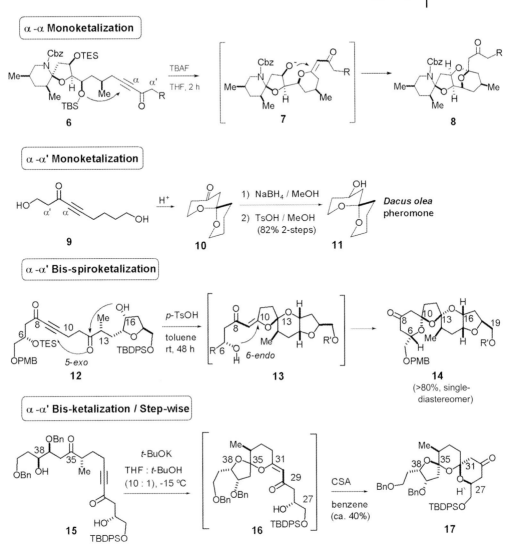

Scheme 2 Variations on DIHMA ketalizations.

Alternatively, we found that a Staudinger reduction of azide **18** performed in the presence of water liberated a primary amine that initiated a DIHMA process to give the spiroaminal **19** in high yield (Scheme 3) [9]. Finally, the fully functionalized azaspiracid spiroaminal could also be assembled from δ-azido-γ-hydroxy-ketone **20**. Treatment of **20** with triethylphosphine under anhydrous conditions led to an intramolecular aza-Wittig reaction and subsequent capture of the imine by the γ-hydroxy group to close the spiroaminal and generate **21**.

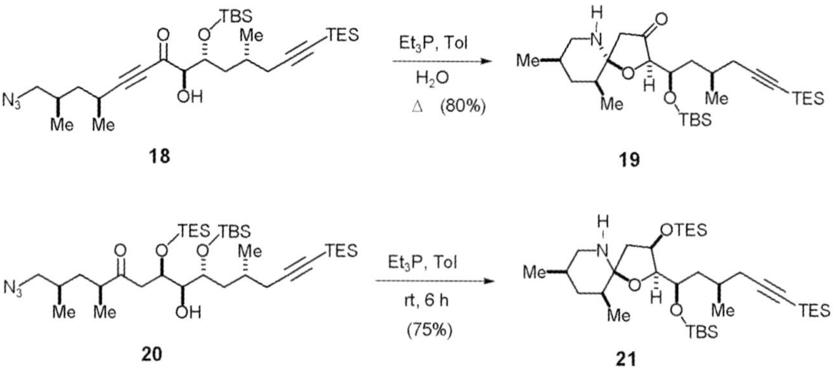

Scheme 3 Two novel asymmetric spiroaminalization methods.

Conclusion and Future Perspectives

The presence of cyclic ketal and aminal moieties in natural product targets inspires the development of efficient methods for their asymmetric synthesis. Approaches involving substrate dehydration, transketalization, or simple isomerization have been exploited to access polycyclic ring systems under thermodynamic control. A remaining challenge in this area is to develop general methods to assemble readily equilibratable ketal and aminal stereogenic centers contra-thermodynamically.

CV of Craig J. Forsyth

Craig J. Forsyth obtained his B.S. at Humboldt State University and a Ph.D. from Cornell University with J. C. Clardy (1989). After postdoctoral research with Y. Kishi, he was appointed at the University of Minnesota, Twin Cities, where he holds the rank of Professor. He received the Zeneca Pharmaceuticals Excellence in Chemistry Award (1997), the BMS Research Award in Synthetic Organic Chemistry (1999), a Japan Society for the Promotion of Science Fellowship (2000), a Novartis Lectureship (2003), and an Eli Lily U.K. Lectureship (2004).

Selected Publications

1. A. B. Dounay, C. J. Forsyth, *Curr. Med. Chem.* **2002**, *9*, 1939–1980. *Okadaic Acid: the Archetypal Serine/Threonine Protein Phosphatase Inhibitor.*
2. C. J. Forsyth, S. F. Sabes, R. A. Urbanek, *J. Am. Chem. Soc.* **1997**, *119*, 8381–8382. *An Efficient Total Synthesis of Okadaic Acid.*
3. A. B. Dounay, R. A. Urbanek, S. F. Sabes, C. J. Forsyth, *Angew. Chem. Int. Ed.* **1999**, *38*, 2258–2262. *Total Synthesis of the Marine Natural Product 7-Deoxy-Okadaic Acid: a Potent Inhibitor of Serine/Threonine-Specific Protein Phosphatases.*
4. A. B. Dounay, C. J. Forsyth, *Org. Lett.* **1999**, *1*, 451–453. *Abbreviated Synthesis of the C3-C14 (Substituted 1,7-Dioxaspiro[5.5]undec-3-ene) System of Okadaic Acid.*
5. C. J. Forsyth, J. Hao, J. Aiguade, *Angew. Chem. Int. Ed.* **2001**, *40*, 3663–3667. *Synthesis of the (+)-C26-C40 Domain of the Azaspiracids by a Novel Double Intramolecular Hetero-Michael Addition Strategy.*

6. J. Hao, C. J. Forsyth, *Tetrahedron Lett.* **2002**, *43*, 1–2. *Application of the Double Intramolecular Hetero-Michael Addition (DIHMA) Approach in Spiroketal Synthesis: Total Synthesis of (±)-(4S*,6S*)-4-Hydroxy-1,7-dioxaspiro[5.5]undecane, a Dacus Oleae Olive Fly Pheromone.*
7. L. K. Geisler, S. Nguyen, C. J. Forsyth, *Org. Lett.* **2004**, *6*, 4159–4162. *Synthesis of the Azaspiracid-1 Trioxadispiroketal.*
8. S. Nguyen, C. Wang, C. J. Forsyth, Abstract of Papers, 229th National Meeting of the American Chemical Society, San Diego, CA; American Chemical Society: Washington, DC, 2005, Abstract ORGN 414. *Synthesis of the Trioxadispiroketal Domain of Spirastrellolide A.*
9. J. Xu, C. J. Forsyth, Abstract of Papers, 229th National Meeting of the American Chemical Society, San Diego, CA; American Chemical Society: Washington, DC, 2005, Abstract ORGN 330. *Synthesis of a Universal Hapten for Antibody Detection of the Azaspiracids.*

Case Studies at the Metathesis/Asymmetric Synthesis Interface

Alois Fürstner, MPI für Kohlenforschung, Mülheim, Germany

Background

During the last decade, the advent of well defined metal alkylidene complexes combining high catalytic activity with a truly remarkable functional group tolerance [1] has revealed the exceptional preparative potential of metathesis in general and ring closing metathesis (RCM) in particular [2]. Most notable amongst them are the ruthenium catalyst **1** pioneered by Grubbs [1] and descendants thereof such as **2–4** (Figure 1). The impact of metathesis on the logic of total synthesis can hardly be overestimated and is maximized when interfaced with asymmetric synthesis regimens. Outlined below are two case studies from our group to illustrate this notion.

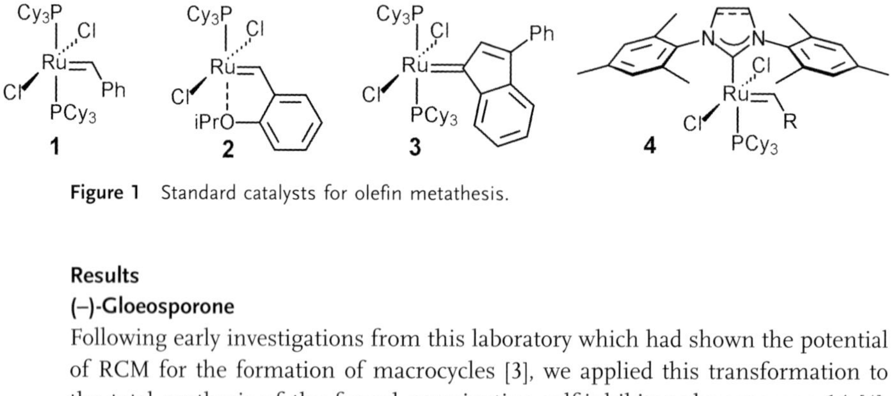

Figure 1 Standard catalysts for olefin metathesis.

Results

(−)-Gloeosporone

Following early investigations from this laboratory which had shown the potential of RCM for the formation of macrocycles [3], we applied this transformation to the total synthesis of the fungal germination self-inhibitor gloeosporone **14** [4]. To this end, cycloheptene was ozonized according to Schreiber's procedure to give the monoprotected dialdehyde **5** in one step. Enantioselective addition of dipentylzinc to this substrate in the presence of Ti(OiPr)$_4$ and catalytic amounts of ligand **6** afforded the secondary alcohol **7** in 88% yield with an *ee* of > 98% on a multigram scale. Subsequent esterification with 4-pentenoyl chloride followed by hydrolysis of the dimethyl acetal furnished aldehyde **8** which was subjected to a highly efficient catalytic asymmetric "Keck allylation" reaction (*de* > 98%). Silyla-

Asymmetric Synthesis – The Essentials.
Edited by Mathias Christmann and Stefan Bräse
Copyright © 2007 WILEY-VCH Verlag GmbH & Co. KGaA, Weinheim
ISBN: 978-3-527-31399-0

tion of the free alcohol in **9** set the stage for the crucial ring closure via RCM which was co-catalysed by the ruthenium carbene **1** (3 mol%) and Ti(OiPr)$_4$ (30 mol%), delivering cycloalkene **12** in 80% isolated yield. The addition of Ti(OiPr)$_4$ was necessary to avoid the formation of stable chelate structures such as **11** between the Lewis acidic metal center in the emerging carbene and the donor sites of the substrate which sequester the metal template in an unreactive form and were previously shown to be deleterious for macrocyclizations using "first generation" Grubbs-type catalysts [3]. (Note that this total synthesis was finished before "second generation" ruthenium carbenes such as **4** became available which might catalyze the cyclization of **10** even in the absence of any further additive.) Oxidation of cycloalkene **12** gave 1,2-diketone **13** which, on cleavage of the silyl group during the aqueous work-up, spontaneously cyclized to (–)-gloeosporone **14** (Scheme 1) [4].

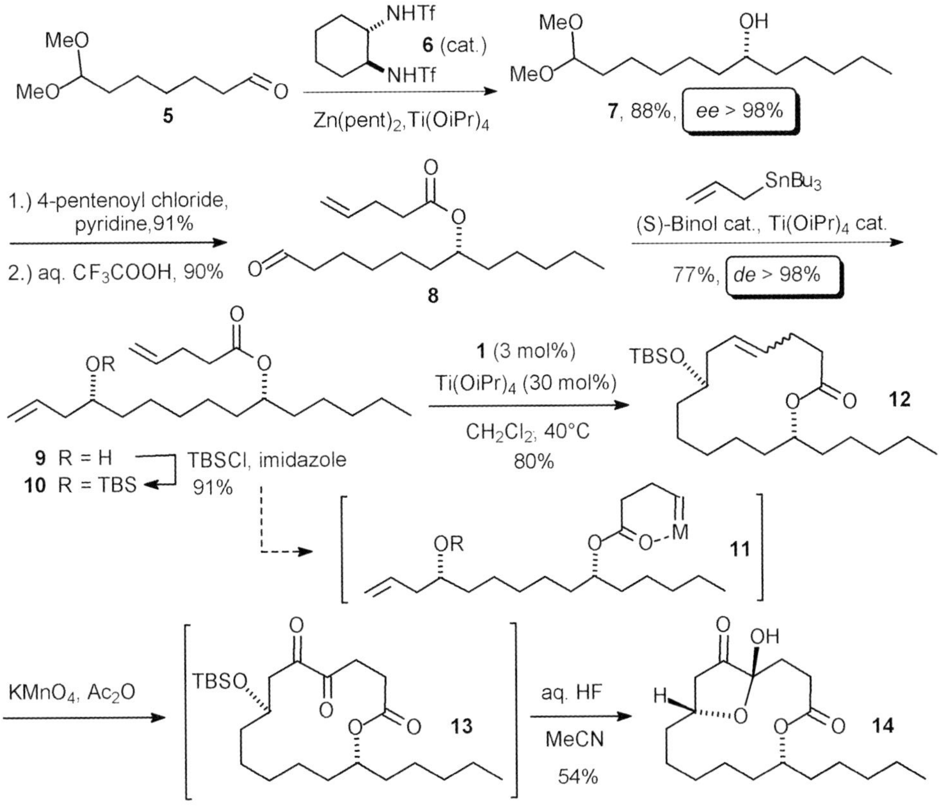

Scheme 1 Total synthesis of (–)-gloeosporone **14** [4].

Even though RCM has been successfully applied in the meanwhile to much more intricate targets [2], the approach outlined above illustrates some aspects of general relevance:

1. Enantiomerically pure **14** was obtained in only 8 steps from cycloheptene, with all C–C-bond formations being transition metal catalyzed. The chosen route is *significantly* shorter and more productive than all previous approaches to this target, not least because the number of protecting group manipulations is kept to a minimum. This is only possible because catalyst **1** allows us to activate alkenes chemoselectively in the presence of almost any functional group, thus avoiding inherently 'unproductive' interconversions.
2. The RCM-based macrocyclization was effective because it was performed at a site *remote* from the pre-existing functional groups. Since conventional bond formations usually follow the reactivity pattern induced by polar substituents and are therefore most effective at or near the functional groups, this aspect illustrates that *metathesis is complementary to established retrosynthetic logic* and hence expands preparative chemistry to a significant extent.

(−)-Isooncinotine

Because the macrocyclic spermidine alkaloid isooncinotine **22** isolated from the stem bark of the tropical trees *Oncinotis nitida* and *O. tenuiloba* is relatively simple in structural terms, any preparative approach considered 'adequate' must be short, practical and efficient. Metathesis combined with other catalytic bond formations holds great promise in this regard [5].

In line with the conclusions reached above, ring closure by RCM at a remote site within the aliphatic loop of **22** should be particularly effective. The required diene precursor could be secured by the innovative asymmetric hydrogenation methodology developed by Glorius [6] using an alkylated pyridine as the substrate which, in turn, might be available by cross coupling using cheap, non-toxic and benign iron salts rather than palladium- or nickel complexes as precatalysts [7].

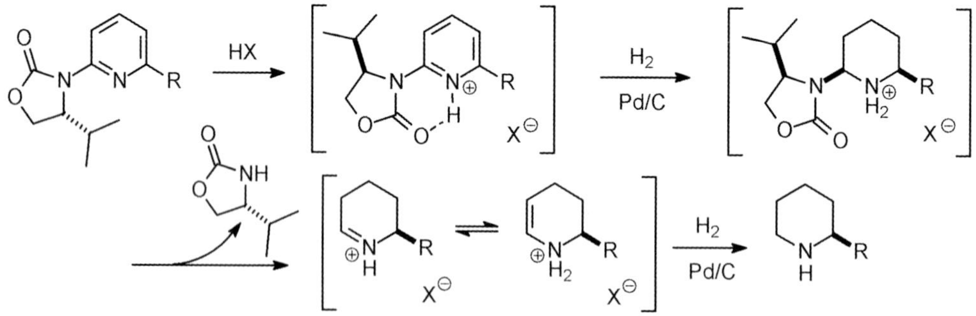

Scheme 2 Concept of the asymmetric hydrogenation of pyridines developed by Glorius et al. [6].

The essence of the Glorius hydrogenation of pyridines is sketched in Scheme 2 [6]. Protonation of a substituted pyridine bearing a chiral oxazolidinone at C-2 activates the heterocycle for hydrogenation and prevents catalyst poisoning. Moreover, the ensuing hydrogen bond locks a conformation in which the ring of the auxiliary is coplanar with the pyridine moiety; this conformational preference results in an effective shielding of one of the π-faces of the aromatic system by the isopropyl substituent and hence allows H_2 to be delivered from the surface of a heterogeneous catalyst to the "open" back side. Traceless cleavage of the resulting aminal under the reaction conditions releases the auxiliary and ultimately leads to the desired piperidine by hydrogenation of the iminium- (or enaminium-) species thus formed.

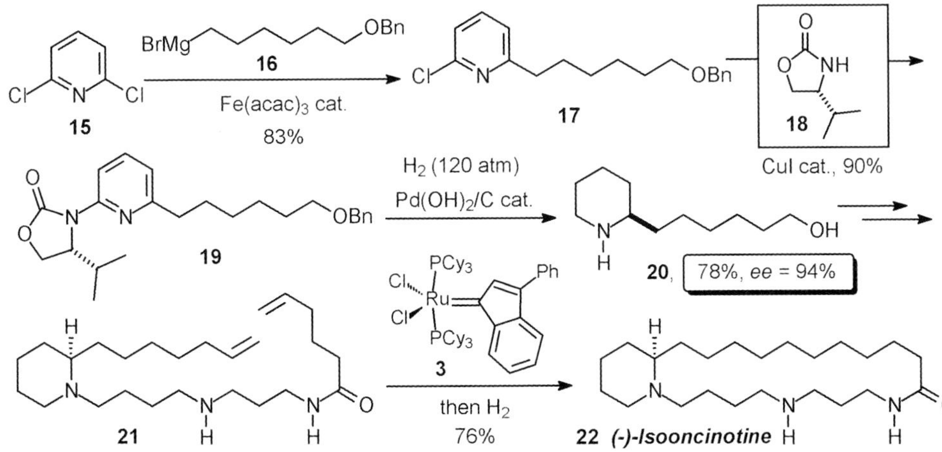

Scheme 3 Total synthesis of isooncinotine [5].

In the forward sense (Scheme 3), a highly efficient iron-catalyzed cross coupling between dichloropyridine **15** and the functionalized Grignard reagent **16** afforded compound **17** in good yield on a multigram scale, which allowed the introduction of the required chiral auxiliary **18** by a copper-catalyzed amidation process. Hydrogenation of the resulting product **19** over $Pd(OH)_2$ on charcoal in HOAc delivered piperidine **20** in 78% yield with an ee = 94%. Subsequent elaboration into diene **21** by standard transformations was streamlined by the fact that the lateral benzyl ether protecting group had also been concomitantly cleaved. RCM of compound **21** proceeded smoothly using the Ru-indenylidene complex **3** [8] which constitutes a cheap, stable and equipotent alternative to the classical Grubbs catalyst **1**. Importantly, stirring of the crude reaction mixture under H_2 engendered the saturation of the newly formed cycloalkene in 'one pot' and delivered (–)-isooncinotine **22** in excellent overall yield, most likely by converting the ruthenium alkylidene species responsible for RCM into a ruthenium hydride serving as the actual catalyst in the hydrogenation step [5].

Overall, this total synthesis is short and productive, largely catalysis-based, and features homogeneous as well as heterogeneous methodology. It illustrates the power of catalytic 'one-pot' metathesis/hydrogenation cascades, the promise of asymmetric hydrogenation of arenes, and showcases recent advances in cross coupling reactions.

Conclusions and Future Perspectives

The awesome power of olefin metathesis is evident from countless examples in the literature which are beyond the scope of this chapter [1, 2]. During the last decade, it has revolutionized preparative organic chemistry and rapidly evolved into a routine transformation. While unabated growth can be expected as a result of the development of even more effective catalysts, related fields such as the metathesis of alkynes are also starting to impact target oriented synthesis [9].

CV of Alois Fürstner

Alois Fürstner (born 1962) obtained his Ph.D. in 1987 at the Technical University of Graz, Austria. After a postdoctoral appointment with the late Prof. Oppolzer in Geneva, he finished his Habilitation in Graz (1992) before joining the Max-Planck-Institut für Kohlenforschung in 1993 as a group leader. Since 1998 he has been Director at that Institute. He has received the Leibniz award of the German Science Foundation (1999), the Thieme-IUPAC Prize (2000), an Arthur C. Cope Scholar Award of the ACS (2002), and the Otto-Bayer-Preis (2006).

Selected Publications

1R. T. M. Trnka, R. H. Grubbs, *Acc. Chem. Res.* **2001**, *34*, 18-29. *The Development of $L_2X_2Ru=CHR$ Olefin Metathesis Catalysts: An Organometallic Success Story.*

2R. A. Fürstner, *Angew. Chem. Int. Ed.* **2000**, *39*, 3012–3043. *Olefin Metathesis and Beyond.*

3. A. Fürstner, K. Langemann, *J. Org. Chem.* **1996**, *61*, 3942. *Conformationally Unbiased Macrocyclization Reactions by Ring Closing Metathesis.*

4. A. Fürstner, K. Langemann, *J. Am. Chem. Soc.* **1997**, *119*, 9130–9136. *Total Synthesis of (+)-Ricinelaidic Acid and (−)-Gloeosporone Based on Transition-Metal-Catalyzed C-C-Bond Formations.*

5. B. Scheiper, F. Glorius, A. Leitner, A. Fürstner, *Proc. Natl. Acad. Sci. USA* **2004**, *101*, 11960–11965. *Catalysis-Based Enantioselective Total Synthesis of the Macrocyclic Spermidine Alkaloid Isooncinotine.*

6. F. Glorius, N. Spielkamp, S. Holle, R. Goddard, C. W. Lehmann, *Angew. Chem. Int. Ed.* **2004**, *43*, 2850–2852. *Efficient Asymmetric Hydrogenation of Pyridines.*

7R. A. Fürstner, R. Martin, *Chem. Lett.* **2005**, *34*, 624–629. *Advances in Iron Catalyzed Cross Coupling Reactions.*

8. A. Fürstner, O. Guth, A. Düffels, G. Seidel, M. Liebl, B. Gabor, R. Mynott, *Chem. Eur. J.* **2001**, *7*, 4811–4820. *Indenylidene Complexes of Ruthenium: Optimized Synthesis, Structure Elucidation, and Performance as Catalysts for Olefin Metathesis. Application to the Synthesis of the ADE-Ring System of Nakadomarin A.*

9R. A. Fürstner, P. W. Davies, *Chem. Commun.* **2005**, 2307–2320. *Alkyne Metathesis.*

Asymmetric Synthesis of Amino Acids by Rhodium and Ruthenium Catalysis

Jean Pierre Genet, Ecole Nationale Supérieure de Chimie de Paris (ENSCP), France

Introduction

Asymmetric syntheses of proteinogenic and non-proteinogenic amino acids **1** have been of great interest due to their wide utility as components of proteins and peptides and as a starting material for the synthesis of biologically active compounds, α-amino, β-hydroxy acids *syn* **2** and *anti* **3** are also key components, of biologically active cyclopeptides. Catalytic asymmetric routes of amino acids are shown in Figure 1.

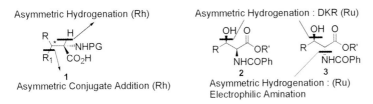

Figure 1 Catalytic asymmetric strategies of amino acids.

Asymmetric Hydrogenation Reactions [1R]

Asymmetric catalysis is certainly among the most challenging and widely investigated area in modern synthetic organometallic chemistry. Such a statement is fully confirmed by the award of the 2001 Nobel Prize in Chemistry to W.S. Knowles and R. Noyori for their work on asymmetric hydrogenation [2] and to K. B. Sharpless for his contribution in asymmetric oxidation reactions. Homogeneous asymmetric hydrogenation was first formally demonstrated through the work of Horner and Knowles. A very important improvement was introduced when Kagan demonstrated that a chiral phosphorus atom is not necessary if a chiral bidendate ligand is used, such as DIOP, a C_2-symmetric diphosphine ligand. In addition, the perception that amino acids can be directly synthesized from dehydroamino acids gave a significant impetus to this technology. The successful development by Knowles of Rh-DIPAMP as a catalyst for L-DOPA with enantiomeric excess up to 96 % for many years enabled Monsanto to be the supplier of the main drug used in stabilizing the effect of Parkinson's disease.

Asymmetric Synthesis – The Essentials.
Edited by Mathias Christmann and Stefan Bräse
Copyright © 2007 WILEY-VCH Verlag GmbH & Co. KGaA, Weinheim
ISBN: 978-3-527-31399-0

Chiral Diphosphine Ligands and Diversity of Chiral Ru Catalysts

Development of optically active phosphine ligands, especially C_2-chiral diphosphines, provided a great advance in asymmetric hydrogenation. In the context, of our long interest in asymmetric hydrogenation, particularly with ruthenium catalysis, we have developed the preparation of new atropisomeric ligands such as SYNPHOS **4** [3] and DIFLUORPHOS **5** [4] in order to induce fine-tuning of the steric and electronic properties of the biaryl skeleton.

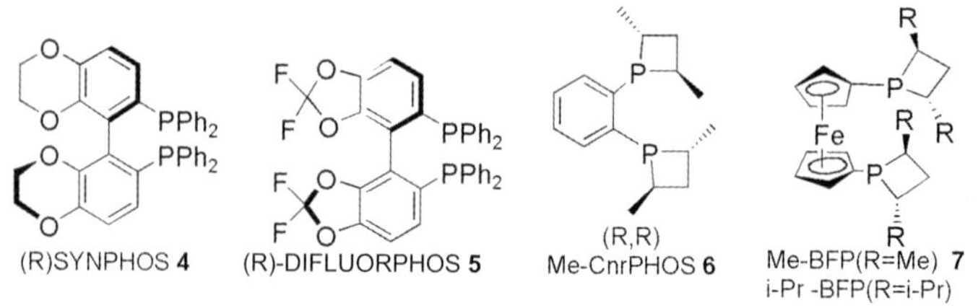

Figure 2 Chiral diphosphines.

Another interesting class of electron-rich C_2-symmetric phosphetane CnrPHOS **6** and BFP **7** (bis ferrocenyl phosphetanes) has been developed in our laboratory (Figure 2) [5].

A suitable combination of metal species and chiral organic ligands is the key factor in the preparation of high performance catalysts for asymmetric hydrogenation reactions. Asymmetric hydrogenation has been dominated for three decades by rhodium complexes. The focus changed from rhodium to ruthenium in the mid 1980s with the outstanding work of Noyori and coworkers [1R]. We also have focused some efforts on the design of a general synthesis of chiral ruthenium catalysts. Our method is based on the use of Ru (COD) (η^3-methylallyl)$_2$, which is an excellent starting material for a facile synthesis of chiral Ru (P*P) X$_2$ complexes [6].

Results

Asymmetric Hydrogenation of Dehydroamino Acids

Since the invention of the well designed Rh-complexes containing chiral disphosphines used in the asymmetric hydrogenation of dehydroamino acids, this reaction has become the model reaction to evaluate the efficiency of a new chiral ligand. In our work, we have considered this reaction for an evaluation of the catalytic potential of our recently reported chiral diphosphines. In this context Rh-BFP catalysts and Ru-SYNPHOS also have revealed high efficiency in asymmetric hydrogenation of α-dehydroamino acids **8** to amino acids **9** [6A] (Figure 3).

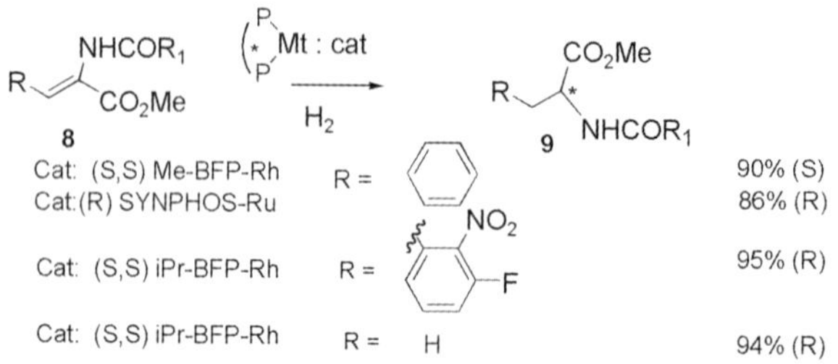

Figure 3 Asymmetric synthesis of α-amino acids.

The practical importance of asymmetric hydrogenation maintains the great interest in the mechanistic aspects of this reaction. It was previously well accepted that the "unsaturated-alkene" mechanism (Halpern-Brown) was operating with a wide range of phosphines. Recently with an electron-rich P-stereogenic ligand such as MINIPHOS the situation has changed. In particular Gridnev and Imamoto, through experimental and computational studies, have established that the so-called "dihydride mechanism" of Rh-catalyzed asymmetric hydrogenation is of greater relevance than previously accepted. They also suggested an approach for the prediction of the sense of enantioselectivity [7].

Asymmetric Syntheses of α-Amino β-Hydroxy Acids (DKR) [1R]

Hydrogenation of racemic α-monosubstituted β-keto esters should, in principle, provide four possible stereomers of hydroxy esters. However it is possible under appropriate conditions with Ru catalyst to obtain a single product *syn* or *anti*. This ruthenium promoted hydrogenation via dynamic kinetic resolution (DKR) has turned out to be an elegant and powerful method of simultaneously controlling two adjacent stereogenic centers with a high level of selectivity in a single chemical operation. This reaction, was first reported independently by Noyori and our group for the synthesis of *syn* β-hydroxy α-amino acid **10** using at the α position the protected β-keto esters **11**. Very recently, we found that the Ru-Synphos catalyzed hydrogenation affords a general and efficient access to *anti* α-amino-β-hydroxy esters **14** by using α-NH$_2$.HCl.β-keto esters **13** with a high level of selectivity (Scheme 1) [8]. Thus, optically active *syn* and *anti* β-hydroxy α-amino acids are efficiently prepared from β-keto esters **12** as common intermediates, Scheme 1.

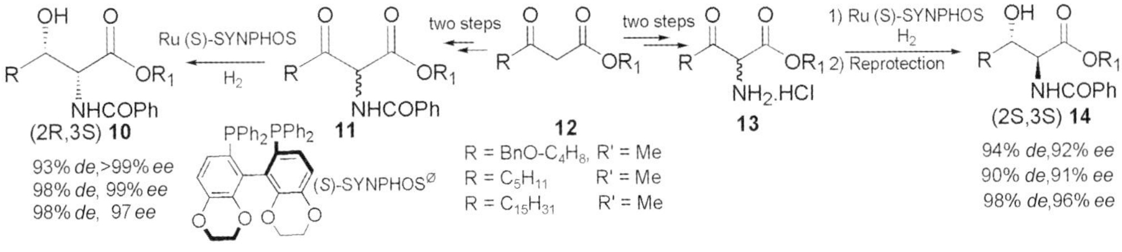

Scheme 1 Asymmetric synthesis of *syn* **10** and *anti* **14** α-amino β-hydroxy acids. Conditions: [Ru(S)-SYNPHOS® Br$_2$ 2mol%, α-benzamido-β-keto esters **11** (0.5 mmol), H$_2$ 130 bar, 80 °C, 4 days, CH$_2$-Cl$_2$[Ru(S)-SYNPHOS®Br$_2$ 2 mol%, α-amino-β-keto esters hydrochlorides **13** (0.5 mmol), H$_2$ 12 bar, 80 °C, 24 h, CH$_2$Cl$_2$ROH then (PhCO)$_2$O, NEt$_3$CH$_2$Cl$_2$.

Asymmetric Syntheses of Natural Products

We have used and are using the described systems for the synthesis of natural products such as dolastatin 10 [9], (+)-lactacystin. In addition, we have used Ru-asymmetric hydrogenation of β-keto esters to yield optically active β-hydroxy esters; a subsequent diastereoselective electrophilic amination provides a direct approach to chiral *anti* α-amino β-hydroxy amino acids of biological interest such as Sulfobacin A (Figure 4) [10].

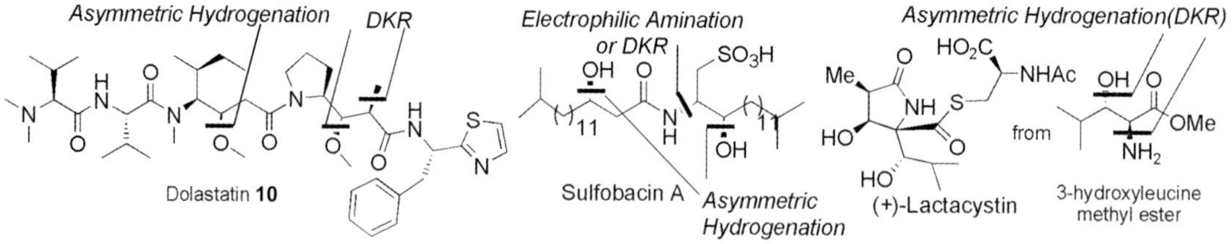

Figure 4 Synthesis of products of biological interest.

Tandem-1,4 addition/Enantioselective Protonation Catalyzed by Rhodium

The 1,4-addition of organometallic reagents (Michael type addition) to electron-deficient alkenes catalyzed by transition metals has emerged as a powerful tool in synthesis for the construction of carbon–carbon bonds. We have found that trifluoro(organo)borates are good organometallic partners to α-acylamidoacrylate in the tandem 1,4-addition / enantioselective protonation using one equivalent of *ortho*-methoxy phenol (Guaiacol) as protonating agent in the presence of [[Rh(Cod)$_2$]PF$_6$] (cod = cyloocta-1,5-diene) and a chiral ligand such as BINAP or electrodeficient atropisomeric DIFLUORPHOS (Scheme 2). Aryl and alkenyl α amino acids are obtained with a high level of enantioselectivity, up to 95% [11].

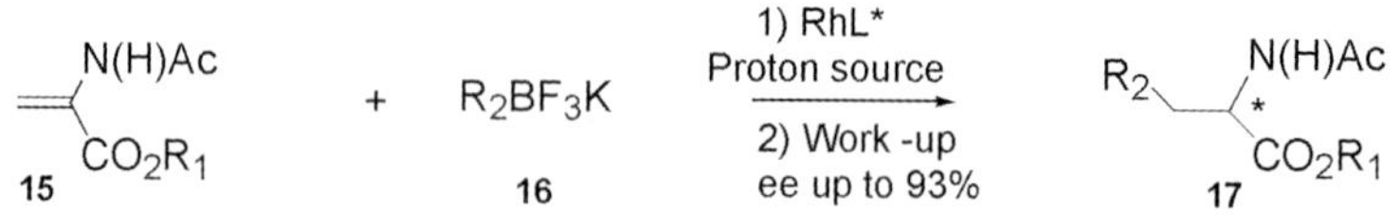

Scheme 2 Enantioselective Addition Organopotassium trifluoroborates **16** to Amidoacrylate **15**. Conditions: methyl-2 acetylaminoacrylate (0.5 mmole)[Rh(cod)$_2$] PF$_6$, 3 mol%, L* 6 mol%, R$_2$BF$_3$K (2 equiv.), 1 equiv. of guaiacol, toluene, 110°C for 20 h.

The best enantioselectivities were obtained with a bulky substituent at the ester moiety such as isopropyl ester [12]. Deuterium-labeling as well as computational studies are currently underway to get insight into the mechanism of this interesting transformation [12].

Conclusions

We have designed and synthesized new chiral ligands and catalysts for the asymmetric synthesis of amino acids. The principle of tandem -1,4 addition/enantioselective protonation catalyzed by rhodium is a highly efficient process for the introduction of an organic substituent in the β-position of an electro-deficient alkene with concomitant control of the chirality of the α-center. This principle provides a new and excellent entry into various chiral α amino acids and has very recently been applied to $β^2$ amino acids [13].

CV of Jean Pierre Genet

Jean-Pierre Genet was born in Tulle, France in 1942. He received his education in Paris with a B. Sc. and a PhD from the University of Pierre and Marie Curie with Professor Jacqueline Ficini. In 1975–1976 he did postdoctoral work with Professor B.M. Trost at the University of Wisconsin, Madison. He was appointed at the University Pierre and Marie Curie as Assistant Professor in 1970, and then full Professor in 1980. In 1988 he moved to the Ecole Nationale Supérieure de Chimie de Paris (ENSCP) where he holds the Chair of Organic Chemistry. Since 1992, he has been Director of the Department of Organic and Bioorganic Chemistry at ENSCP and the Laboratory associated to Centre National de la Recherche Scientifique (C.N.R.S) U.M.R 7573. From 1994 to 1998 he was President of the Organic Division of the French Chemical Society. He has served as Chairman for OMCOS 10 in Versailles (1999). He received the Award of Académie des Sciences Institut de France (1988), a Japanese Society for the Promotion of Science Fellowship (1998), Lady Davis Lecturer Technion, Haifa, Israel (1999); the Innovation Award of University Pierre and Marie Curie-ENSCP (ADFAC-2003), Novartis Lecturer (2003) and Le Bel Award of the French Chemical Society (2004).

Selected Publications

1R. O. Takeshi, M. Kitamura, R. Noyori, in *Catalytic Asymmetric Synthesis*, 2nd Edn., I. Ojima (Ed.), Wiley-VCH, Weinheim, **2000**, pp. 1–101. Asymmetric Hydrogenation.

2. W. S. Knowles, *Angew. Chem., Int. Ed.* **2002**, *41*, 1998–2007. Asymmetric hydrogenations (Nobel lecture). R. Noyori., *Angew. Chem., Int. Ed.* **2002**, *41*, 2008–2019. Asymmetric catalysis: Science and opportunities (Nobel lecture).

3. S. Duprat de Paule, S. Jeulin, V. Ratovelomanana-Vidal, J. P. Genet, N. Champion, P. Dellis, *Eur. J. Org. Chem.* **2003**.1931–1941. Synthesis of Synphos®, a new efficient diphosphine ligand in ruthenium-catalyzed asymmetric hydrogenation and molecular modeling studies.

4. S. Jeulin, S. Duprat de Paule, V. Ratovelomanana-Vidal, J. P. Genet, N. Champion, P. Dellis *Angew. Chem., Int. Ed.* **2004**, *43*, 320–325. Difluorphos®, an Electron-Poor Diphosphane: a Good Match Between Electronic and Steric Features.

5. A. Marinetti, F. Labrue, J. P. Genet, *Synlett* **1999**, *12*, 1975–1977 Synthesis of 1,1'-bis(phosphetano) ferrocenes, a new class of chiral ligands for asymmetric catalysis.

6A. J. P. Genet, *Acc. Chem. Res.* **2003**, *36*, 908–918. Asymmetric Catalytic Hydrogenation. Design of New Ru-Catalysts and Chiral Ligands: From Laboratory to Industrial Applications.

7. I. D. Gridnev, T. Imamoto, *Acc. Chem. Res.* **2004**, *37*, 633–644. On the mechanism in Rh-catalyzed asymmetric hydrogenation: a general approach for predicting the sense of enantioselectivity.

8. C. Mordant, P. Dünkelman, V. Ratovelomanana-Vidal, J. P. Genet, *Chem. Commun.* **2004**, 1296–1297; *Eur. J. Org. Chem.* **2004**, 3017–3026. Dynamic kinetic resolution: an efficient route to anti α–amino-β-hydroxy esters via Ru-Synphos® catalyzed hydrogenation.

9. C. Mordant, S. Reymond, V. Ratovelomanana-Vidal, J. P. Genet *Tetrahedron*, **2004**, *60*, 9715–9723, Stereoselective synthesis of the stereoisomers of Boc-dola-proline from (S)-Boc-proline through dynamic kinetic resolution.

10. O. Labeeuw, P. Phansavath, J. P. Genet, *Tetrahedron Lett.* **2003**, *44*, 2405–2409; *Tetrahedron : Asymmetry* **2004**, *15*, 1899–1905. *A short synthesis of Sulfobacin A*.
11. L. Navarre, S. Darses, J. P. Genet, *Angew. Chem. Int. Ed.* **2004**, *43*, 719–723. *Tandem 1,4-addition/enantioselective protonation catalyzed by rhodium complexes: efficient access to α-amino acids*.
12. L. Navarre, S. Darses, J. P. Genet, Manuscript in preparation. *Rhodium-catalyzed tandem carbometallation/enantioselective protonation Asymmetric synthesis of α-amino acids and mechanistic studies*.
13. M. P. Sibi, H. Tatamidani, K. Patel, *Org. Lett.* **2005**, *7*, 2571–2573. *Enantioselective rhodium enolate protonation. A new methodology for the synthesis of β^2-amino acids*.

Asymmetric Syntheses of Pheromones

Kenji Mori, University of Tokyo, Japan

Background

In 1973 when I started my work on asymmetric syntheses of pheromones, a few chiral pheromones such as *exo*-brevicomin (**1**) had already been identified. Of course nothing was known yet about the complexity of the genes encoding olfactory receptors. I wanted to solve the following two questions. (i) What is the absolute configuration of the chiral pheromones? Their oily nature and scarce availability precluded the use of conventional methods for that purpose such as X-ray analysis and degradative studies. (ii) Is only one of the enantiomers of a chiral pheromone responsible for the bioactivity, or are both of them bioactive [1R]?

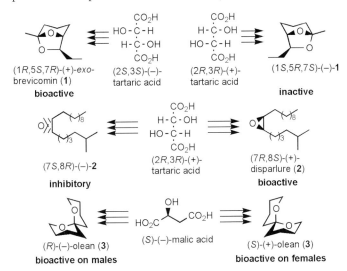

Scheme 1

Asymmetric Synthesis – The Essentials.
Edited by Mathias Christmann and Stefan Bräse
Copyright © 2007 WILEY-VCH Verlag GmbH & Co. KGaA, Weinheim
ISBN: 978-3-527-31399-0

Strategy and Results

In my pheromone research, my friends in the Entomology Department were waiting for the arrival of my samples. I had to send out sufficient amounts of chemically and enantiomerically pure samples as soon as possible. I therefore employed any kind of asymmetric reactions and/or separation techniques, if they were simple, efficient and selective [2R].

Derivation from Optically Active Starting Materials

In 1974, both the enantiomers of *exo*-brevicomin (1), a bark beetle pheromone, were synthesized, and only (1R,5S,7R)-1 showed bioactivity (Scheme 1) [3]. In 1976, the enantiomers of disparlure (2), the gypsy moth pheromone, were synthesized from (+)-tartaric acid. The bioactive enantiomer was (7R,8S)-2, while (7S,8R)-2 was inhibitory [4]. In 1985, the enantiomers of olean (3), the olive fruit fly pheromone, was synthesized from (−)-malic acid [5]. Male flies were activated by (R)-3, while (S)-3 activated the females.

Enantiomer Separation

Although classical optical resolution by recrystallization of diastereomeric salts was useful [2R], enantiomer separation could be facilitated by liquid chromatography and enzymatic kinetic resolution, as shown in Scheme 2. The enantiomers of lineatin (4), the pheromone of an ambrosia beetle, were synthesized from 6 by separating 5 and its diastereomer by medium pressure liquid chromatography. The absolute configuration of 5 was determined by X-ray analysis [6]. The bioactive enantiomer of lineatin was (1R,,4S,5R,7R)-4. The enantiomers of the pheromone 9 of the Colorado potato beetle were synthesized by employing enzymatic kinetic resolution of (±)-7 [7]. (S)-9 was bioactive.

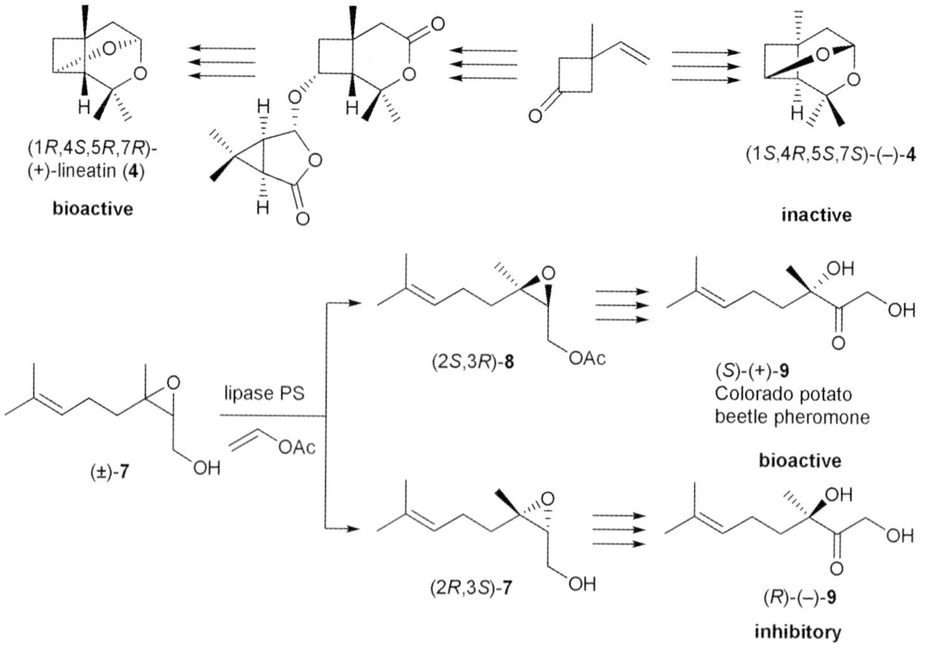

Scheme 2 Synthesis of enantiopure pheromones. Separation of the racemic intermediates.

Asymmetric Synthesis

Both chemical and enzymatic asymmetric reactions could be used efficiently in pheromone synthesis, as illustrated in Scheme 3. In the synthesis of the pheromone (R)-13 of *Janus integer*, a fly, Sharpless asymmetric dihydroxylation of 10 gave diol 11, which was converted to epoxide 12 (87% ee). Further purification of this epoxide 12 by Jacobsen's hydrolytic kinetic resolution (HKR) yielded 12 of 96% ee, which was converted to the pheromone lactone (R)-13 [8]. Lipase PS was employed for the synthesis of leucomalure [(3Z,6S,7R,9S,10R)-16], the pheromone of the Satin moth [9]. Lipase-catalyzed asymmetric acetylation of (±)-14 gave (2R,3S)-14 and (2S,3R)-15. They were converted to the four stereoisomers of 16 [9]. Only (3Z,6S,7R,9S,10R)-16 was bioactive.

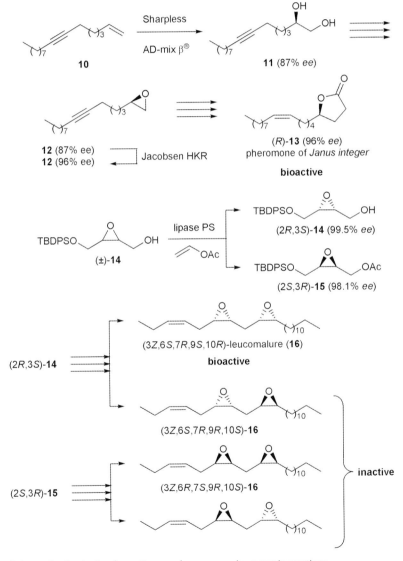

Scheme 3 Synthesis of enantiopure pheromones. Asymmetric reactions.

Conclusion and Future Perspective

Synthesis of the pure enantiomers of pheromones allowed us to examine the relationships between stereochemistry and pheromone activity, as shown in Figure 1 [1R, 10]. The relationships turned out to be complicated and diverse. Chirality was thus shown to be of key importance in pheromone perception. Many new pheromones with stereogenic centers will be identified in future. We chemists must synthesize them to enable their further study by biologists.

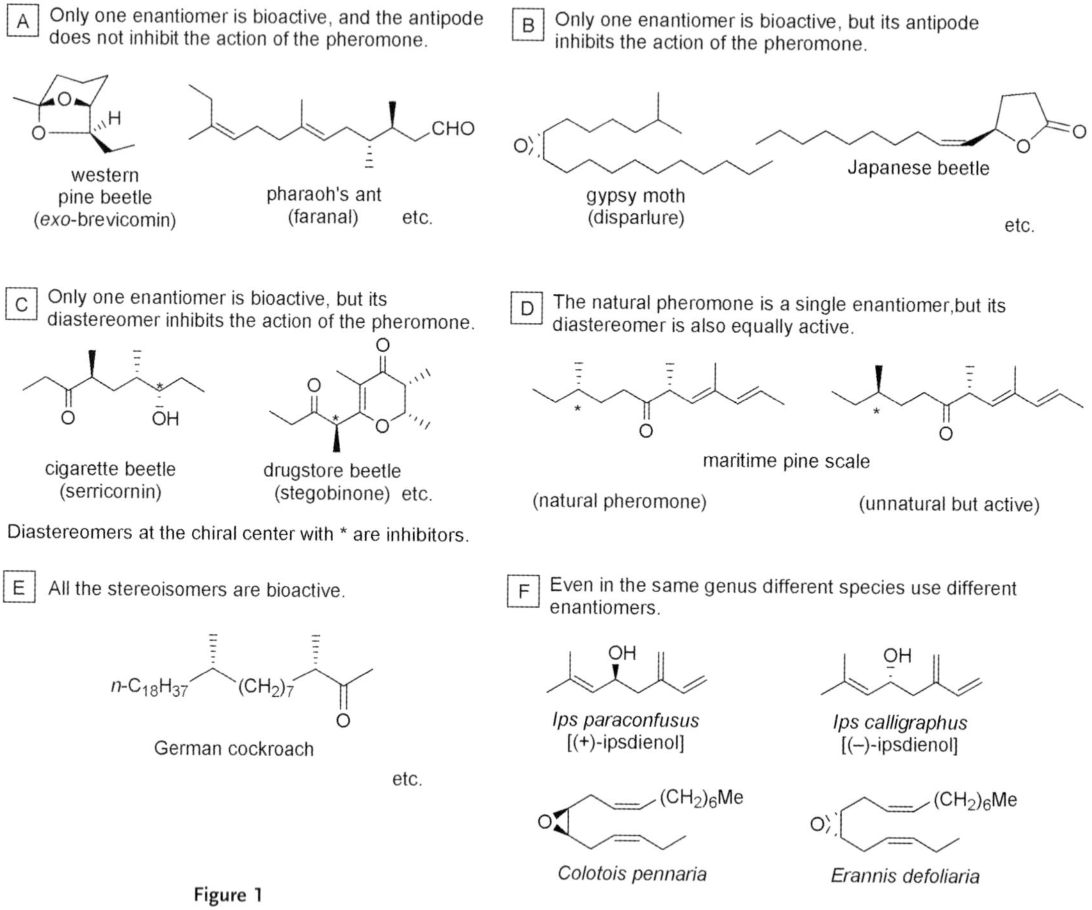

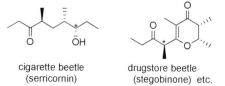

Figure 1

CV of Kenji Mori

Kenji Mori was born in 1935, and obtained his B.Sc. (1957), M.Sc. (1959) and Ph.D. (1962) degrees from the University of Tokyo. He remained there until March 1995 (1962–1968, assistant; 1968–1978, Associate Professor; 1978–1995, Professor), and then worked at the Science University of Tokyo (1995–2001, Professor). He is now a Research Consultant at RIKEN (Institute of Physical and Chemical Research) and also at Toyo Gozei Co. He has been awarded the Japan Academy Prize (1981), the Silver Medal of the International Society of

Chemical Ecology (1998), the American Chemical Society's Ernest Guenther Award in the Chemistry of Natural Products (1999), the Special Prize of the Society of Synthetic Organic Chemistry, Japan (2003), and the Frantisek Sorm Memorial Award of the Academy of Sciences of the Czech Republic (2003).

Selected Publications

1R. K. Mori, in *Chirality in Natural and Applied Science*, W. J. Lough, I. W. Wainer (Eds), Blackwell Science, CRC Press, **2002**, Ch. 9, pp. 241–259. *Chirality in the Natual World: Chemical Communications.*

2R. K. Mori, *Tetrahedron* **1989**, *45*, 3233–3298. *Synthesis of Optically Active Pheromones.*

3. K. Mori, *Tetrahedron* **1974**, *30*, 4223–4227. *Synthesis of exo-Brevicomin, the Pheromone of Western Pine Beetle to Obtain Optically Active Forms of Known Absolute Configuration.*

4. K. Mori, T. Takigawa, M. Matsui, *Tetrahedron Lett.* **1976**, 3553–3556. *Stereoselective Synthesis of Optically Active Disparlure, the Pheromone of the Gypsy Moth. (Porthetria dispar L.).*

5. K. Mori, T. Uematsu, K. Yanagi, M. Minobe, *Tetrahedron* **1985**, *41*, 2751–2758. *Synthesis of the Optically Active Forms of 4,10-Dihydroxy-1,7-dioxaspiro[5.5]undecane, and their Conversion to the Enantiomers of 1,7-Dioxaspiro[5.5]undecane, the Olive Fly Pheromone.*

6. K. Mori, T. Uematsu, M. Minobe, K. Yanagi, *Tetrahedron* **1983**, *39*, 1735–1743. *Synthesis and Absolute Configuration of Both the Enantiomers of Lineatin, the Pheromone of Trypodendron lineatum.*

7. T. Tashiro, K. Mori, *Tetrahedron: Asymmetry* **2005**, *16*, 1801–1806. *Enzyme-assisted Synthesis of (S)-1,3-Dihydroxy-3,7-dimethyl-6-octen-2-one, the Male-produced Aggregation Pheromone of the Colorado Potato Beetle, and its (R)-Enantiomer.*

8. K. Mori, *Eur. J. Org. Chem.* **2005**, 2040–2044. *Concise Synthesis of (4R,9Z)-Octadec-9-en-4-olide, the Female Sex Pheromone of Janus integer.*

9. S. Muto, K. Mori, *Eur. J. Org. Chem.* **2003**, 1300–1307. *Synthesis of all Four Stereoisomers of Leucomalure, Components of the Female Sex Pheromone of the Satin Moth, Leucoma salicis.*

10. K. Mori, *Acc. Chem. Res.* **2000**, *33*, 102–110. *Organic Synthesis and Chemical Ecology.*

Total Synthesis of Polyketides Using Asymmetric Aldol Reactions

Ian Paterson, University of Cambridge, UK

Background

As stereochemically complex bioactive natural products, the polyketides represent challenging target molecules for adventurous practitioners of organic synthesis. The therapeutic utility of many polyketides, isolated initially from soil microorganisms, is well established in human and veterinary medicine, the classic example being the erythromycin family of macrolide antibiotics first reported in 1952. Following on from the landmark total synthesis of erythromycin A (Figure 1) completed by the Woodward group in 1981, the goal of assembling such elaborate 3D structures in a more concise fashion has been a key driver for the rapid progress made in new methodology for acyclic stereocontrol.

More recently, the isolation of highly bioactive polyketides from marine organisms, particularly sponges, has generated important lead structures for the development of new anticancer drugs, for example discodermolide [1] and swinholide A [2]. Altogether, this enticing combination of complex structures and multiple stereocentres, with significant biological activity, and, for marine polyketides, extremely limited natural abundance, calls for ever more powerful synthetic strategies and methodology. For the total synthesis of such polyketides, we were attracted to developing experimentally straightforward methodology based on the boron-mediated aldol reactions of ketones [3].

Results

Stereocontrol in the Boron Aldol Reactions of Ketones

Inspired by the pioneering studies on the directed aldol reaction performed in the early 1980s by Evans, Heathcock, Masamune and Mukaiyama, we first concentrated on the development of general procedures for the asymmetric synthesis of β-hydroxyketones, as in **1** and **2** (Figure 2). Initially, the target selected was the characteristic polypropionate sequences, like **3** and **4**, found embedded in many polyketide structures [4A].

Pursuing this strategy, led us to introduce a range of chiral ketones [5, 6], *inter alia* derived from the Roche ester (as for **5**), that undergo *syn* and *anti* aldol reactions with aldehydes *via* their kinetically generated Z- or E-configured boron en-

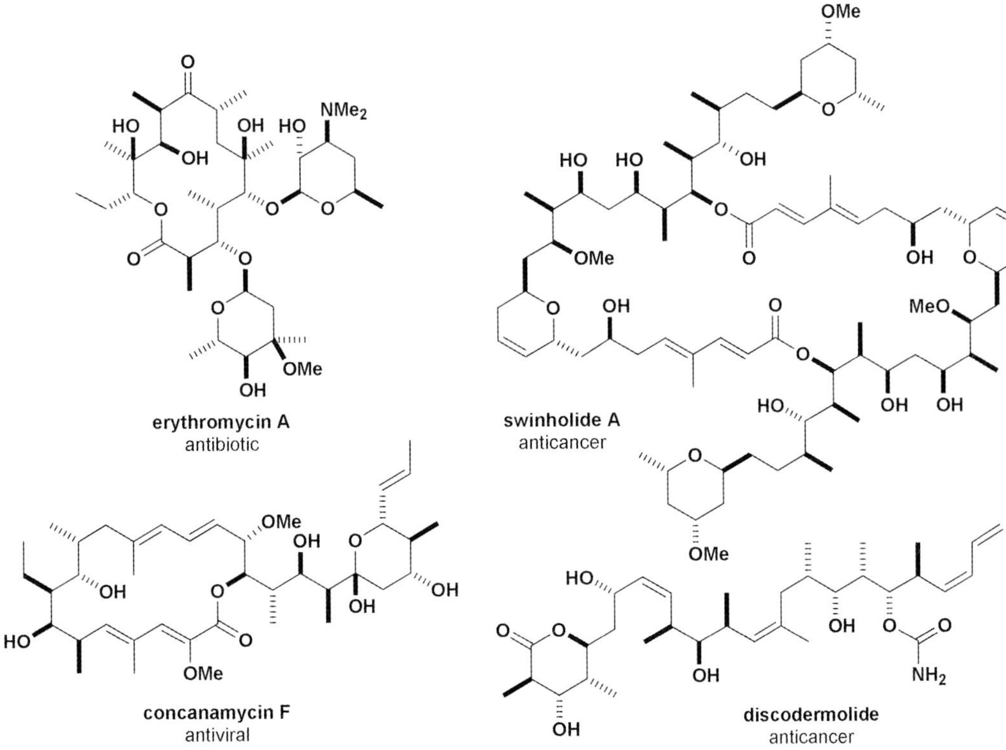

Figure 1 Some representative polyketide natural products synthesised in the Paterson group using aldol reactions of chiral boron enolates as key carbon–carbon bond-forming and stereo-defining steps.

olates (**6–9** and **10–14**, Figure 3) with reliable stereochemical results. Here π-facial discrimination arises from the minimisation of unfavorable steric and electronic interactions in the 6-membered cyclic transition state. Asymmetric aldol additions for simple ethyl and methyl ketones can be performed via their corresponding diisopinocampheyl boron enolates, **15** and **16** respectively, formed by enolisation with Ipc$_2$BX (X = OTf or Cl) prepared from (+)- or (−)-α-pinene [7]. In certain cases (**17–19**), 1,4-, 1,5- and, remarkably, even 1,6-induction is possible, enabling an expedient solution to achieving remote stereocontrol in certain acyclic systems.

Application to the Total Synthesis of Altohyrtin A (Spongistatin 1)

To demonstrate the scope of this aldol methodology, we applied it to the synthesis of altohyrtin A (**20**, Scheme 1), a rare marine macrolide that represents a novel class of tubulin-binding anticancer agent. This endeavor nicely showcased the power of chiral boron enolates for constructing such highly oxygenated and stereochemically elaborate polyketides [8], being used here in 10 separate instances (aldols #1–10, selected examples are shown) involving highly functiona-

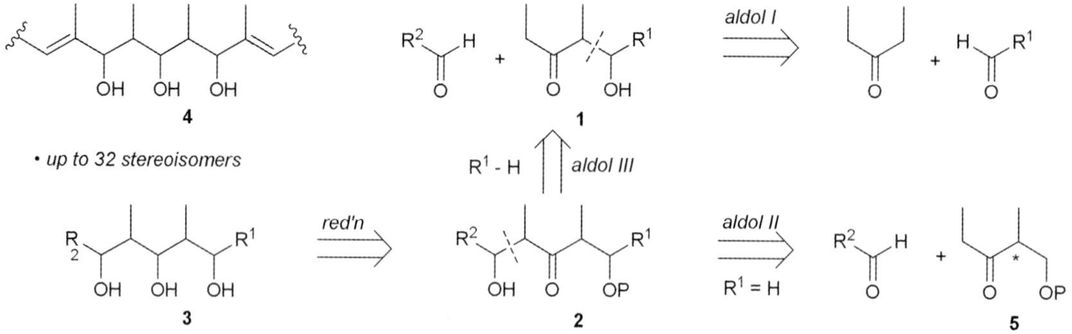

Figure 2 Aldol-based retrosynthetic analysis for a generalised stereopentad sequence leading to the development of reagent and substrate controlled boron aldol reactions of ethyl ketones with aldehydes, enabling selective access to all 32 stereoisomers.

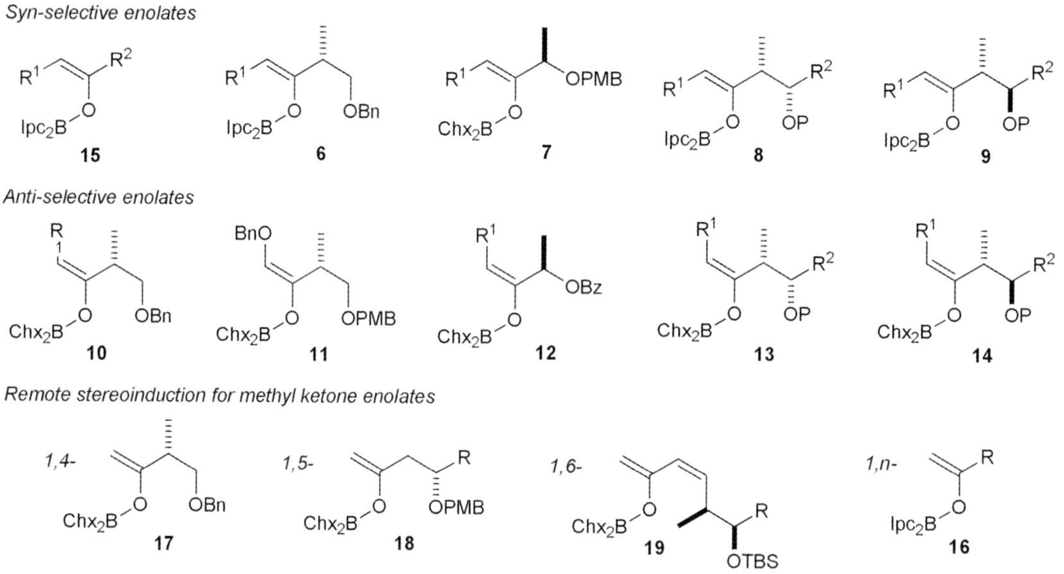

Figure 3 Representative kinetically generated chiral boron enolates obtained from ketones by soft enolization using (+)- and (−)-Ipc$_2$BCl, or Ipc$_2$BOTf, and Chx$_2$BCl in the presence of a tertiary amine base, e.g. Et$_3$N. All of these enolates add to aldehydes with synthetically useful levels of π-facial discrimination through a highly ordered chair (or occasionally boat) transition state.

lised and sensitive intermediates, as well as leading to the discovery of the 1,5-*anti* aldol reaction [9].

Conclusions and Future Perspectives

The boron-mediated aldol reaction has amply proved its versatility in the asymmetric construction of structurally diverse polyketides. For example, enabling the scalable synthesis of useful quantities of both discodermolide (in industry)

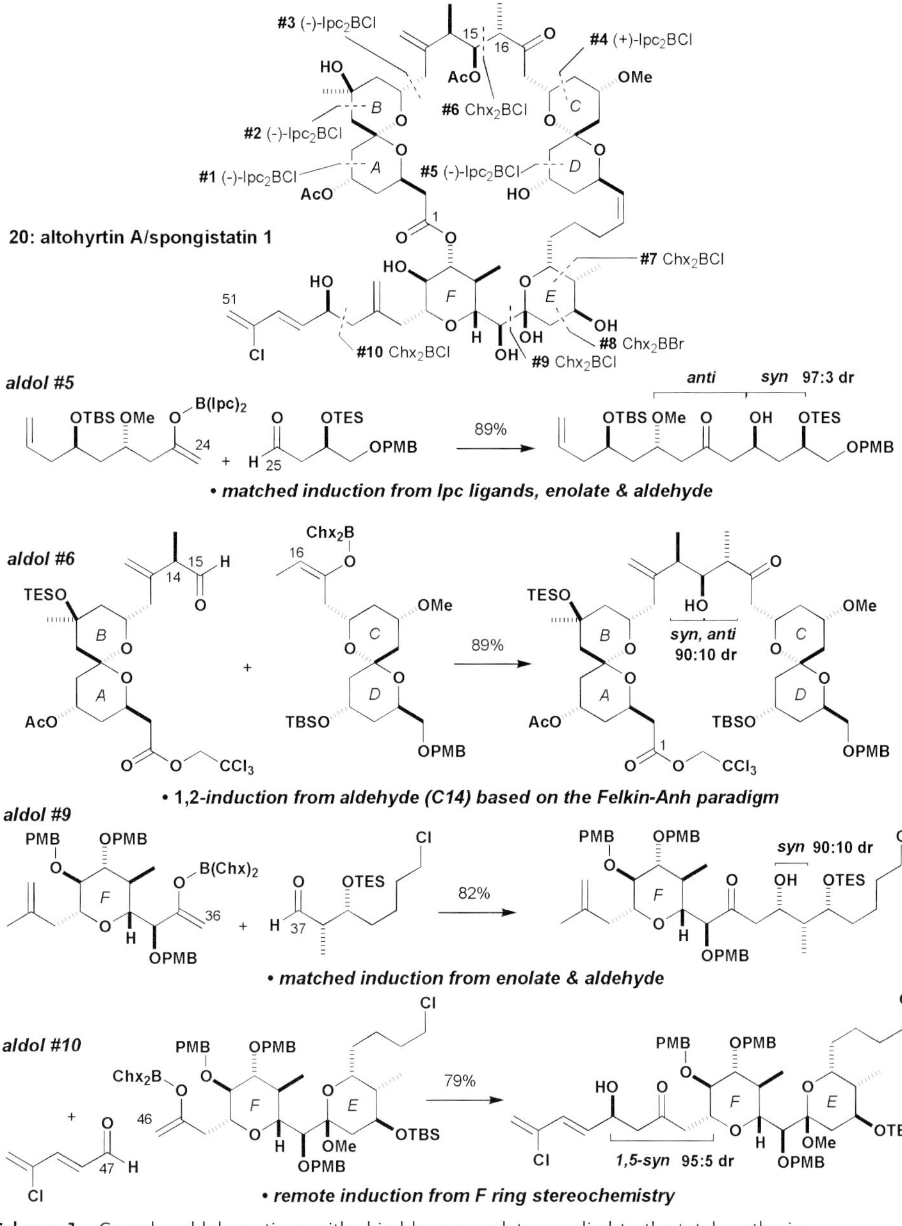

Scheme 1 Complex aldol reactions with chiral boron enolates applied to the total synthesis of altohyrtin A (spongistatin 1). This strategy led to the highly selective introduction of 13 of the 24 stereocentres present.

and altohyrtin A. Over the last 25 years, this powerful carbon–carbon bond-forming reaction has evolved into an established and experimentally straightforward method for achieving useful levels of substrate and reagent based stereocontrol in a multitude of demanding situations. More recently, it has been adapted to solid phase synthesis, with the objective of expanding polyketide diversity by combinatorial chemistry [10]. The development and application of useful new aldol methodology to the total synthesis of important polyketide targets continues to be pursued in our laboratories.

CV of Ian Paterson

Ian Paterson was born in Dundee, Scotland, in 1954. Following his BSc at the University of St Andrews (1976), he obtained his Ph.D. in 1979 from the University of Cambridge working with Ian Fleming. After a NATO postdoctoral fellowship at Columbia University with Gilbert Stork, he took up a lectureship at University College London. In 1983, he moved back to Cambridge, where he is currently Professor of Organic Chemistry and a Professorial Fellow of Jesus College. He was the recipient of the Royal Society of Chemistry's Robert Robinson Lectureship Award (2004), Synthetic Organic Chemistry Award (2001), Bader Prize (1996), Hickinbottom Fellowship (1989) and Meldola Medal (1983). In 2005, he was elected as a Fellow of the Royal Society.

Selected Publications

1. I. Paterson, G. J. Florence, K. Gerlach, J. P. Scott, N. Sereinig, *J. Am. Chem. Soc.* **2001**, *123*, 9535–9544. *A practical synthesis of (+)-discodermolide and analogues: fragment union by complex aldol reactions.*
2. I. Paterson, K.-S. Yeung, R. A. Ward, J. D. Smith, J. G. Cumming, S. Lamboley, *Tetrahedron* **1995**, *51*, 9467–9486. *The total synthesis of swinholide A. Part 4: Synthesis of swinholide A and isoswinholide A from a protected version of the monomeric seco acid, pre-swinholide A.*
3R. C. J. Cowden, I. Paterson, in *Organic Reactions 51*, L. A. Paquette (Ed.), pp. 1–200, Wiley, New York, **1997**. *Asymmetric aldol reactions using boron enolates.*
4A. I. Paterson, *Pure Appl. Chem.* **1992**, *64*, 1821–1830. *New methods and strategies for the stereocontrolled synthesis of polypropionate-derived natural products.*
5. I. Paterson, J. M. Goodman, M. Isaka, *Tetrahedron Lett.* **1989**, *30*, 7121–7124. *Aldol reactions in polypropionate synthesis: high π-face selectivity of enol borinates from α-chiral methyl and ethyl ketones under substrate control.*
6. I. Paterson, D. J. Wallace, S. M. Velazquez, *Tetrahedron Lett.* **1994**, *35*, 9083–9086. *Studies in polypropionate synthesis: high π-face selectivity in syn and anti aldol reactions of chiral boron enolates of lactate-derived ketones.*
7. I. Paterson, J. M. Goodman, M. A, Lister, R. C. Schumann, C. K. McClure, R. D. Norcross, *Tetrahedron* **1990**, *46*, 4663–4684. *Enantio- and diastereoselective aldol reactions of achiral ethyl and methyl ketones with aldehydes: the use of enol diisopinocampheylborinates.*
8. I. Paterson, D.Y.-K. Chen, M. J. Coster, J. L. Acena, J. Bach, D. J. Wallace, *Org. Biomol. Chem.* **2005**, *3*, 2431–2440. *The total synthesis of altohyrtin A/spongistatin 1. Fragment couplings, completion of the synthesis, analogue generation and biological evaluation.*
9. I. Paterson, K. R. Gibson, R. M. Oballa, *Tetrahedron Lett.* **1996**, *37*, 8585–8588. *Remote 1,5-anti stereoinduction in the boron-mediated aldol reactions of β-oxygenated methyl ketones.*
10. I. Paterson, M. Donghi, K. A. Gerlach, *Angew. Chem.* **2000**, *112*, 3453–3457. *A combinatorial approach to polyketide-type libraries via iterative asymmetric aldol reactions performed on solid support.*

Asymmetric Synthesis on the Solid Phase

Torben Leßmann and Herbert Waldmann, Max-Planck-Institut für Molekulare Physiologie, Dortmund, Germany

Introduction

The synthesis of compound collections has gained steadily increasing interest in both industry and academia. In the past, typically, compound libraries were synthesized which contain hardly any stereocenter, and asymmetric synthesis was applied only occasionally in compound library development.

However, newer developments in library design that focus on either natural product guided [1R] or diversity-oriented synthesis [2R] try to access more complex structures. In such endeavors, asymmetric synthesis is needed as an integral method of the library development effort.

Results

Asymmetric Aldol Reactions

Initial studies by Reggelin et al. and Paterson et al. that demonstrated compliance of asymmetric aldol reactions with solid-support-bound substrates were extended to library synthesis by Schreiber et al. [3]. Immobilized furaldehydes **1** (Scheme 1) reacted with the chiral Z-enolate **2** in an Evans *syn*-aldol reaction with a diastereomeric ratio > 20:1. The reaction products **3** were subsequently transformed into several compound classes with different scaffold geometries. Following this route with an extended set of starting materials, a 1260-member library was obtained.

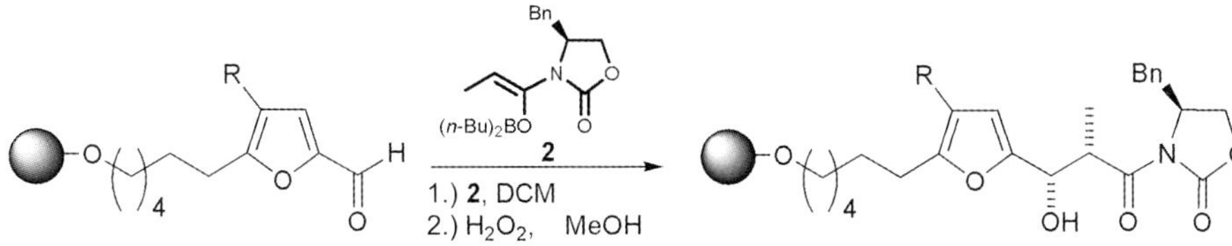

Scheme 1 Asymmetric aldol reactions in diversity-oriented synthesis.

Asymmetric Synthesis – The Essentials.
Edited by Mathias Christmann and Stefan Bräse
Copyright © 2007 WILEY-VCH Verlag GmbH & Co. KGaA, Weinheim
ISBN: 978-3-527-31399-0

Waldmann et al. used chiral boron enolates for the synthesis of a library of 6,6′-spiroketals, which are a substructure of many natural products [4]. The immobilized aldehyde 4 (Scheme 2) and the preformed chiral Z-enolate 5 gave the enantioenriched product 6 after two steps. Stereocontrolled formation of a boron-E-enolate on the solid phase was a prerequisite to control the course of a second, *anti*-selective aldol reaction with different aldehydes 7 and furnished bis-β-hydroxyketones 8. The final products 9 were obtained upon oxidative release from the carrier with very high stereoselectivity.

Based on earlier studies, Paterson et al. published a similar strategy for the synthesis of a fragment of the natural product spongistatin [5].

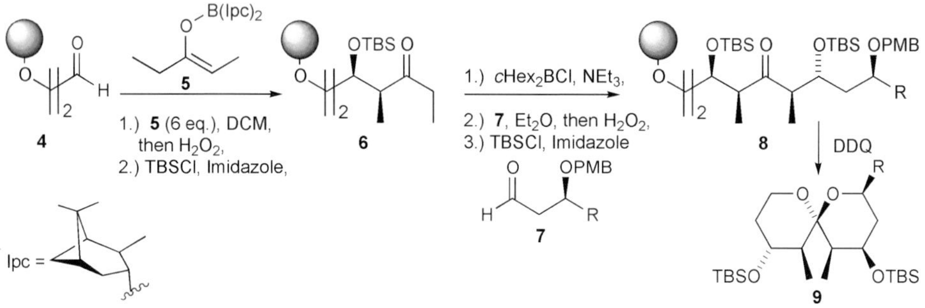

Scheme 2 Asymmetric synthesis of 6,6′-spiroketals.

Asymmetric Cycloadditions

Jiang et al. reported the synthesis of a set of isoxazolines by applying an asymmetric 1,3-dipolar cycloaddition of immobilized nitrile oxides to allyl alcohol under mediation of (–)-diisopropyl tartrate ((–)-DIPT) in 50–70 % yield with up to 95 % enantiomeric excess [6].

Schreiber et al. succeeded in the synthesis of a 4320-member library via an asymmetric inverse-electron-demand hetero-Diels-Alder reaction [7]. In the presence of 20 mol % of chiral Cu(II)-bisoxazoline catalyst 10 (Scheme 3) several immobilized enol ethers 11 underwent enantioselective cycloaddition with ten heterodienes 12, yielding dihydropyrans with enantiomeric excesses ranging from 80 to 98 % ee. Subsequent transformations led to a set of dihydropyrancarboxamides 13.

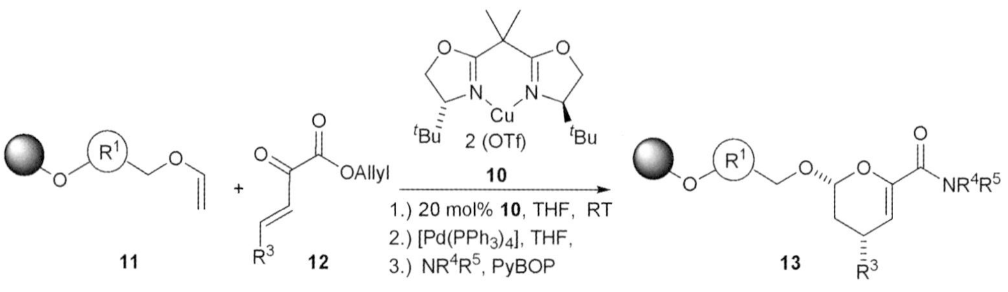

Scheme 3 Asymmetric hetero Diels-Alder reactions on solid-phase bound enol ethers.

In a study to identify biologically active compounds derived from the natural phosphatase inhibitor dysidiolide, Waldmann et al. reported the use of the chiral dienophile **14** (Scheme 4) to enhance the directing force of a resin-bound chiral substrate [8].

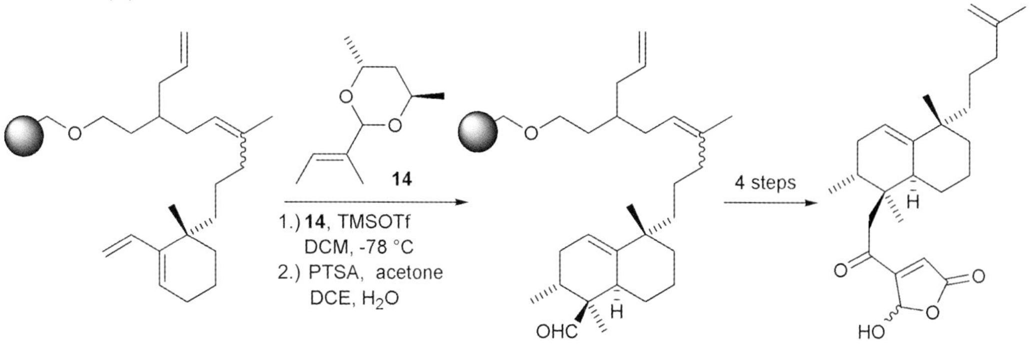

15 **16** 87% of desired isomer **17**

Scheme 4 Asymmetric synthesis of dysidiolide derivatives.

In the crucial Diels-Alder reaction to build up the bicyclic scaffold, diene **15** was treated with the tiglic aldehyde derived acetal **14** to yield compound **16**. The use of the chiral auxiliary raised the ratio of the desired *endo*-isomer from 67 % to 87 % in the isomer mixture of the reaction product.

Asymmetric Epoxide Openings

Jacobsen et al. used the chiral (*salen*)-chromium catalyst **18** (Scheme 5) to promote the ring opening of resin-bound *meso*-epoxides like **19** with $TMSN_3$ with up to 94 % enantiomeric excess in the primary products **20** [9]. Further transformations led to a series of cyclic peptides that contained the pharmacophoric Arg-Gly-Asp (RGD) sequence.

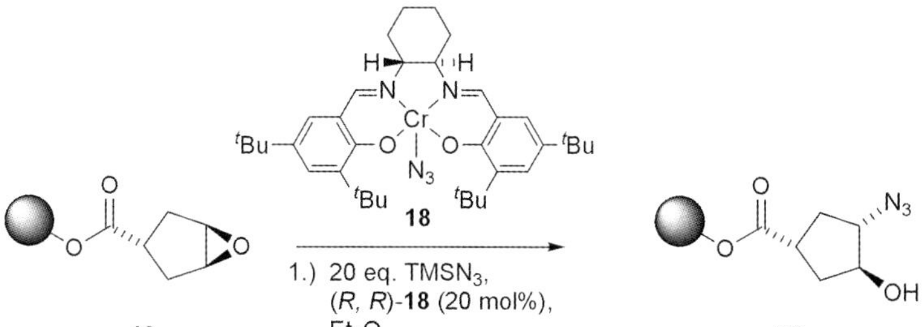

19
6 examples

20
92 %*ee*

Scheme 5 Catalytic asymmetric epoxide openings on solid support.

Other Reactions

Davies et al. reported an asymmetric catalytic cyclopropanation reaction of an immobilized 1,1-diarylethylene with a set of 7 different aryldiazoacetates [10]. In this transformation, 1 mol % of the chiral catalyst $Rh_2(S\text{-}DOSP)_4$ was sufficient to achieve yields of more than 80 % and to induce good diastereo- and enantioselectivities ($E:Z > 3:1$, up to 93 % ee).

Janda's finding that the asymmetric dihydroxylation of immobilized ethyl (E)-cinnamate proceeds with high enantioselectivity (up to 97 % ee) [11] was complemented by a more thorough study by Berkessel et al. [12]. Application to library synthesis was successfully performed by Lee et al. who transformed an immobilized dihydroxylated coumarin derivative (91 % ee) to a set of khellactones [13].

Bräse et al. reported the addition of organozinc compounds to a solid-phase bound aldehyde by using a chiral [2.2]paracyclophane ligand [14].

Conclusion

The examples discussed above demonstrate that asymmetric synthesis on the solid support is possible and comparable to results obtained in solution-phase studies. Solid phase protocols for asymmetric synthesis should be further developed and improved, since they allow the generation of compound libraries with high complexity.

CV of Herbert Waldmann

Herbert Waldmann received his PhD in chemistry from the University of Mainz (Germany) with H. Kunz. After postdoctoral studies with George Whitesides and habilitation (University of Mainz) he was appointed Associate Professor at the University of Bonn, from where he moved to the University of Karlsruhe as Full Professor in Organic Chemistry in 1993. Since 1999 he has held a position as Director of the Department of Chemical Biology at the Max-Planck-Institute of Molecular Physiology in Dortmund and a Professorship in Biochemistry at the University of Dortmund. He received the Friedrich Weygand Award for the advancement of peptide chemistry, the Carl Duisberg Award, the Otto-Bayer-Award and the Steinhofer Award. He is a member of the "Deutsche Akademie der Naturforscher Leopoldina" and a fellow of the Royal Society of Chemistry.

Selected Publications

1R. R. Breinbauer, I. R. Vetter, H. Waldmann, Angew. Chem. Int. Ed. **2002**, 41, 2878–2890. *From Protein Domains to Drug Candidates – Natural Products as Guiding Principles in the Design and Synthesis of Compound Libraries.*

2R. M. D. Burke, S. L. Schreiber, Angew. Chem. Int. Ed. **2004**, 43, 46–58. *A Planning Strategy for Diversity-Oriented Synthesis.*

3. M. D. Burke, E. M. Berger, S. L. Schreiber, Science **2003**, 302, 613–618. *Genrating Diverse Skeletons of Small Molecules Combinatorially.*

4. O. Barun, S. Sommer, H. Waldmann, Angew. Chem Int. Ed. **2004**, 43, 3195–3199. *Asymmetric Synthesis of 6,6-Spiroketals.*

5. I. Paterson, D. Gottschling, D. Menche, Chem. Commun. **2005**, 3568–3570. *Towards the Combinatorial Synthesis of Spongistatin Fragment Libraries by Using Asymmetric Aldol Reactions on Solid Support.*

6. N. Zou, B. Jiang, J. Comb. Chem. **2000**, 2, 6–7. *Solid Phase Asymmetric Synthesis of Isoxazolines.*

7. R. A. Stavenger, S. L. Schreiber, Angew. Chem. Int. Ed. **2001**, 40, 3417–3421. *Asymmetric Catalysis in Diversity-Oriented Organic Synthesis: Enantioselective Synthesis of 43230 Encoded and Spatially Segregated Dihydropyrancarboxamides.*

8. D. Brohm, N. Philippe, S. Metzger, A. Bhargava, O. Müller, F. Lieb, H. Waldmann, *J. Am. Chem. Soc.* **2002**, *124*, 13172–13178. Solid-Phase Synthesis of Dysidiolide-Derived Protein Phosphatase Inhibitors.
9. D. A. Annis, O. Helluin, E. N. Jacobsen, *Angew. Chem. Int. Ed.* **1998**, *37*, 1907–1909. Stereochemistry as a Diversity Element: Solid-Phase Synthesis of Cyclic RGD Peptide Derivatives by Asymmetric Catalysis.
10. T. Nagashima, H. M. L. Davies, *J. Am. Chem. Soc.* **2001**, *123*, 2695–2696. Catalytic Asymmetric Solid-Phase Ccyclopropanation.
11. H. Han, K. D. Janda, *Angew. Chem. Int. Ed.* **1997**, *36*, 1731–1733. Multipolymer-Supported Substrate and Ligand Approach to the Sharpless Asymmetric Dihydroxylation.
12. R. Riedl, R. Tappe, A. Berkessel, *J. Am. Chem. Soc.* **1998**, *120*, 8994–9000. Probing the Scope of the Asymmetric Dihydroxylation of Polymer-Bound Olefins. Monitoring by HRMAS NMR allows for Reaction Control and On-Bead Measurement of Enantiomeric Excess.
13. Y. Xia, Z.-Y. Yang, A. Brossi, K.H. Lee, *Org. Lett.* **1999**, *1*, 2113–2115. Asymmetric Solid-Phase Synthesis of (3′R, 4′R)-Di-O-cis-acyl 3-Carboxyl Khellactones.
14. K. Knepper, R. E. Ziegert, S. Bräse, *Tetrahedron* **2004**, *60*, 8591–8603. Solid-Phase Synthesis of Isoindolinones and Naturally-Occuring Benzobutyrolactones (Phthalides) Using a Cyclative-Cleavage Approach.

Part V

Asymmetric Synthesis in Industry

Herbert Hugl, Lanxess Deutschland GmbH, Germany

Background

The economic importance of chiral compounds is significant and increasing. A majority of pharmaceuticals, especially of those registered within the last ten years, many flavors, fragrances, food and feed additives, as well as some important agrochemicals, are produced in enantio-pure or enantio-enriched forms.

For their production chemical industry uses a well established portfolio of methods:
- asymmetric chemical synthesis and enzymatic synthesis
- chiral building blocks either from naturally occurring chiral pool compounds or synthesized

and, still very important
- the separation of racemates by chemical, enzymatic or physical (SMB) methods.

Industrial Production of Chiral Compounds

For many of the large scale productions enzymatic processes are in use [1–3]. L-Aspartic acid, the artificial sweetener aspartame, riboflavin, niacinamide and several D- and L-amino acids are produced on a greater than 1000 t year^{-1} scale. These compounds, especially the amino acids, also contribute to the portfolio of chiral building blocks for the synthesis of pharmaceuticals and other follow up products. Enzymatic processes often need several years for development, process engineering and the construction of dedicated plants or suitable equipment. If time is a critical factor for the process development of, for example, a new pharmaceutical compound, chemical syntheses is often preferred.

During the last 20 years a broad variety of well performing methods for enantioselective chemical synthesis has been developed. By far the most important is asymmetric hydrogenation. This method enables the introduction of chiral centers into a molecule using homogeneous metal complexes with chiral ligands as catalysts. Among the industrially most important substrates for asymmetric hydrogenation are: enamides, itaconic acid derivatives, allylic alcohols, α,β-unsaturated acids, dehydro-amino acid derivatives, ketones, β-ketoesters, and various C=C and C=N systems [4].

Pioneers of this technology were Monsanto with the L-Dopa process, described in this chapter and Takasago using Noyori's Binap-ligand for several large scale applications: intermediates for vitamin E, (S)-oxfloxacin and carbapenem [5]. With greater than 10000 t year^{-1} Syngenta's (Solvias) intermediate for the (S)-metolachlor herbicide is currently the largest industrial application of the technology [6]. β-Ketoesters as chiral building blocks for pharmaceuticals have been enantioselectivity reduced using Ru/Binap as catalyst by NSC Technologies [4] and with Ru/Cl-MeO-BIPHEP by Lanxess [7, 8] on a greater than 70 t production scale. The so-called "statin side chain" of the HMG-CoA reductase inhibitors, a 3,5-dihydroxy hexanoic acid was, and is still, another challenge for enantioselective synthesis. Job and Stolle have prepared a brief summary for this Part.

Firmenich has set up a multi-ton production process for (+)-cis methyl dihydrojasmonate, an important perfume ingredient [9].

As reported by Knowles in this Part, Takasago introduced a large scale, greater than 1000 t year^{-1}, process for the asymmetric isomerization of an L-menthol intermediate using Rh (Binap)$_2$ in the 1980s. The same process is used for the production of citronellal, D- and L-citronellol and methoprene juvenile hormone.

Following the invention of his scientifically and industrially very successful Binap-catalysts, Noyori also discovered Ru/diphosphine/diamine systems (shown in Scheme 1) and Ru-based transfer hydrogenation catalysts for hydrogenation of aromatic ketones [10].

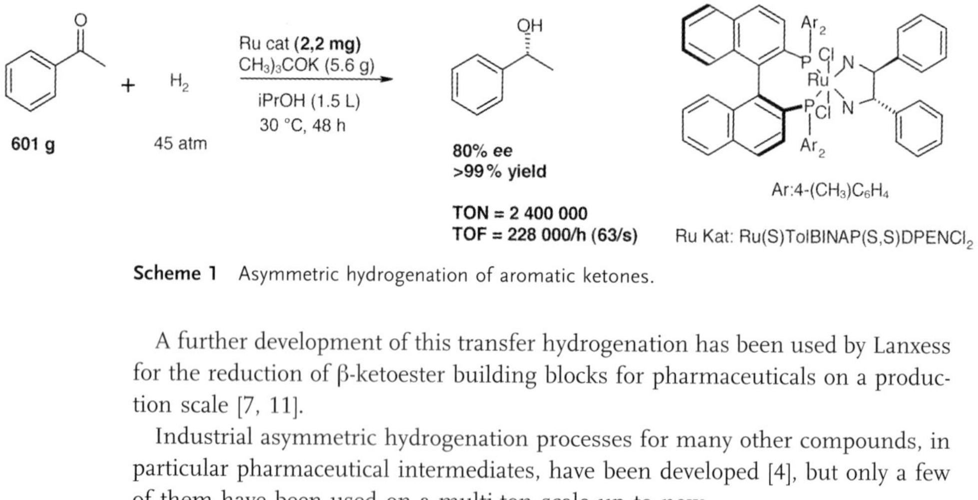

Scheme 1 Asymmetric hydrogenation of aromatic ketones.

A further development of this transfer hydrogenation has been used by Lanxess for the reduction of β-ketoester building blocks for pharmaceuticals on a production scale [7, 11].

Industrial asymmetric hydrogenation processes for many other compounds, in particular pharmaceutical intermediates, have been developed [4], but only a few of them have been used on a multi-ton scale up to now.

Oxidation processes are very often used for the production of bulk chemicals but even for achiral or racemic compounds only a few methods are in use in the area of fine chemicals and, in particular, for pharmaceutical intermediates. Nevertheless the PPG-Sipsy epoxidation using Ti-disopropyltartrate as catalyst and tert-BuOOH as oxidant, the epoxidation of indene resulting in an intermediate for Crixivan® by Rhodia/Chirex using Mn-salen as catalyst and 3-phenyl-propyl-pyridine N-Oxide as oxidant [4] should be mentioned. Furthermore Jacobsen's hydrolytic kinetic resolution, using Co-Salen catalysts (Scheme 2) is used by Rhodia for the production of several chiral epoxides on the ton scale [12].

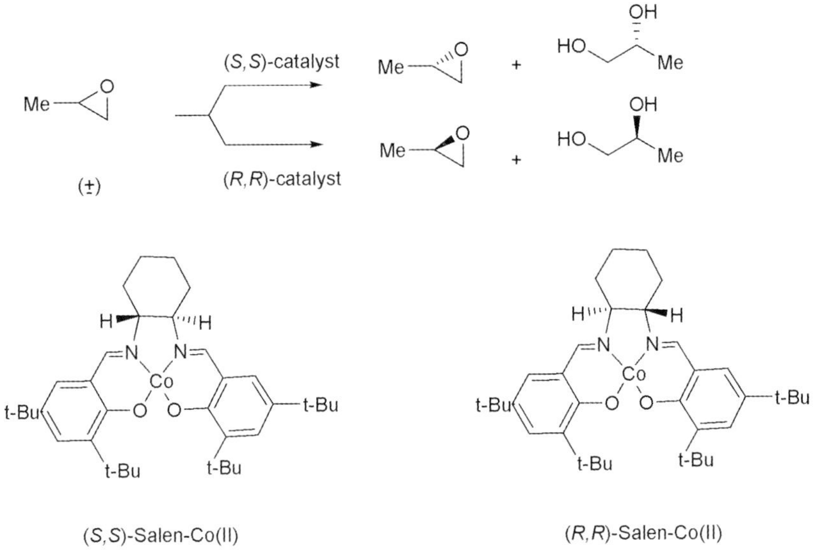

Scheme 2 Jacobsen's hydrolytic kinetic resolution.

These expoxides are used as building blocks for various chiral molecules. In addition to these epoxidations the Sumitomo cyclopropanation using a Cu-complex as catalyst and diazoaceticacid ethylester as oxidant [13], resulting in an intermediate for Cilastatin® marked a significant step of industrial development. To my knowledge asymmetric dihydroxylations have not been performed on a production scale (> 1 t) up to now.

Presumably a majority of the chiral active pharmaceutical ingredients (APIs) synthesized on an annual production scale of 1–100 t are still produced by separation of the racemate. In many cases this has been the only opportunity to get the enantiomerically pure compound within a short time for the process development. In addition it should be mentioned that patent protection of innovative methods for enantioselective synthesis often acts as an obstacle for the industrial use of the particular method. Time-consuming licensing negotiations between the owner of the method and potential applicants could extend the time to market for a new product. For the separation of racemates following traditional chemical synthesis in an increasing number of cases enzymes, often lipases, are used. Also stimulated moving bed chromatography (SMB) enables short development times and excellent optical purities.

Outlook

Enantioselective synthesis went through two decades of very intensive development of methods. Three pioneers of this development, R. Noyori, K.B. Sharpless and W. Knowles were awarded the Nobel prize for their outstanding contributions. This portfolio of new methods will enable chemical and pharmaceutical industries to produce chiral compounds routinely generating less waste at reasonable costs.

CV of Herbert Hugl

Herbert Hugl was born in Vienna, Austria, in 1945. He studied Chemistry at the University of Vienna and earned his Ph.D. working on steroid synthesis with Erich Zbiral in 1973. Thereafter he joined the Corporate Research of Bayer AG in Leverkusen. Main working areas over the years have been dyestuffs, medical plastics, diagnostic tests and kits, process research and catalysis. His research efforts led to a dozen new commercial products. Between 1997 and 2002 he headed the Synthesis Department of Bayer's Corporate Research. Since 2002 he worked with the Fine Chemicals Business Unit of Bayer Chemicals AG, later Lanxess Deutschland GmbH. Herbert Hugl is lecturer at the RWTH-Aachen and Honorary Professor of that University since 2004.

Selected Publications

1. A. Schmid, F. Hollmann, J. B. Park, B. Bühler, *Curr. Opin. Biotechnol.* **2002**, *13*, 359–366. *The use of enzymes in the chemical industry in Europe*
2. J. Ogawa, S. Shimizu, *Curr. Opin. Biotechnol.* **2002**, *13*, 367–375. *Industrial microbial enzymes: their discovery by screening and use in large-scale production of useful chemicals in Japan.*
3. A. Liese, K. Seelbach, C. Wandrey, *Industrial Biotransformations*, Wiley-VCH, Weinheim, **2000**.
4. H. U. Blaser, F. Spindler, M. Studer, *Appl. Catal. A* **2001**, *221*, 119–143. *Enantioselective catalysis in fine chemicals production.*
5. H. Kumobayashi, *Recl. Trav. Chim. Pays-Bas* **1996**, *115*, 201–210. *Industrial application of asymmetric reactions catalyzed by BINAP-metal complexes.*
6. H. U. Blaser, H. P. Buser, K. Coers, R. Hanreich, H. P. Jalett, E. Jelsch, B. Pugin, H. D. Schneider, F. Spindler, A. Wegmann, *Chimia* **1999**, *53*, 275–280. *The chiral switch of metolachlor. The development of a large-scale enantioselective catalytic process.*
7. F. Rampf, *Processing Chiral USA 2004*, Scientific Update, Mayfeld, UK.
8. A. Gerlach, U. Scholz, *Speciality Chemicals Mag.*, **2004**, *24*(4), 37–38. *Industrial application of chiral phosphines*
9. D. A. Dobbs, K. P. M. Vanhesche, E. Brazi, V. Rautenstrauch, J.-Y. Lenoir, J.-P. Genet, J. Wiles, S. H. Bergens, *Angew. Chem. Int. Ed.* **2000**, *39*, 1992–1995. *Industrial synthesis of (+)-cis-methyl dihydrojasmonate by enantioselective catalytic hydrogenation; identification of the precatalyst [Ru((-)-Me-DuPHOS)(H)-(h6-1,3,5-cyclooctatriene)](BF_4).*
10. R. Noyori, M. Kitamura, T. Ohkuma, *Proc. Natl. Acad. Sci.* **2004**, *101*, 5356–5362. *Toward efficient asymmetric hydrogenation: architectural and functional engineering of chiral molecular catalysts.*
11. B. E. Bosch, M. Eckert, H.-C. Militzer, C. Dreisbach, EP 1340746 (Lanxess). *Process for the preparation of stereoisomer-enriched 3-heteroaryl-3-hydroxypropanoates via transfer hydrogenation of 3-heteroaryl-3-oxopropanoates using ruthenium catalysts, formic acid, and amines.*

12. E.N. Jacobsen, M. H. Wu, in *Comprehensive Asymmetric Catalysis*, E. N. Jacobsen, H. Yamamoto, A. Pfaltz (Eds.), Springer, Berlin, **1999**, pp. 1309–1326. *Ring opening of epoxides and related reactions.*

13. T. Aratani, *Pure Appl. Chem.* **1985**, *57*, 1839–1844. *Catalytic asymmetric synthesis of cyclopropanecarboxylic acids: an application of chiral copper carbenoid reaction.*

Industrial Application of Enantioselective Catalysis

Hans-Ulrich Blaser, Solvias AG, Switzerland

The Potential of Enantioselective Catalysis

Over the years, homogeneous metal complexes with chiral, usually bidentate ligands with a chiral backbone carrying two coordinating heteroatoms have been shown to be the most useful asymmetric catalysts. For noble metal complexes, the preferred coordinating atoms are tertiary P or N, for metals such as Ti, B, Zn, Co, Mn or Cu ligands with O or N atoms are favored. This methodology has received its due recognition with the 2001 Nobel Prize to W. S. Knowles, R. Noyori, and K. B. Sharpless [1]. Very considerable efforts, both in academia and in industry, have led to a wide variety of catalytic transformations with *ee*s often reaching >99% [2].

Despite this spectacular scientific progress with literally hundreds of catalytic transformations with very high enantioselectivity and the recognition of the importance of enantioselective catalysis, a recent survey has revealed that relatively few enantioselective catalytic reactions are used on an industrial scale today (see Table 1) [3]. An analysis of the processes shows that hydrogenation of C=C and C=O is by far the predominant transformation applied for industrial processes. If one examines the structures of the starting materials, many of these compounds are complex and multifunctional (for selected examples see Schemes 1–8), i.e., the successful catalytic systems must be not only enantioselective but also tolerate many functional groups.

Hurdles on the Way to an Industrial Process

A major reason for the relative scarcity of industrial processes is that large scale applications of enantioselective catalysts present some very special challenges and problems [4]. Whether a synthetic route containing an enantioselective catalytic step can be considered for a particular product is usually determined by the answer to two questions:
- Can the costs for the overall manufacturing process compete with alternative routes?
- Can the catalytic step be developed in the given time and cost frame?

Asymmetric Synthesis – The Essentials.
Edited by Mathias Christmann and Stefan Bräse
Copyright © 2007 WILEY-VCH Verlag GmbH & Co. KGaA, Weinheim
ISBN: 978-3-527-31399-0

Table 1 Statistics for the various types of industrial processes [3]

Transformation	Production >5 t y⁻¹	Production <5 t y⁻¹	Pilot >50 kg	Pilot <50 kg	Bench scale
Hydrogenation of enamides	1	1	2	6	4
Hydrogenation of C=C-COOR and C=C–CH–OH	2	0	3	4	6
Hydrogenation of other C=C systems	1	0	1	1	2
Hydrogenation of α and β functionalized ketones	2	3	3	2	4
Hydrogenation / reduction of other keto groups	0	0	2	2	4
Hydrogenation of C=N	1	0	1	0	0
Dihydroxylation of C=C	0	1	0	0	4
Epoxidation of C=C, oxidation of sulfide	2	2	1	0	2
Isomerization, epoxide opening, addition reactions	2	4	2	0	1
Total	11	11	15	15	27

In our experience, the following critical factors determine the technical feasibility of an enantioselective process step:

Catalyst Performance
The enantioselectivity expressed as enantiomeric excess should be >99% for pharmaceuticals if no purification is possible. This case is quite rare and ees >90% are often acceptable. Functional group tolerance will be very important for multifunctional substrates. The catalyst productivity, given as turnover number (ton, mol product/mol catalyst), determines catalyst costs. For hydrogenation reactions the ton ought to be >1000 for high value and >50 000 for large scale and/or less expensive products (catalyst re-use increases the productivity). The catalyst activity given as average turnover frequency (tof = mol product/mol catalyst/reaction time, h⁻¹), affects the production capacity. For hydrogenations, tofs ought to be >500 h⁻¹ for small and >10 000 h⁻¹ for large scale products. Due to lower catalyst costs and often higher added values, lower ton and tof values are acceptable for enantioselective oxidation and C–C bond-forming reactions.

Availability and Cost of the Catalyst
Chiral ligands and many metal precursors are expensive and/or not easily available. Typical costs for chiral diphosphines are 100–500 $ per gram for laboratory quantities and 2000–100000 $ per kilogram on a larger scale. Chiral ligands used for early transition such as salen or amino alcohols metals are usually much cheaper. At this time, only selected chiral ligands (mostly diphosphines) are available commercially. If ligands are patent protected, licensing models must be available which are easy to handle and not prohibitive.

Development Time

When developing a process for a new chemical entity (NCE) in the pharmaceutical or agrochemical industry, time restraints can be severe. For so-called second generation processes, e.g., for generic pharmaceuticals or the manufacture of other fine chemicals the time factor is usually not so important.

Selected Milestones in Industrial Application

Schemes 1–8 show selected industrial processes which were or still are used for the manufacture of chiral intermediates and which are considered to be significant either for historical reasons or due to their commercial importance. The selection is of course somewhat subjective, the reasons for their selection and some further information are given in the captions. In particular, the selected processes do not reflect the fact that the vast majority of the processes which are in the development pipeline are enantioselective hydrogenation of functionalized C=C systems. For all processes, the following information is provided: catalytic system (metal and chiral ligand), reaction conditions, ee, tonnage and tof. More details as well as information on all further processes included in Table 1 can be found in Ref. [3] and in a recently published monograph of case histories which also includes biocatalytic processes using enzymes or whole cells [5].

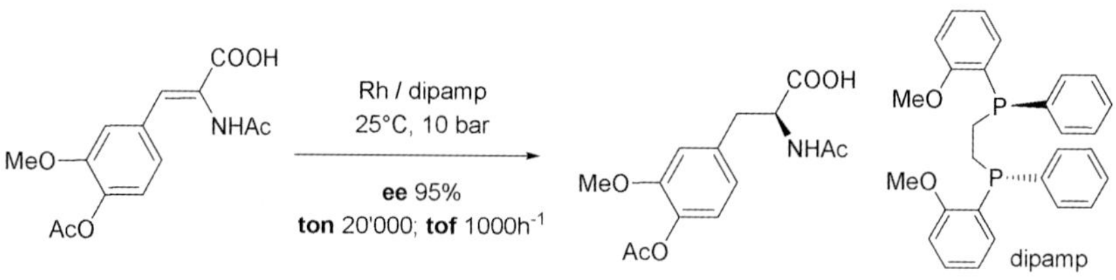

Scheme 1 Intermediate for L-dopa (Parkinson disease); scale ca. 1 t y^{-1}. First commercial process using a chiral chemocatalyst. Developed by Knowles at Monsanto. Discontinued.

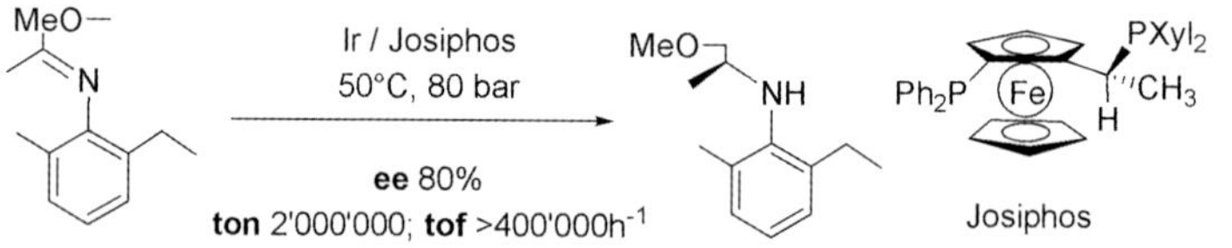

Scheme 2 Intermediate for (S)-metolachlor (herbicide). Produced since 1996, with a volume of >10 000 t y^{-1} the largest commercial process and with exceptionally high tonnage. Developed by Ciba-Geigy (now Syngenta/Solvias).

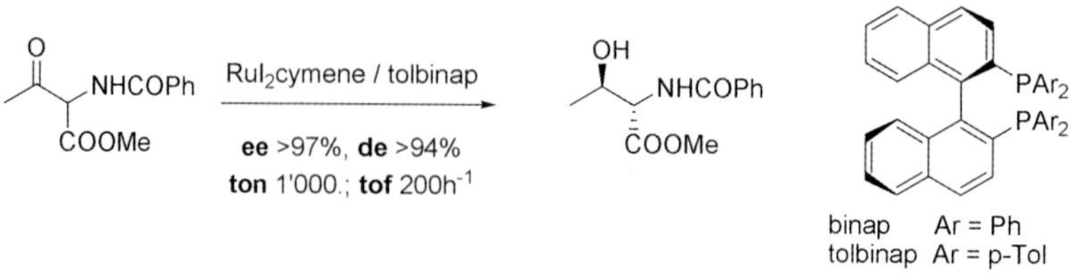

Scheme 3 Intermediate for carbapenem (antibiotic); scale 50–120 t y^{-1}. The first process involving dynamic kinetic resolution. Developed by Takasago.

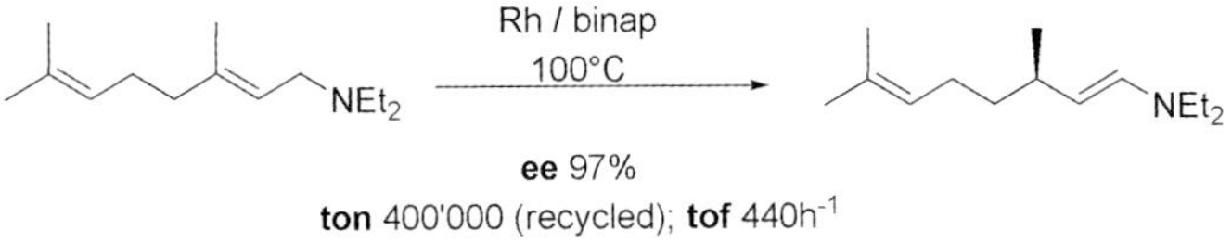

Scheme 4 Intermediate for L-menthol (>1000–t y^{-1}); hydroxy-dihydro-citronellal (40 t y^{-1}); D- and L- citronellol (20 t y^{-1}); methoprene (juvenile hormone) (20 t y^{-1}). The second largest enantioselective process and the first asymmetric isomerization. Developed by Takasago.

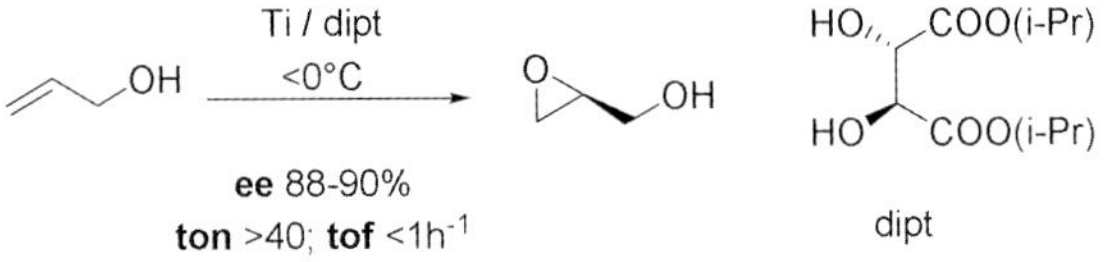

Scheme 5 Chiral building block for various applications, scale multi- t y^{-1}. Developed by Arco and operated by PPG-Sipsy. The largest application of the Sharpless epoxidation catalyst. Discontinued.

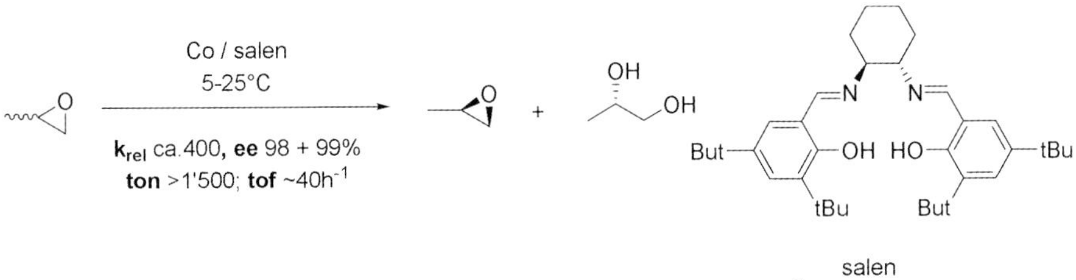

Scheme 6 Chiral building block for various applications, scale multi- t y^{-1}. The first application of the Jacobsen epoxide ring opening catalyst. Developed by Chirex.

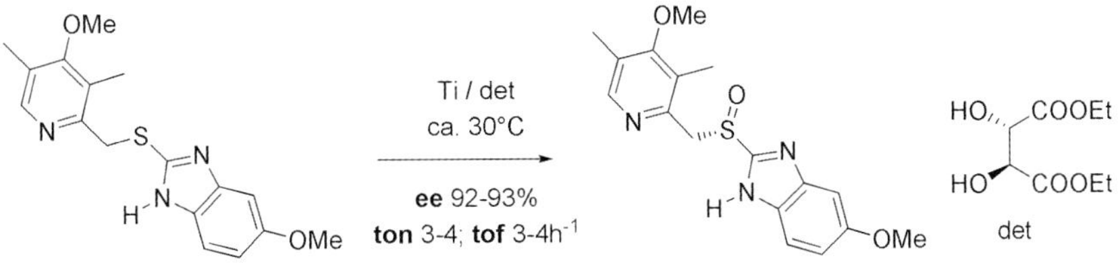

Scheme 7 Intermediate for Esomeprazole (anti-ulcer), multi-ton scale. The first large scale sulfide oxidation. Developed by AstraZeneca.

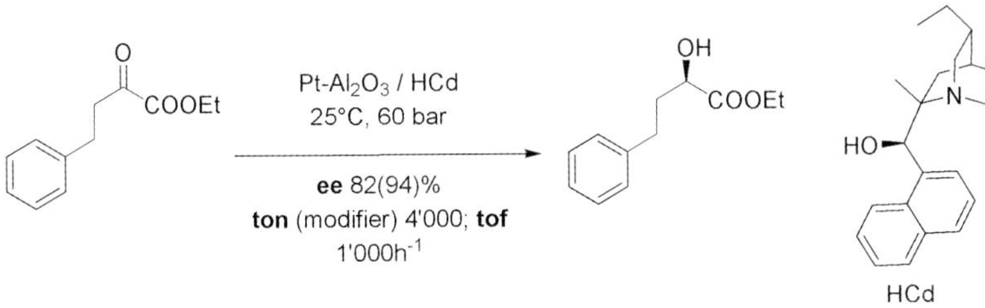

Scheme 8 Intermediate for various ACE inhibitors, produced on a multi-ton scale by Ciba-Geigy. The first application of a chirally modified heterogeneous catalyst. Discontinued.

Conclusions and Future Perspectives

In my personal opinion, enantioselective catalysis has not yet attained its appropriate position in the production of fine chemicals. By "appropriate position" I mean in accord with its potential for economically as well as ecologically superior production processes. Even if we acknowledge that it takes time for a new technology to be adapted by the notoriously conservative production managers (conservative for good reasons!) there are additional hurdles responsible for this unsatisfactory situation. Some of these hurdles are technical, others may be more psychological. However, much is moving in this exciting field of chemical technology. I am convinced that the Nobel Prizes for Knowles, Noyori and Sharpless has led to a better visibility of enantioselective catalysis and will give new impetus to its application to fine chemicals production.

CV of Hans-Ulrich Blaser

Hans-Ulrich Blaser carried out his doctoral research with A. Eschenmoser at the ETH Zürich, where he received a Ph.D. degree in 1971. Between 1971 and 1975 he held postdoctoral positions at the University of Chicago (J. Halpern), Harvard University (J.A. Osborn), and Monsanto (Zürich). During 20 years at Ciba-Geigy (1976–1996) he gained practical experience in R&D in the fine chemicals and pharmaceutical industry, which continued at Novartis (1996–1999) and at Solvias where he presently is chief technology officer. His main interest is selective catalysis with emphasis on enantioselective catalysts. During his industrial carrier he has developed and implemented numerous catalytic routes for agrochemicals, pharmaceuticals and fine chemicals, both as project leader and section head.

References

1. W. S. Knowles, R. Noyori, K. B. Sharpless, Nobel Lectures, *Angew. Chem. Int. Ed.* **2002**, *41*, 1998–2022.
2. H. Brunner, W. Zettlmeier, *Handbook of Enantioselective Catalysis*, VCH, Weinheim, **1993**; E.N. Jacobsen, A. Pfaltz, H. Yamamoto (Eds.) *Comprehensive Asymmetric Catalysis*, Springer, Berlin, **1999**.
3. H. U. Blaser, F. Spindler, M. Studer, *Appl. Catal. A: Gen.* **2001**, *221*, 119. *Enantioselective Catalysis in Fine Chemicals Production*.
4. H. U. Blaser, B. Pugin, F. Spindler, *J. Mol. Catal. A: Chem.* **2005**, *231*, 1. *Progress in Enantioselective Catalysis Assessed from an Industrial Point of View*.
5. H. U. Blaser, E. Schmidt (Eds.), *Large Scale Asymmetric Catalysis*, Wiley-VCH, Weinheim, 2003.

Crystallization-Induced Diastereoselection for the Synthesis of Aprepitant

J. Brands and P. Pye, Merck & Co, USA, and K. Rossen, Sanofi-Aventis, Germany

Background

The development of organic synthesis, including the discovery of truly useful asymmetric catalysis, has been driven largely by the challenges posed by the complexity of natural products and the desire for the development of efficient methods for their preparation. While the progress over the last 50 years has been astounding, the challenge for the large scale synthesis of increasingly complex pharmaceuticals is equally daunting. As an example of one such structure, Merck & Co has discovered and commercialized aprepitant (**1**), a potent human NK-1 receptor antagonist, which is the active ingredient in Emend™, a novel medication for the treatment of chemotherapy-induced nausea and vomiting (CINV). The introduction of fast and practical laboratory chromatography techniques has had a great impact on the practical mode of operation for the organic chemist: the classical purification by crystallization or distillation has been largely supplanted by the routine chromatography of crude reaction products. While chromatography is rapid and can be automated, crystallizations offer opportunities far beyond simple isolations, such as the crystallization of a single isomer from a solution of equilibrating diastereomers. The latter provides a powerful tool in practical asymmetric synthesis. The design and execution of these crystallization-induced diastereoselective transformations frequently involves the attachment – and subsequent removal – of a chiral auxiliary. Useful as this approach may be, it should also be possible to make use of a crystallization-induced diastereoselective transformation in which the inducing stereocenter is an integral part of the desired target and which thus does not need to be removed at the end of the transformation. Furthermore, additional stereocenters could then be set up starting from the existing stereocenter. Interestingly, two Merck approaches towards aprepitant owe their efficiency to two very different crystallization-driven diastereoselective transformations.

1, Aprepitant (Emend™)

Asymmetric Synthesis – The Essentials.
Edited by Mathias Christmann and Stefan Bräse
Copyright © 2007 WILEY-VCH Verlag GmbH & Co. KGaA, Weinheim
ISBN: 978-3-527-31399-0

Results

Bicyclic Acetal Approach

The key synthetic challenge in synthesizing aprepitant is to control the assembly of the two adjacent *cis*-stereocenters on the morpholine core. The first approach by Merck [1] targets *cis*-enolether **5**, whose diastereoselective hydrogenation was known [2] (Scheme 1). The synthesis relies upon the selective formation of bicyclic acetal **3** from *trans*-lactol **2** followed by quaternization of **3** with benzyl iodide. Fragmentation of the bicyclic core of **4** through a highly regioselective Hofmann elimination then affords *cis*-enolether **5**. As all these transformations work well, the synthetic challenge amounts to the preparation of *trans*-lactol **2**.

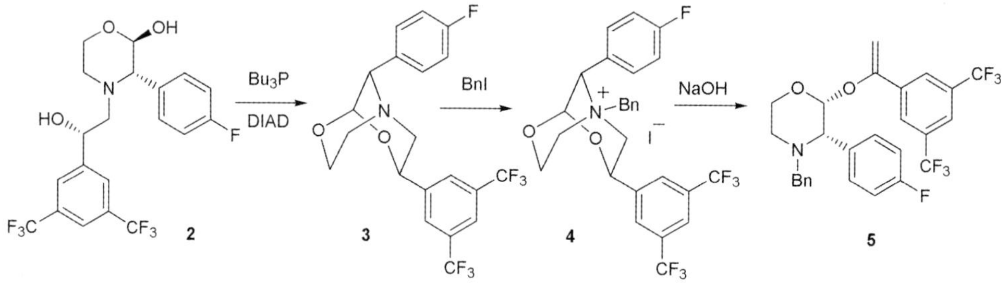

Scheme 1 Bicyclic acetal approach to aprepitant.

Lactol **2** is assembled in a novel, multi-component condensation between glyoxal, 4-fluorophenylboronic acid and aminodiol **6** which yields a complex mixture of isomers (Scheme 2). On closer analysis, all these species can be made to interconvert quite readily, presumably through the open aldehyde, so that the same equilibrium mixture is obtained from each of the isolated, individual compounds. In the racemic series, inducing **2** to crystallize from the reaction mixture via seeding funnels the complex mixture into pure *rac*-**2** in 65 % isolated yield!

Crystal properties of enantiomerically pure compounds are often different from those of the racemate. This frequently quite inconsequential fact becomes a key issue for any crystallization-induced asymmetric transformation. Indeed, the analogous crystallization-induced process with enantiomerically pure **6** leads to the isolation of a minor constituent in the reaction mixture: *cis*-lactol **9** crystallizes in 86 % yield! While it is remarkable that a component that constitutes < 2 % in the reaction mixture can be isolated essentially pure in excellent yield, the *cis*-lactol **9** is not directly usable in the synthesis – clearly the enantiomerically pure material and not the racemate is required and this detracts from the attractiveness of this approach.

While this example nicely demonstrates the power of crystallization-induced asymmetric transformations, it also shows its pitfall: crystal properties are a physical property that cannot be easily influenced or predicted. Thus, a synthesis will only become viable, if the required diastereomer possesses the energetically most favorable crystal packing. As this property is very substrate dependent, success is difficult to design.

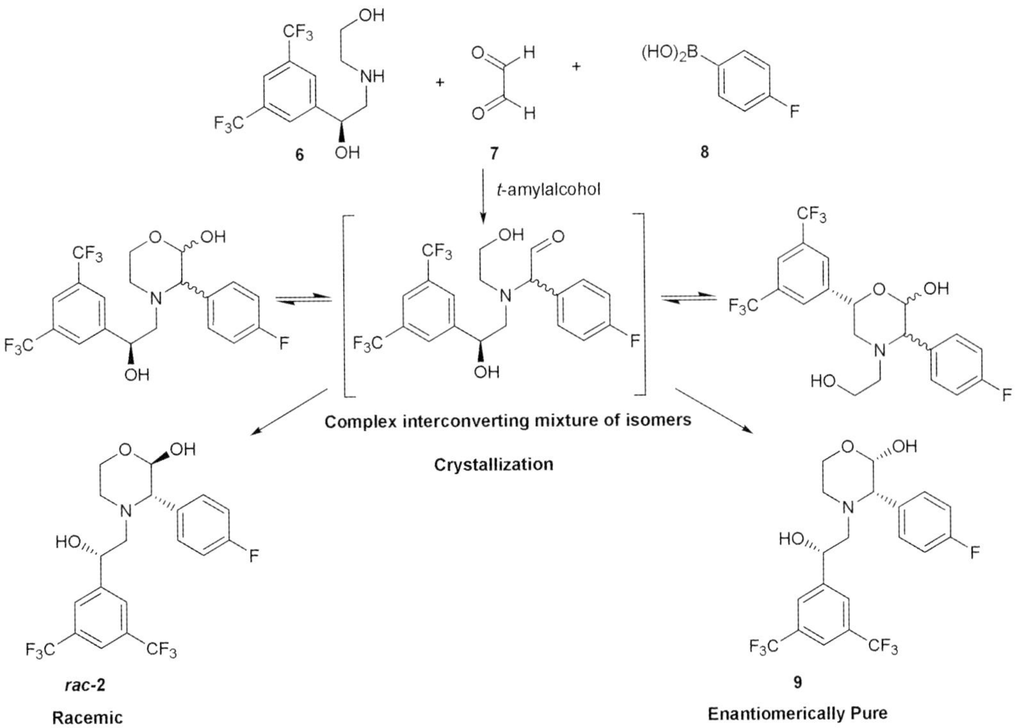

Scheme 2 Diastereoselective transformation: racemic versus enantiomerically pure.

Lactam Approach

The synthetic route (Scheme 3) that is the basis for the commercial synthesis of aprepitant introduces the stereogenic centers in a consecutive manner starting with enantiopure 3,5-bis(trifluoromethyl)phenethylalcohol **10**. This stereocenter is used to introduce the *cis*-oriented substituents on the morpholine core in a sequential manner with near perfect control of stereochemistry. The lactam functionality in **11** performs a critical role in this strategy by facilitating the substitution of the lactol hydroxyl, while preventing β-elimination. Consequently, displacement of the trifluoroacetate derivative of **11** with **10** catalyzed by $BF_3 \cdot Et_2O$ gives an essentially quantitative yield of the two possible diastereomeric acetals. As expected, little induction is observed and **12** and **13** are formed in a 55:45 ratio, respectively.

The key feature of the synthetic plan is the expectation that it is possible to epimerize the acetal center in combination with a crystallization of the less soluble and desired acetal diastereomer **12**. Gratifyingly, this epimerization can be accomplished using a tertiary potassium alkoxide in the presence of the corresponding alcohol. The next key challenge requires a tertiary alkoxide/alcohol pair which does not significantly increase the polarity of the highly nonpolar crystallization medium. In the end, dissolving the mixture of diastereomers obtained in the coupling reaction in heptane, adding 0.9 equiv. of the readily available 3,7-dimethyl-3-

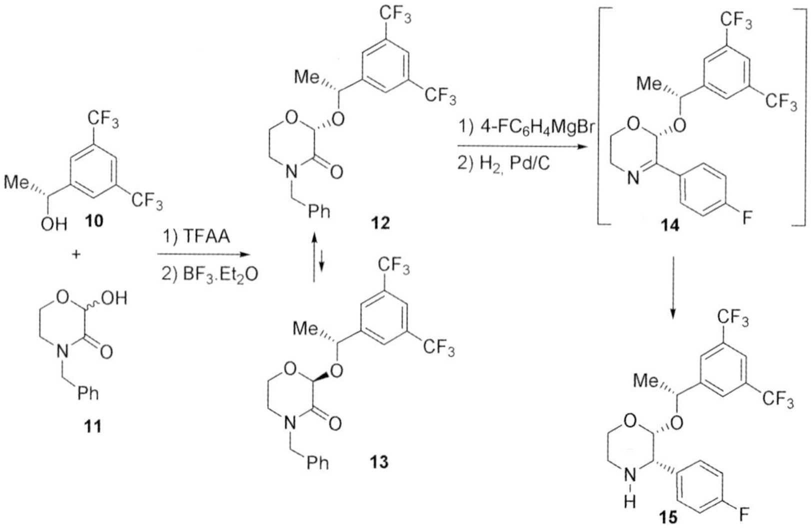

Scheme 3 Commercial route to aprepitant.

octanol as well as 0.3 equiv. of the corresponding potassium salt and ageing the resulting mixture at approximately −5 °C produces a 85 % isolated yield of diastereomerically and enantiomerically pure **12**. Thus, the chirality of **10**, which is to become an integral part of the target molecule, in combination with the crystalline properties of the required diastereomer are effectively used to introduce the second stereogenic center in a highly efficient and practical manner.

The synthesis culminates in an innovative and expedient conversion of oxazinone **12** to morpholine **15** via a nucleophilic addition of 4-fluorophenylmagnesium bromide followed by a double reduction with hydrogen. Detailed kinetic studies of the latter reaction showed that imine **14** is an initial intermediate. The reduction of **14** occurs with > 300:1 selectivity to set up the third stereogenic center in 91 % overall yield.

Conclusion

We report two contemporary examples of the power of so-called crystallization-induced asymmetric transformations in the context of two very different syntheses of the same target molecule. In the first example, we show how this technique can be effectively used to isolate the minor but desired product from a complex mixture of equilibrating stereoisomers in high yield, thus overcoming the limitations of a nonselective reaction. In the second example, we demonstrate that this technique can also be effectively used to relay stereochemical information from one stereocenter to the next in the build-up of a complex target. Many examples of crystallization-induced transformations are reported in the literature, mostly as the result of a serendipitous discovery. However, this phenomenon is rarely used as a design element in modern asymmetric synthesis. We hope that our work shows that this type of transformation should be seriously considered in synthetic planning, particularly when the chemistry has to be performed on a larger scale. However, to this end synthetic organic chemists will have to rediscover the value of the often ignored crystal properties of organic compounds rather than thoughtlessly resorting to the use of chromatographic techniques for purification.

CV of J. Brands
Karel M. Jos Brands completed his Ph.D at the University of Amsterdam with Professor U.K. Pandit. After a postdoctoral stay at the University of Rochester with Professor A.S. Kende he joined the Process Research department at Merck in 1992.

CV of P. Pye
Philip Pye completed his Ph.D. at the University of Texas at Austin with Professor P.D. Magnus and has been a member of the Process Research department at Merck since 1995.

CV of K. Rossen
Kai Rossen received his Ph.D. working with Professor Ganem at Cornell University. He worked in the Process Research department at Merck for 10 years and has been employed by Degussa in Germany from 2000 to 2005 before joining Sanofi-Aventis.

Selected Publications

1. P. J. Pye, K. Rossen, S. A. Weissman, A. Maliakal, R. A. Reamer, R. Ball, N. N. Tsou, R. P. Volante, P. J. Reider, *Chem. Eur. J.* **2002**, *8*, 1372–1376. Crystallization-Induced Diastereoselection: Asymmetric Synthesis of Substance P Inhibitors.

2. J. J. Hale, S. G. Mills, M. MacCoss, P. E. Finke, M. A. Cascieri, S. Sadowski, E. Ber, G. G. Chicchi, M. Kurtz, J. Metzger, G. Eiermann, N. N. Tsou, F. D. Tattersall, N. M. J. Rupniak, A. R. Williams, W. Rycroft, R. Hargreaves, D. E. MacIntyre, *J. Med. Chem.* **1998**, *41*, 4607–4614. Structural Optimization Affording 2-(R)-(1-(R)-3,5-Bis(trifluoromethyl)phenylethoxy)-3-(S)-(4-fluoro)phenyl-4-(3-oxo-1,2,4-triazol-5-yl)methylmorpholine, a Potent, Orally Active, Long-acting Morpholine Acetal Human NK-1 Receptor Antagonist.

3. K. M. J. Brands, J. F. Payack, J. D. Rosen, T. D. Nelson, A. Candelario, M. A. Huffman, M. M. Zhao, J. Li, B. Craig, Z. J. Song, D. M. Tschaen, K. Hansen, P. N. Devine, P. J. Pye, K. Rossen, P. G. Dormer, R. A. Reamer, C. J. Welch, D. J. Mathre, N. N. Tsou, J. M. McNamara, Paul J. Reider *J. Am. Chem. Soc.* **2003**, *125*, 2129–2135. Efficient Synthesis of NK-1 Receptor Antagonist Aprepitant Using a Crystallization-Induced Diastereoselective Transformation.

Combinatorial Methods in Asymmetric Syntheses

Stefan Dahmen, cynora GmbH, Germany

Background

Asymmetric catalysis is mostly regarded as a discipline, which is driven by mindful planning of substrates, reagents and chiral catalysts. Despite the efforts to achieve a rational access to catalytic reactions, the development of most processes is nowadays still accompanied by massive experimental expenditure. Reaction conditions, solvents and additives are as important for the (enantio)-selectivity and catalytic turnover as is the choice of the right chiral ligand or catalyst. Additives have also been found to have a dramatic effect in catalysis. An indispensable tool of modern catalysis is high-throughput screening. State of the art automation equipment enables the testing of hundreds of reactions per day. Such automation techniques are usually referred to as combinatorial catalysis [1R].

Results

Ligand Screening

Often ligand screening is the most obvious application of automation techniques in asymmetric catalysis. When important parameters of a given reaction such as metal, solvent, reagents and further experimental parameters – a protocol – have been established and optimized, the accomplishment of high enantioselectivity strongly depends of the availability of suitable chiral ligands.

A good example is the addition of diethylzinc to imines (Scheme 1). While there are hundreds of known amino alcohol ligands that, over the years, have been shown to promote the diethylzinc addition to aldehydes, a corresponding

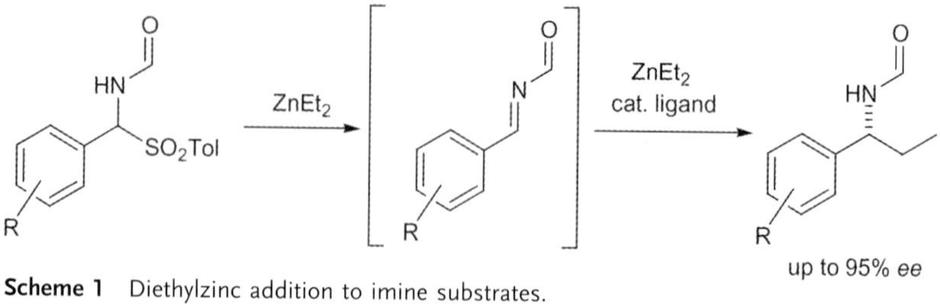

Scheme 1 Diethylzinc addition to imine substrates.

Asymmetric Synthesis – The Essentials.
Edited by Mathias Christmann and Stefan Bräse
Copyright © 2007 WILEY-VCH Verlag GmbH & Co. KGaA, Weinheim
ISBN: 978-3-527-31399-0

catalytic system for imines was developed only recently [2, 3]. When we developed the reaction, our initial problems were to find a suitable substrate that would be easily accessible, stable and prone to undergo the addition reaction with a minimum of (undirected) background alkylation.

After establishing N-formyl imines, formed *in situ* from the corresponding sulfinyl amides by reaction with diethylzinc, as a new class of imine electrophiles, we were able to optimize the enantioselectivity by using our [2.2]-paracyclophane-based imine ligands [4, 5]. However, a broad ligand screening using cynora's library of N,O-ligands (Figure 1) revealed that exactly these ligands have an outstanding performance in this reaction. Up to now, no other ligand class has been found to deliver comparable performance.

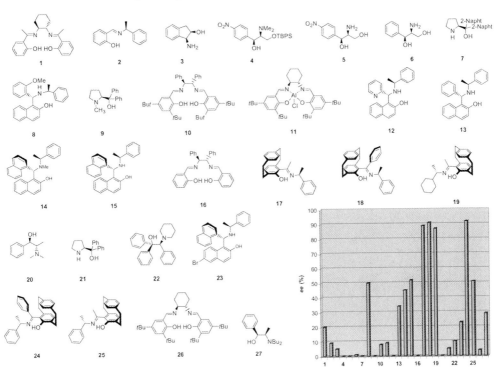

Figure 1 Ligand screening in the diethylzinc addition to imines according to Scheme 1.

A limited set of ligands was used to optimize the alkynylzinc addition to aldehydes. Different protocols for the enantioselective addition of alkynylzinc reagents to aldehydes were compared. Crucial at this point was *in situ* formation of the active zinc reagent **28**. The concurrent formation of symmetric dialkynylzinc reagents **29** drastically diminishes the stereoselectivity of the reaction (Scheme 2). Based on the protocol that allows the lowest catalyst loading, a screening of different commercially available N,O-ligands as well as several [2.2]paracyclophane-based ketimine ligands was conducted. Due to their high activity, the [2.2]para-

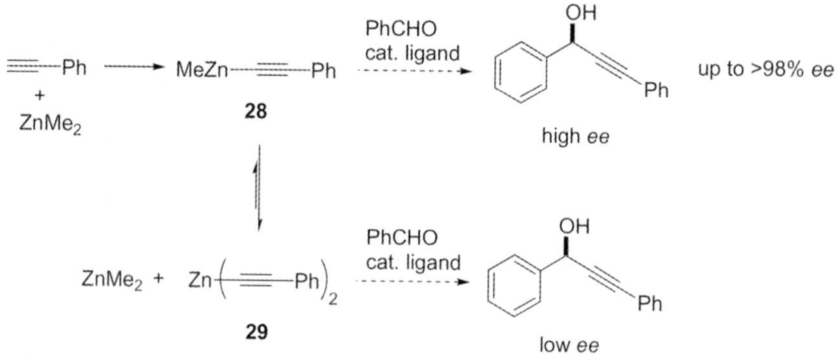

308 | Combinatorial Methods in Asymmetric Syntheses

Scheme 2 Alkynylzinc addition to aldehydes.

cyclophane-based ligands proved to be superior to conventional amino alcohol ligands, especially when employed at low catalyst loading (2–5 mol%).

Additive Screening

In an additive screening using automation techniques we demonstrated that several additives have positive effects in the phenylzinc addition to 2-bromobenzaldehyde (Figure 2) [8]. It also became apparent that the additive amount is equally important. Polyethyleneglycol derivatives generally improve the enantioselectivity, presumably by complexing zinc salts and thereby reducing the importance of the background reaction. Alcohols, in contrast, react with the zinc reagent and influence the selectivity by formation of phenylzinc alkoxides. The most interesting results were obtained with imidazol as additive, which led to a reversal in enantioselectivity. Overall, DiMPEG 2000 proved to be the best additive, giving an improvement of about 20% ee compared to the catalyzed reaction without additive.

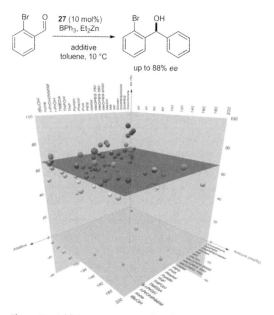

Figure 2 Additive screening in the phenylzinc addition to 2-bromobenzaldehyde.

Arylzinc Precursors

The enantioselective addition of arylzinc reagents to aldehydes has matured to a reliable process giving rise to diarylmethanols, some of which are important precursors for pharmacologically active compounds. The use of boronic acids as aryl source enabled the transfer of functionalized aryls. Triphenylborane was demonstrated to be a viable phenyl source and additives enabled the use of simple, commercially available ligands for the triphenylborane protocol. Serious drawbacks of all these protocols however, were price and availability of the aryl sources.

We therefore searched for other aryl sources for the transmetallation step yielding active arylzinc reagents and were able to demonstrate that triarylborane ammonia complexes are outstandingly stable, versatile and economic precursors for arylzinc reagents in asymmetric catalysis (Scheme 3).

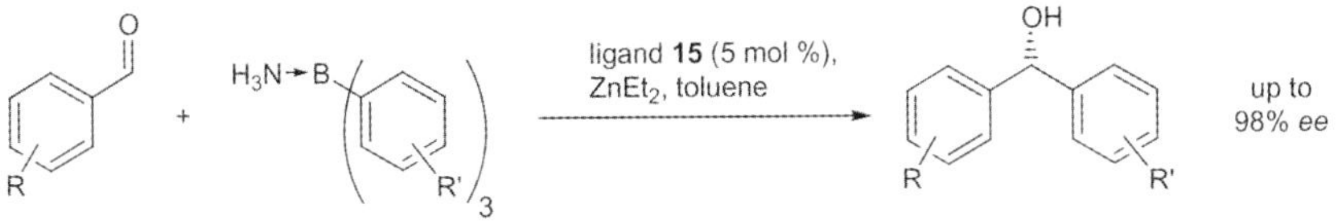

Scheme 3 Triarylborane ammonia complexes as precursors in the arylzinc addition to aldehydes.

In a kinetic approach we were able to monitor the transmetallation reaction and rationalize the observed differences of three classes of arylboranes in catalytic applications. Using this method, we synthesized an array of chiral diarylmethanols in high yield and enantioselectivity. The simple reaction protocol as well as the ready availability of substrates and ligand provided for the first time an excellent opportunity for technical applications.

CV of Stefan Dahmen

Stefan Dahmen was born in Rheydt, Germany in 1971 and studied chemistry at the RWTH Aachen, Germany and University of York, UK. He received his Diploma in Chemistry in 1999 and his Ph.D. from RWTH Aachen in 2002 working with Stefan Bräse. In 2003, he co-founded cynora GmbH, where he currently heads R&D. He is recipient of the Borchers medal and his interests focus on combinatorial methods, asymmetric catalysis and OLED materials.

Selected Publications

1. R. S. Dahmen, S. Bräse, *Synthesis* **2001**, 1431. *Combinatorial Methods for the Discovery and Optimisation of Homogeneous Catalysts*.
2. S. Dahmen, S. Bräse, *J. Am. Chem. Soc.* **2002**, *124*, 5940. *The Asymmetric Dialkylzinc Addition to Imines Catalyzed by [2.2]Paracyclophane-Based N,O-Ligands*.
3. N. Hermanns, S. Dahmen, C. Bolm, S. Bräse, *Angew. Chem.* **2002**, *114*, 3844; *Angew. Chem. Int. Ed.* **2002**, *40*, 3692. *Asymmetric, Catalytic Phenyl Transfer to Imines: Highly Enantioselective Synthesis of Diarylmethylamines*.
4. S. Dahmen, S. Bräse, *Chem. Commun.* **2002**, 26. *Planar and central chiral [2.2]paracyclophane-based N,O-ligands as highly active catalysts in the diethylzinc addition to aldehydes*.

5. S. Dahmen, S. Bräse, *Org. Lett.* **2001**, *3*, 1119. *[2, 2]Paracyclophane-based N,O Ligands in Alkenylzinc Additions to Aldehydes.*
6. S. Dahmen, *Org. Lett.* **2004**, *6*, 2113. *Enantioselective alkynylation of aldehydes catalyzed by [2.2]paracyclophane-based ligands.*
7. S. Dahmen, M. Lormann, *Org. Lett.* **2005**, *7*, 4597. *Triarylborane ammonia complexes as ideal precursors for arylzinc reagents in asymmetric catalysis.*
8. J. Rudolph, M. Lormann, C. Bolm, S. Dahmen, *Adv. Synth. Catal.* **2005**, *347*, 1361. *A high-throughput screening approach for the determination of additive effects in organozinc addition reactions to aldehydes.*

Biocatalytic Production of Optically Active Amines

Klaus Ditrich, BASF AG, Ludwigshafen, Germany

Background
With the trend towards single-enantiomer drugs and drug candidates, industrially feasible methods for the production of optically active intermediates and building blocks are of increasing importance. Of central interest are optically active amines which are frequently used as building blocks, resolving agents and chiral auxiliaries. Established commercial routes to the synthesis of this class of compounds have been reviewed in a recently published paper [1R].

Results
In the course of a program aimed at the identification of new biocatalytic reactions, initiated at BASF in the early 90s, a new enzyme-catalyzed kinetic resolution of chiral amines was discovered [2] (Scheme 1).

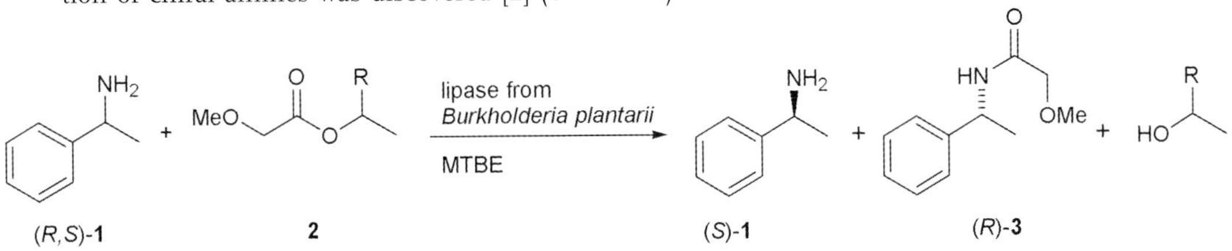

Scheme 1 Enzyme-catalyzed kinetic resolution of 1-phenylethylamine (R,S)-1.

A bacterial lipase from *Burkholderia plantarii* turned out to be a very efficient catalyst in the enantioselective acylation of 1-phenylethylamine (R,S)-1. The most effective acylation reagents were esters 2 of methoxyacetic acid which, compared to the isosteric butyric acid esters, showed a remarkably rate acceleration in the amidation reaction. With esters of primary alcohols a selectivity $E = 150$ was achieved, esters originating from secondary alcohols were even more selective [3]. E increased to >2000 which corresponds to an energy difference $\Delta\Delta G^*$ of >4.5 kcal mol^{-1} in the transition states! Under these conditions both products,

Asymmetric Synthesis – The Essentials.
Edited by Mathias Christmann and Stefan Bräse
Copyright © 2007 WILEY-VCH Verlag GmbH & Co. KGaA, Weinheim
ISBN: 978-3-527-31399-0

the unreacted (S)-enantiomer (S)-1 and the (R)-amide (R)-3 were isolated in nearly optically pure form. Due to the differences in their physico-chemical behavior amine (S)-1 and amide (R)-3 were easily separated by extraction or distillation.

A remaining task was the liberation of the (R)-amine (R)-1 from the R-configurated amide (R)-3, again phenylethylamine was chosen as an example (Scheme 2). Despite the plethora of publications dealing with base-catalyzed racemizations during hydrolysis of optically active amides, we found that, by applying the right conditions, optically active amides like (R)-3 could be hydrolyzed without any undesired side reaction. The most important factor was the addition of co-solvents like glycols or aminoalcohols; in the presence of these compounds, amides like (R)-3 were hydrolyzed in nearly quantitative yield without racemization using inexpensive 50% aqueous NaOH solution as the hydrolyzing reagent [4]:

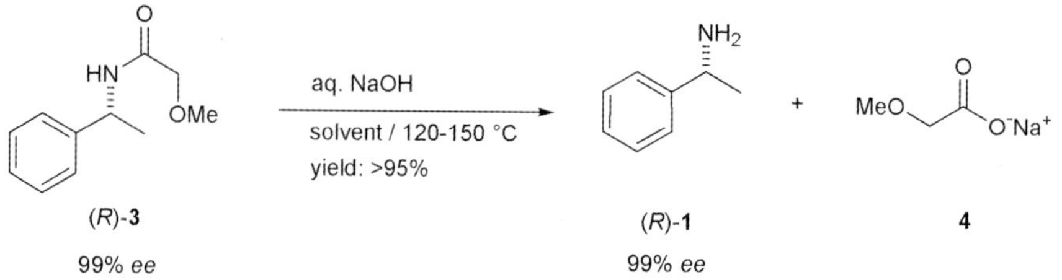

Scheme 2 Hydrolysis of (R)-amides to the corresponding (R)-amines.

The liberated amine (R)-1 was isolated by distillation or extraction, as a by-product, the sodium salt of methoxyacetic acid 4 was formed. In terms of cost-efficiency of the overall process, recycling of the acylation reagent 2 starting from the sodium salt 4 seemed to be very attractive. The desired conversion was realized by acidification (concentrated sulfuric acid) of the aqueous solution of 4, followed by an extraction/esterification procedure [5].

The most impressive feature of the new resolution process is the enzyme's unique flexibility regarding substrates. In contrast to the well established "classical" resolution processes for amines based upon precipitation of diastereomeric ammonium salts with optically active acids, the new enzyme-catalyzed kinetic resolution is applicable to a wide variety of amine substrates (Figure 1).

Aryl alkyl amines can be resolved without any restriction, heterocyclic systems are also accessible. In the case of alkyl amines, the selectivity of the resolution process increases with increasing difference in the steric bulkiness of the substituents at the asymmetric center: whereas differentiation of an ethyl group vs. a methyl substituent is not very pronounced ($E = 8$), a higher selectivity ($E = 50$) is observed by increasing the chain length of one substituent by a methylene unit; a secondary substituent is well recognized over a methyl group ($E = 80$), introduction of a tertiary substituent leads to a perfect selectivity ($E \geq 1000$). Amino alcohols are also accessible, in most cases protection of the hydroxy function as a benzyl- or methyl ether is advantageous [6].

Based upon the results of research and process development we planned a new dedicated plant with tailor-made equipment. For maximal exploitation of the en-

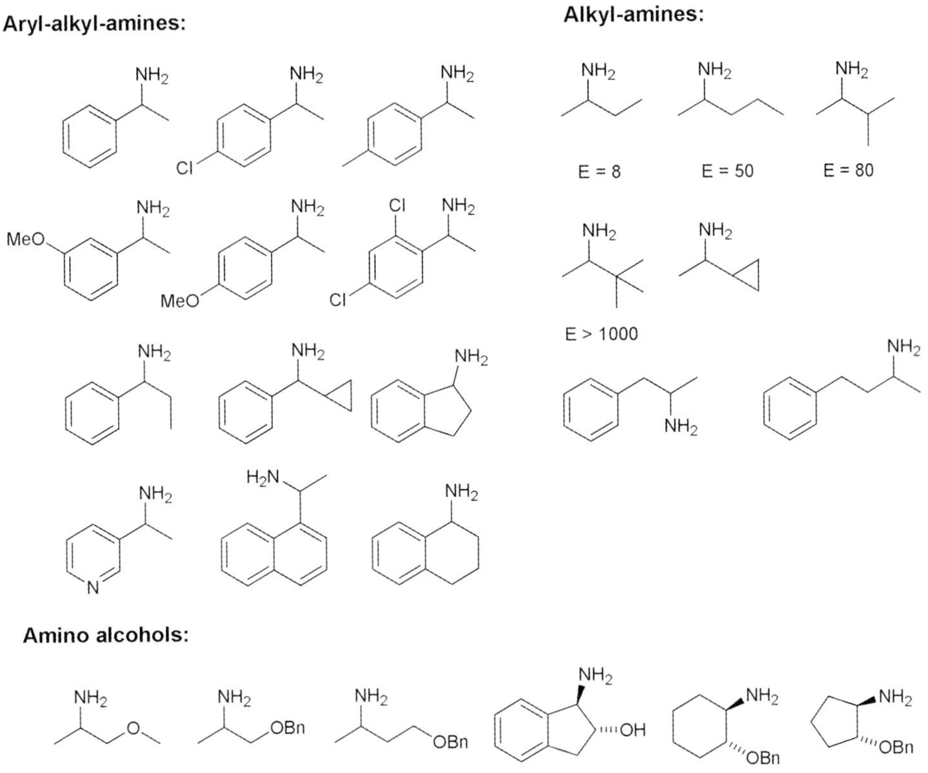

Aryl-alkyl-amines:

Alkyl-amines:

Figure 1 Some typical examples of chiral amines which can be resolved by the enzyme-catalyzed kinetic resolution (see Scheme 1). Some of them are resolved on a multiton scale.

zyme's selectivity, the new production facility, which went on stream in 2001, was designed as a multi-product facility. In total the equipment enables us to produce > 1000 t a^{-1} of optically active amines in full accordance with cGMP-guidelines.

In a second single-product plant built in the US, "S-MOIPA" (S)-**5**, a key intermediate for the production of an optically active corn herbicide (Outlook® (**7**)), is produced on a scale of > 2000 t a^{-1} (Scheme 3) [7]:

Since in most cases only one enantiomer of a chiral amine can be commercialized, recycling of the undesired enantiomer by racemization would contribute to the overall cost-efficiency of the process. With standard substrates the desired conversion could be realized by a well-known metal (e.g. Ni or Co)-catalyzed dehydrogenation-hydrogenation procedure [8]. Due to dehalogenation reactions, the method was not applicable to halogenated amines. For these compounds, we developed a new catalytic racemization procedure based on Schiff bases (Scheme 4) [9]. Condensation of a chiral amine with a catalytic amount of the appropriate ketone **8** leads to the formation of an optically active Schiff base **9** which can be deprotonated by a strong base like DBU to form a symmetrical aza-allyl anion **10**. By nonspecific protonation of **10**, racemic **9** is formed. An amine exchange reaction with

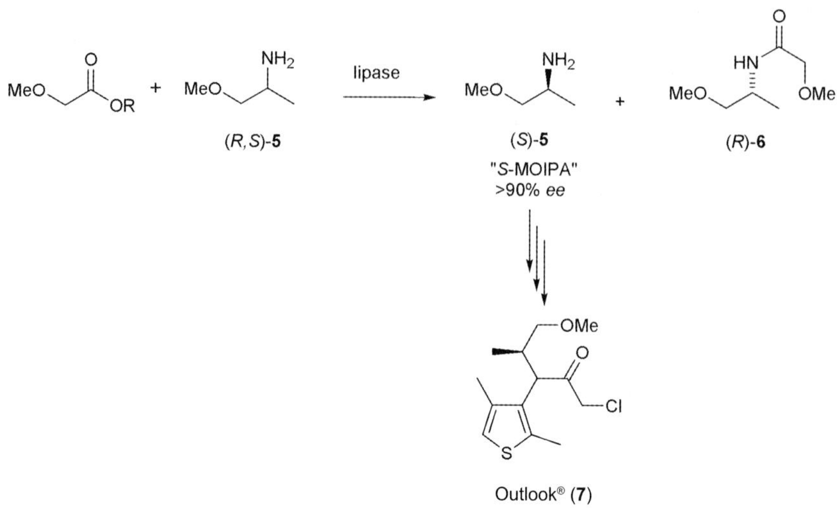

314 Biocatalytic Production of Optically Active Amines

Scheme 3 Resolution of 1-methoxyisopropylamine (R,S)-**5**; production of (S)-MOIPA (S)-**5**, a key intermediate for the synthesis of the corn herbicide "Outlook®" (**7**).

optically active amine via the unstable aminal **11** completes the catalytic cycle, and will accumulate the racemic amine. After completion, the racemic amine is distilled off and the nonboiling residues (Schiff base **9** and DBU) are used in the next racemization batch.

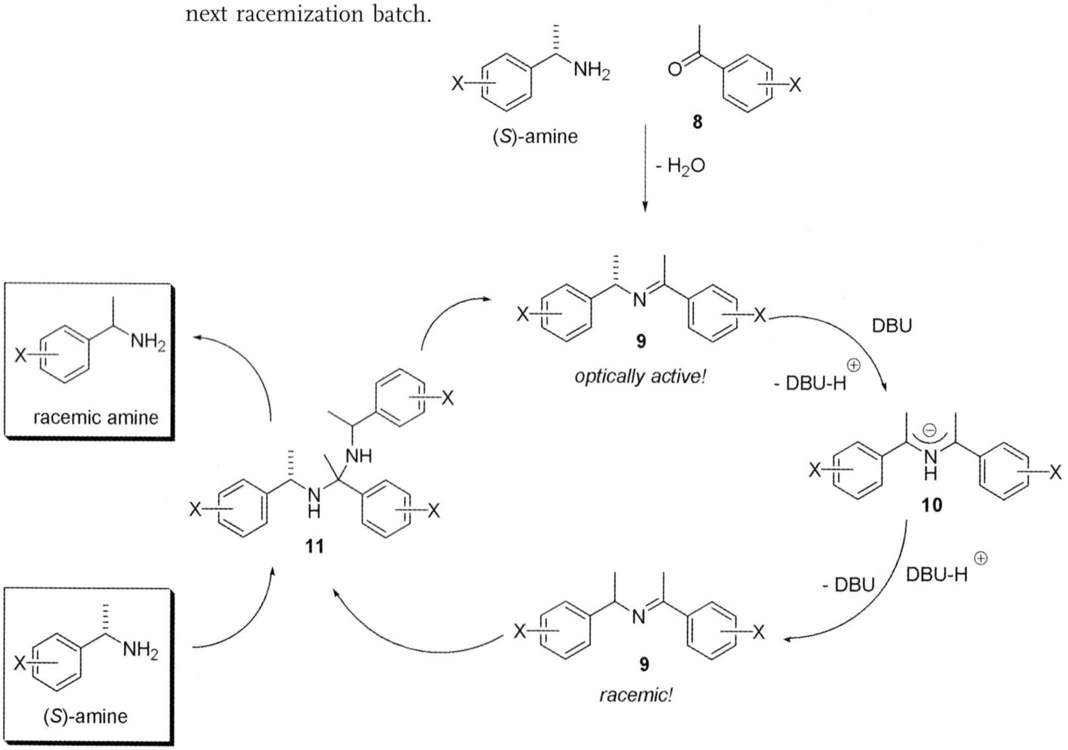

Scheme 4 Racemization of halogenated amines using catalytic amounts of Schiff bases.

Conclusions and Future Perspectives

The new BASF-process based upon a biocatalytic enantioselective amidation turned out to be a generally applicable method for the resolution of chiral amines. Having two dedicated production facilities on hand, optically active amines are produced at BASF on a commercial scale. A dynamic kinetic resolution would be even more attractive but, up to now, only a few examples of a truly dynamic kinetic resolution of chiral amines have been reported [10]. Before becoming attractive for commercial applications, more active racemization catalysts have to be developed.

CV of Klaus Ditrich

Klaus Ditrich was born in Rhina, Germany in 1956. He studied chemistry at the University of Marburg where he joined Professor R.W. Hoffmann's group working in the field of natural product synthesis. After receiving his Ph.D. in 1986, he joined BASF AG in 1987, starting his career in the field of agrochemicals research. In 1992 he moved to the research unit for fine chemicals and biocatalysis, since 1995 he has been responsible for the process development for the production of optically active amines. In 1996 he received the "BASF Innovations Award". He is a board member of the "Liebig-Vereinigung für Organische Chemie" (since 2004), in 2006 he was appointed as a professor at the Albert-Ludwigs-University of Freiburg.

Selected Publications

1R. M. Breuer, K. Ditrich, T. Habicher, B. Hauer, M. Keßeler, R. Stürmer, *Angew. Chem. Int. Ed.* **2004**, *43*, 788–824. *Industrial Methods for the Production of Optically Active Intermediates.*

2. F. Balkenhohl, K. Ditrich, B. Hauer, W. Ladner, *J. Prakt. Chem./Chem.-Ztg.* **1997**, *339*, 381. *Optically Active Amines via Lipase-catalyzed Methoxyacetylation.*

3. K. Ditrich, F. Balkenhohl, W. Ladner (BASF AG), DE 19531116, **1995**; *Chem. Abstr.* **1997**, *126*, 277259. *Hydrolysis of Optically Active Amides in Enzymatic Resolutions.*

4. K. Ditrich, W. Ladner, J.-P. Melder (BASF AG), DE 19913256, **1999**; *Chem. Abstr.* **2000**, *133*, 252151. *Stereoretentative Hydrolysis Method and Solvents for the Conversion of Optically Active Carboxamides into Optically Active Amines.*

5. K. Ditrich, U. Block (BASF AG), WO 2000046177, **1999**; *Chem. Abstr.* **2000**, *133*, 165414. *Method for the Production of Esters.*

6. F. Balkenhohl, K. Ditrich, C. Nübling (BASF AG), WO 9623894, **1995**; *Chem. Abstr.* **1996**, *125*, 219729. *Racemate Separation of Primary and Secondary Heteroatom-substituted Amines by Enzyme-catalyzed Acylation.*

7. C. Nübling, K. Ditrich, C. Dully (BASF AG), DE 19837745, **1998**; *Chem. Abstr.* **2000**, *132*, 165214. *Enzyme-catalyzed separation of Primary Amines.*

8. G. Vitt, H. Siegel, M. Schneider (BASF AG), DE 2851039, **1978**; *Chem. Abstr.* **1983**, *98*, 185928. *Method for the Racemization of Optically Active 1-Aryl-alkylamines.*

9. K. Ditrich (BASF AG), DE 19606124, **1996**; *Chem. Abstr.* **1997**, *127*, 234169. *Method for the Racemization of OpticallyActive Amines.*

10. M. T. Reetz, K. Schimossek, *Chimia* **1996**, *50*, 668–669. *Lipase-Catalyzed Dynamic Kinetic Resolution of Chiral Amines: Use of Palladium as the Racemization Catalyst;* J. Paetzold, J. E. Bäckvall, *J. Am. Chem. Soc.* **2005**, *127*, 17260–17261. *Chemoenzymatic Dynamic Kinetic Resolution of Primary Amines.*

The Monsanto L-Dopa Process

William S. Knowles, Monsanto Co. (Retired), St. Louis, USA

Background

It has been pointed out that one of the charms of doing exploratory research is that one doesn't know where it is going to lead or who the leaders will be. Our program on asymmetric hydrogenations, which has led to a direct synthesis of L-α-amino acids without making the unwanted D-isomer, provides an excellent example of what a purely exploratory effort can accomplish. We did not know where our work would lead us and we were definitely not experts in stereochemistry or coordination compounds. Our success, which resulted in Nobel recognition in 2001, certainly surprised many of the experts in these fields.

I should like to point out that in any research organization budgets are always tight, and it is very difficult to get much purely exploratory effort. Obviously, most research has to be on safe projects, but somehow time must be found to do undirected effort where you let the results lead wherever they will. Thus, we did not approach this problem looking for an α-amino acid synthesis, but let our research lead us in that direction. When success came our way we immediately exploited the area, and being in an industrial laboratory we had the thrill of seeing our technology run on a plant scale almost immediately. Our work provides an excellent example of how a modest and inexpensive effort in industry can produce significant results.

L-DOPA Process

Our discovery in 1969 of an efficient L-amino acid synthesis by hydrogenating an enamide precursor using as catalyst a chiral phosphine, CAMP (cyclohexyl-o-anisylmethylphosphine) complexed with Rhodium coincided very nicely with another seemingly unrelated development. This was the discovery that a fairly massive dose of the rare amino acid, l-3,4-dihydroxyphenylalanine (L-DOPA), was useful in treating Parkinson's disease. We learned that Monsanto, because of its position in vanillin, was custom manufacturing a racemic intermediate, which Hoffmann-La Roche was resolving and de-blocking to L-DOPA. The synthesis followed closely the Erlenmeyer azlactone synthesis described in Organic Syntheses [1] using vanillin instead of benzaldehyde. It went by way of a hydro-

Asymmetric Synthesis – The Essentials.
Edited by Mathias Christmann and Stefan Bräse
Copyright © 2007 WILEY-VCH Verlag GmbH & Co. KGaA, Weinheim
ISBN: 978-3-527-31399-0

genation of a properly substituted enamide. The symmetrically hydrogenated product was then resolved and de-blocked to L-DOPA. This enamide offered a golden opportunity for us to commercialize this burgeoning technology. Seldom has an invention been so closely timed with an emerging need. Our chemistry followed very closely the Organic Syntheses preparation substituting our chiral hydrogenation catalyst for their symmetrical one (Figure 1). It turned out that this was the simplest step in the sequence.

To develop a workable process we had a number of important considerations, which differ from the usual laboratory preparation. First we had to choose between benzoyl or acetyl as an enamide blocking group. The former did give better aldehyde condensation yields, but its great bulk and greater difficulty of removal tipped the balance in favor of the acetyl, especially since the *ee* (enantiomeric efficiency) was the same for both using our chiral phosphines. The choice of vanillin as a starting aldehyde was fairly straightforward since it was about the only molecule containing a 3,4-dihydroxy moiety, which enjoyed an economy of scale because of its flavor use. The closely related veratraldehyde or piperonal could have been used, but were much too expensive.

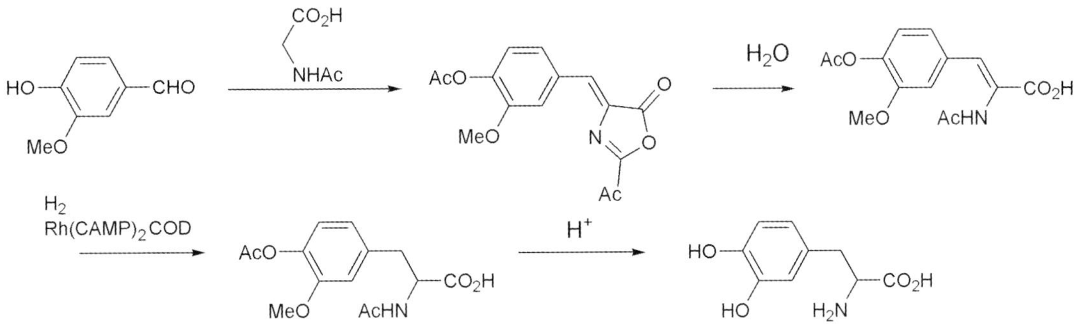

Scheme 1 Monsanto L-DOPA process.

At the time we did this work virtually all hydrogenations were heterogeneous. The substrate was dissolved in a solvent and contacted with hydrogen in the presence of a metal catalyst. When complete, the catalyst could be filtered and in many cases reused. With these soluble chiral catalysts the blocked L-isomer (3-methoxy, 4-acetoxy-N-acetylphenylalanine) had to be separated from spent catalyst by crystallization. Since both substrate and product were sparingly soluble in alcohol we could start with a slurry of substrate and end up with a slurry of product. We then filtered off the product leaving behind the spent catalyst and the small amount of racemate that was formed. We could even add up to 30% water to reduce solubility losses even further. Thus, our pay-load was limited not by solubility, but by how heavy a slurry we could stir. This technique would not be possible for a typical heterogeneous catalysis, where a slurry would coat and inactivate the catalyst.

Originally we used aqueous methanol as a solvent, but the long cycles incurred in the plant caused a trace of esterification, throwing our L-Dopa slightly off spec.

This problem was readily solved by shifting to isopropanol. We later found that aqueous *tert*-butanol worked even better, but it was not worth making the change.

Even in the best case some racemic product is made and must be separated. This separation is easy or hard depending on the nature of the racemate crystals. If the racemate has a different crystalline form than pure D or L, then separation of the excess enantiomer will be inefficient. If one obtains 90% *ee* then it is possible to get out easily only 75 to 80% pure enantiomer. With lower *ees* losses become prohibitve. For such a system a catalyst of high efficiency must be used. Unfortunately, most compounds are of this type. If on the other hand the racemic modification is a conglomerate or an equal mix of D and L crystals, then recovery of excess L can be achieved with no losses. Since the L and DL are not independently soluble, a 90% *ee* gives a 90% recovery of pure isomer. In our L-DOPA process the intermediate is such a conglomerate and separations are efficient. This lucky break was most welcome and indeed it was the same luck that Pasteur encountered in his classical tartaric acid separations 150 years ago. Incidentally, the closely related phenylalanine intermediate does not have easy separations and needs a much more efficient catalyst than CAMP. This factor coupled with the high cost of benzaldehyde, which did not enjoy an economy of scale, tipped the balance in favor of a one step biochemical route from glucose for the manufacture of this important amino acid.

A very important consideration was catalyst loading. Laboratory preparations were usually done at a mole ratio of substrate to catalyst of 1000/1 expressed as turnover number (TON), taking an hour or two. Although, in principle, it was possible to recover and recycle the spent catalyst it would have been messy and impractical. It was much better to use a lot less and take a much longer time, working on the premise that time was much cheaper than rhodium. At 60–70 °C we were able to achieve a TON of 20 000 with a 15–20 h cycle. We found that these chiral catalysts were not poisoned by the usual sulfur compounds so damaging to heterogeneous hydrogenations but were poisoned by oxygen or peroxides, since they did not catalyze the reaction of hydrogen and oxygen to harmless water. Surprisingly, this was less of a problem in the plant, since on a large scale the hydrogen contained less oxygen. We learned that our laboratory cylinders had to be carefully chosen to get one low in oxygen. Another plus on the large scale was the use of wet substrate. This was done to avoid an expensive drying step, but it turned out to be easier to purge out the air, presumably since none was trapped in the interstices of the crystals. Even so we had to carefully purge the system first with nitrogen and then with hydrogen to remove residual air. In addition, we had to be sure the solvent was peroxide free.

Another advantage in the use of these soluble catalysts is that they are not pyrophoric. With heterogenous catalysts using a flammable solvent, fire is always a problem when you filter off catalyst and one must take appropriate precautions. With more than ten years of production we never had a fire problem and never had a poisoned batch.

The key to our success was of course the catalyst. It was made by a multistep synthesis [1, 2], but only a small amount was needed. We achieved some economy of scale by making a ten-year supply in a single plant batch. We could combine the phosphine with rhodium *in situ*, but found it was better to use an air stable solid complex of the type [Rh(Bisligand)(COD)]$^+$ (BF4)$^-$. We used L-menthol to give the phosphine its chirality. It could be recovered for reuse, but was not worth the effort. We started using our first successful ligand, CAMP. Later we changed to a more efficient bidentate ligand called DiPAMP [1]. This ligand was more efficient and easier to make and did not change the general working of the process. De-blocking to L-DOPA used conventional chemistry.

Other Industrial Applications

For five to ten years, L-DOPA manufacture was the only large scale use of this new chemistry. Finally, in the 1980s Takasago [3] commercialized a L-menthol synthesis at perhaps a ten times larger scale. The catalyst was a rhodium (BINAP)$_2$ complex. This was not a hydrogenation at all, but an asymmetric isomerization of an olefin, which was then converted to L-menthol. A key feature of the synthesis was that the catalyst achieved a TON of 100 000, which is necessary for such an expensive complex.

A number of other chiral catalyses have been reported, but it is hard to tell how many are even in small scale production [4]. A notable exception is a process for making the herbicide metolachlor, the active ingredient in Dual4. The key step is the hydrogenation of a prochiral imine precursor. The chiral catalyst xylophos was derived from ferrocene complexed with Iridium. An impressive TON of 1 000 000 was reported. This is an elegant application and shows the potential of these soluble, molecularly dispersed catalysts. For such a high cost catalyst for a relatively cheap herbicide, a very high TON is imperative.

Conclusions

The reported development of ligands [5, 6] with both high TON numbers as well as 99–100 % efficiency will be a great help in future commercial processes, since a conglomerate type racemate with its easy separations will no longer be needed. Even so, though they can be a nice alternative, they will by no means replace biochemical processes, where the problem of dilute solutions and difficult isolations are often less than the problem of a multi-step synthesis. Chiral catalysis will be the favored route to non-natural products, where fermentation routes are not available.

Perhaps the most important use of these catalysts will be to provide an easy way to increase the chiral pool for life sciences studies with very little effort. We can look at them as a labor saving device in the laboratory. For this they will have impact as long as chemists run reactions.

CV of William Knowles

William Knowles, born in 1917, earned a Bachelor's degree in chemistry at Harvard in 1939 and a Ph.D. in steroid chemistry at Columbia University in 1942. He accepted a position at Monsanto, in St. Louis, immediately after graduating from Columbia. In 1951, he studied the total synthesis of steroids while on a company-sponsored postdoctoral fellowship in the laboratory of Harvard Chemistry Professor and Nobel Laureate, Robert B. Woodward. In the late 1960s, Knowles headed a three-man team that set out to develop a catalyst that could be used to synthesize individual enantiomers of chiral compounds directly.

A resident of Kirkwood, MO, he has been retired from Monsanto since 1986. Among the awards he received were the IR 100 Award for Asymmetric Hydrogenation (1974), the St. Louis ACS Section Award (1978), the Monsanto Thomas and Hochwalt Award (1981), the ACS Award for Creative Invention (1982) and the The Organic Reactions Catalysis Society - Paul N. Rylander Award (1996). In 2001 he shared the Nobel prize for chemistry with R. Noyori and K. B. Sharpless.

Selected Publications

1. W. S. Knowles, *Acc. Chem. Res.* **1983**, *16*, 106–112 and references cited therein. See also W. S. Knowles, Angew. Chem. Int. Ed. **2002**, *41*, 1998–2007. Asymmetric hydrogenations (Nobel Lecture).
2. W. S. Knowles, *J. Chem. Ed.* **1986**, *63*, 222–225. Application of organometallic catalysis to the commercial production of L-DOPA
3R. R. Noyori, *Chemtech* **1992**, 360–367. Asymmetric catalysis by chiral metal complexes.
4R. *Asymmetric Catalysis on an Industrial Scale*, H. U. Blaser, E. Schmidt (Eds.) Wiley-VCH, **2004**, (especially Chap. 3).
5. M. J. Burk, J. E. Feaster, W. A. Nugent, R. I. Harlow, *J. Am. Chem. Soc.* **1993**, *115*, 10125–10138, Preparation and use of C2-symmetric bis(phospholanes): production of a-amino acid derivatives via highly enantioselective hydrogenation reactions. M. J. Burk, M. F. Gross, J. P. Martinez, *J. Am. Chem. Soc.* **1995**, *117*, 9375–9376. Asymmetric catalytic synthesis of β-branched amino acids via highly enantioselective hydrogenation of a-enamides.
6R. *Principles and Applications of Asymmetric Synthesis*, G. Q. Lin, Y. M. Li, A. S. C. Chan, Wiley-Interscience, 2001, Chap. 6, Asymmetric Catalytic Hydrogenation and Other Reduction Reactions.

Asymmetric Hydrogenation through Metal–Ligand Bifunctional Catalysis

Ryoji Noyori, Takeshi Ohkuma, Christian A. Sandoval, and Kilian Muñiz, Nagoya University, Japan

Background

Dihydrogen is a clean, abundant molecule with an enormous potential from both scientific and technical viewpoints. Its key involvement in asymmetric hydrogenation renders it an ideal compound for the production of a wide range of chiral, non-racemic compounds. Ideally, such transformations are produced by efficient asymmetric catalytic reactions, which allow rapid and kinetically controlled hydrogenation processes without formation of waste or byproducts.

Strategy

Early research by our group revealed that BINAP–RuX$_2$ complexes [BINAP = 2,2'-bis(diphenylphosphino)-1,1'-binaphthyl; X = halogen] catalyse the efficient asymmetric hydrogenation of functionalised ketones such as β-keto esters or β-hydroxy esters to yield the corresponding chiral alcohols in high enantiomeric excess (*ee*) of up to 100 %. Here, the BINAP–Ru architecture provides a powerful enantio-differentiating template for the coordinating ester/olefin moieties, resulting in efficient diastereomeric discrimination. Unfunctionalised ketones such as acetone, however, do not react under such conditions [1–3].

Results

The problem of asymmetric hydrogenation of simple, unfunctionalised ketones was finally solved through the development of a new catalytic system [1, 2]. In 1995, we found that a combination of BINAP–RuCl$_2$ complexes with a suitable 1,2-diamine in the presence of an alkaline base in 2-propanol affords the enantioselective hydrogenation of various aromatic, heteroaromatic and unsaturated ketones. Various functional groups are tolerated while alkene functionalities remain largely unreactive under standard conditions. [4, 5] The isolation of bench-stable (diphosphine)RuCl$_2$(1,2-diamine) complexes and the introduction of 1,1-dianisyl-2-isopropyl-1,2-ethylenediamine (DAIPEN) greatly improved both productivity and enantioselectivity [6, 7]. For example, hydrogenation in the presence of catalyst precursor **1** leads to smooth reaction even at a substrate-to-catalyst (S/C) ratio of 2.4 million, while alcohol product of 99 % *ee* is obtained when complex **4**

(*R,R,R*)-**1**: Ar = C_6H_5,
(*R,R,R*)-**2**: Ar = 4-$CH_3OC_6H_4$,
(*R,R,R*)-**3**: Ar = 3,5-$(CH_3)_2C_6H_3$

(*R,R*)-**4**: R = 4-$CH_3OC_6H_4$

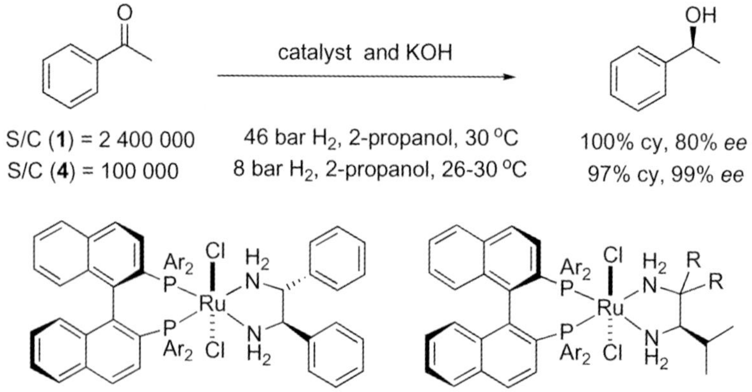

Scheme 1 Highly enantioselective hydrogenation of simple ketones with (binap)RuX_2(diamine) catalysts.

(Ar = 3,5-dimethylphenyl) is utilised at S/C of 100 000 (Scheme 1). This highly enantioselective reaction proved general for a variety of different ketones [8–10]. Efficient hydrogenation in the absence of base cocatalyst addition was realised using stable hydride-borohydride complexes (Figure 1), which in turn are obtained from treatment of **1–4** with sodium borohydride [11]. When [(*S*)-xyl-binap] RuH(η^1-BH_4)[(*S,S*)-1,2-diamine] **6** was used (S/C of 100 000), acetophenone (102 g) was quantitatively hydrogenated to (*R*)-1-phenylethanol in 99% *ee* in 2-propanol under 8 atm of hydrogen employing as little as 9 mg of precatalyst.

(*R,R,R*)-**5**: Ar = C_6H_5,
(*R,R,R*)-**6**: Ar = 3,5-$(CH_3)_2C_6H_3$

5 (X-Ray)

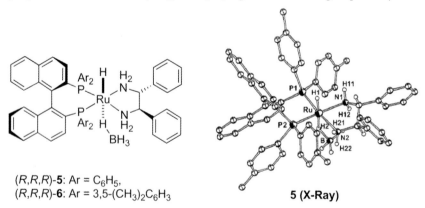

Figure 1 Catalyst precursor of (binap)RuH(η^1-BH_4)(1,2-diamine) composition.

Different substrates require structurally more diverse 1,2-diamine ligands for efficient hydrogenation. For example, hydrogenation of 1-tetralones is better effected by the use of a preformed ruthenium complex **7** (Figure 2) bearing a

1,4-diamine ligand [12]. With this precatalyst (S/C = 3000), under 9 atm of hydrogen, α-tetralone is converted to the corresponding (*R*)-configured alcohol with 99% ee and in 99.6% yield. A practical method for asymmetric hydrogenation of *tert*-alkyl ketones was recently discovered using the RuX_2(binap) motif together with the NH_2/pyridine hybrid α-picolylamine as ligand [13]. Here, when *tert*-butyl-methyl ketone (21.1 g) was reduced at S/C of 100 000 in ethanol in the presence of 83 mg of $KOC(CH_3)_3$ and under 20 atm of hydrogen with as little as 2 mg of (*S*)-BINAP-derived **8** (Figure 2), the corresponding (*S*)-configured alcohol was quantitatively produced in 98% ee.

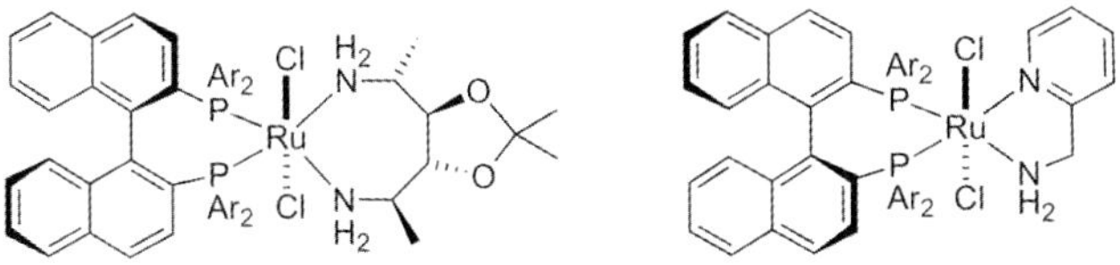

7: Ar = 3,5-$(CH_3)_2$-C_6H_3

8: Ar = 4-$CH_3C_6H_4$

Figure 2 Catalyst structures for hydrogenation of tetralones and *tert*-alkyl ketones.

Catalytic Cycle

The exact working mode of this uniquely enantioselective catalysis was uncovered in 2003 [14]. The structurally defined hydride/borohydride ruthenium complex **5** served as precursor for the unambiguous determination of underlying steady-state kinetics, base and hydrogen pressure dependence, solvent effects, and elucidation of the active catalytic species and mode of enantioselection. Two catalytic cycles with the common reactive intermediate complex **9** were found to be of importance (Figure 3). In protic solvents such as the usual 2-propanol (cycle II), the amido-complex **10** generated by ketone reduction abstracts a proton from the ambient solvent yielding cationic **12**. This electrophilic complex then accomodates molecular dihydrogen yielding the dihydrogen complex **13**, which functions as the resting state and represents the immediate catalyst precursor. Active catalyst **9** originates from **13** via loss of a proton, a process highly accelerated by the presence of ambient base. For reactions in aprotic solvents, **10** generates the dihydride catalyst **9** by intramolecular heterolytic cleavage of H_2 through short-lived **11** (cycle I). Under optimum turn-over frequency conditions this metal-ligand bifunctional hydrogenation functions through sustainable, well-balanced overall neutral pathways with only local appearance of acidic and basic species.

The enantiofacial differentiation of the substrate is kinetically accomplished on the molecular surface of the chirally modified RuH_2 catalyst. The final hydrogen transfer proceeds through a six-membered transition state involving simultaneous transfer of a proton from an amino moiety of the chiral diamine ligand to the carbonyl oxygen and hydride transfer from ruthenium to the carbonyl. No direct interaction between the ruthenium metal center and the ketone substrate is involved. A molecular model of the active *trans*-RuH_2 complex **9** displaying the hydride and amine proton (H-Ru-N-H_{ax}) involved in the transition state is depicted in Figure 3. The contribution to the chiral environment of the catalyst,

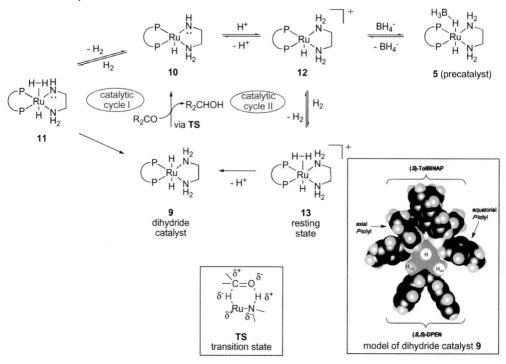

Figure 3 Mechanism for metal–ligand bifunctional hydrogenation of ketones.

and active performance in proton transfer during ketone reduction and electrophilic metal-center generation, constitute the pivotal roles of the prerequisite diamine ligand.

CV of Ryoji Noyori

Ryoji Noyori (born 1938) was educated at Kyoto University and became an instructor in Hitosi Nozakis group at the same university in 1963. He was appointed Associate Professor at Nagoya University in 1968, spent a postdoctoral year at Harvard with E. J. Corey in 1969–1970 and, shortly after returning to Nagoya, was promoted to Professor in 1972. In 2003, he was appointed President of RIKEN and University Professor at Nagoya. Noyori is a Member of the Japan Academy and the Pontifical Academy of Sciences, a Foreign Member of the National Academy of Sciences, USA, the Russian Academy of Sciences and the Royal Society of Chemistry. His research has long focused on the fundamentals and applications of molecular catalysis based on organometallic chemistry, particularly asymmetric catalysis. In 2001, he shared the Nobel Prize in Chemistry with W. S. Knowles and K. B. Sharpless.

Selected Publications

1. R. Noyori, *Angew. Chem.* **2002**, *114*, 2108; *Angew. Chem. Int. Ed.* **2002**, *41*, 2008.
2. R. Noyori, T. Ohkuma, *Angew. Chem.* **2001**, *113*, 40; *Angew. Chem. Int. Ed.* **2001**, *40*, 40.
3. R. Noyori, *Chem. Commun.* **2005**, 1807.
4. T. Ohkuma, H. Ooka, S. Hashiguchi, T. Ikariya, R. Noyori, *J. Am. Chem. Soc.* **1995**, *117*, 2675.
5. T. Ohkuma, H. Ooka, T. Ikariya, R. Noyori, *J. Am. Chem. Soc.* **1995**, *117*, 10417.
6. H. Doucet, T. Ohkuma, K. Murata, T. Yokozawa, M. Kozawa, E. Katayama, A. F. England, T. Ikariya, R. Noyori, *Angew. Chem.* **1998**, *110*, 1792; *Angew. Chem. Int. Ed.* **1998**, *37*, 1703.
7. T. Ohkuma, M. Koizumi, H. Doucet, T. Pham, M. Kozawa, K. Murata, E. Katayama, T. Yokozawa, T. Ikariya, R. Noyori, *J. Am. Chem. Soc.* **1998**, *120*, 13529.
8. T. Ohkuma, M. Koizumi, M. Yoshida, R. Noyori, *Org. Lett.* **2000**, *2*, 1749.
9. T. Ohkuma, M. Koizumi, H. Ikehira, T. Yokozawa, R. Noyori, *Org. Lett.* **2000**, *2*, 659.
10. T. Ohkuma, D. Ishii, H. Takeno, R. Noyori, *J. Am. Chem. Soc.* **2000**, *122*, 6510.
11. T. Ohkuma, M. Koizumi, K. Muñiz, G. Hilt, C. Kabuto, R. Noyori, *J. Am. Chem. Soc.* **2002**, *124*, 6508.
12. T. Ohkuma, T. Hattori, H. Ooka, T. Inoue, R. Noyori, *Org. Lett.* **2004**, *6*, 2681.
13. T. Ohkuma, C. A. Sandoval, R. Srinivasan, Q. Lin, Y. Wei, K. Muñiz, R. Noyori, *J. Am. Chem. Soc.* **2005**, *127*, 8288.
14. C. A. Sandoval, T. Ohkuma, K. Muñiz, R. Noyori, *J. Am. Chem. Soc.* **2003**, *125*, 13490.

Many Ways are Leading to Rome – Today's Variety of Competing Synthetic Methods in Industry

Andreas Job and Andreas Stolle, Saltigo Deutschland GmbH, Germany

Introduction

Statins represent the current top selling class of drugs worldwide. These HMG-CoA reductase inhibitors (shown in Figure 1) effect the regulation of the cholesterol synthesis and thus increase blood levels of LDL cholesterol and triglycerides and decrease HDL cholesterol levels. Clinical trials have shown that decreased levels of low-density lipoproteins, generally known as LDL cholesterol, reduce morbidity and mortality from cardiovascular diseases [1].

Since the introduction of Lovastatin by Merck in 1987, the market for statins has grown tremendously reaching $30 billion in 2004 for cholesterol and triglyceride lowering drugs. Pfizer's Atorvastatin is the current market leader with sales of more than $10 billion in 2004. Further growth is expected due to lifestyle changes

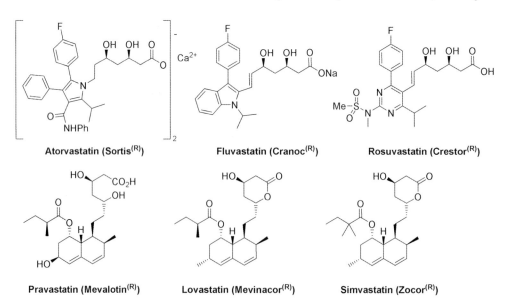

Figure 1 Examples of HMG-CoA reductase inhibitors (statins).

Asymmetric Synthesis – The Essentials.
Edited by Mathias Christmann and Stefan Bräse
Copyright © 2007 WILEY-VCH Verlag GmbH & Co. KGaA, Weinheim
ISBN: 978-3-527-31399-0

and growing numbers of cardiovascular diseases in the fast growing economies like India and China.

Due to the importance of these drugs, various concepts for the synthesis of statins and particularly for the "statin side chain" have been accomplished by the pharmaceutical, the biotechnological and the fine chemical industry.

Synthetic Concepts

A common pharmacophore in all statins is the so-called "statin side chain" (Figure 2) which constitutes a 3,5-dihydroxy hexanoic acid derivative connected to a heteroaromatic ring system or bicyclic residue. The chiral building block needs to be available in an enantio- and diastereomerically pure form and needs to be produced on a multi-ton scale.

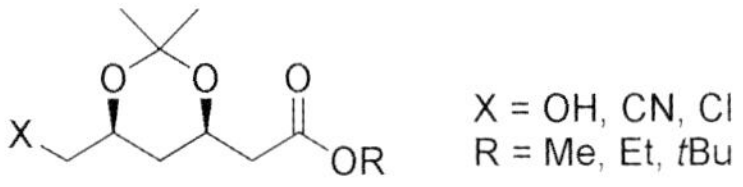

X = OH, CN, Cl
R = Me, Et, tBu

Figure 2 Statin side chain as key intermediate for the syntheses of statins.

With respect to scalability to industrial scale, the most promising concepts for the synthesis of the statin side chain are featured: biochemical, catalytic and chiral pool based syntheses.

Biochemical Approach

Scheme 1 highlights a typical biochemical approach to the statin side chain precursor. The key step is the incorporation of the chirality by reduction of a β-keto ester **1** to obtain (S)-**2**.

Kaneka has developed this enzymatic process using cell extracts of an *E. coli* transformant co-producing carbonyl reductase S1 and glucose dehydrogenase as biocatalyst to yield the hydroxyester (S)-**2** in 95 % yield and > 99 % enantiomeric purity [2]. The major disadvantage of this reaction is the high amount of gluconate obtained as a side product from the NADPH-recycling. Chain extension to

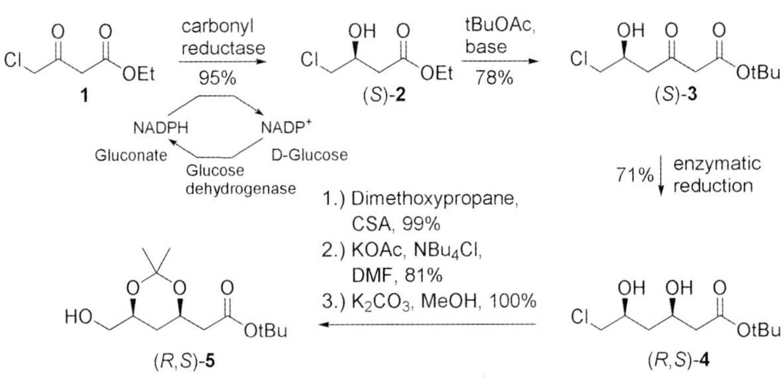

Scheme 1 Biochemical approach by Kaneka via enzymatic reduction of ethyl-4-chloro-3-oxo-butanoate [2].

(S)-3 and subsequent enzymatic reduction [3] finally provides (R,S)-5 as a key building block.

A similar procedure to generate (S)-3 in a single step from *tert*-butyl 6-chloro-3,5-dioxohexanoate was published by Müller in 2000 [4]. Applying a recombinant *Lactobacillus brevis* alcohol dehydrogenase (recLBADH) (S)-3 could be obtained with 99,5% ee and 72% yield. The recycling of NADPH was accomplished with the redox-system acetone/isopropanol to avoid further side-products that had to be separated.

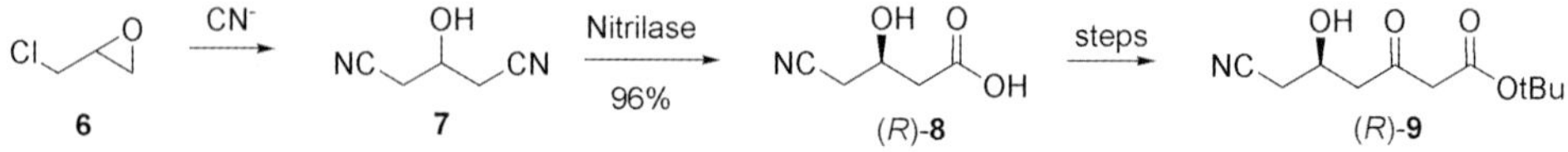

Scheme 2 Biochemical approach by Diversa [5].

An alternative enzymatic synthesis has been introduced by Diversa using a nitrilase for the desymmetrization of 3-hydroxyglutaronitrile [5] to yield (R)-9 in 96% yield and 98.5% ee (Scheme 2).

Based on an enzyme catalyzed C–C bond-formation a new synthetic route was developed by Wong and DSM [6]. Simple aldehydes **10** and **11** are transformed with 2-desoxyribose-5-phosphate aldolase (DERA) to pyran (R,S)-**12** in 70% yield (Scheme 3). This versatile pyran building block can be transferred into nitrile substituted acetal (R,R)-**13**.

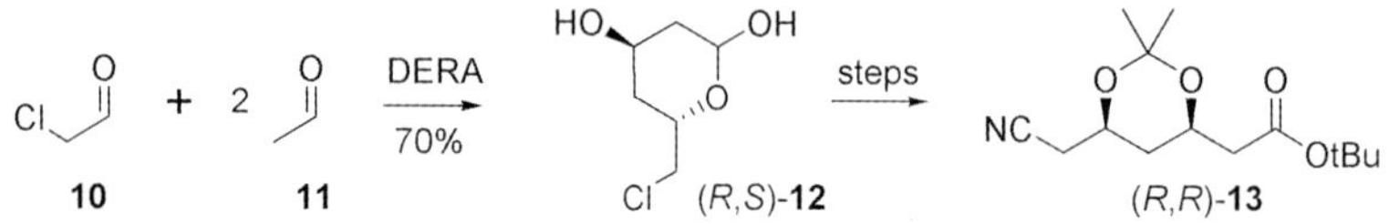

Scheme 3 Biochemical approach by Wong and DSM via enzymatic C–C bond-formation [6].

Catalytic Hydrogenation Approach

Various companies have published syntheses of statin side chain intermediates via asymmetric hydrogenation. Scheme 4 demonstrates a synthesis developed by Hoechst [7] using a ruthenium BINAP catalyst in a low pressure hydrogenation to obtain hydroxyester (S)-**15** in 96% yield and 97.5% ee.

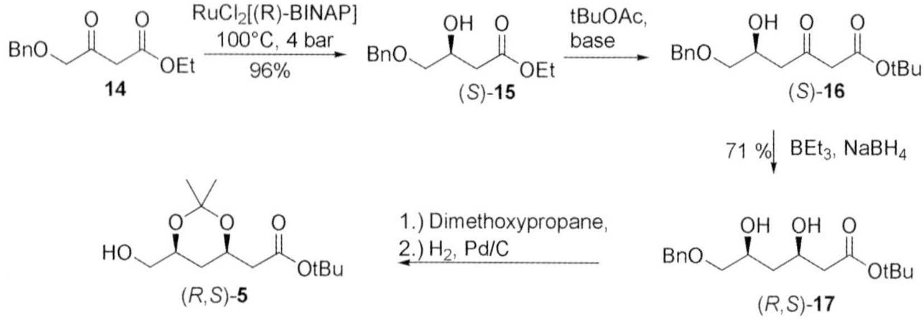

Scheme 4 Asymmetric hydrogenation approach published by Hoechst [7].

Saltigo has further extended this methodology to a multi-ton process for the manufacture of a hydroxyester using its proprietary Cl-MeOBIPHEP ligand in a high pressure hydrogenation [8].

Chiral Pool Synthesis
Various syntheses of statin side chains via the (S)-3-hydroxybutyrolactone (**19**) have also been described.

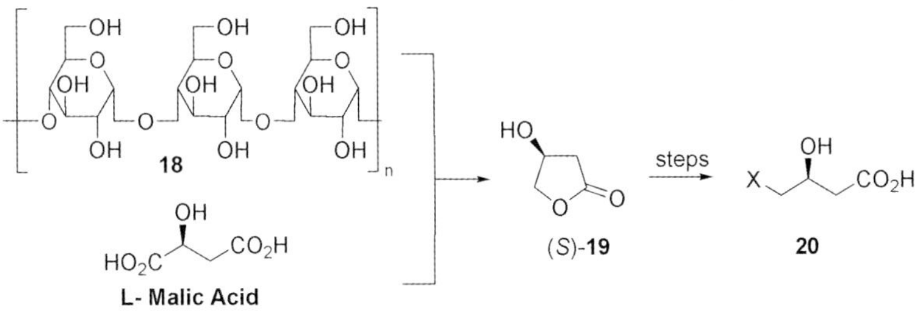

Scheme 5 Chiral Pool approach by Samsung [9] and SK Energy [10].

Hydroxylactone **19** can be obtained with an overall yield of > 50 % and excellent ee (99.9 %) from [1,4]-linked oligosaccharides **18** applying amylopectin to cleave the saccharide units [9]. A completely continuous process from L-malic acid via an esterification/hydrogenation/cyclization sequence leading to (S)-**19** has been developed by SK Energy [10].

Conclusion
The importance and high demands of statin drugs combined with increasing price pressure led to the development of several syntheses for the statin side chain. Within each approach a mature and scalable technology for production at technical scale is available today.

CV of Andreas Job
Andreas Job was born in Neuwied am Rhein in 1972. He received his diploma in chemistry from the RWTH Aachen in 1998. He earned his Ph.D. working on natural product synthesis with Dieter Enders. In 2001, as a Feodor Lynen Fellow (Alexander von Humboldt Foundation) he moved to The Scripps Research Institute, La Jolla, USA for postdoctoral studies with Julius Rebek, Jr. in supramolecular chemistry. In 2002 Andreas Job joined the process development division at Bayer Chemicals, Business Unit Fine Chemicals, now Saltigo Deutschland GmbH.

CV of Andreas Stolle
Andreas Stolle was born in Lüneburg, Germany in 1965. He studied Chemistry at the University of Hamburg, the Georg-August University in Göttingen, Germany and Paris-Sud, France. He received his Ph.D. in 1992 after working with Armin de

Meijere in Hamburg and Göttingen. After a post-doctoral appointment as a Feodor Lynen Fellow at the University of Illinois at Urbana-Champaign, US working with Scott E. Denmark, Andreas Stolle joined the Bayer AG where he was working in Medicinal Chemistry in Germany and West Haven, US. Since 2003, he has been a Section Head in Process R&D of Saltigo, formerly Bayer Chemicals AG.

Selected Publications

1. J. A. Tobert, *Nat. Rev. Drug Discovery* **2003**, *2*, 517–526. *Lovastatin and Beyond: the History of the HMG-CoA Reductase Inhibitors.* A. A. Daemmrich, M. E. Bowden, *Chem. Eng. News*, **2005**, *June 20*, 82. *Top Pharmaceuticals, Chapter: Lovastatin.*
2. Y. Yasohara, N. Kizaki, J. Hasegawa, M. Wada, M. Kataoka, S. Shimizu, *Tetrahedron: Asymm.* **2001**, *12*, 1713–1718. *Stereoselective Reduction of Alkyl 3-oxobutanoate by Carbonyl Reductase from Candida Magnoliae.*
3. A. Nishiyama, M. Horikawa, Y. Yasohara, N. Ueyama, K. Inoue, EP 1288213, **2001**. *Process for Preparing Optically Active 2-[6-(Hydroxymethyl)-1,3-dioxan-4-yl] Acetic Acid Derivatives.*
4. M. Wolberg, W. Hummel, C. Wandrey, M. Müller, *Angew. Chem., Int. Ed. Engl.* **2000**, *39*, 4306–4308. *Highly Regio- and Enantioselective Reduction of 3,5-Dioxocarboxylates.*
5. G. DeSantis, K. Wong, B. Farwell, K. Chatman, Z. Zhu, G. Tomlinson, H. Huang, X. Tan, L. Bibbs, P. Chen, K. Kretz, M.J. Burk, *J. Am. Chem. Soc.* **2003**, *125*, 11476–11477. *Creation of a Productive, Highly Enantioselective Nitrilase Through Gene Site Saturation Mutagenesis (GSSM).*
6. H. J. M. Gijsen, C.-H. Wong, *J. Am. Chem. Soc.* **1994**, *116*, 8422–8423. *Unprecedented Asymmetric Aldol Reactions with Three Aldehyde Substrates Catalyzed by 2-Desoxyribose-5-phosphate Aldolase.* J. H. M. H. Kooistra, H. J. M. Zeegers, D. Mink, J. M. C. A. Mulders, WO 02/06266, **2000**. *Process for the Preparation of 2-(6-Substituted-1,3-dioxanone-4-yl)acetic Acid Derivatives.*
7. G. Beck, J.-H. Jendralla, K. Kessler, EP 0577040, **1993**, *Process for the Preparation of (3R, 5S)6-6-Hydroxy-3,5-O-isopropylidene-3,5-dihydroxy-hexanoic acid, tert.-butyl ester.*
8. A. Gerlach, U. Scholz, *Specialty Chemicals Mag.*, April, **2004**, 37–38. *Industrial Application of Chiral Phosphines.*
9. J. Chun, Y. Cho, K. R. Roh, YM. Park, H. Yu, D. Hwang, II, US 6251642, **1999**. *Continuous Process for Preparing Optically Pure (S)-3-Hydroxy-γ-butyrolactone.*
10. B.-S. Kwak, K.-N.Chung, T.Y. Kim, K.-K. Koh, J.-W. Kim, S.-I. Lee, WO 04/026223, **2004**. *Continuous Process for Production of Optically Pure (S)-3-Hydroxy-γ-butyrolactone from Carboxylic Acid Ester Derivatives.* Review: A.M. Roughi, *Chem. Eng. News*, **2005**, *83*, 41–51. *Custom Chemicals.*

Index

a

AC133 179
acquisition of stem cell 57
β-actin promoter 134
acute ischemia 151
acute lymphoid leukemia (ALL) 222, 240
acute myeloid leukemia (AML) 222 ff., 239
ADA-SCID gene therapy clinical trial 110
adenosine deaminase (ADA) 110
adhesion 184, 234
– cell-cell 63
– cell-ECM 63
– molecule 62
adipocyte 77, 150
adipose tissue 94, 233
adoptive immunotherapy 221 ff.
adult stem cell (ASC) 6 ff., 237
– plasticity potential 6 ff.
– trans-differentiation 6 ff.
alkaline phosphatase (ALP)
– bone-specific 77
ALL, see acute lymphoid leukemia
allogeneic stem cell transplantation
  (allo-SCT) 221, 234, 247 ff.
alloreactivity 237
AMD-3100 63 f.
AML, see acute myeloid leukemia
amniotic fluid 94
angiogenesis 158 f., 185, 202
animal model 109 f., 147 ff.
– clonality analysis 109 ff.
– fetal sheep 80,123 ff.
– mouse 109, 112, 152
– non-human primate 110
– non-injury model 121
– potential of MAPC 147 ff.
antibody 221 ff.
– bispecific 227
antigen expression 92
– hematopoietic 92

antithymocyte globulin 222
antitumor
– activity 29
– agent 158
Ara-C (cytarabine) 226
Argonaute protein 46
arterial and venous oxygen content (AVDO$_2$) 198
arteriogenesis 150, 203
arylsulfatase A 239
ASC, see adult stem cell
asymmetric cell division 12, 28
autoimmune disorder 234
autologous stem cell transplantation 249
5-azacytidine (5azaD) 29 ff., 206
5-aza-2'-desoxycytidine 29

b

B-cell 151, 247 ff.
– lymphoma 247
basic fibroblast growth factor (bFGF) 134, 158
Berashi cell 73
blastocyst 133 ff.
– transfer 135
blood cell 6
blood stem cell
– mobilization 57 f.
bone 77 ff.
bone marrow (BM) 61, 73 ff., 122, 147, 159, 202 ff.
– bone marrow-derived mononuclear cell (BMC) 189, 209
– mesenchymal stem cell 83, 128, 157
– stem cell 210
– transplantants 7
– transplantation 7, 122
bone morphogenetic protein 34, 238
bypass grafting 197 ff.

Asymmetric Synthesis – The Essentials.
Edited by Mathias Christmann and Stefan Bräse
Copyright © 2007 WILEY-VCH Verlag GmbH & Co. KGaA, Weinheim
ISBN: 978-3-527-31399-0

c

C3aR antagonist 66
CABG, see coronary artery bypass grafting
CAD, see coronary artery disease
CAFC, see cobblestone area-forming cell
cancer, see tumor
cardiac regeneration 179 ff.
cardiomyocyte 151, 205 f.
cardiomyopathy 191
cartilage 77 ff.
CD14$^+$ mononuclear cell 180
CD26 62
CD34 11, 28 ff., 210, 228, 234
– CD34$^+$ cell 100 ff., 128, 179
– CD34$^+$ cell differentiation 28 ff.
– CD34$^+$ cell mobilization 63
– CD34$^+$ stem/progenitor cell 60
CD38 11
CD44
CD45$^-$ 142, 234
CD45$^+$ 102
CD133 11, 179, 210
cell culture
– trans-differentiated 9
cell fusion 137 ff., 205
cell membrane raft formation 62
cell plating density 94
cell therapy 200 ff., 247 ff.
cellular microenvironment 15
central nervous system (CNS) 12
CFC, see colony-forming cell
chemoattractant 142, 181
chemokine 142, 186
chemotaxis 186
chimerism 135 ff., 229
chondroblast 150
chromatin modification 9
– chromatin-modifying agent 27 ff.
chromosome instability 36
chronic ischemia 151
chronic lymphocytic leukemia (CLL) 227, 247 ff.
chronic myelogenous leukemia (CML) 221 ff.
circulating system 202
clinical studies 18
clinical scale production 91 ff.
coating 168
cobblestone area-forming cell (CAFC) 31
colony-forming cell (CFC) 33
colony-forming unit fibroblast (CFU-Fs) 91
concanavalin A 235
connexion 43 (Cx43) 207
cord blood (CB) 33, 61, 73 ff., 93
– cryoconserved 76
– mesenchymal stem cell 85
– umbilical cord blood 73 ff., 93, 126
coronary artery bypass grafting (CABG) 197 ff.
– off-pump 201
– surgery 199 ff., 214
coronary artery disease (CAD) 197 f.
Creutzfeld-Jakob disease 95
CTCE0021 65
CTCE0214 65
culture condition 180
– ex-vivo 150
– in-vitro 142
CXCL2Δ4 66
CXCR4 57 ff.
– pathway 58
cytarabine (Ara-C) 226
cytokine 221 ff.
– delivery 165
– production 82 f.
cytomegalovirus (CMV) 73
– enhancer 134
cytotoxicity 29 ff.
cytotoxic T-lymphocyte (CTL) 235

d

decitabine 29
delivery
– coating 168
– MSC-based 167
– therapeutic gene 165
– vehicle 157 ff.
Delta 24 virus 169
desmoplastic reaction 164
development potential 133 ff.
dextran sulphate (DEX) 61
Dicer 46 f.
differentiation 7 ff., 150 f., 183, 206 ff., 233 f.
– dopaminergic 150
– multilineage 7
– multipotent 147
– pluripotent 147
– potential 35, 126, 150 ff.
– stage 137
– trans-differentiation 6 ff.
division 12
– asymmetric 12
– symmetric 12
DNA (cytosine-5)-methyltransferase (*dnmt1*) 16, 29
Dolly 3
donor lymphocyte infusion (DLI) 249
donor lymphocyte transfusion (DLT) 221 f.
dopaminergic differentiation 150
Drosha 46

e

E-selectin 185
ECM, see extracellular matrix
ectodermal derivation 121
embryo
– murine 141
– permanent embryo 5
embryogenesis 121
embryonal carcinoma (EC) cell 143
embryonic antigen 148
– stage-specific (SSEA) 148
embryonic hematopoietic tissue 142
embryonic progenitor cell homing 185
embryonic state 143
– pluripotent 143
embryonic stem cell (ESC) 3 ff.
– differentiation 14
– pluripotency 28
endoderm 122
endothelial cell 158
endothelial differentiation 180
endothelial growth factor (EGF) 134
endothelial progenitor cell (EPC) 151, 179 ff.
– circulating bone marrow-derived (CEPC) 179
– heterogeneity 180
endothelium 151
enzyme replacement therapy 112
epigenetic alteration 133 ff., 206
epigenetic mechanism 28
Epstein-Barr virus (EBV) 73
erythroid-like cell 134
erythroleukemia 143
ESC, see embryonic stem cell
ethics 143
expansion 33 ff.
– ex-vivo 33, 182
– malignant expansion 112
Exportin 46
extracellular matrix (ECM) 63, 158 ff.

f

FDCPmix cell 142
fetal sheep
– model 123 ff.
– preimmune 81
– xenograft model 80
fibroblast 158
Flt3 ligand (FL) 60
fludarabine 252
follicular lymphoma (FL) 253
fumarylacetoacetate hydrolase (FAH) 152
fusion 183

g

G-CSF, see granulocyte colony-stimulating-factor
G6PD, see glucose-6-phosphate dehydrogenase
GAG, see glycosaminoglycan
ganglion mother cell (GMC) 12
gap junction protein 207
gene deletion 143
gene expression 34
– dynamics 28
gene therapy 109
– clinical trial 110 f.
– clonality analysis 109 ff.
genetic targeting 157
genotype 14
germ layer 121, 149 f.
germline stem cell (GSC) 14
glial fibrillary acidic protein (GFAP) 80
globulin 222
– antithymocyte 222
glucose-6-phosphate dehydrogenase (G6PD) 108
glycosaminoglycan (GAG) 151
GMC, see ganglion mother cell
graft versus host disease (GVDH) 63, 221 f., 249
graft versus leukemia (GVL) effect 224, 249
granulocyte colony-stimulating-factor (G-CSF) 57 ff.
GROβT 66
growth factor 142

h

heart failure 190
hemangioblast 202
hematopoiesis 36, 82 ff.
– in vivo 80
hematopoietic cell 151
hematopoietic chimerism 138, 247
hematopoietic engraftment 126
hematopoietic stem cell (HSC) 6 ff., 27 ff., 107 ff., 133 f.
– acquisition 57 ff.
– asymmetric cell division 28
– 5azaD/TSA treatment 30 ff.
– chromatin-modifying agent 27 ff.
– clonal activity 107 ff.
– differentiation 234
– ex-vivo expansion 33
– expression profile 14
– fast-dividing fraction (FDF) 14
– fetal sheep 124
– genetic marker 107

- genotype 14
- human cord blood-derived 138
- malignant expansion 112
- marked HSC 109 ff.
- myogenic potential 216
- slow-dividing fraction (SDF) 14
- symmetric cell division 35
- transplantation 163

hematopoietic system 73
hepatic cell 81
hepatocyte 126, 133, 151
- hepatocyte-like cell 152

hereditary tyrosinemia type I (HT1) 152
heterogeneity 180
histone acetylation 28
histone deacetylase (HDAC) 29
- HDAC inhibitor 30 ff.

HLA, see human leukocyte antigen
HMG-CoA reductase 60
homeostasis 157
homing 61 f., 162, 184
- mechanism 184

HoxB4 33 ff.
HSC, see hematopoietic stem cell
human gene therapy 109 ff.
- clonality analysis 109 ff.

human leukocyte antigen (HLA) 228, 234
- HLA-DR 11
- HLA-haploidentical transplantation 228
- HLA-identical related donor (ID-SCT) 251
- HLA-identical sibling transplants 240, 252 ff.
- HLA-matched unrelated donor (UD-SCT) 256

human umbilical cord perivascular cell (HUCPV) 85
Hurler's disease 239
hyaluronic acid (HA) 60
hypomethylation 29
hypoxia-inducible factor-1 (HIF-1) 186

i

α-L-iduronidase 239
IFN, see interferon
IL, see interleukin
immune escape 233 f.
immune system 6 ff., 150 f.
immunosuppression 233 ff.
immunotherapy 247 ff.
- adoptive 221 ff.

implantable cardioverter defibrillator (ICD) 207
indoleamine 2,3-dioxygenase (IDO) 238

information
- cell-autonomous 12
- intrinsic 12

initiating cell
- long-term (LTC-IC) 11
- myeloid-lymphoid initiating cell (ML-IC) 11

inner cell mass 4
β_2-integrin 184
interferon-alpha (IFN-α) 225
interferon-beta (IFN-β) 165
interferon-gamma (IFN-γ) 234 ff.
interleukin (IL) 59, 186, 225
- IL-2 receptor gamma common chain (γc) 111

international bone marrow transplant registry (IBMTR) 253
invasion 186
ischemia 151, 198
- myocardial 198

ischemic cardiomyopathy 191
ischemic heart failure 190
ischemic myocardium 189

j

JAK-STAT signalling cascade 14
junction sequences
- proviral-genomic 108

k

kinase inhibitor 143

l

left ventricular (LV) function 200
leukemia 8, 221 ff.
- acute lymphoid leukemia (ALL) 222
- acute myeloid leukemia (AML) 139 ff., 222 ff.
- chronic lymphocytic leukemia (CLL) 227
- chronic myelogenous leukemia (CML) 221 ff.
- erythroleukemia 143

leukemic cell 138
lineage commitment 11
linear amplification-mediated PCR (LAM-PCR) 109
LMO2 insertion 112
long-term initiating cell (LTC-IC) 11
LV, see left ventricular function
lymphocyte 221
- donor lymphocyte infusion (DLI) 249
- donor lymphocyte transfusion (DLT) 221 f.

lymphoma 222, 247 ff.
– lymphoplasmacytic lymphoma 255

m
major histocompatibility complex (MHC) 164, 234
MAPC, *see* multipotent adult progenitor cell
matrix-degrading protease 162
MCP-1, *see* monocyte chemoattractant protein-1
mesenchymal lineage 150, 210
mesenchymal stem cell (MSC) 16 ff., 73 ff., 122 ff., 157 ff., 233
– adult 237
– bone marrow (BMSC) 190
– clinical-scale production 91 ff.
– differentiation 128
– fetal 237
– homing 162
– immune escape 233
– immunosuppression 235
– lineage 77
– MSC-based cellular mini-pump 158
– precursor cell 81
– quality assurance 99
– quality control 99
– starting material 92
– stroma cell-derived 206
mesenchymal tissue 163
mesodermal cell 152
mesodermal derivation 122
metachromatic leukodystrophy (MLD) 239
metastase 157
methyl-β-cyclodextrin 62
micro RNA (miRNA) 43 ff.
– biogenesis 44
– potential function mode 48 f.
micro ribonucleoprotein (miRNP) 46
microenvironment 15, 142 f.
– influence 126
– nonmalignant 158
– tumor 161 ff.
migration 186
mini-pump 165
– MSC-based cellular mini-pump 158
mitogen 235
mixed lymphocyte culture (MLC) 235
MLD, *see* metachromatic leukodystrophy
MLV, *see* murine leukemia virus
mobilization 62 f.
mobilized peripheral blood (mPB) 126
model
– animal model 109 f., 147 ff.
– animal non-injury model 121
– fumarylacetoacetate hydrolase (FAH) 152
– fetal sheep 80, 123 ff.
– *in-vivo* 123
– LMO2 transgenic mouse model 112
– mouse model 109, 112, 152
– non-human primate model 110
monocyte chemoattractant protein-1 (MCP-1) 186
mouse model 109, 112, 152
MSC, *see* mesenchymal stem cell
multilineage differentiation 7
multipotent adult progenitor cell (MAPC) 16, 35, 92, 142, 148 ff.
– characterization 148
– culture condition 148 ff.
– cynomolgus monkey-derived (pMAPC) 149
– differentiation potential 150 ff.
– *ex-vivo* culture 153
– mechanism 152 f.
– phenotype 148
– pluripotency 152
– potential in animal model 147 ff.
– proliferative capacity 148
multipotent differentiation 147
multipotent stem cell 210
multipotentiality 28
murine blastocyst 133 ff.
murine embryo 141
murine leukemia virus (MLV)-based retroviral vector 111
murine neurosphere cell 142
muscle
– fiber 151
– regeneration 180
– smooth muscle 151
myeloid cell 180
myeloid-lymphoid initiating cell (ML-IC) 11
myoblast 150, 207
myocardial cell 81
– therapy 209
myocardial regeneration 197 ff.
myocardium 189, 205 ff.
– ischemic 189, 198
myocyte 181
myofibroblast 160 ff.
myogenesis 205
myogenic potential 216

n
Nanog 148
natural killer (NK) cell 151, 221 ff., 235
– natural killer T (NK-T) cell 228

neoangiogenesis 203
neovascularization 182
neural cell 80, 151
neural stem cell (NSC) 123 ff., 133 f.
neuroblast (NB) 12
neuroectoderm 152
neuron 150
neurosphere cell 134 ff.
2-(2-nitro-4-trifluoro-methylbenzoyl)-1,3-cyclohexanedione 152
NOD-SCID 150
non-human primate model 110
non-obese diabetic (NOD) 150
nonhematopoietic cell 122, 234
– regenerative capacity 85
nonhematopoietic differentiation potential 35
nonhematopoietic lineage 210
nonhematopoietic stem cell 85
– multipotent 85
notch family 34
NTBC, see 2-(2-nitro-4-trifluoro-methyl-benzoyl)-1,3-cyclohexanedione
nuclear size 137 ff.

o

oct3a 148
oct4 148
ontogeny 121, 157
Onyx-15 virus 168
osteoblast 150

p

P-selectin 185
paracrine effect 183
pcna, see proliferating cell nuclear antigen
PCR
– linear amplification-mediated (LAM-PCR) 109
PEG-ADA, see polyethylene glycol adenosine deaminase
percutaneous transluminal coronary angioplasty (PTCA) 199
peripheral blood
– mobilized (mPB) 126
peripheral blood mononuclear cell (PBMC) 180
peripheral blood progenitor cell (PBPC) 62
– mobilization 63
peripheral blood stem cell (PBSC) 239, 256
peripheral vascular disease 151
perivasculature cell 158
– human umbilical cord perivascular cell (HUCPV) 85

phytohemagglutinin (PHA) 235
plasmacytic lymphocyte 255
plasminogen-activator receptor
– urokinase-type (uPAR) 63
plasticity 9, 121
pluripotency 28, 143, 152
pluripotent embryonic state 143
pluripotent stem cell 73 ff., 180
– expansion 74
– generation 74
pMAPC, see multipotent adult progenitor cell, cynomolgus monkey derived
polyethylene glycol adenosine deaminase (PEG-ADA) 112
progenitor cell
– homing 181
– multipotent adult progenitor cell (MAPC) 16, 35, 92, 142, 148 ff.
– therapy 190
– vascular progenitor 151
– vasculogenesis 184
proliferating cell nuclear antigen (pcna) 16
proliferation potential 147
prominin 179
protease
– matrix-degrading 162
proviral-genomic junction sequences 108
Purkinje fiber 81

q

quality assurance 55, 99
quality control 99

r

RANTES 186
reduced-intensity conditioning transplantation (RICT) 249 ff.
regeneration 4 ff., 179 ff.
– cardiac 179 ff.
– myocardial 197 ff.
– tissue 157 ff.
– vascular 197 ff.
replacement therapy 18, 112, 128, 157
replicating oncoselective virus 168
retroviral integration site 113
– analysis 108
retroviral vector 111
Rex1 148
RNA
– micro RNA 43 ff.
– RNA-induced silencing complex (RISC) 44 f.
RS phenotype 167

s

SB 290157 66
SCID (severe compromised immunodeficient)
- ADA-SCID 110
- mouse model 12, 164
- SCID repopulating cell (SRC) 33
- SCID-X1 111

secretory leukocyte protease inhibitor (SLPI) 59
self-renewal 147
- ability 7 ff.
senescence 149
side population 180
simvastatin 60
somatic cell 140
sonic hedgehog 34
sphingosine-1-phosphate 186
stage-specific embryonic antigen (SSEA)-1 148
stem cell 3 ff., 197 ff.
- acquisition 57
- adult stem cell (ASC) 6 ff.
- allogeneic stem cell transplantation (allo-SCT) 221, 234, 247 ff.
- clinical potential 3 ff.
- embryonic stem cell (ESC) 3 ff.
- germline stem cell (GSC) 14
- hematopoietic stem cell (HSC) 6 ff., 133 f.
- human stem cell plasticity 121
- homing 181
- mesenchymal stem cell (MSC) 16 ff., 73 ff.
- multipotent nonhematopoietic stem cell 85
- multipotent stem cell 210
- neural stem cell (NSC) 123, 133 f.
- peripheral blood stem cell (PBSC) 239
- pluripotent 73 ff., 180
- preparation 55 ff.
- regeneration 4 ff.
- resource 73
- somatic stem cell (SSC) 9, 133 ff.
- therapy 190, 207
- trans-differentiation 8
- transplantation 247 ff.
- unrestricted somatic stem cell (USSC) 16, 36, 73 ff.
- vasculogenesis 184
stemness 11 ff.
stroma 157 ff.

stromal cell-derived factor-1 (SDF-1) 58 ff., 186
symmetric cell division 12, 28 ff.

t

T-cell 151, 252
- depletion (TCD) 252
- T-cell inhibitory effector mechanism 238
T-lymphycyte 235
TAF, see tumor-associated fibroblast
telomere 148
teratocarcinoma 11
therapeutic agent delivery 162 ff.
therapeutic cloning 10
Thy-1 11
tissue
- engineering 163
- regeneration 157
- replacement therapy 128
total body irradiation (TBI) 59
total nuclear cell (TNC) 32
totipotent 3
toxicity 36
trans-differentiation 6 ff., 142
transcription factor 33
transforming growth factor-beta (TGF-β) 158 ff.
transgene expression 113 f.
translational control
- miRNA-mediated 47
transplantation 7, 163, 221 ff., 247 ff.
- allogeneic stem cell transplantation (allo-SCT) 221, 234, 247 ff.
- bone marrow 7
- fetal sheep 124
- HLA-identical transplantation 228, 240, 251 ff.
- in-vivo 123
- reduced-intensity conditioning transplantation (RICT) 249 ff.
transposon 46
treatment-related mortality (TRM) 248
trichostatin A (TSA) 29 ff.
Trojan-Horse concept 158
tryptophan depletion 238
tumor 11
- development 161
- genetic targeting 157
- invasion 161
- microenvironment 161
- progression 158 ff.
- stroma 161
- tropism 162

u

– tumor-associated fibroblast (TAF) 158 ff.
– tumor-associated macrophage 158
tumorigenesis 158

u

umbilical cord blood 73 ff.
umbilical cord perivascular cell 85
– human (HUCPV) 85
unpaired 14
unrestricted somatic stem cell (USSC) 16, 36, 73 ff.
– differentiation potential 77
– expansion 74
– generation 74
– hepatic cell 81
– immunophenotype 77
– *in-vitro* differentiation 77
– *in-vivo* differentiation 77
– liver cell-specific differentiation 82
– myocardial cell 81
– Purkinje fiber 81
urokinase-type plasminogen-activator receptor (uPAR) 63

v

valproic acid 33
vascular endothelial growth factor (VEGF) 158, 186
vascular progenitor 151
vascular regeneration 197 ff.
vascular remodelling 202 f.
vasculature cell 158
vasculogenesis 151, 184, 202 f.
very late antigen-4 (VLA-4) 57
virus 168
von Willebrand factor (vWF) 179

w

Waldenström's disease 255
Waldenström's macroglobulinemia (WM) 255
wnt family 34
wound healing 158 ff.

Author Index

Abe, H. 246
Aggarwal, V. 52, 176
Ahlbrecht, H. 103
Alexakis, A. 80, 97
Arai, S. 178
Armstrong, A. 177
Armstrong, R. W. 230
Bach, T. 166a
Bäckvall, J.-E. 171a
Baeyer, A. 57
Bakshi, R. K. 235
BASF 315
Beak, P. 103
Belokon, Y. N. 64, 67, 69
Berkessel, A. 63, 176a, 286
Bettray, W. 21a
Blackmond, D. G. 181a, 197, 199
Blaser, H.-U. 296
Blechert, S. 241a
Bolm, C. 57a
Brands, K. M. J. 301a
Bräse, S. 62a, 286
Bringmann, G. 246a
Brown, H. C. 28, 230
Brown, J. M. 181, 269
Brückner, R. 10a
Brunner, H. 133
Bulger, P. G. 225a
Caro, H. 57
Carreira, E. M. 163
Chirex 294, 299
Ciba-Geigy 298f
Consiglio, G. 91
Corey, E. J. 4, 21, 132, 178, 235f
Cram, D. J. 62
Crimmins, M. T. 5
Curran, D. P. 227
Dahmen, S. 306a
Dai, L.-X. 132
Danishefsky, S. J. 34, 251a

Davies, H. M. L. 285
Davies, S. G. 5
Davis, F. A. 16
De Meijere, A. 67a
Dittrich, K. 311a
Diversa 328
Drueckhammer, D. G. 172
DSM 328
Duthaler, R. O. 29
Eder, U. 21
Enders, D. 4, 21a, 162, 177, 226
Eschenmoser, A. 131
Etter, M. C. 136
Evans, D. A. 3a, 33, 72a, 132, 142, 162, 225, 227, 278
Feringa, B. L. 78a, 97
Forsyth, C. J. 256a
Frank, F. C. 181
Fu, G. C. 73, 186a
Fujita, E. 5
Fujita, J. 116
Fürstner, A. 262a
Furstoss, R. 58
Furukawa, N. 53
Gais, H.-J. 84a
Gates, M. 232
Genet, J. P. 267a
Ghosh, A. K. 5
Glorius, F. 264
Goodman, J. M. 53
Greenberg, W. A. 217a
Gridnev, I. D. 269
Grubbs, R. H. 262, 265
Gulder, T. A. M. 246a
Gulder, T. 246a
Hall, D. 28
Halpern, J. 269
Harayama, T. 248
Hassfeld, J. 105a
Hayashi, T 80, 90a

Asymmetric Synthesis – The Essentials.
Edited by Mathias Christmann and Stefan Bräse
Copyright © 2007 WILEY-VCH Verlag GmbH & Co. KGaA, Weinheim
ISBN: 978-3-527-31399-0

Heathcock, C. H.   278
Helmchen, G.   3a, 4f, 63, 95a, 103, 132
Hoechst   328
Hoffmann, R. W.   4, 27a
Hoffmann-La Roche   316
Holub, N.   241a
Hoppe, D.   100a
Horner, L.   267
Hoveyda, A. H.   79f
Hugl, H.   291
Imada, Y.   40a
Imamoto, T.   269
Jackson, R. F. W.   177
Jacobsen, E. N.   116, 229, 275, 285, 294
Jiang, B.   284
Job, A.   292, 326a
Jorgensen, K. A.   191a
Jung, M. E.   108
Kagan, H. B.   196a
Kalesse, M.   105a
Kaminsky, W.   126
Kaneka   327
Kanger, T.   59
Katsuki, T.   59, 116a, 233
Kawasaki, T.   212a
Kazlauskas, R. J.   174
Kemp, D. S.   167
Kinugasa, M.   72
Kleinbeck, F.   72a
Knowles, W. S.   236, 267, 292, 294, 296, 298, 300, 316a
Kobayashi, S.   110a
Kochi, J. K.   116
Kumada, M.   91
Kunz, H.   4f, 32a
Lanxess   292
Larionov, O. V.   67a
Lee, K. H.   286
Lehn, J.-M.   132
Leighton, J. L.   29, 80
Lerner, R. A.   187
Leßmann, T.   283a
Ley, S. V.   201a
Liang, B.   126a
Lippard, S. J.   78
List, B.   141, 161a
López, F.   78a
Lotz, F.   147a
Maruoka, K.   121a, 178
Masamune, S.   132, 278
Matsumoto, K.   116a
Matsunaga, S.   47a
Merck   301f, 326
Metzner, P.   53

Meyers, A. I.   4, 21f, 37
Mihara, J.   118
Miura, M.   72
Miyaura, N.   80
Monsanto   267, 292, 298, 316
Mori, K.   273a
Morken, J. P.   142
Mukaiyama, T.   278
Müller, M.   328
Muñiz, K.   321a
Murahashi, S.-I.   40a
Myers, A. G.   4, 227
Nagao, Y.   5
Negishi, E.-I.   126a, 155
Nicolaou, K. C.   225a
Nishiyama, H.   133
Novak, T.   126a
Noyori, R.   172, 197, 236, 267ff, 292, 294, 296, 300, 321a
Ogawa, A. K.   230
Ogawa, C.   110a
Oguni, N.   197
Ohkuma, T.   321a
Ohshima, T.   141a
Omura, S.   237
Onishi, M.   132
Ooi, T.   121a
Oppolzer, W.   4f
Overman, L. E.   231
Parke-Davis   37
Pasteur, L.   318
Paterson, I.   108, 278a, 283f
Pauling, L.   136
Pfaltz, A.   63, 79, 96, 131a
Pfizer   326
Pye, P. J.   62, 301a
Rawal, V. H.   136a
Reetz, M. T.   207a
Reggelin, M.   283
Reich, H. J.   62
Rhodia   294
Roberts, S. M.   58
Rossen, K.   62, 301a
Rozenberg, V.   63f
Rüping, M.   3a, 72a
Saltigo   329
Sammakia, T.   79
Samsung   329
Sandoval, C. A.   321a
Sauer, G.   21
Schering   21
Schöllkopf, U.   4
Schreiber, S. L.   11f, 262, 283f
Schwab, J. M.   58

Seebach, D 4f, 79, 138
Shair, M. D. 235
Sharpless, K. B. 116, 267, 275, 294, 296, 299f
Shi, M. 80
Shi, Y. 118, 177
Shibasaki, M. 47a, 141a, 163f, 177, 231
Shibata, S. 235
Sibi, M. P. 5
Sih, C. J. 172
SK Energy 329
Soai, K. 181, 199, 212a
Solvias 292, 298
Somitomo 293
Spescha, M. 79
Stolle, A. 292, 326a
Strukul, G. 58f
Suzuki, K. 249
Syngenta 292, 298
Tai, A. 171
Takasago 292, 298f
Tan, Z. 126a

Tang, Y. 73
Thadani, A. N. 136a
Thomas, E. J. 103
Tietze, L. F. 147a
Trost, B. M. 97, 141, 232
van Koten, G. 79
Villiger, V. 57
Waldmann, H. 283a
Walsh, C. T. 58
Weinreb, S. M. 19
Whitesell, J. K. 4
Wiechert, R. 21
Williams, J. M. J. 172
Williams, J. M. 63, 96
Wong, C.-H. 217a, 328
Woodward, R. B. 278
Würthwein, E.-U. 103
Yamada, S. 4, 21
Yamamoto, H 153a
Yan, T. H. 5
Yang, D. 177

Subject Index

absolute asymmetric synthesis   215
1,4-addition, see conjugate addition
ALAPHOS   91
aldolase   161, 217
alkylation   4, 23
π-allyl complexes   95
allylation   24ff, 122, 230
allylboration   27
allylboronate   29
allyldiisopinocampheyl-borane   28
allylic alcohol   84ff, 116
allylic strain   30
allylic substitution   48, 84, 95ff, 150, 231f, 242
allylsilane   91, 147
altohyrtin, see spongistatin
ambruticin   229
aminalization   256
amination reaction   73
α-aminoacid   54, 67ff, 87, 267ff, 291, 316ff
β-aminoacid   34, 41, 87
α-aminonitrile   34
anabasine   35
anhydride opening   60
anomeric effect   34, 257
antibiotic   40
aphidinolide   107
apicularen A   230
aprepitand   301ff
aqueous media   110ff
archaea cell wall lipid   97
arylation   148
aspartame   236, 291
astrophylline   244
asymmetric amplification   183
atorvastatin   219f, 326
atropisomers   246
attenol A   24
autocatalysis   181ff
autocatalysis   212
automation   306

auxiliary   3ff, 22, 28, 32ff, 162, 285
axial chirality   48, 90, 93, 246ff
azaenolate   21ff
azaspiracid   256, 258
aziridination   54
aziridine   54
aelactosin A   67ff
aenanomycin B   249
aetaine   52
aiaryl synthesis 92, 246ff
bioisostere   201
bisoxazoline   72ff, 131ff
borrelidin   128, 237
BPPFA   91
arevetoxin B   232f
arevicomin   273, 276
callystatin   24
calyculin A   230
camphor   4
carbamate   101
carbapenem   41, 298
carbene   24, 52, 263
carboalumination   126ff
carbohydrate   24, 32ff
CDP-840   55
cetirizine   55
chalcone   178
chelation   12
chiral cooperativity   63
chiral pool synthesis   329
cilastatin   293
cinchona alkaloids   204
circularily polarized light   212
citronellol   292, 299
colombiasin A   228f
combinatorial chemistry   48, 306ff
concanamycin F   279
coniine   23, 35
conjugate addition 24, 32, 35, 38, 48, 64, 78ff, 98, 197, 202

Asymmetric Synthesis – The Essentials.
Edited by Mathias Christmann and Stefan Bräse
Copyright © 2007 WILEY-VCH Verlag GmbH & Co. KGaA, Weinheim
ISBN: 978-3-527-31399-0

conjugate reduction 132
CP-263,114 235f
Cranoc® 326
Crestor® 326
crixivan 293
cross coupling 90ff, 97f, 232, 265
cross metathesis 241
cuprate 78ff
cuscohygrine 244
cycloaddition 32, 73, 132, 136ff, 168, 284
cyclopropanation 47, 54, 67ff, 133, 204, 285
cytovaricin 226
dehydropinidine 35
dendrochrysine 244
desymmetrization 10f, 134, 237, 252
diazo compound 52, 285
dictyostatin 227
DIFLUORPHOS 268, 270
dihydroxylation 47, 286, 293, 297ff
dioncopeltine A 247f
dioxanone 23
dioxirane 177
[3+2] dipolar cycloaddition 73, 123, 284
directed evolution 207ff
discodermolide 278
disparlure 274, 276
diversity oriented synthesis 283ff
diversonol 65
dolastatin 269
doliculide 129
domino 1,4-addition-α-chlorination 33
domino conjugate addition enatioselective
 protonation 270
domino conjugate addition-Aldol 79, 192
domino Mannich-Michael 34
domino reaction 147ff, 241ff
dynamic kinetic resolution (DKR) 48, 84ff,
 171ff, 246, 269
dysidiolide 285
ecteinascidin 236f
emend 301ff
enamine catalysis 195
enolate 17ff, 21ff
enzyme 161, 172ff, 207ff, 217, 311ff
epothilone 217, 219f
epoxidation 10ff, 47, 52ff, 116ff, 176ff,
 232f, 253f, 297ff
epoxide opening 285, 297
epoxyalcohol 11
erythromycin A 278
esomeprazole 299
faranal 276
ferrocene 91f
fluorination 18

fluvastatin 326
FR-182877 7
galactose 35
garsubellin A 251
gloeosporone 236, 262f
glucose 32
glycin equivalent 67
heterobimetallic catalysis 141
hexahelicene 214
high-troughput screening 207, 306ff
himachalene 7
homoaldol reaction 100ff
hormaomycin 67ff
hydrazone 21ff, 53, 162
hydroformylation 48
hydrogenation 41ff, 47, 97f, 236f, 264, 292,
 296ff, 321ff, 328
hydrosilylation 42
hydroxymethylation 112
imidazolidine catalysis 194
imines 34
iminium catalysis 191
iminoether 37
indolizidine 19
iodolactonization 252
ionomycin 128
ipsdienol 276
isooncinotine 264
isoquinoline 43
jiadifenin 251
JOSIPHOS 80
kelsoene 166
Kemp's triacid 166
ketals 256, 284
ketene 72, 187
β-ketoester 321ff
khellactones 286
kinetic resolution 10, 208, 235, 274, 311ff
lactacystin 269
β-lactam 40, 72ff, 187
lactone 57ff, 193, 209, 252
γ-lactone 102
lactone method 246ff
lardolure 81
lasubine II 243
leucomalure 275
library 208, 283
ligand screening 306
lineatin 274
lipase 172, 207, 274, 311ff
lithium carbanions 101ff
lovastatin 326
macrolactonization 7, 263
mastigophorene B 249

Subject Index

menaquinone 178
menthol 299
merrilactone A 251
metolachlor 292, 298, 319
metathesis 262ff
methoprene juvenile hormone 292, 299
Mevalotin® 326
Mevinacor® 326
moipa 313f
monooxygenase 209
morphine 232
mutagenesis 207
neobenodine 55
neurotoxin 232
NGA0187 251
N-hydroxylamine 40
niacinamide 291
nitroalkene 24, 203
nitrone 40, 72ff, 123
nitrosamine 21
nonlinear effect 196ff
N-Sulfonyloxaziridines 16
nucleophilic carbene, see carbene
okadaic acid 256
olean 273
oleandolide 106
organocatalysis 186ff, 201ff, 235
Outlook® 313
oxaziridine 17
oxazolidinone 6, 33, 162
oxazoline 37, 96, 133
oxidation 292
Oxone® 177
paracyclophane 62, 286
Parkinson's disease 236, 267, 316
peptide 177
PHANEPHOS 62
PHEPHOS 91
pheromone 23, 81, 256, 258, 273ff
phomoidride B 7
phophazene 55
phosphoramidite 79, 97
photochemical reaction 118, 166ff
photoinduced electron transfer (PET) 169
pinacol rearrangement 107f
pinene 279
piperidine 19, 35, 241, 265
planar chirality 48, 62, 186ff
pochonin E 65
polyalanine 178
polyketide 105, 276ff
polymerase chain reaction (PCR) 207
polypropionate 128, 278
PPFA 91

pravastatin 326
proline 21, 162, 201
prostacyclin 87
prostaglandin 79, 87, 235f
pumiliotoxin C 35
pyrrolidine 19
pyrrolizidine 241
quadrigenin 231
quaternary stereogenic center 48
quinidine 205
quinine 179, 205
quinolizidine 19
racemization 84, 171ff, 314
radical reaction 169
radical reaction 252
RAMP 22ff
ratjadone 105
Red-Al 14
rhamnose 32
riboflavin 291
ring closing metathesis (RCM) 97, 241f, 262ff
ring opening metathesis 241ff
rosuvastatin 326
salen 116ff
SAMP 22ff
scrabronine G methyl ester 251
selenoxide 17
self-replication 181
semicorrin 131ff
serricornin 276
silyl ketene acetal 41, 163, 253
simvastatin 326
solid phase synthesis 283ff
sordidin 24
Sortis® 326
sparteine 78, 102f
spirastrelloride A 256, 258
spiroaminal 256
spongistatin 226, 281, 284
statin 217
steganone 248
stegobinone 276
stenosine 23
sulfenamide 16
sulfenimines 16
sulfide oxidation#60
sulfinimines 16
sulfobacin 269
sulfonimines 16
sulfonium ylide 52f
sulfoxidation 116, 118
sulfoxide 17
swinholide A 226, 278

SYNPHOS 268
tabersonine 136
TANIAPHOS 80
taxol 54
tedanolide 107
TEI 9826 98
TentaGel® 178
tetrahydrocannabinol 65
tetrahydroquinoline 147ff
tetraline 148
tetraponerine 242
tetrazole 201
THC, tetrahydrocannabinol

thioester 81
thiostrepton 233f
tobacco alkaloid 98
trans-195A 243
tricycloillicinone 251
valinol 249
VALPHOS 91
vancomycin 234f
vitamin E 129
vitamin K 129
vitamin K_3 178
ylide 204
Zocor® 326

Related Titles

Berkessel, A., Gröger, H.

Asymmetric Organocatalysis

From Biomimetic Concepts
to Applications in Asymmetric Synthesis

2005
ISBN 3-527-30517-3

Blaser, H. U., Schmidt, E. (eds.)

**Asymmetric Catalysis
on Industrial Scale**

Challenges, Approaches and Solutions

2004
ISBN 3-527-30631-5

Nicolaou, K. C., Snyder, S. A.

Classics in Total Synthesis II

More Targets, Strategies, Methods

2003
ISBN 3-527-30685-4

Lin, G.-Q., Li, Y.-M., Chan, A. S. C.

**Principles and Applications
of Asymmetric Synthesis**

2001
ISBN 0-471-40027-0

Ojima, I. (ed.)

Catalytic Asymmetric Synthesis, 2nd ed.

2000
ISBN 0-471-29805-0